D0221451

An Atlas of *Drosophila* Genes

An Atlas of *Drosophila* Genes

Sequences and Molecular Features

GUSTAVO MARONI

Department of Biology
University of North Carolina
Chapel Hill, NC 27599

With Contributions by

Stephen M. Mount
Douglas R. Cavener and Beth A. Cavener
Paul M. Sharp and Andrew T. Lloyd

New York Oxford
Oxford University Press
1993

Oxford University Press

Oxford New York Toronto
Delhi Bombay Calcutta Madras Karachi
Kuala Lumpur Singapore Hong Kong Tokyo
Nairobi Dar es Salaam Cape Town
Melbourne Auckland Madrid

and associated companies in
Berlin Ibadan

Published by Oxford University Press, Inc.,
200 Madison Avenue, New York, New York 10016

Library of Congress Cataloging in Publication Data
Maroni, Gustavo.
An atlas of Drosophila genes : sequences and molecular features /
Gustavo Maroni with contributions by Stephen M. Mount . . . [et al.].
p. cm. Includes bibliographical references and index.
ISBN 0-19-507116-6
1. Drosophila—Genetics—Atlases.
2. Gene mapping—Atlases.
I. Title.
QL537.D76M37 1993
595.77′4—dc20 92-35001

9 8 7 6 5 4 3 2 1
Printed in the United States of America
on acid-free paper

Preface

The time is long past when all workers in the field knew the main characteristics of all of the *Drosophila* genes that have been sequenced. My objective in preparing this book was to bring together the available molecular information concerning *Drosophila melanogaster* genes and thereby to make that information more readily accessible.

In Part I of this volume, I describe the main molecular features of genes for which the sequence of the entire transcription unit is available (with a special dispensation for *Ubx*). This sample includes 90 genes, approximately half of all the *Drosophila* genes that fulfill the condition mentioned above and that were listed in the GenBank and EMBO databases early in 1992. In organizing the voluminous data, I have tried to develop a form that would facilitate, and perhaps encourage, a comparative approach for future studies.

Part II includes four chapters that consider different aspects of gene organization as they occur in the *Drosophila* genome. These chapters cover: (1) size correlations among various genetic elements; (2) splicing signals (by S. M. Mount); (3) translation initiation signals (by D. R. Cavener and B. A. Cavener); and (4) codon bias (by P. M. Sharp and A. T. Lloyd). These last three chapters are not restricted to the genes covered in the first part of the book. On the contrary, the authors' analyses cover as much of the available data as possible.

Many people helped me by reviewing individual chapters, pointing out deficiencies and suggesting improvements. Some of these colleagues also made unpublished material available. For such help, I am very grateful to Paul D. Boyer, Carlos V. Cabrera, Sean B. Carroll, Robert S. Cohen, Allan Comer, Victor G. Corces, Winifred W. Doane, Wolfgang Driever, Marshal Edgell, James Fristrom, Eric Fyrberg, Donal A. Hickey, Jay Hirsh, Dan Hultmark, David Ish-Horowicz, Clyde Hutchison, Herbert Jäckle, Allen S. Laughon, Judith A. Lengyel, Michael Levine, John T. Lis, John C. Lucchesi, J. Lawrence March, Elliot M. Meyerowitz, Markus Noll, Christiane Nüsslein-Volhard, Mark Peifer, William H. Petri, Michael Rosbash, Georgette Sass, Lillie L. Searles, Stephen Small, Wayne Steinhauer, Alain Vincent, Gail L. Waring, Pieter Wensink, Theodore R. F. Wright and Ray Wu.

The internal consistency of the material in this book, as well as the clarity of its presentation benefited greatly from the editing of my wife,

Donna Maroni. I am grateful to her for her patience and generosity and for her support.

Format and Conventions

I have tried to be consistent in presenting equivalent data for different genes using the same format. All chapters in Part I are arranged according to the following plan:

- Product
  - Structure
  - Function
  - Tissue distribution
  - Mutant phenotype
- Gene organization and expression
  - Developmental pattern
  - Promoter

The sections *Tissue distribution* and *Developmental pattern* contain comparable information, except that the former reflects results obtained from studies of the protein product and the latter from studies at the RNA level. In some cases, when a group of genes are considered as part of a cluster or a gene family, there may be other sections within the chapter.

The section *Promoter* includes information on all *cis*-acting regulatory regions.

Some of the conventions I used are the following: *Nomenclature*, *cytogenetic, and genetic map position* follow *The Genome of* Drosophila melanogaster by Lindsley and Zinn (New York: NY: Academic Press, 1992). The names of proteins are abbreviated by using the same letters of the corresponding gene, capitalized and non-underlined, i.e. ADH for *Adh* and ACT5C for *Act5C*.

Sequences

All nucleotide sequences are numbered with A at the proposed site of translation initiation as position 1. The position immediately upstream of the initiation ATG is 0. Dots above the sequence mark the decades. Positions in the polypeptide chain obtained by virtual translation are indicated along the right-hand margin in parentheses.

The sequence figures were prepared using programs of the Genetics Computer Group of the University of Wisconsin (Madison, Wisconsin). Most of the sequence data were obtained from the GenBank and EMBL databases and the Accession numbers are given. In some cases, segments with no defined function at the 5′ and 3′ ends of a published sequence were omitted in the interest of space.

The site of transcription initiation is identified by the first dash of a three-character arrow (-->); it should be remembered that the resolution in defining this site experimentally is usually no better than $\pm$ 2 nucleotides.

The Hogness–Goldberg box and the polyadenylation signal are marked with double underlining (------). If a segment exists that matches the CAAT box sequence (or its reverse complement) 60–100 bp upstream of the transcription initiation site, it is also doubly underlined.

The polyadenylation site is marked by $|(A)_n$ below the sequence, where | indicates the last transcribed position or the last nucleotide before a string of A's? Introns in non-coding regions are delimited by brackets, ⌉⌋ and ⌈⌊ marking the end of one exon and the beginning of the next. Introns in coding regions can be identified by discontinuities in the amino acid sequence.

Short segments of interest such as promoter and enhancer elements are marked by dashes below the sequence (---). Arrowheads are often used to distinguish a certain sequence from its reverse complement (---> = 5′TAA3′, <--- = 5′TTA3′).

Longer segments are delimited by |--| below the sequence line, usually with some designation or label between the delimiters or after the second vertical line.

Base substitutions are indicated above the line followed by = followed by the designation of the mutant allele (e.g., A = n11 marks the position where an A for G substitution is found in the *Adh*^{n11} allele). Larger rearrangements are delimited by |--| above the sequence with a label describing the type of mutation (deletion, duplication, etc.).

Amino acid sequences are always the outcome of virtual translation. The initiation ATG is chosen according to the proposal of the original investigators. When confirmation of the amino acid sequence is available from direct protein sequencing, this fact is noted in the "product" section. In most cases, the positions of introns are derived exclusively from the comparison of cDNA and genomic sequences. TATA boxes and polyadenylation signals are indicated according to the proposals of the original investigators; these are usually based on sequence data alone. For other features, transcription initiation and termination sites, regulatory regions, etc., I indicate in the text the methods used to ascertain those features.

Gene Diagrams

The transcription initiation sites are marked by ⌈ for units in which transcription is from left to right, and by ⌉ for units in which transcription is from right to left. The boxes downstream of these symbols represent exons with the black boxes representing coding regions. The lines between exons represent introns.

Sequence Comparison Figures

In the case of some gene families, a comparison of polypeptide sequences is included to highlight differences or similarities between different members of the family. When the sequence of putatively homologous proteins from distant groups, mammals in particular, were available, a sequence comparison figure is included. The sequence alignments were done with the program *Pileup* of the Genetics Computer Group.

Contents

I

Contributors

Beth A. Cavener
Department of Molecular Biology
Vanderbilt University
Nashville, Tennessee

Douglas R. Cavener
Department of Molecular Biology
Vanderbilt University
Nashville, Tennessee

Andrew T. Lloyd
Department of Genetics
Trinity College
Dublin, Ireland

Stephen M. Mount
Department of Biological Sciences
Columbia University
New York, New York

Paul M. Sharp
Department of Genetics
Trinity College
Dublin, Ireland

I

1

The *achaete-scute* Complex: *ac, sc, lsc, ase*

Chromosomal Location:		**Map Position:**
ac	X, 1B2-3	1-0.0
sc; lsc	X, 1B3-4	1-0.0
ase	X, 1B3-4	1-0.0

Organization of the Complex

The *achaete-scute* complex is proximal to *yellow* (*y*) in a 90 kb segment that includes eight or nine transcription units; the units have been designated *T1* through *T9* (*T6* corresponds to *y*) (Fig. 1.1). Four of these are thought to be responsible for the *ac-sc* genetic function, the *scute* family. Within the *sc* family the following correspondence has been suggested: *T5* = *ac*; *T4* = *sc*; *T3* = *lsc* and *T8* = *ase*. Each of these four genes is transcribed toward the centromere. (Campuzano et al. 1985; Villares and Cabrera 1987; Alonso and Cabrera 1988; González et al. 1989, and references therein; see Ghysen and Dambly-Chaudière 1988 for a review).

Products

DNA-binding regulatory proteins of the basic helix-loop-helix (bHLH) type that promote neuroblast differentiation.

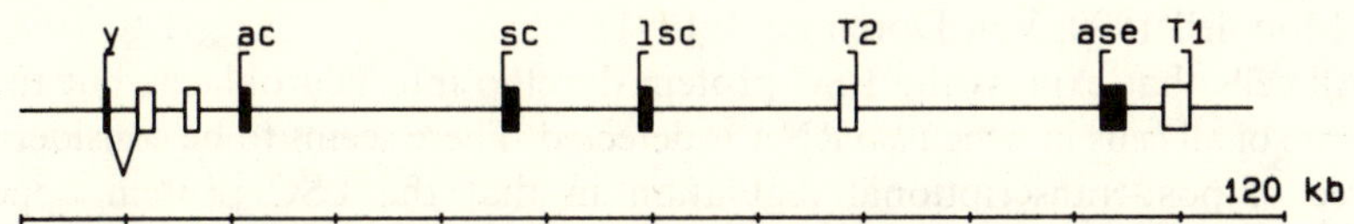

FIG. 1.1. The *ac-sc* complex and *y*. The open box to the left of *ac* corresponds to unidentified embryonic transcripts. *T7* (immediately to the right of *sc*) and *T9* (between *ase* and *T1*) have been omitted; there are conflicting reports on the existence of *T9* (Alonso and Cabrera 1988; González et al. 1989)

Structure

Sequence comparisons show that, in the region of the HLH domain, the products of *ac*, *sc*, *lsc* and *ase* are similar to each other and to the products of the mammalian oncogene *myc*, the myogenic gene *MyoD*, and the *Drosophila* genes *daughterless* (*da*), *Enhancer of split*, *extramacrochaetae* (*emc*), *hairy* (*h*), and *twist* (Fig. 1.2). In these proteins, the hydrophobic surface of each helix is involved in dimer formation; the amino acids in these regions are particularly well conserved. The basic amino acids in the vicinity of the helices, which effect DNA binding, are also conserved (Villares and Cabrera 1987; Alonso and Cabrera 1988; Murre et al. 1989a, 1989b; Harrison 1991). PEST elements, regions rich in Pro, Glu, Ser and Thr and thought to be important in protein degradation, are common to the various proteins; however, these are not correlated with sequence similarities (González et al. 1989).

Three genes in the complex, *ac*, *sc* and *lsc* share certain sequence elements that distinguish them from *ase*. Particularly noteworthy is the occurrence of a Tyr at the end of a run of acidic amino acids (position 394 in Fig. 1.2; a similar arrangement is found at position 222 of *ase*). A Tyr so associated with acidic residues is reminiscent of a motif found in substrates for protein tyrosine kinases (Villares and Cabrera 1987; Alonso and Cabrera 1988; González et al. 1989).

Function

Products of the *scute family* are transcriptional activators that promote transcription of genes involved in neuroblast differentiation. They act by binding to regulatory DNA sequences in association with ubiquitous helix-loop-helix proteins such as DA, the product of *da*. *In vitro*, AC, SC and LSC form heterodimers with DA. These complexes bind with high affinity to a DNA segment with the core sequence CANNTG, a sequence that is also found in the immunoglobulin kappa chain enhancer (Murre et al. 1989b), in the *hunchback* (*hb*) zygotic (proximal) promoter and at three positions in the *ac* promoter (Cabrera and Alonso 1991; Van Doren et al. 1991). In yeast cells, LSC/DA heterodimers induce transcription of a reporter gene bearing the *hb* target sequence in its promoter (Cabrera and Alonso 1991).

ac-sc function is counteracted by EMC, the product of *emc*. EMC, an HLH protein lacking the basic DNA-binding region, competes with the *ac-sc* products for DA binding. Thus, deficiency of EMC leads to excessive *ac-sc* function and the occurrence of ectopic sensory organs (see below; Ellis et al. 1990; Garrell and Modolell 1990; Van Doren et al. 1991).

All cells that express the LSC protein develop into neuroblasts, but this is not true of all cells in which *lsc* RNA is detected. There seems to be considerable degree of post-transcriptional regulation in that the LSC protein appears significantly later than the corresponding transcript and in a much more restricted subset of cells. Mutations in the neurogenic genes *Notch* and *Delta* (whose normal function is to limit neuronal differentiation to a single cell in a cluster of potential precursors) lead to the presence of LSC in all cells with *lsc*

```
     1                                                 50                                                100
lsc .........M TSICSSKF.Q QQHYQLT.NS NIFLLQHQH. ......HHQT QQHQLIAPKI PLGTSQL..Q NMQQSQQ... ....SNVGPM LSSQKKKFNY
 sc MKNNNNTTKS TTMSSSVLST NETFPTTINS ATKIFRYQHI MPAPSPLIPG GNQNQPAGTM PIKTRKY.TP RGMALTR... ...CSESVSS LSPGSSPAPY
 ac .......... .......... .......... .......... .......... .......... .......... ..MALGSE.. ....NHSVFN DDEESSSA.F
ase IRKIRDFGML GAVQSAAAST TNT..TPISS QRK.....RP LGESQKQNRH NQQNQQLSKT SVPAKKCKTN KKLAVERPPK AGTISHPHKS QSDQSFGTP.
CON ---------- ----S----- ---------S ---------- ---------- ---------- ---------- ---A------ ----S----- -S--S-----

    101                                                150                                               200
lsc NNMPYGEQLP SVARRNARER NRVKQVNNGF VNLRQHLPQT VVNSLS.... NGGRGSSKKL SKVDTLRIAV EYIRGLQDML D......DGT ASSTRHIYN.
 sc N....VDQSQ SVQRRNARER NRVKQVNNSF ARLRQHIPQS IITDLT...K GGGRGPHKKI SKVDTLRIAV EYIRSLQDLV D......DLN GGSNIGANNA
 ac N.......GP SVIRRNARER NRVKQVNNGF SQLRQHIPAA VIADLSNGRR GIGPGANKKL SKVSTLKMAV EYIRRLQKVL H......E.. .......NDQ
ase .GRKGLPLPQ AVARRNARER NRVKQVNNGF ALLREKIPEE VSEAFE..AQ GAGRGASKKL SKVETLRMAV EYIRSLEKLL GFDFPPLNSQ GNSSGSGDDS
CON N--------- SV-RRNARER NRVKQVNNGF --LRQHIP-- V---L----- G-GRG--KKL SKV-TLR-AV EYIR-LQ--L ---------- --S-------
                                 --*---* --*--- Helix I                 -*-**--** -**--*- Helix II

    201                                                250                                               300
lsc ......SADE SSNDGSSYND YNDS...... ........LD SSQQ...... .......... ......F... .......... .....LTGAT QSAQSRSYHS
 sc VTQLQLCLDE SSSHSSSSST CSSSGHNTYY QNRIS...VS PVQQQQQLQR .......... ........QQ FNHQPLTALS LNTNLVGTSV PGGDA.GCVS
 ac QKQKQLHLQ. .......... .......... .......... ..QQHLHFQQ .......... ........QQ .QHQHLYAWH QELQL..... ..........
ase FMFIKDEFDC LDEHFDDSLS NYEMDEQQTV QQTLSEDMLN PPQASDLLPS LTTLNGLQYI RIPGTNTYQL LTTDLLGDLS HEQKLEETAA SGQLSRSPVP
CON --------D- ---------- ---------- ---------- --QQ------ ---------- --------Q- -----L---- ----L----- ----------

    301                                                350                                               400
lsc A........S PTPSYSGSEI S........G GG........ .........Y IKQELQEQ.. .D.LKFDSFD SFSDEQ.... PDDEELL... ..DYISSWQE
 sc TSKNQQTCHS PTSSFNSS.M SFDSGTYEGV PQQ....... ISTHLDRLDH LDNELHTHSQ LQ.LKFEPYE HFQLDEEDCT PDDEEIL... ..DYISLWQE
 ac ....QSPTGS TSSC..NSIS SYCKPATSTI PGA....... TPP....... ..NNFHTKLE ......ASFE DYRNNSCSSG TEDEDIL... ..DYISLWQD
ase QKVVRSPCSS PVSPVASTEL LLQTQTCATP LQQQVIKQEY VSTNISSSSN AQTSPQQQQQ VQNLGSSPIL PAFYDQEPVS FYDNVVLPGF KKEFSDILQQ
CON ---------S P-S----S-- S--------- ---------- ---------- ---------- ---L------ ---------- --DE--L--- --DYIS-WQ-

    401                               439
lsc Q*........ .......... .......... .........
 sc Q*........ .......... .......... .........
 ac DL*....... .......... .......... .........
ase DQPNNTTAGC LSDESMIDAI DWWEAHAPKS NGACTNLSV
CON ---------- ---------- ---------- ---------
```

FIG. 1.2. *sc* family polypeptide sequences. The residues involved in the two helices are underlined, the conserved hydrophobic positions are marked with asterisks. The CON(sensus) sequence indicates positions where at least three of the sequences are identical. Alignment was done using the University of Wisconsin Genetics Computer Group *Gap* program. The first residue shown in this figure corresponds to amino acid 29 in the *ase* sequence.

mRNA and, thus, to the development of ectopic neural derivatives (Cabrera 1990).

SC distinguishes itself from the three other products in this family in that it plays a role in sexual development. This function was indicated by the ability of sc^+ to complement *sisterless b* (*sisb*) mutations (*sisb* is one of the "numerators" used to measure the X:autosome ratio that controls sex determination and dosage compensation early in embryonic development). This prompted the realization that *sc* and *sisb* are one and the same gene. The role of SC in sex determination is likely to involve the formation of a heterodimer between SC and DA. In embryos with two copies of *sc*, enough product is generated to form heterodimers capable of inducing transcription of *sex lethal* (*sxl*) and thus leading to female development. In embryos with only one *sc* copy (males), not enough heterodimer exists to induce *sxl* expression (Parkhurst et al. 1990; Erickson and Cline 1991).

Mutant Phenotypes

Mutations in the complex affect the development of sensory organs and central nervous system: *ac* and *ase* affect different subsets of larval and adult sensory organs while *sc* affects only a subset of adult sensory organs. Amorphic mutations that involve both *ac* and *sc* (sc^{10-1}) lead to the absence of all macro- and microchaetae except for those of the wing margin and eye. *lsc* mutations are embryonic lethals that lead to degeneration of the larval peripheral and central nervous systems; chaetae are, however, present. Also, in *lsc* mutant embryos, there is reduced expression of *hb* in neuroblasts (see *Function*). Mutations that increase expression of *ac* and *sc*, such as the dominant gain-of-function allele *Hairy-wing* (*Hw*), are associated with supernumerary chaetae at ectopic sites (Campuzano et al. 1986, and references therein). Amorphic mutations of *ase* cause abnormalities in the development of the adult optic lobes as well as alterations in peripheral neurons and chaetae (González et al. 1989).

ac
(*achaete*)

Synonym: *T5*

Gene Organization and Expression

Open reading frame, 201 aa; expected mRNA length, 912 bases. The 5′ end was determined by primer extension, RNase protection and sequencing of a cDNA clone; the 3′ end was determined from the sequence of the cDNA clone. There are no introns (*ac* Sequence) (Villares and Cabrera 1987).

ac

```
               .         .         .         .         .         .         .         .         .
-939  GAATTCTGAAATAATGGGACCTCCTAAATGCTTTCAAAATGCTTTCGGCTGAGAGGAACAACTGATACGTTGGGCATAAAGGCCCCGGGG   -850

               .         .         .         .         .         .         .         .         .
-849  CATTAGAAGTGTTAATAGAAAAGTCCTCCGGCTGATCAGGTTTCGTTGCAGGACCGAATGGATCGCCGCCTGAGGTGTTGATGAGCTGGC   -760

               .         .         .         .         .         .         .         .         .
-759  CTTGAAAATTCCTACGACTTTGGAGTCGAGCGACAATGGTCTAGTGTTTAAGATAATGTCCGAATGATCCAGGGATCGGAAGGTCATCAG   -670

               .         .         .         .         .         .         .         .         .
-669  TACATAAAATAAATTAAATTAAATGTATTAACATAAAATTAAAGATTTTTAAAAGTCTAAAATACCTAGCCTTGTTAATTAAAGATTAT   -580

               .         .         .         .         .         .         .         .         .
-579  TTTTTCGTAAACACTTTTGGTAGTGTATAAATTGTAAATGTCCCCATTTTTATAATTGTAATGACAGTCTATTCCACTAATTTTGTTGTA   -490

               .         .         .         .         .         .         .         .         .
-489  TTTTGTTAGTTATAAAAATTGGATGGCCACTTTCAATAGGAGATACAGCTTTTTACTTCGGAGGTGTTTTTACTTGGCTCTGATGTCTGG   -400

               .         .         .         .         .         .         .         .         .
-399  ACCTTGTTGCCTTTTTAAACCGGTTGGCAGCCGGCACGCGACAGGGCCAGGTTTTCGTTTGGGGACGACAGGCAGCTGAAAATGAACAAA   -310
                                                                              ------ e3

               .         .         .         .         .         .         .         .         .
-309  AACACTCAGAAACTCTTCCCACTCGACAACGGGAACACTCAGGTCACCAACAGCTGCGTTTTACAGAGAGAACGAGAGATAATATTACTA   -220
                                                        ------ e2

               .         .         .         .         .         .         .         .         .
-219  CCTCTCTATTAAAATCAGAGAAAACACTCATCTCAAGAGACGATCCTTCAGTGATGATGCTGTTGCACCTTTTCCAGGGGCAGGTAGGTA   -130

               .         .         .         .         .         .       -->-62      .         .
-129  GTCACGCAGGTGGGATCCCTAGGCCCTGATACCTATAAATAGCCTGAACGGAACGGGGAAGGGCATCAGAACAGAGCCAGCGCTGAAGCA   -40
            ------ e1                  ------

               .         .         .         .         .         .         .         .         .
 -39  AGGAGCATCGTCACACAATAACGTTATACTATCTCTTAAAATGGCTTTGGGCAGCGAAAATCACTCTGTTTTCAACGACGACGAGGAGTC   50
                                              MetAlaLeuGlySerGluAsnHisSerValPheAsnAspAspGluGluSe   (17)

               .         .         .         .         .         .         .         .         .
  51  ATCTTCGGCCTTTAATGGACCCTCTGTTATCCGGAGAAATGCCCGGGAACGCAACCGCGTAAAGCAGGTCAACAATGGCTTCAGCCAACT   140
      rSerSerAlaPheAsnGlyProSerValIleArgArgAsnAlaArgGluArgAsnArgValLysGlnValAsnAsnGlyPheSerGlnLe   (47)

               .         .         .         .         .         .         .         .         .
 141  ACGACAACATATCCCTGCGGCCGTAATAGCCGATTTAAGCAATGGTCGCCGGGGAATTGGTCCCGGCGCCAATAAAAAACTGAGCAAAGT   230
      uArgGlnHisIleProAlaAlaValIleAlaAspLeuSerAsnGlyArgArgGlyIleGlyProGlyAlaAsnLysLysLeuSerLysVa   (77)

               .         .         .         .         .         .         .         .         .
 231  TAGCACACTGAAAATGGCAGTAGAGTACATACGGCGCTTGCAGAAAGTTCTTCATGAAAACGACCAGCAGAAACAGAAACAGTTGCATTT   320
      lSerThrLeuLysMetAlaValGluTyrIleArgArgLeuGlnLysValLeuHisGluAsnAspGlnGlnLysGlnLysGlnLeuHisLe   (107)

               .         .         .         .        ||=Hw-1    .         .         .         .
 321  GCAGCAGCAACATTTGCACTTTCAGCAGCAGCAACAGCATCAACACTTATACGCCTGGCACCAAGAGTTGCAGTTGCAATCTCCAACTGG   410
      uGlnGlnGlnHisLeuHisPheGlnGlnGlnGlnGlnHisGlnHisLeuTyrAlaTrpHisGlnGluLeuGlnLeuGlnSerProThrGl   (137)

               .         .         .         .         .         .         .         .         .
 411  CAGCACAAGTTCCTGCAACAGCATTAGCTCTTATTGCAAGCCAGCAACATCGACGATTCCGGGAGCAACACCTCCTAACAATTTTCATAC   500
      ySerThrSerSerCysAsnSerIleSerSerTyrCysLysProAlaThrSerThrIleProGlyAlaThrProProAsnAsnPheHisTh   (167)

               .         .         .         .         .         .         .         .         .
 501  CAAGTTGGAAGCCAGTTTTGAAGACTACCGTAACAATTCCTGCAGTTCTGGTACTGAAGATGAGGACATCCTCGACTATATATCACTCTG   590
      rLysLeuGluAlaSerPheGluAspTyrArgAsnAsnSerCysSerSerGlyThrGluAspGluAspIleLeuAspTyrIleSerLeuTr   (197)

               .         .         .         .         .         .         .         .         .
 591  GCAGGACGACCTGTAAAAAAACAGATCAAATCTTCAGCTATTGCTAGTCGCACCCAACCATAACACACATCAAACCATTGATTGGCCAAC   680
      pGlnAspAspLeuEnd                                                                             (201)

               .         .         .         .         .         .         .         .         .
 681  AAGTATTACCTCAGCCACAAAGTATTTATATTCCCTAGAACTACCTTTTTGCCTTATAAATTAGTATTTAAGGTTTTATATAGTTTCTAA   770

               .         .         .         .         .         .         .         .         .
 771  GGATAGTTTCTAATGGAAGACAATTTATATTTAAGTTTTTTTTTATAGCATACATTCAGGACATTAAACTGATATATATAAAATTTTAAA   860
                                                                    ------          |(A)n
```

(continued)

```
                 .         .         .         .         .         .         .         .         .
861  TGAATTTTTATTGTAAACAAAATTAAACGGTAATTAAAGTGAAACAAATTTATGTACAAAAGGAGTAAAATTCAGAAAAGTTTTAATGAA   950
                 .         .         .         .         .         .         .         .         .
951  CAAATGCTTTATGAATATGGGCGTAGCAATGTTTTGATACAAACTTGATCCTGTCCTGTATACCACAGGACACGCTTCCTTTTACCTGGT   1040
                 .         .         .         .         .         .         .         .         .
1041 ACATTCCTTTAAACGATCCTAGTATACGCTTTATTCGGGGTAAGCCCGAAAAAAGTATTCGAAACTGTAACCGTTAAGTATTTACAGATC   1130
                 .         .         .         .         .         .         .         .         .
1131 ACTAGCCAATGAAGATAAATTACAATAACATTTTGTAAACACTTTTGATCGAAAACGCCGATTTGCATAAATAAAGTTGGATTGAGTAGG   1220
                 .         .         .         .         .         .         .
1221 GTGAAAAAGGAAAATATTTACCTGCTGCATTTTTGCATATGAACCGGTCAAGGTAATAAGATCCTGAGAATTC  1293
```

ac SEQUENCE. Strain, *Canton S*. Accession M17120 (DROASC1). e1, e2 and e3, AC/DA binding sites (Van Doren et al. 1991). The dominant allele Hw^1 is caused by insertion of a *gypsy* element after nucleotide 368; termination occurs within the transposon's terminal repeat, one codon after the insertion (R. Villares and C. V. Cabrera, personal communication).

Developmental Pattern

The expression patterns of *ac* and *sc* and *lsc* are very similar. Before blastoderm formation, expression is uniform throughout the embryo. Later, in early gastrula, transcripts begin to accumulate in stripes restricted to ectodermal cells. During the period of fast germ-band extension (stages 8 and 9), a pattern of two stripes per metamere develops; soon thereafter, when neuroblasts segregate from the ectoderm, transcription is restricted to the neurogenic cells and ceases in epidermal precursors. At the end of stage 9, when neuroblasts begin to divide, transcripts fade (Cabrera et al. 1987).

As development proceeds, expression appears restricted to small clusters of cells that are distributed in a more complex pattern. Even so, the general design outlined above persists: as waves of neuroblast differentiation occur throughout the embryo, transcripts appear immediately before and during the segregation of neuroblasts from the ectoderm; then, the transcripts disappear again, first from the epidermal precursor cells, and finally from the dividing neuroblasts. During germ-band shortening, as differentiation of the neural precursors is completed, expression ceases in the segmented portion of the embryo. After germ-band shortening, expression persists in the primordia for the optic lobes and stomatogastric nervous system (Cabrera et al. 1987). In third instar larvae and early pupae, these genes are expressed in imaginal discs in groups of cells from which the sensory organ mother cells will develop (Romaní et al. 1989). In wing imaginal discs, *ac* and *sc* are expressed with very similar distributions, although mutations affect different sensory organs. Experiments with a reporter gene in transgenic flies indicate that *ac* and *sc* are initially expressed in different clusters of cells; but their products stimulate transcription of each other, so that the ranges of expression soon overlap. As a consequence, in mutants for only one of the two genes, expression of both genes is affected, albeit in different subsets of clusters (Martínez and Modolell 1991).

Differences in expression among the genes are: (1) *ac* stripes are slightly offset from those of *lsc* and *sc*; and (2) during the later stages of expression (stages 10, 11 and 12), transcription of *ac* is more intense than that of *sc* and *lsc*, but *lsc* RNA occurs in more cells.

sc

```
               .         .         .         .         .         .         .         .         .
-659  AAAAAATTTTGATCCTTTTGATAATTTAATTGGAGAAATAAGTGAAATTGTTTGAACACCTTTAGGGAGCGTACTCCGAATGTCTAATAA  -570
               .         .         .         .         .         .         .         .         .
-569  GGAGGATCCCAGGATCGGCTGTCGATCCCTTGGATCCGTCCGGCGCTAATGAATAGAAGCGTGCGTGAGCTGCACATAAAATTGGCGATC  -480
               .         .         .         .         .         .         .         .         .
-479  GCGACTTTTGCTAAGTTAATTAACACAGAAATCAAATTCCTGGCGTGCCGTAGCAAAAAGAGCCCTCACTCAGATACCTTGATCGTTTTT  -390
               .         .         .         .         .         .         .         .         .
-389  CGATATTTCGAGTTGATATTTTGAGTTTAAAATTTGAGTGTTTCTTTTGGACTGTCGAGTGAGAACAGTTTTCCTGTGGGATACTCGAGT  -300
               .         .         .         .         .         .         .         .         .
-299  ACCTGAGACAGAGAAAGAGAGAGAGACTACCTGTGGCTCACTCACTTCGAGTTCCCTACCTGTGCAGGCAGCTCTTGCCGTCACTCTCTC  -210
               .         .         .         .         .         .         .         .         .
-209  TCTCTCTTTCTCTCCGATTCTCTCGCCCGTTTCTCTGCCTGAGTGTTGTGCAGAGAGTTGCATAAAGGGTACATAACGCGAGGGTTTAGG  -120
                                                                  ------
         -->  --> -116/-111                        .         .         .         .         .         .
-119  ACGAAGGGACTCATTCTTGTGTAAGGTGTCAAACGATCAAGTTCAAGTATTGTACTCTGTTCATTTATTTTTTTCTGTTGATCGTTATCC  -30
               .         .         .         .         .         .         .         .         .
 -29  GGAAAGTGAAAGAAAGCTCCGAGTGTGTTAATGAAAAACAATAATAATACAACGAAAAGCACTACCATGTCATCGAGTGTGCTGTCCACC  60
                                    MetLysAsnAsnAsnAsnThrThrLysSerThrThrMetSerSerSerValLeuSerThr  (20)
               .         .         .         .         .         .         .         .         .
  61  AACGAAACGTTTCCAACGACCATCAATTCGGCAACGAAGATCTTTCGTTATCAGCACATAATGCCAGCCCCTAGTCCATTAATTCCCGGT  150
      AsnGluThrPheProThrThrIleAsnSerAlaThrLysIlePheArgTyrGlnHisIleMetProAlaProSerProLeuIleProGly  (50)
               .         .         .         .         .         .         .         .         .
 151  GGCAATCAAAATCAACCCGCTGGCACAATGCCAATTAAGACTCGCAAGTATACACCAAGGGGTATGGCACTGACCAGATGCTCTGAATCA  240
      GlyAsnGlnAsnGlnProAlaGlyThrMetProIleLysThrArgLysTyrThrProArgGlyMetAlaLeuThrArgCysSerGluSer  (80)
               .         .         .         .         .         .         .         .         .
 241  GTATCATCTCTATCGCCTGGTTCCTCGCCGGCTCCATATAATGTAGACCAATCCCAGTCGGTCCAAAGGCGCAATGCTAGAGAACGAAAT  330
      ValSerSerLeuSerProGlySerSerProAlaProTyrAsnValAspGlnSerGlnSerValGlnArgArgAsnAlaArgGluArgAsn  (110)
               .         .         .         .         .         .         .         .         .
 331  CGTGTAAAGCAGGTGAACAACAGCTTCGCCAGGTTGCGGCAACATATACCACAATCCATAATCACGGATTTGACAAAGGGTGGTGGTCGA  420
      ArgValLysGlnValAsnAsnSerPheAlaArgLeuArgGlnHisIleProGlnSerIleIleThrAspLeuThrLysGlyGlyGlyArg  (140)

                                                                       T=sc-10.1
               .         .         .         .         .         .         .          .        .
 421  GGACCTCACAAAAAGATCTCCAAAGTAGACACACTGCGCATTGCCGTCGAGTACATCCGGAGCCTTCAGGATCTGGTGGATGACCTAAAT  510
      GlyProHisLysLysIleSerLysValAspThrLeuArgIleAlaValGluTyrIleArgSerLeuGlnAspLeuValAspAspLeuAsn  (170)
                                                                        End

               .         .         .         .         .         .         .         .         .
 511  GGGGGCAGCAATATTGGTGCCAACAATGCAGTCACCCAGCTTCAACTTTGTTTGGATGAGTCCAGCAGTCACAGTTCGAGCAGCAGTACT  600
      GlyGlySerAsnIleGlyAlaAsnAsnAlaValThrGlnLeuGlnLeuCysLeuAspGluSerSerSerHisSerSerSerSerSerThr  (200)
               .         .         .         .         .         .         .         .         .
 601  TGCAGTTCCTCAGGGCATAATACCTACTATCAAAACAGGATCTCTGTCAGTCCTGTGCAACAACAGCAGCAGCTACAGAGGCAGCAGTTC  690
      CysSerSerSerGlyHisAsnThrTyrTyrGlnAsnArgIleSerValSerProValGlnGlnGlnGlnGlnLeuGlnArgGlnGlnPhe  (230)
               .         .         .         .         .         .         .         .         .
 691  AATCACCAACCGCTGACAGCGCTCTCATTAAATACCAACTTGGTGGGCACATCCGTACCAGGTGGAGATGCAGGATGCGTATCCACCAGC  780
      AsnHisGlnProLeuThrAlaLeuSerLeuAsnThrAsnLeuValGlyThrSerValProGlyGlyAspAlaGlyCysValSerThrSer  (260)
               .         .         .         .         .         .         .         .         .
 781  AAAAACCAGCAAACCTGCCACTCGCCAACATCATCATTCAACTCCAGCATGTCCTTTGATTCAGGCACCTACGAAGGAGTTCCCCAACAA  870
      LysAsnGlnGlnThrCysHisSerProThrSerSerPheAsnSerSerMetSerPheAspSerGlyThrTyrGluGlyValProGlnGln  (290)

                                  ||=Hw-Ua
               .         .                 .         .         .         .         .         .
 871  ATATCCACCCACCTGGATCGTCTGGATCATCTGGACAACGAATTACACACGCACTCCCAACTTCAGCTAAAATTTGAACCGTACGAACAT  960
      IleSerThrHisLeuAspArgLeuAspHisLeuAspAsnGluLeuHisThrHisSerGlnLeuGlnLeuLysPheGluProTyrGluHis  (320)
               .         .         .         .         .         .         .         .         .
 961  TTTCAATTAGACGAGGAGGACTGCACCCCCGACGACGAGGAGATTTTGGACTACATCTCTCTATGGCAGGAGCAGTGACTTAATCCCCAA  1050
      PheGlnLeuAspGluGluAspCysThrProAspAspGluGluIleLeuAspTyrIleSerLeuTrpGlnGluGlnEnd              (345)
```

(*continued*)

```
                 .         .         .         .         .         .         .         .         .
1051  AATTTACCACCACGCCCTATTTTCTTCTAGTCAATGTTGAGTTGAACCAAGTGCCTCAAATTGTAAATAACACTAATACAAAAACAACAT  1140

                 .         .         .         .         .         .         .         .         .
1141  ACCCCCAATTTTTTTTTCTTACTTTAAGCTATTTTTTTACATTGTTAAGAACCACGAGACCAGTTTCAAATTTATATATTTATGAAATAA  1230

                 .         .         .         .         .         .         .         .         .
1231  CTATAGCATGGAAACGAAAACATATTTTTTTGGCTAATACAATTTTATGTTAATTAGTTTTGGTGGAAAAATAAAATGAAAAAATTAAAC  1320
                                                                             ------
                                                                             ======
                 .         .         .         .         .         .         .         .         .
1321  GAAAAATAATATTTAAGTTTTTTTGTACAAAGGGGATCCATCTATTGCATCAGGTTTGTAAAACATTCGGGTACTACTTGCATTGCCTTG  1410
      |(A)n
                 .         .         .         .         .         .
1411  CAGTGCCGATGGGACCATGTGCAGCCGTTATGTACATTGGTTGCTTTGCATTGGTTTTCCA  1471
```

sc SEQUENCE. Strain, *Canton S.* Accession M17119 (DROASC2). The base substitution at 487 in the null allele sc^{10-1} is indicated; this mutation also involves a breakpoint that inactivates *ac*. The dominant allele Hw^{Ua} is caused by insertion of a *copia* element after nucleotide 899; termination occurs within the transposon's terminal repeat, 21 codons after the insertion (R. Villares and C. V. Cabrera, personal communication).

Promoter

A segment of 0.9 kb upstream of the transcription initiation site is sufficient for nearly normal expression of *ac* (Ruiz-Gómez and Modollel 1987). Within that segment, there are three binding sites apparently responsible for autocatalysis: binding of heterodimers of *ac-sc* and *da* products has been detected at three copies of the element CANNTG (sites marked e in the *ac* sequence at -327, -259 and -123). This binding is blocked by the simultaneous presence of EMC (Van Doren et al. 1991).

sc
(*scute*)

Synonyms: *T4* and *sisb*

Gene Organization and Expression

Open reading frame, 345 aa; expected mRNA length, 1,437 and 1,432 bases. The 5′ ends were determined by primer extension; sequencing of a cDNA clone provided the 3′ end. There are no introns. Translation might initiate at any of five in-frame AUGs in the mRNA. In the *sc* Sequence, translation is depicted as starting at the first of those ATGs, but the best fit to the initiation of translation consensus is next to the fifth ATG (Villares and Cabrera 1987).

Developmental Pattern (see *ac*)

Product from the blastoderm period of *sc* expression is probably associated with the *sisb* function.

lsc

```
-302  CTGAGTAGGAATAGAGGCACCCACCACAGAAAAAGAACCCCTAGAAAGAGAGGAAAAATGTACGATCACTTGTGCAAAGGACTTAGGTCC  -213

-212  CGGTTTTTCGAGGGCAGGTAGCCAGGATCCGACCCCGTACCAACCCCTGTAGCTCCTCTGCCGAAGTCGCTGCCTCTGTCGCGGCGCGTT  -123

-122  TCCCTCTGCCACTGGCCGGGTATTTAAAGCCCTAGATCAGAACAGCAATTATCATTGCGGAATCTGATTCCACACAGTCAACATCTGTAA  -33
          !-26
 -32  ACTAAATCTTAGAAAACTCTCACAAGGATTACCATGACGAGCATTTGCAGCAGCAAATTCCAGCAGCAGCATTACCAGCTGACCAACAGT  57
                                     MetThrSerIleCysSerSerLysPheGlnGlnGlnHisTyrGlnLeuThrAsnSer  (19)

  58  AACATTTTCTTGCTGCAACATCAGCATCACCATCAAACGCAGCAGCACCAGTTGATTGCTCCGAAAATACCTTTGGGTACCAGCCAACTG  147
      AsnIlePheLeuLeuGlnHisGlnHisHisHisGlnThrGlnGlnHisGlnLeuIleAlaProLysIleProLeuGlyThrSerGlnLeu  (49)

 148  CAGAATATGCAGCAGAGTCAACAGTCCAATGTTGGACCCATGTTGTCCTCCCAGAAGAAGAAGTTCAACTACAATAACATGCCCTATGGC  237
      GlnAsnMetGlnGlnSerGlnGlnSerAsnValGlyProMetLeuSerSerGlnLysLysLysPheAsnTyrAsnAsnMetProTyrGly  (79)

 238  GAGCAATTGCCATCGGTAGCCAGACGAAATGCCCGTGAACGCAATCGCGTGAAGCAGGTGAACAATGGATTCGTCAATCTCCGCCAGCAT  327
      GluGlnLeuProSerValAlaArgArgAsnAlaArgGluArgAsnArgValLysGlnValAsnAsnGlyPheValAsnLeuArgGlnHis  (109)

 328  TTGCCTCAAACTGTGGTAAACTCGCTGTCCAATGGAGGACGTGGTAGCAGCAAGAAGTTATCCAAGGTGGACACACTGCGAATCGCCGTT  417
      LeuProGlnThrValValAsnSerLeuSerAsnGlyGlyArgGlySerSerLysLysLeuSerLysValAspThrLeuArgIleAlaVal  (139)

 418  GAATATATTCGAGGACTACAGGACATGCTTGATGATGGCACTGCTTCATCAACTCGTCACATCTACAATTCCGCCGATGAAAGTAGCAAC  507
      GluTyrIleArgGlyLeuGlnAspMetLeuAspAspGlyThrAlaSerSerThrArgHisIleTyrAsnSerAlaAspGluSerSerAsn  (169)

 508  GATGGCAGCAGCTATAACGATTACAACGATAGTTTGGACAGTTCGCAACAGTTCTTGACGGGAGCCACCCAGTCTGCCCAATCCCGCTCG  597
      AspGlySerSerTyrAsnAspTyrAsnAspSerLeuAspSerSerGlnGlnPheLeuThrGlyAlaThrGlnSerAlaGlnSerArgSer  (199)

 598  TACCACTCCGCCTCGCCCACGCCGTCGTACTCCGGATCCGAGATTTCCGGAGGTGGCTATATCAAACAGGAACTACAAGAGCAGGACCTC  687
      TyrHisSerAlaSerProThrProSerTyrSerGlySerGluIleSerGlyGlyGlyTyrIleLysGlnGluLeuGlnGluGlnAspLeu  (229)

 688  AAATTCGACTCCTTTGATAGCTTCAGTGACGAGCAGCCAGATGACGAGGAGCTACTCGATTATATTTCATCTTGGCAAGAGCAGTGAAGG  777
      LysPheAspSerPheAspSerPheSerAspGluGlnProAspAspGluGluLeuLeuAspTyrIleSerSerTrpGlnGluGlnEnd     (257)

 778  GGTCTTACTAAAAGTCCCAAACAAACAAATATTGTACAAACTGTAAATACCCTAAATTGTTGCCTTAGTGAGTGTAAAACCAAGTCTCAA  867

 868  ATTTCACATTAGCCTCTAAGTTACCCCATATTTTTTTTTATTATATTTTAACGCAATGGAAGACAATGATAGAAAACCACATATTTTTTT  957

 958  CATAGTTATAAGTTTGTTATAAGCATGGAAGACACTAAACTAACTACTTTTAAGCCAAAATAAAAACATATTGATAAATTTAATTCCAAA  1047

1048  TGTTTTTTACTGAAATCACTTACTCGTAAATATATTCAGATCGTCATGTAGGGTAATTACAACGAGTTCTCGTTCTCATACCAGCATCAG  1137

1138  AGCCAAAAAGGTTTTTAAACAATCTGCATTTTGAAGCATTGCTTTGACTATATATATGTTATGATATTCGTTTTTAATTTATGTATTTTT  1227
                      |(A)n

1228  ATATTATTATTATTATTTTTTAGCTTAGCTGTTTTGGCCTCAGGCTTAATAATGGTACTAGCGATAGAAATAATAATTTCACAAAAAAGT  1317

1318  TACCCAATTTATTATTTATATTCAATTACTTTTGGAGCGTGGACATGACTCACTCAAAATTCGTAACCAACATAGAGTTAAGCACCTGAC  1407

1408  AGGAAACAACAGCGAATATTTTCATGATTGGTTCCCTAACGAGCTACAATTCGGCCGGGAATTGTTAATGGCGCGTAAATAGCCCGGAAA  1497

1498  TAGGCAGTCACGCCTGAGAGGATGAAATTGTCCTAGTCCAAGG  1540
```

lsc SEQUENCE. Strain, *Canton S*. Accession, X12549, Y00846 (DROASCA).
The exclamation mark at −26 indicates the 5′ end of a cDNA.

Promoter

An *sc* construction with approximately 1 kb of DNA upstream of the transcribed region and 3 kb downstream is sufficient to provide *sisb* function but not *sc* function (Erickson and Cline 1991). The *cis*-acting regulatory region of *sc* is likely to extend for tens of kilobases.

lsc
(*lethal of scute*)

Synonym: *T3*

Gene Organization and Expression

Open reading frame, 257 amino acids; expected mRNA length, ca. 1,184 bases. A cDNA sequence and low resolution S1 mapping were used to define the 5′ and 3′ ends. There are no introns (*lsc* sequence) (Alonso and Cabrera 1988).

Developmental Pattern

lsc expression follows the general pattern of *ac* and *sc* expression (see *ac*) except that the expression of *lsc* seems to be more extensive than that of the other two genes and persists longer in both epidermal precursors and neuroblasts (Cabrera et al. 1987).

ase
(*asense*)

Synonym: *T8*

Gene Organization and Expression

Open reading frame, 486 amino acids; expected mRNA, 2,263 bases. The 5′ end was defined by primer extension and the 3′ end by a cDNA sequence. There are no introns (*ase* sequence) (Alonso and Cabrera 1988, partial sequence; González et al. 1989).

ase

```
-1942  GGATCCAGTATGTTTCCACGCTAGCGTCAATTCCGTTTACTCATCTGTTTCATTACCATTTGGCGTTTCTCTCGTCGAAAGATATTTTCC  -1853

-1852  CATTGAAATCAATGCGTTTTTAAAATGCAAATAAACCAGAAACCAGAAACCATTCATAAATTGTATTTGCCTAGATTGGAACATTTCGAT  -1763

-1762  CCGCCAAAGATAACAGCCAAAAAAATATATAAAAAAAAGAGTGACAGGACTACAACGCAGGTTCTTAATTTTACGAACGTGGGAGTAAAT  -1673

-1672  TCGAAAAGGTATGCCGCGTCTTGGGGCAAAACTTTTTTGAAACCGTTTACATGTAATATTTTTGGAATCGCTACTTTTATGTATGGTTAA  -1583

-1582  TTTAATTATGAACTATTTTCTTGCAGTGCACGAAAGGCGTGGCTGGGGCAAGGAACAGTTCCTTGAGATGAGTGCTGCGATGCGTCGTCA  -1493

-1492  AAGTGGGACGCAACCGAGTCAAATCCTCTAGGACAACAAAGGACGCCGAGCAGTACTTCCCAGTACTCAAATACTCCTCAGTACGCACAA  -1403

-1402  GCGTTGACTCCTTTTTCTTTGAGAGCTCGTCTGCATAATGAGGAATGAGGACGTGGCATCCTGGATCAAAAACCGGTAGTCGGTCGTCGC  -1313

-1312  AAGTTTCTTCTCCCCGCCGGTTATCCTGCGCTCAAGTCCTTTTCCTTGAACCTTTTAAGTGAACTCAAGTTTATAAATTGTGCAGCAAGT  -1223

-1222  ACAACACACACACATATGTATACTCCTCTATTTACTCAGGTTTGTTGGAAACTACCTGAGAGGAAGGATCAAAAAAGTTTATTGTAGCAC  -1133

-1132  TCGAAAACTTGTTTGCACAGATAGACCTAAGCTCCAAAAAAAAGAATAGATTATAGAAATAAAGCACTGGTAAAATTATTTTCCCACCGG  -1043

-1042  ACTAATTGCCAGAAATTTTTTGCCAGACGTAGAAAAAACAAGAATGTAGAGAAGGATGGGTGATTTCTCACCCCTTAAAGGATTTAAATT  -953

 -952  GGCTCTCTGGCATCTTGTCAATTTCCAACATAAATTGTAGCCCTGTGAATTACCTAAGACATTACTTTCGCAGTATATACTTGCTGTTTA  -863

 -862  TTAGCTTAACAATAGGAAAATGTTTTTGCCAATGCAAGTCCTGGTAAATATATGTATATATTATCCAGTACGAGTTTTTGAAAAGTTAAC  -773

 -772  AATAGGTGATCCCGAGACATTTTTCGAATGAAGTAGAAACCAGTCCTTGGTTTTAGCTATAAGCTAAAAATAAAGATTCGATGCATTTCT  -683

 -682  GCGTTTTACATGACGAATATTGGAGGTCTAAGGTGATCTATTAGGATATTTTGCAAATTCCTAGGTGGTAGGGCATTCCTTGAAAACCAG  -593

 -592  GGCTGAAAAAGCTCCCCAGGGAATATACTTTTTATTATATGCATACGTATATGGTTATTATAATGTCCATATATTAATAGGGGCGGTATA  -503

                                                      -->-455
 -502  TAAGCATAATGTGTTTGCTGCCGATAAATAATGAGAGAGAGCGTGGCGAAATCCACCCTTGAACCAGGTGGACTTTTTGGCTCGAGTTGA  -413
          -----           ------

 -412  TCAGATGTTAGTTTTCCCAAAAGCCGTACTGTATATAATATATATATATGATAAAGGTGTATGTGTGAGCGATCGAGAAGGGACACCCAG  -323

 -322  TCTAGGGGATGAAAAGTCAGGCCCTTTACATAAGGGATACGCAGGACCTCAAATGCCTTCTGTTTTGTATGTGTGTGCTCTGTGTTGTGT  -233

 -232  ATGTCAGTCAACGAAGTCACTTCCGTTGGGTTTGCGTTTTAGTTTGAGTTCGGAGTTTAGGGGCACGCGACACAGAGCGCCAGCAGCTGT  -143

 -142  CCTGATGCAAGGACACGGAAACCATATTACATCAGTCACCAGTTAACATTCACTCAAGAAGGACTAACTTGCTAAAAGTACACCCGCAAT  -53

  -52  CGCCACCAGTTTTTCTCCCGCCCTCAAAAAGCCACGAATCAAAAAACTTAATTATGGCCGCCTTAAGCTTCAGCCCATCACCTCCTCCAA  37
                                                              MetAlaAlaLeuSerPheSerProSerProProProL  (13)

   38  AAGAAAACCCCAAGGAAAACCCCAATCCAGGAATAAAAACCACGTTGAAACCTTTTGGAAAGATTACCGTTCACAATGTTTTAAGTGAGA  127
       ysGluAsnProLysGluAsnProAsnProGlyIleLysThrThrLeuLysProPheGlyLysIleThrValHisAsnValLeuSerGluS  (43)

  128  GTGGCGCCAACGCCTTGCAACAGCATATAGCCAATCAGAACACCATTATTCGAAAGATCCGGGACTTTGGCATGCTGGGCGCTGTTCAAA  217
       erGlyAlaAsnAlaLeuGlnGlnHisIleAlaAsnGlnAsnThrIleIleArgLysIleArgAspPheGlyMetLeuGlyAlaValGlnS  (73)
```

(continued)

```
218  GTGCCGCAGCCAGCACAACTAACACCACACCCATATCCAGTCAACGGAAGAGGCCCCTGGGAGAATCCCAAAAGCAGAACCGGCACAACC   307
     erAlaAlaAlaSerThrThrAsnThrThrProIleSerSerGlnArgLysArgProLeuGlyGluSerGlnLysGlnAsnArgHisAsnG  (103)

308  AGCAGAATCAACAGCTTAGTAAAACATCAGTGCCTGCTAAAAAATGCAAGACCAACAAGAAGTTGGCGGTTGAAAGGCCCCCAAAAGCAG   397
     lnGlnAsnGlnGlnLeuSerLysThrSerValProAlaLysLysCysLysThrAsnLysLysLeuAlaValGluArgProProLysAlaG  (133)

398  GAACTATAAGCCACCCTCATAAAAGCCAAAGCGATCAGAGTTTTGGGACTCCTGGAAGAAAGGGTTTGCCTTTGCCACAAGCCGTTGCCC   487
     lyThrIleSerHisProHisLysSerGlnSerAspGlnSerPheGlyThrProGlyArgLysGlyLeuProLeuProGlnAlaValAlaA  (163)

488  GTAGAAACGCTAGGGAAAGAAATCGCGTGAAGCAGGTTAACAATGGATTTGCTTTACTCCGGGAGAAGATCCCAGAAGAAGTATCTGAGG   577
     rgArgAsnAlaArgGluArgAsnArgValLysGlnValAsnAsnGlyPheAlaLeuLeuArgGluLysIleProGluGluValSerGluA  (193)

578  CTTTTGAGGCCCAGGGGGCGGGTAGAGGAGCAAGCAAGAAGCTATCCAAAGTGGAGACCCTCCGCATGGCCGTAGAGTACATAAGAAGTT   667
     laPheGluAlaGlnGlyAlaGlyArgGlyAlaSerLysLysLeuSerLysValGluThrLeuArgMetAlaValGluTyrIleArgSerL  (223)

668  TGGAAAAACTGCTGGGATTTGATTTTCCACCTCTCAACAGTCAGGGGAATAGTTCTGGTTCCGGCGATGATAGCTTTATGTTTATTAAGG   757
     euGluLysLeuLeuGlyPheAspPheProProLeuAsnSerGlnGlyAsnSerSerGlySerGlyAspAspSerPheMetPheIleLysA  (253)

758  ACGAATTCGATTGTCTGGATGAACATTTCGACGACTCGCTGAGCAACTACGAAATGGATGAGCAACAGACTGTCCAACAAACTTTATCCG   847
     spGluPheAspCysLeuAspGluHisPheAspAspSerLeuSerAsnTyrGluMetAspGluGlnGlnThrValGlnGlnThrLeuSerG  (283)

848  AGGATATGCTAAACCCTCCGCAAGCCAGTGATCTCCTGCCTAGTTTGACTACATTAAATGGGTTGCAATACATCAGAATACCAGGAACCA   937
     luAspMetLeuAsnProProGlnAlaSerAspLeuLeuProSerLeuThrThrLeuAsnGlyLeuGlnTyrIleArgIleProGlyThrA  (313)

938  ACACCTACCAACTGCTGACGACTGACTTATTGGGCGATTTGAGTCACGAGCAAAAACTTGAAGAAACAGCTGCTTCGGGCCAGTTATCGC  1027
     snThrTyrGlnLeuLeuThrThrAspLeuLeuGlyAspLeuSerHisGluGlnLysLeuGluGluThrAlaAlaSerGlyGlnLeuSerA  (343)

1028 GATCGCCCGTGCCACAAAAGGTGGTAAGAAGTCCCTGCTCTTCTCCAGTTTCACCTGTCGCCTCGACTGAATTGCTGTTACAGACACAGA  1117
     rgSerProValProGlnLysValValArgSerProCysSerSerProValSerProValAlaSerThrGluLeuLeuLeuGlnThrGlnT  (373)

1118 CGTGTGCCACACCGCTGCAACAGCAAGTAATCAAACAGGAATACGTCAGTACCAACATTAGCAGCAGCAGCAACGCACAGACTTCCCCGC  1207
     hrCysAlaThrProLeuGlnGlnGlnValIleLysGlnGluTyrValSerThrAsnIleSerSerSerSerAsnAlaGlnThrSerProG  (403)

1208 AGCAGCAGCAGCAAGTTCAGAACCTGGGATCGTCGCCTATTTTACCCGCGTTCTACGACCAGGAGCCCGTGAGCTTCTACGACAACGTAG  1297
     lnGlnGlnGlnGlnValGlnAsnLeuGlySerSerProIleLeuProAlaPheTyrAspGlnGluProValSerPheTyrAspAsnValV  (433)

1298 TCCTTCCCGGATTCAAGAAGGAATTCAGCGATATTTTGCAGCAAGATCAGCCCAACAATACAACCGCTGGCTGCCTTTCGGACGAGAGCA  1387
     alLeuProGlyPheLysLysGluPheSerAspIleLeuGlnGlnAspGlnProAsnAsnThrThrAlaGlyCysLeuSerAspGluSerM  (463)

1388 TGATCGATGCCATTGACTGGTGGGAGGCACATGCACCTAAATCTAATGGTGCATGCACCAATCTGTCCGTTTAGCCGAATTTTTTCACAT  1477
     etIleAspAlaIleAspTrpTrpGluAlaHisAlaProLysSerAsnGlyAlaCysThrAsnLeuSerValEnd                 (486)

1478 CACGCATCTCGGAAAAGCCGATTGCATTTTTTGGCATACTTTTTAAATGATTTTAAATCCTCACAGCATAAGTCTGTGGCAGGCCATTCT  1567

1568 ATCTAAAGTTTTTTTTAATCAAGCCATGACTGAGTCATTGTGTAAATATCAATTTAAGCCGAGAAAGGAGGATAACTTCGGCCAGCCGAA  1657

1658 GCTTATATACCTTTGCTGTTAAAACCATGTATTTAATATGAAAGTTCGCACAATTTCGATGAAGTTTATCACAAATTTACGATTTCATCA  1747

1748 AGATTTGTATATTCTCCAAATTCTATAAAATATATGTACATTTTTGATTCTTGCTATGGTACTTGTACGTATGATATTGTTGATCGATCC  1837
                          ------                                        |(A)n

1838 TGCCCGAGTCACCTTTTATATCACCAGACATGCCGATCATGAATATTTATTGATGTGGGGCTGGAA  1903
```

ase SEQUENCE. Strain, *Canton S*. Accession X51532 (DROASE).

Developmental Pattern

The pattern of *ase* expression is very different from that of the other three genes in the *sc* family. Expression does not initiate until the extending germ-band stage (late stage 8 embryos); then *ase* transcripts occur in neuroblasts after they have segregated from the ectoderm. After germ-band retraction (stage 13), expression in the segmented region of the embryo ceases, but *ase* transcripts persist in the presumptive optic and procephalic lobes (Alonso and Cabrera 1988; González et al. 1989). Expression of *ase* is also evident in late third instar larvae, occurring throughout the central nervous system in many of its actively proliferating cells (González et al. 1989).

References

Alonso, M. C. and Cabrera, C. V. (1988). The *achaete-scute* gene complex of *Drosophila melanogaster* comprises four homologous genes. *EMBO J.* **7**:2585–2591.

Cabrera, C. V. and Alonso, M. C. (1991). Transcriptional activation by heterodimers of the *achaete-scute* and *daughterless* gene products of *Drosophila. EMBO J.* **10**:2965–2973.

Cabrera, C. V., Martínez-Arias, A. and Bate, M. (1987). The expression of three members of the *achaete-scute* gene complex correlates with neuroblast segregation in *Drosophila. Cell* **50**:425–433.

Cabrera, C. V. (1990). Lateral inhibition and cell fate during neurogenesis in *Drosophila*: the interactions between *scute*, *Notch* and *Delta. Development* **109**:733–742 [Reprinted in **110**(1)].

Campuzano, S., Balcells, L., Villares, R., Carramolino, L., García-Alonso, L. and Modolell, J. (1986). Excess function *Hairy-wing* mutations caused by *gypsy* and *copia* insertions within structural genes of the *achaete-scute* locus of Drosophila. *Cell* **44**:303–312.

Campuzano, S., Carramolino, L., Cabrera, C. V., Ruiz-Gómez, M., Villares, R., Boronat, A. and Modolell, J. (1985). Molecular genetics of the *ac-sc* gene complex of *D. melanogaster. Cell* **40**:327–338.

Ellis, H. M., Spann, D. R. and Possakony, J. W. (1990). *extramacrochaetae*, a negative regulator of sensory organ development in *Drosophila*, defines a new class of helix-loop-helix proteins. *Cell* **61**:27–38.

Erickson, J. W. and Cline, T. W. (1991). Molecular nature of the *Drosophila* sex determination signal and its link to neurogenesis. *Science* **251**:1071–1074.

Garrell, J. and Modolell, J. (1990). The *Drosophila extramacrochaetae* locus, an antagonist of proneural genes that, like these genes, encodes a helix-loop-helix protein. *Cell* **61**:39–48.

Ghysen, A. and Dambly-Chaudière, C. (1988). From DNA to form: The *achaete-scute* complex. *Genes Dev.* **2**:495–501.

González, F., Romaní, S., Cubas, P., Modolell, J. and Campuzano, S. (1989). Molecular analysis of the *asense* gene, a member of the *achaete-scute* complex of *Drosophila melanogaster*, and its novel role in optic lobe development. *EMBO J.* **8**:3553–3562.

Harrison, S. C. (1991). A structural taxonomy of DNA-binding domains. *Nature* **353**:715–719.

Martínez, C. and Modolell, J. (1991). Cross-regulatory interactions between the proneural *achaete* and *scute* genes of *Drosophila. Science* **251**:1485–1487.

Murre, C., McCaw, S. and Baltimore, D. (1989a). A new DNA binding and dimerization motif in immunoglobulin enhancer binding, *daughterless*, *MyoD*, and *myc* proteins. *Cell* **56**:777–783.

Murre, C., McCaw, S. P., Vaessin, H., Caudy, M., Jan, L. Y., Cabrera, C. V., Buskin, J. N., Hauschka, S. D., Lassar, A. B., Weintraub, H. and Baltimore, D. (1989b). Interactions between heterologous helix-loop-helix proteins generate complexes that bind specifically to a common DNA sequence. *Cell* **58**:537–544.

Parkhurst, S. M., Bopp, D. and Ish-Horowicz, D. (1990). X:A ratio, the primary sex-determining signal in *Drosophila*, is transduced by helix-loop-helix proteins. *Cell* **63**:1179–1191.

Romaní, S., Campuzano, S., Macagno, E. R. and Modolell, J. (1989). Expression of *achaete* and *scute* genes in *Drosophila* imaginal discs and their function in sensory organ development. *Genes Dev.* **3**:997–1007.

Ruiz-Gómez, M. and Modolell, J. (1987). Deletion analysis of the *achaete-scute* locus of *Drosophila melanogaster. Genes Dev.* **1**:1238–1246.

Van Doren, M., Ellis, H. M. and Posakony, J. W. (1991). The *Drosophila extramacrochaetae* protein antagonizes sequence-specific DNA binding by *daughterless/achaete-scute* protein complexes. *Development* **113**:245–255.

Villares, R. and Cabrera, C. V. (1987). The achaete-scute gene complex of *D. melanogaster*: conserved domains in a subset of genes required for neurogenesis and their homology to *myc. Cell* **50**:415–424.

2

The Actin Genes: *Act5C, Act42A, Act57B, Act79B, Act87E, Act88F*

Chromosomal Location:			**Map Position:**
Act5C	X,	5C3-4	1-[14]
Act42A	2R,	42A	2-[55.4]
Act57B	2R,	57B	2-[97]
Act79B	3L,	79B	3-[47.5]
Act87E	3R,	87E9-12	3-[52.3]
Act88F	3R,	88F	3-57.1

Products

Actins, cytoskeletal and contractile proteins.

Structure

There is great similarity between *Drosophila* and mammalian actin amino acid sequences. Vertebrates have two distinct families of actins, one family as cytoplasmic filaments and the other occurring in muscle fibers. All *Drosophila* actins are more similar to vertebrate cytoskeletal actins than to muscle actins, but *Act5C* and *Act42A* are especially so (Fig. 2.1) (Fyrberg et al. 1981; Sanchez et al. 1983).

Tissue Distribution and Function

Act5C and *Act42A* encode cytoplasmic actins present in all tissues; *Act57B* and *Act87E* encode larval and adult intersegmental muscle actins; *Act79B* encodes thoracic and leg muscle actins and *Act88F* flight muscle actin (Fyrberg et al. 1983; Sanchez et al. 1983; see aslo Fyrberg et al. 1991 and Sparrow et al. 1991).

```
Act88F     C DD.AG    I     MC                                      S              T          I         I
Act79B     C EE.AS    V     MC                                      C              S          I         V
Act57B     C DE.VA    V     MC                                      S              T          I         I
Act87E     C DE.VA    V     MC                                      S              T          I         I
 Act5C     C EE.VA    V     MC                                      S              T          V         I
Act42A     C EE.VA    V     MC                                      S              T          V         I
Muscyt     D XX.IA    V     MC                                      S              T          V         I
Musmus     C EDETT    C     LV                                      S              T          I         I
   CON    M-D-----AL V-DNGSG--K AGFAGDDAPR AVFPSIVGRP RHQGVMVGMG QKD-YVGDEA QSKRGIL-LK YPIEHGI-TN WDDMEK-WHH TFYNELRVAP
          1                                                   50                                                    100

Act88F        V                            S                         L       S T       F      L        D
Act79B        V                            S                         L       S T       Y      L        H
Act57B        V                            S                         L       S T       Y      L        D
Act87E        V                            A                         L       S T       Y      L        D
 Act5C        V                            T                         L       S T       Y      L        D
Act42A        V                            T                         L       S T       Y      L        D
Muscyt        V                            T                         M       T T       Y      L        D
Musmus        T                            V                         L       T N       Y      M        D
   CON    EEHP-LLTEA PLNPKANREK MTQIMFETFN -PAMYVAIQA VLSLYASGRT TGIV-DSGDG V-H-VPIYEG -ALPHAI-RL DLAGR-LTDY LMKILTERGY
          101                                                150                                                    200
```

```
Act88F  T T           T             D       A T                        C   A  Q   L   SC I    V YN      V          S
Act79B  S T           I             Q       A T                        T   A  Q   L   SC I    V YQ      V          N
Act57B  S T           I             Q       A T                        C   S  Q   L   SC I    V YN      V          I
Act87E  S T           I             Q       A T                        C   S  Q   L   SC I    V YN      V          I
 Act5C  S T           I             Q       S S                        C   A  H   L   SC I    T YN      V          T
Act42A  S T           I             Q       S S                        C   S  Q   L   AC L    T YN      V          T
Muscyt  S T           I             Q       S S                        C   A  Q   L   SC I    T FN      V          T
Musmus  S V           I             N       S S                        C   T  Q   I   SA I    T YN      I          N
   CON  -F-TTAEREI VRD-KEKLCY VALDFE-EMA TAA-S-SLEK SYELPDGQVI TIGNERFR-P E-LF-PSF-G ME--G-HET- --SIMKCD-D IRKDLYAN-V
        201                                                250                                                    300

Act88F  L                     T     I     I              L     L       IS  Q    S  S        *
Act79B  L                     A     M     I              S     L       IS  Q    S  G        *
Act57B  M                     S     I     I              S     L       IS  E    S  G        *
Act87E  M                     A     I     I              S     L       IS  Q    S  G        *
 Act5C  L                     A     M     I              S     S       TS  Q    S  S        *
Act42A  L                     A     M     V              S     L       IS  Q    S  S        *
Muscyt  L                     A     M     I              S     L       IS  Q    S  S        *
Musmus  M                     A     M     I              S     L       IT  Q    A  S        *
   CON  -SGGTTMYPG IADRMQKEIT -LAPST-KIK I-APPERKYS VWIGG-ILAS -STFQQMW-- K-EYDE-GP- IVHRKCF-
        301                                                350                            378
```

FIG. 2.1. Comparison of the six *Drosophila* actins to the mouse striated muscle and cytoskeletal actins. The CON(sensus) line displays all positions for which there is total agreement among the sequences. Where there is no such agreement, the residues occupying that position in each sequence is indicated. The sequence of *Act57B* is known from a cDNA. There is 98% overall identity between the *Drosophila* and mouse cytoskeletal proteins.

Mutant Phenotype

Mutations in *Act88F* affect only the development of indirect flight muscles, and mutants are viable (Karlik et al. 1984; Mahaffey et al. 1985; Okamoto et al. 1986). Some mutations, such as *Act88F*^{KM88} and *Act88F*^{KM129}, are recessive hypomorphs producing severely altered proteins that fail to accumulate. Other alleles, those with more subtle changes such as *Act88F*^{KM75}, are antimorphs; they are dominant even in the presence of two normal alleles and often result in the expression of heat-shock genes, probably induced by the accumulation of denatured muscle proteins (Okamoto et al. 1986; Drummond et al. 1991).

Common Features of Gene Organization and Expression

Open reading frame, 376 amino acids. Although coding sequences are 85–95% conserved among all *Drosophila* actins, the position of introns is not constant (Fyrberg et al. 1981; Fig. 2.2). Transcription from the six genes is differentially modulated during development, in accordance with the tissue distribution of their products (Bond-Matthews and Davidson 1988; Burn et al. 1989; Tobin et al. 1990).

Act5C

Gene Organization and Expression

Determination of 5′ and 3′ ends was by S1 mapping and by RNase protection studies, primer extension and sequencing of several cDNAs. Transcription occurs from two main initiation sites. The upstream site is preceded by a putative TATA box, and the position of the 5′ end seems to be quite invariant. The downstream initiation site lacks a canonical TATA box, and there is some microheterogeneity in the 5′ end, although the main site seems to be −712. Both leaders have introns with donor sites at −1,675 and −602 and a common acceptor site at −7 (*Act5C* Sequence and Fig. 2.2). Three major and two minor alternative poly-A sites exist, and it is probable that all possible combinations of initiation and polyadenylation sites are used. The major classes of mRNAs would range from 1,524 to 1,919 bases. Three mRNA bands resolved by northern analysis are 1.8 kb, 2.0 kb and 2.3 kb long (Fyrberg et al. 1981; Bond and Davidson 1986; Vigoreaux and Tobin 1987; Chung and Keller 1990a).

Developmental Pattern

The gene for the cytoplasmic actin 5C is, as would be expected, transcribed in all tissues. Its maternal mRNA is uniformly distributed in preblastoderm embryos. During blastoderm formation this mRNA becomes localized in a peripheral layer; and, as tissue differentiation proceeds, it remains present in

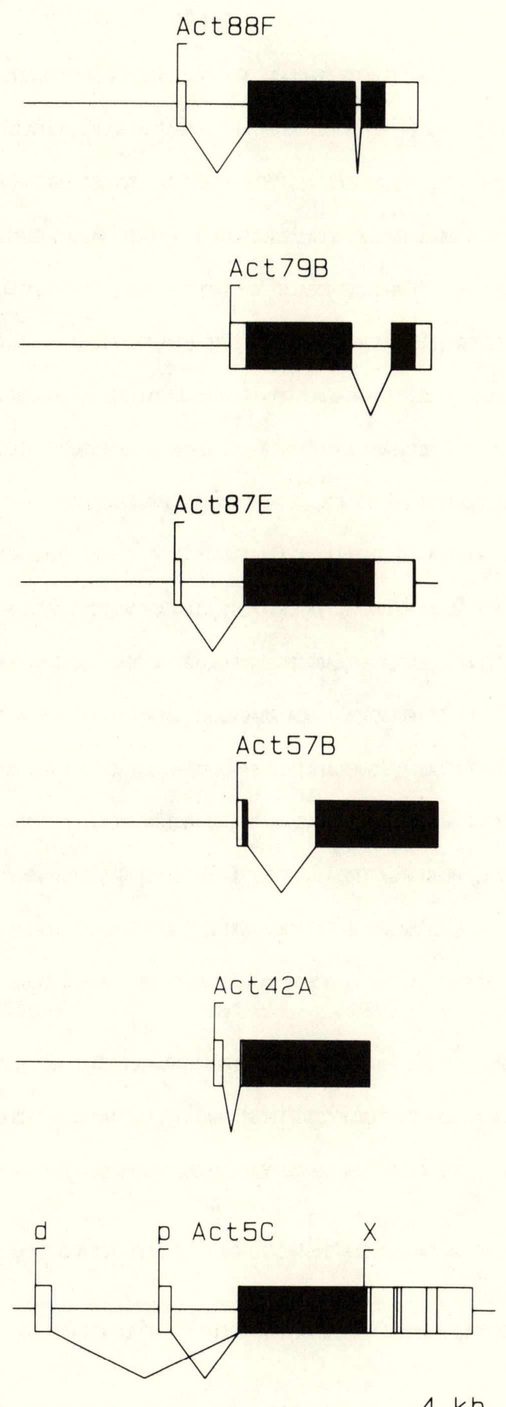

FIG. 2.2. Organization of the six actin genes.

Act5C

```
-3735 ATTTTCTACAAAAACATGTTATCTATAGATAATTTTGTTGCAAAATATGTTGACTATGACAAAGATTGTATGTATATACCTTTAATGTAT -3646
-3645 TCTCATTTTCTTATGTATTTATAATGGCAATGATGATACTGATGATATTTTAAGATGATGCCAGACCACAGGCTGATTTCTGCGTCTTTT -3556
-3555 GCCGAACGCAGTGCATGTGCGGTTGTTGTTTTTTGGAATAGTTTCAATTTTCGGACTGTCCGCTTTGATTTCAGTTTCTTGGCTTATTCA -3466
-3465 AAAAGCAAAGTAAAGCCAAAAAAGCGAGATGGCAATACCAAATGCGGCAAAACGGTAGTGGAAGGAAAGGGGTGCGGGGCAGCGGAAGGA -3376
-3375 AGGGTGGGGCGGGGCGTGGCGGGGTCTGTGGCTGGGCGCGACGTCACCGACGTTGGAGCCACTCCTTTGACCATGTGTGCGTGTGTGTAT -3286
-3285 TATTCGTGTCTCGCCACTCGCCGGTTGTTTTTTTCTTTTTATCTCGCTCTCTCTAGCGCCATCTCGTACGCATGCTCAACGCACCGCATG -3196
-3195 TTGCCGTGTCCTTTATGCGTCATTTTGGCTCGAAATAGGCAATTATTTAAACAAAGATTAGTCAACGAAAACGCTAAAATAAATAAGTCT -3106
-3105 ACAATATGGTTACTTATTGCCATGTGTGTGCAGCCAACGATAGCAACAAAAGCAACAACACAGTGGCTTTCCCTCTTTCACTTTTTGTTT -3016
-3015 GCAAGCGCGTGCGAGCAAGACGGCACGACCGGCAAACGCAATTACGCTGACAAAGAGCAGACGAAGTTTTGGCCGAAAAACATCAAGGCG -2926
-2925 CCTGATACGAATGCATTTGCAATAACAATTGCGATATTTAATATTGTTTATGAAGCTGTTTGACTTCAAAACACACAAAAAAAAAAATAA -2836
-2835 AACAAATTATTTGAAAGAGAATTAGGAATCGGACAGCTTATCGTTACGGGCTAACAGCACACCGAGACGAAATAGCTTACCTGACGTCAC -2746
-2745 AGCCTCTGGAAGAACTGCCGCCAAGCAGACGATGCAGAGGACGACACATAGAGTAGCGGAGTAGGCCAGCGTAGTACGCATGTGCTTGTG -2656
-2655 TGTGAGGCGTCTCTCTCTTCGTCTCCTGTTTGCGCAAACGCATAGACTGCACTGAGAAAATCGATTACCTATTTTTTATGAATGAATATT -2566
-2565 TGCACTATTACTATTCAAAACTATTAAGATAGCAATCACATTCAATAGCCAAATACTATACCACCTGAGCGATGCAACGAAATGATCAAT -2476
-2475 TTGAGCAAAAATGCTGCATATTTAGGACGGCATCATTATAGAAATGCTTCTTGCTGTGTACTTTTCTCTCGTCTGGCAGCTGTTTCGCCG -2386
-2385 TTATTGTTAAAACCGGCTTAAGTTAGGTGTGTTTTCTACGACTAGTGATGCCCCTACTAGAAGATGTGTGTTGCACAAATGTCCCTGAAT -2296
-2295 AACCAATTTGAAGTGCAGATAGCAGTAAACGTAAGCTAATATGAATATTATTTAACTGTAATGTTTTAATATCGCTGGACATTACTAATA -2206
-2205 AACCCACTATAAACACATGTACATATGTATTGTTTTGGCATACAATGAGTAGTTGGGGAAAAAATGTGTAAAAGCACCGTGACCATCACA -2116
                              --------A5Ce3 ------cA5Ce3           --------A5Ce2
-2115 GCATAAAGATAACCAGCTGAAGTATCGAATATGAGTAACCCCCAAATTGAATCACATGCCGCAACTGATAGGACCCATGGAAGTACACTC -2026
-2025 TCATGGCGATATACAAGACACACACAAGCACGAACACCCAGTTGCGGAGGAAATTCTCCGTAAATGAAAACCCAATCGGCGAACAATTCA -1936
-1935 TACCCATATATGGTAAAAGTTTTGAACGCGACTTGAGAGCGGAGAGCATTGCGGCTGATAAGGTTTTAGCGCTAAGCGGGCTTTAATAAA -1846
                                                     ----                                -------
                              -->-1821
-1845 ACGGGCTGCGGGACCAGTTTTCATATCACTACCGTTTGAGTTCTTGTGCTGTGTGGATACTCCTCCCGACACAAAGCCGCTCCATCAGCC -1756
                                                                                         -
                                                                                          |
-1755 AGCAGTCGTCTAATCCAGAGACACCAAACCGAAAGACTTAATTTATATTTATTTAATTAATTTTAATAAAACACACCAAATGTAAGTAGC -1666
                                                                                         _|
-1665 TTTCCCCTTCCCAACAACAAAACACCATCGAACCACTCCCACCAAGAAAAAGCAATAATCGAGAAAAGCCGCGGAAAATGTGTGATTTTT -1576
-1575 TTTGTAAACAAAATTTTTTTATGTGCCAGTGCTGAAAGTGATCAAAAAATACTAGCCACGAGCTAAAGAGTTATTGTATTGACCAAAACT -1486
```

```
-1485 CCAAAAATACCCAAGTTTGGCCCTAAATTGTCAATCAAAATACCAATAGGTCGAAAGACATCAAAATTAACAAAACCAGGGTTTCAAATA -1396

-1395 CCATAACTCAAGAATCAGGATTACAACTGCAGATTTCAGGATATATACATACAAATTATAGCAAATTATAAAAACCAAAGCAATTCATAG -1306

-1305 CCCCAACTCAAATGTTAGGATCTAATATAGTGTTTAAAGCCAAGCTCGCTGATGTGGGCGTGTCACGATTTCACCCAAAGATATGCCAAA -1216

-1215 TTACGAATTGCAAATCAATTCGCCAACACTTCTTTTTTTCCCACGCCTAAAACACAGATCATCATAAATGTACATACATACAGTATATGC -1126

-1125 ATATTATAATCTGTAAACTAGATCAGGTTCTTGAAAATAGTGACGTAGGAGCCGTTTTGGCTGAAGCAGAAATTTTTTGCCGGTTTTTCAA -1036
                                                                                         ---

-1035 AGTTGTAGTTGCAAAAATGGAGAAAACCTTCGAGCATTCGTTCATATACACACACTCACGCGCAAAATAACGAGAGAGAGTGTATGTGTG -946
      -----------A5Ce1

-945  TGTGAGAGAGCGAAAGCCAGACGACGGTTTGCTTTTCGCCTCGAAACATGACCATATATGGTCACAAAACTTGGCCGCCGCAATTCAACA -856

-855  CACCAGCGCTCTCCTTCGCACCCATAGCGACCATGGCGCGGAGCGAGCGAGATGGCGAGAGCGAGCGACGCTATGGCGACGTCGACGCAG -766
            <----                                    --->  ---->
                                                   -->-712 -->-704 (minor)
-765  GCAGCGATTGAAAAACGCAGTTAACTGGCATTCAACATTCACCAGCCACTTTCAGTCGGTTTATTCCAGTCATTCCTTTCAAACCGTGCG -676

                                                                          _
                                                                          |
-675  GTCGCTTAGCTCAGCCTCGCCACTTGCGTTTACAGTAGTTTTCACGCCTTGAATTTGTTAAATCGAACAAAAAGGTAAAGTTTAACTAGC -586
                                                                          _|

-585  TTTGAAAAGTTTCGTGGCTCTTAATTGTTAAATTTTCTAGAGTGCGTTTAGTGTTTTTTTTTTTTTTTATTTTGTAATGTTAATTTCGGG -496

-495  TTCCAATTCGAGTTTTAGGCAGCCGCCATTTTAAGGGCGCATACACACAGGCAACTGTGCTCTCTTTGCGGCTTTCTTTTGCACCGGCAT -406

-405  TCGTTAAGCTGTCGTCTAGAAGCTTCTCCCCTCCCTTTTCGGCATATTCGTATTGTGGTTTTAATTTTTCGGGGCGGGGCTTCTATTTTG -316

-315  TAACTGTTCTTTTAATTTCTTATTACAATTCGATCGCAAGTGAAAATCAGTTTTCAATCGGAAAAGTATTTTTTTATGAAATTTTTTTTT -226

-225  GTCCAAGATTAAAATTTTGTACTAAAAAAACGTACATTGCATTGAGTGATTTTTAATTGTACACGAAAAACAAGTTAGTTTGTTATGACA -136

-135  ATTGTACTTTGGTAGACCAGCGCAGTCCAAGGAGACCACGCAAATTCTCAGTTTTTTTTTTGCCATTTCTACATTACCAAATAAGGTAAC -46

                                   _
                                   |
-45   CAAAAACTAATGGGAAATCCGCATTCTTTCCATTGCAGCTTACAAAATGTGTGACGAAGAAGTTGCTGCTCTGGTTGTCGACAACGGCTC 44
                                   |_          MetCysAspGluGluValAlaAlaLeuValValAspAsnGlySe (15)

45    TGGCATGTGCAAGGCCGGATTTGCCGGAGACGATGCTCCCCGCGCCGTCTTCCCATCGATTGTGGGACGTCCCCGTCACCAGGGTGTGAT 134
      rGlyMetCysLysAlaGlyPheAlaGlyAspAspAlaProArgAlaValPheProSerIleValGlyArgProArgHisGlnGlyValMe (45)

135   GGTCGGCATGGGCCAGAAGGACTCGTACGTGGGTGATGAGGCGCAGAGCAAGCGTGGTATCCTCACCCTGAAGTACCCCATTGAGCACGG 224
      tValGlyMetGlyGlnLysAspSerTyrValGlyAspGluAlaGlnSerLysArgGlyIleLeuThrLeuLysTyrProIleGluHisGl (75)

225   TATCGTGACCAACTGGGACGATATGGAGAAGATCTGGCACCACACCTTCTACAATGAGCTGCGTGTGGCACCCGAGGAGCACCCCGTGCT 314
      yIleValThrAsnTrpAspAspMetGluLysIleTrpHisHisThrPheTyrAsnGluLeuArgValAlaProGluGluHisProValLe (105)

315   GCTGACCGAGGCCCCGCTGAACCCCAAGGCCAACCGTGAGAAGATGACCCAGATCATGTTCGAGACCTTCAACACACCCGCCATGTATGT 404
      uLeuThrGluAlaProLeuAsnProLysAlaAsnArgGluLysMetThrGlnIleMetPheGluThrPheAsnThrProAlaMetTyrVa (135)
```

(continued)

```
405  GGCCATCCAGGCTGTGCTCTCGCTGTACGCTTCGGGTCGTACCACCGGTATCGTTCTGGACTCCGGCGATGGTGTCTCCCACACCGTGCC  494
     lAlaIleGlnAlaValLeuSerLeuTyrAlaSerGlyArgThrThrGlyIleValLeuAspSerGlyAspGlyValSerHisThrValPr  (165)

495  CATCTACGAGGGTTATGCCCTCCCCCATGCCATCCTGCGTCTGGATCTGGCTGGTCGCGATTTGACCGACTACCTGATGAAGATCCTGAC  584
     oIleTyrGluGlyTyrAlaLeuProHisAlaIleLeuArgLeuAspLeuAlaGlyArgAspLeuThrAspTyrLeuMetLysIleLeuTh  (195)

585  CGAGCGCGGTTACTCTTTCACCACCACCGCTGAGCGTGAAATCGTCCGTGACATCAAGGAGAAGCTGTGCTATGTTGCCCTCGACTTTGA  674
     rGluArgGlyTyrSerPheThrThrThrAlaGluArgGluIleValArgAspIleLysGluLysLeuCysTyrValAlaLeuAspPheGl  (225)

675  GCAGGAGATGGCCACCGCTGCCAGCAGCTCCTCGTTGGAGAAGTCCTACGAGCTGCCCGACGGACAGGTGATCACCATCGGCAACGAGCG  764
     uGlnGluMetAlaThrAlaAlaSerSerSerSerLeuGluLysSerTyrGluLeuProAspGlyGlnValIleThrIleGlyAsnGluAr  (255)

765  TTTCCGCTGCCCCGAGGCCCTGTTCCATCCCTCGTTCCTTGGGATGGAGTCTTGCGGCATCCACGAGACCACCTACAACTCCATCATGAA  854
     gPheArgCysProGluAlaLeuPheHisProSerPheLeuGlyMetGluSerCysGlyIleHisGluThrThrTyrAsnSerIleMetLy  (285)

855  GTGTGATGTGGATATCCGTAAGGATCTGTATGCCAACACCGTGCTGTCCGGTGGCACCACCATGTACCCTGGCATCGCCGACCGTATGCA  944
     sCysAspValAspIleArgLysAspLeuTyrAlaAsnThrValLeuSerGlyGlyThrThrMetTyrProGlyIleAlaAspArgMetGl  (315)

945  GAAGGAGATCACCGCCCTGGCACCGTCGACCATGAAGATCAAGATCATTGCCCCGCCAGAGCGCAAGTACTCTGTCTGGATCGGTGGCTC  1034
     nLysGluIleThrAlaLeuAlaProSerThrMetLysIleLysIleIleAlaProProGluArgLysTyrSerValTrpIleGlyGlySe  (345)

                                                                                 -->1108 (X)
1035 CATCCTGGCTTCGCTGTCCACCTTCCAGCAGATGTGGATCTCCAAGCAGGAGTACGACGAGTCCGGCCCCTCCATTGTGCACCGCAAGTG  1124
     rIleLeuAlaSerLeuSerThrPheGlnGlnMetTrpIleSerLysGlnGluTyrAspGluSerGlyProSerIleValHisArgLysCy  (375)

1125 CTTCTAAGAAGGATCGCTTGTCTGGGCAAGAGGATCAGGATCGGGATGGTCTTGATTCTGCTGGCAGGAGGAGGAGGAGAAGTCGAGGAA  1214
     sPheEnd   (376)                                  MetValLeuIleLeuLeuAlaGlyGlyGlyGlyGluValGluGlu  (15)

1215 GCAGCAGCGAAAGTGCAAGTGCGAGTGGTGGAAGTTTGGAGTGCAGCACAACAAAATCAACAACAACACCAACTACAAGATGAAAAGAGC  1304
     AlaAlaAlaLysValGlnValArgValValGluValTrpSerAlaAlaGlnGlnAsnGlnGlnGlnHisGlnLeuGlnAspGluLysSer  (45)

1305 GGAACCACCTGCCACACCATCATCACTATCATCATCGTTTTGGGCGCATGTTGTGTGGTTCCAGCGTATTAATATAATTAATTTATTCCA  1394
     GlyThrThrCysHisThrIleIleThrIleIleIleValLeuGlyAlaCysCysValValProAlaTyrEnd                   (68)
                                                                                    -------

1395 TGAGATATGATATGATATACTATGTATTTTTTGTTTTTTTTTTATTTGTAAACCTTTAATATAACAAGAACTACAAAAAAGTGAAAATGA  1484
           |(A)n                                        |(A)n   ------- (X)          |(A)n (X)

1485 GCGAAAATGCATATTCTGCCATTCCACACACACACCAACAACACCCAACACACGCACACCCACAAGCTTACACACACACACATTCGCGGC  1574

1575 ATGACAAGGACATCAAGATAAAGAAGAACTTAAAGAAGATATTTCCCAAAGCGCAAAAAGAACACACACACATTGCAAAACACAAACAAC  1664

1665 ACACTAGCGTTTTGTACAATTCGTCAGCAACCTTATGTATTATTTTTTAATTATGATGTAATTATAAACAAAGTGAAACAAAAATATGAA  1754
        |(A)n                                                           -----

1755 AACAAAAAGGAAAATCAAATCTGTCTTCTCTTTCTCCCGCTCTCCTCGCTCTCTGCTGCTAACCTCGCCCTCTCCTCTCTCATCTTTTTG  1844
           |(A)n

1845 TCTGTCTCTCTTCACATTTTTGCCGGCCGGCAAAATAATAACCCACACACACTCACACTTGGCTGCAGTTTCGCGTGCGATATTCACACA  1934

1935 CATTCAAGCATACATACATATGTATTTTTTTTTTATTTGTACACTTTTCTAATTGCATGCGTATCGATTGATAAGTTTACGCTGAAAATG  2024

2025 TTAATTAAAATGTGAAAATGCAACTGAAAAACTGATGAAATGAAACAACAACAAGCGAACAA  2086
```

all organs. There is a slightly greater accumulation of *Act5C* mRNA in the anterior and posterior segments of the prospective midgut, apparently due to increased transcription from the distal initiation site (Burn et al. 1989). Both cytoplasmic actin genes, *Act5C* and *Act42A*, are the only actin genes transcribed in Kc cells, with *Act5C* transcripts being 6–8 fold more abundant. The level of *Act5C* transcript increases 3–5 fold in response to 20-hydroxyecdysone treatment (Couderc et al. 1987). Most *Act5C* mRNA is associated with polysomes (Rao et al. 1988).

Promoter

The two transcription initiation sites respond to independent regulatory regions, as shown by the expression of a reporter gene in cultured cells (Bond-Matthews and Davidson 1988). The distal promoter is the stronger and is developmentally regulated; the proximal promoter is uniformly expressed in all cell types (Vigoreaux and Tobin 1987; Burn et al. 1989).

Distal Promoter The controlling elements of the distal promoter include one that extends between 2,071 and 1,866 bp upstream of the transcription initiation site and several others that lie within 540 bp of the 5′ end. These were identified by reporter gene expression essays performed in cultured cells (Bond-Matthews and Davidson 1988; Chung and Keller 1990b). A bipartite element between −2,343 and −2,182 strongly represses expression, and three elements with a positive effect on expression are found between −2,182 and −2,099, between −2,068 and −2,040 and between −1,911 and −1,864. The segment between −2,182 and −2,099 has the strongest effect, and footprinting and mutational analysis identified A5Ce2 (*Act5C* Sequence) as the main regulatory element in this region. *In vitro* mutagenesis identified two other elements, A5Ce3 and cA5Ce3 (Chung and Keller 1990b).

Proximal Promoter The proximal promoter contains three elements involved in the control of transcription, which were identified by band-shift assays, footprinting and expression of a reporter gene (Chung and Keller 1990a): (1) a 14-bp segment between −1,038 and −1,025 (A5Ce1 in the *Act5C* Sequence) that is necessary for full expression; (2) the 98 base pairs between −872 and −774 whose effect is probably due to the presence of three copies of the GAGA transcription factor binding sites (Biggin and Tjian 1988); and (3) the segment

---

Act5C Sequence (*opposite*). Mostly from *Canton S.* Accession, X15730 (DROACT5CB), X06382 (DRO5CACT1), X06383 (DRO5CACT2), X06384 (DRO5CACT3), M13586 (DROACT5C2). Two bases, −819 and −820, were corrected as suggested by Chung and Keller (1990a). Arrows between −855 and −766 underline potential binding sites for the GAGA factor. The initiation and termination of the 3′ transcriptional unit are marked by X.

Act42A

```
-513  TCGAATTTTGAGAACACTGCATAATTTTTAAATGCATTTTCAAGGATTCTTAGATCATTTCTAATTTGTTGATAACACGTCAGTATACCA  -424

-423  ATGAATAAAAAATTTTAAAAAAAGTCCGCTCTCCAGTCTTCACCGTTTCCAACTTATCGCACATTTATTGTTGGTGGAGTCACTTCGGAA  -334
                                                                                  ----
                                                                                           -->-257
-333  GTAAAAAAGACCATAATTTTATGCGTATATGGTCACACTACTTTTCAACACTTTAACTCGAAAAGTAGCGTCGTCAATTCAATCTTAAAG  -244

-243  CGTCTGTCATTGTGCTAAGTGTGTGCAGCGGATAACTAGAAACTACTCCTACATATTTCCATAAAAGGTAAGACTCCTGCCCAACACTTT  -154

-153  TTTTTTGTCTGTGCGGTCATTATTATTCCTTTCTGGAAGGGGTCGGTCCCGTCTGCGTTCTTTTTTACGTTAGCCGCTGTCCTGTCTCCT  -64

-63   TGTTTTTTTAGTGTACACATCCAGATTTCTTTTCTCTTGCAGATCCAAATAAAATTTCTACAAAATGTGTGACGAAGAGGTTGCAGCTTT  26
                                                                  MetCysAspGluGluValAlaAlaLe  (9)

27    AGTGGTCGACAACGGATCCGGCATGTGCAAAGCCGGCTTTGCCGGTGATGACGCACCGCGTGCAGTTTTTCCTTCTATTGTCGGCCGTCC  116
      uValValAspAsnGlySerGlyMetCysLysAlaGlyPheAlaGlyAspAspAlaProArgAlaValPheProSerIleValGlyArgPr  (39)

117   ACGTCACCAGGGCGTAATGGTAGGAATGGGACAAAAGGACTCTTATGTCGGCGATGAGGCACAGAGCAAACGTGGTATCCTTACCCTGAA  206
      oArgHisGlnGlyValMetValGlyMetGlyGlnLysAspSerTyrValGlyAspGluAlaGlnSerLysArgGlyIleLeuThrLeuLy  (69)

207   GTACCCCATTGAGCACGGTATCGTGACTAACTGGGACGACATGGAGAAGATCTGGCATCACACTTTCTACAACGAGCTTCGTGTGGCCCC  296
      sTyrProIleGluHisGlyIleValThrAsnTrpAspAspMetGluLysIleTrpHisHisThrPheTyrAsnGluLeuArgValAlaPr  (99)

297   GGAGGAGCACCCCGTCTTGCTTACTGAGGCTCCTTTGAACCCCAAGGCTAATCGCGAAAAGATGACTCAGATTATGTTTGAAACCTTCAA  386
      oGluGluHisProValLeuLeuThrGluAlaProLeuAsnProLysAlaAsnArgGluLysMetThrGlnIleMetPheGluThrPheAs  (129)

387   CACTCCGGCCATGTATGTTGCCATCCAAGCGGTGCTTTCTCTCTACGCCTCCGGCCGTACCACAGGTATCGTGTTGGACTCCGGGGACGG  476
      nThrProAlaMetTyrValAlaIleGlnAlaValLeuSerLeuTyrAlaSerGlyArgThrThrGlyIleValLeuAspSerGlyAspGl  (159)

477   TGTCTCCCATACCGTGCCCATCTATGAGGGCTACGCTCTGCCGCACGCTATCCTCCGCTTGGATCTAGCCGGTCGCGATTTAACCGACTA  566
      yValSerHisThrValProIleTyrGluGlyTyrAlaLeuProHisAlaIleLeuArgLeuAspLeuAlaGlyArgAspLeuThrAspTy  (189)

567   CCTGATGAAGATTCTTACTGAGCGCGGTTACAGCTTCACCACCACCGCCGAGCGTGAAATTGTGCGCGACATCAAGGAGAAGCTGTGCTA  656
      rLeuMetLysIleLeuThrGluArgGlyTyrSerPheThrThrThrAlaGluArgGluIleValArgAspIleLysGluLysLeuCysTy  (219)

657   CGTGGCCTTGGACTTCGAGCAGGAGATGGCCACGGCCGCTTCAAGCTCGTCCCTGGAGAAGTCGTACGAGTTGCCCGATGGACAGGTCAT  746
      rValAlaLeuAspPheGluGlnGluMetAlaThrAlaAlaSerSerSerSerLeuGluLysSerTyrGluLeuProAspGlyGlnValIl  (249)

747   CACCATCGGAAATGAGCGATTCCGTTGCCCCGAATCGCTGTTCCAGCCGTCGTTCCTCGGCATGGAGGCCTGTGGACTTCACGAGACCAC  836
      eThrIleGlyAsnGluArgPheArgCysProGluSerLeuPheGlnProSerPheLeuGlyMetGluAlaCysGlyLeuHisGluThrTh  (279)

837   CTACAACTCAATCATGAAGTGTGACGTCGACATCCGTAAGGATCTGTACGCCAACACTGTGCTGTCCGGCGGCACCACCATGTACCCGGG  926
      rTyrAsnSerIleMetLysCysAspValAspIleArgLysAspLeuTyrAlaAsnThrValLeuSerGlyGlyThrThrMetTyrProGl  (309)

927   AATCGCTGACCGCATGCAAAAGGAAATCACGGCGTTGGCTCCGTCCACCATGAAGATTAAGATTGTTGCCCCGCCAGAACGCAAGTACTC  1016
      yIleAlaAspArgMetGlnLysGluIleThrAlaLeuAlaProSerThrMetLysIleLysIleValAlaProProGluArgLysTyrSe  (339)

1017  TGTTTGGATCGGCGGCTCCATCCTAGCTTCGCTGTCTACTTTCCAGCAGATGTGGATCTCGAAGCAAGAGTACGACGAGTCGGGCCCCTC  1106
      rValTrpIleGlyGlySerIleLeuAlaSerLeuSerThrPheGlnGlnMetTrpIleSerLysGlnGluTyrAspGluSerGlyProSe  (369)

1107  CATTGTTCACCGCAAGTGCTTCTAA                                                                   1131
      rIleValHisArgLysCysPheEnd                                                                   (376)
```

between −770 and −744, the position that a TATA box would normally occupy.

Transcription unit X

This transcription unit overlaps the last few codons and 3′ untranslated region of *Act5C*.

Gene Organization and Expression

Open reading frame, 68 amino acids; mRNA, 368 bases, in agreement with a 0.45 kb band detected by northern analysis. S1 mapping and primer extension were used to determine the 5′ end. S1 mapping was used to determine the 3′ end (see *Act5C* sequence). This mRNA is found in polysomes and has the same tissue and developmental distribution as *Act5C* mRNA. Its function is unknown (Rao et al. 1988).

Act42A

Gene Organization and Expression

The 5′ end was determined by S1 mapping; there is no obvious TATA box in its neighborhood. The 3′ end has not been determined. There is a leader intron with a donor site at −177 and an acceptor site at −21. Because most of the coding sequence was determined from a cDNA, the presence of other small introns cannot be ruled out (*Act42A* sequence) (Fyrberg et al. 1981; Couderc et al. 1987).

Developmental Pattern

During embryonic development, *Act42A* transcription follows a pattern similar to that of *Act5C*. The accumulation of transcripts is greatest in the midgut, central nervous system and gonads (Tobin et al. 1990). *Act42A* is expressed in

---

Act42A Sequence (*opposite*). Mostly from *Canton S*. Accession, K00670, K00671 (DROACT2A), X05176 (DROACT42A).

Act79B

```
-517  AGCTTACAAGTGTGTGCGGACCAAAATTCTAACAATATAACAAGACTTACAACTTACAAAACAACTTATTTTATATCGAAATCCAGTACC  -428

-427  AATTTAGTTGCTCTAAGTTGTGGCTTAACTAGGGTTCTTTAATTCGTAATCCAACTTGTTGCCGTAGGCATACCCGAAATCGGAACAATT  -338

-337  TTTGTGAAATCGAAATGATGTCGATCCGACCACCCTCCCCGGAAACGCCTGATCCCCAGCCAGCTTACATATCGCGGAATTCATCAACAT  -248

-247  GTTACTAGATGAACAATTGTTCGAGATGACAGGGACATGGGCGTGGGGCCGGCGGGGCGGGACAGAACTTATTTAAATGCAGCTGCCGGA  -158
                ----                                                            ------

         -->-146
-157  GCGCATAACGAATCACTCTGATCGCTGTCGCTGTTGGATTTACACGTCGTGAGTGTAGTCTTGTCCGCCCATCCGAAATCCGTAACCCGC  -68

 -67  ATAAGGGATAACCGATTCTGTTGTACCCTTGTACCCTTGTGTACCGCCCCGCACCAAACTAACCAAACATGTGTGACGAAGAAGCATCAG  22
                                                                                MetCysAspGluGluAlaSerA  (8)

  23  CCCTGGTCGTAGACAACGGCTCCGGCATGTGCAAGGCCGGATTCGCCGGAGACGACGCGCCCCGCGCGGTATTCCCCTCGATCGTAGGCC  112
      laLeuValValAspAsnGlySerGlyMetCysLysAlaGlyPheAlaGlyAspAspAlaProArgAlaValPheProSerIleValGlyA  (38)

 113  GTCCCCGTCACCAGGGCGTGATGGTGGGTATGGGTCAGAAGGACTGCTACGTGGGCGACGAGGCGCAAAGCAAGCGCGGTATCCTGTCGC  202
      rgProArgHisGlnGlyValMetValGlyMetGlyGlnLysAspCysTyrValGlyAspGluAlaGlnSerLysArgGlyIleLeuSerL  (68)

 203  TGAAGTACCCCATCGAACACGGCATTATCACCAACTGGGATGACATGGAGAAGGTCTGGCACCACACCTTCTACAACGAGCTGCGTGTGG  292
      euLysTyrProIleGluHisGlyIleIleThrAsnTrpAspAspMetGluLysValTrpHisHisThrPheTyrAsnGluLeuArgValA  (98)

 293  CCCCCGAGGAGCACCCCGTTCTGCTGACCGAGGCTCCCTTGAACCCCAAGGCCAACCGCGAGAAGATGACCCAGATCATGTTCGAGACGT  382
      laProGluGluHisProValLeuLeuThrGluAlaProLeuAsnProLysAlaAsnArgGluLysMetThrGlnIleMetPheGluThrP  (128)

 383  TCAACTCCCCGGCCATGTACGTGGCCATCCAGGCCGTGCTCTCCCTGTATGCTTCCGGCCGTACCACCGGTATCGTCCTGGACTCCGGTG  472
      heAsnSerProAlaMetTyrValAlaIleGlnAlaValLeuSerLeuTyrAlaSerGlyArgThrThrGlyIleValLeuAspSerGlyA  (158)

 473  ACGGTGTCTCCCACACCGTGCCCATCTATGAGGGCTATGCCCTGCCCCACGCCATCCTTCGTCTAGATCTGGCCGGTCGCCATCTAACCG  562
      spGlyValSerHisThrValProIleTyrGluGlyTyrAlaLeuProHisAlaIleLeuArgLeuAspLeuAlaGlyArgHisLeuThrA  (188)

 563  ACTACCTGATGAAGATCCTCACCGAGCGCGGCTACAGCTTCACCACCACCGCCGAGCGCGAGATTGTGCGCGACATCAAGGAGAAGCTGT  652
      spTyrLeuMetLysIleLeuThrGluArgGlyTyrSerPheThrThrThrAlaGluArgGluIleValArgAspIleLysGluLysLeuC  (218)

 653  GCTACGTCGCCCTGGACTTCGAGCAGGAGATGGCCACTGCCGCCGCCTCCACCTCCCTGGAGAAGTCTTACGAGCTGCCCGATGGCCAGG  742
      ysTyrValAlaLeuAspPheGluGlnGluMetAlaThrAlaAlaAlaSerThrSerLeuGluLysSerTyrGluLeuProAspGlyGlnV  (248)

 743  TAATCACCATCGGCAACGAGCGCTTCCGCACCCCGGAGGCCCTCTTCCAGCCATCGTTCCTAGGCATGGAGTCCTGCGGCATCCACGAGA  832
      alIleThrIleGlyAsnGluArgPheArgThrProGluAlaLeuPheGlnProSerPheLeuGlyMetGluSerCysGlyIleHisGluT  (278)

 833  CCGTCTACCAGTCCATCATGAAGTGCGACGTGGACATCCGCAAGGATCTGTATGCCAACAATGTGCTGTCTGGCGGCACTACCATGTATC  922
      hrValTyrGlnSerIleMetLysCysAspValAspIleArgLysAspLeuTyrAlaAsnAsnValLeuSerGlyGlyThrThrMetTyrP  (308)

 923  CAGGTACGTAGTCTTAATTATTTAGGACCATAAAGTTCAGAGGAAATTCTTCCGAGGGAATGGGATCAAAACTATGCGGGATACTTAAAA  1012
      roG                                                                                        (309)

1013  AAAAAAACAAGTGTTACTTTATACATTCATTTGGCAGAGAGCAAATCTTTAAATAATAAAGCCTAAATATTTAGCTGAGGTTTGTAATAA  1102

1103  CAGTTAAAAAAAATCTTATGGAAAGTAGTATTACAAAAAAAAAAAAAAGAATTCACCTAACGGGTTATATCCTTTCCCTATATCTCATAT  1192

1193  TCATGCATGCTATTATTAAAATGTCATGTAATGAGTACACCAAAGCTCCACGGTCCGTAGCACCACCAATGGATTCTATTTCCGCCTCTT  1282
```

```
1283 CAGGTATCGCTGACCGTATGCAAAAGGAAATCACCGCACTTGCCCCGTCCACCATGAAGATCAAGATCATCGCCCCGCCAGAGCGCAAGT  1372
        lyIleAlaAspArgMetGlnLysGluIleThrAlaLeuAlaProSerThrMetLysIleLysIleIleAlaProProGluArgLysT  (338)

1373 ACTCCGTCTGGATCGGTGGCTCCATCCTGGCTTCGTTGTCCACCTTTCAGCAGATGTGGATCTCCAAGCAAGAGTATGACGAGTCCGGTC  1462
     yrSerValTrpIleGlyGlySerIleLeuAlaSerLeuSerThrPheGlnGlnMetTrpIleSerLysGlnGluTyrAspGluSerGlyP  (368)

1463 CCGGCATCGTCCACCGCAAGTGCTTCTAAGCATCCAGGCCACCCAAACCAGGTCAACATCTCCTCGAGGCGCGGCCCTGGTGTTTGTCTC  1552
     roGlyIleValHisArgLysCysPheEnd                                                               (376)

1553 CAGCGTAAGACATCCGACTAGGCGTCGGCGCACAGGGTCCGAGGACCGCAGTTCACTGAAAAGATCCTTAAATAACATTTAGTCGATGAA  1642

1643 GAAGTTTTAACA  1654
```

Act79B SEQUENCE. Strain, *Canton S*. Accession, M18829 (DROACT79B).

Kc cells and transcription is enhanced 6–8 fold in the presence of 20-hydroxyecdysone (see *Act5C*; Couderc et al. 1987).

Act57B

Gene Organization and Expression

The 5′ and 3′ ends were not determined. There is an intron in the Gly-14 codon. Most of the coding sequence was determined from cDNA clones only; and the presence of other small introns cannot be ruled out [Fyrberg et al. 1981; Accession, K00672 (DROACT7A1) and K00673 (DROACT7A3)]. The amino acid sequence is shown in Fig. 1. In embryos, transcripts are detectable in the developing musculature of the future larval body wall (Tobin et al. 1990).

Act79B

Gene Organization and Expression

The 5′ end was determined by S1 mapping. The 3′ end has not been determined. There is an intron within the Gly-309 codon (*Act79B* Sequence) (Fyrberg et al. 1981; Sanchez et al. 1983).

Developmental Pattern

Transcription is undetectable in embryos (Tobin et al. 1990), it increases during the first larval instar, peaks during the second instar and diminishes in the third instar and in prepupae. Another small burst of transcription occurs during pupation (Sanchez et al. 1983). Studies of transcript distribution and the pattern of expression of a reporter gene controlled by 4 kb of the *Act79B* promoter region showed that transcription starts in midpupae (at 168 h) and continues

Act87E

```
-981  TATTAGAAAACCATCACACAATAGAAAATAGGTACAAAATAGATAATTTTCATTCCATCATATGCGCTTTACAAAAATCTATATTTTCTC  -892

-891  ATAACATATTTTGAGCCATCTTTCCTGCAGTGCACCATCTGGGAAATTATGAACGAAGCGAGCAGAAGTCCAAAAGCAAAAATCCTACGA  -802

-801  AAACAAATTATTTTTAAAAGAAACTCAGAATCTCCCCCCGCCGGCGCAATGTGCATCCATGTGCACATGTGTGCCGAGAGGCGATTGAGT  -712

                                                                                     -->-637.
-711  GTGCGTGCGGAAAATATCTAAAACGACTGAGGGTCGCCAGAATGGTATAAATATTAGCGCATCTCGGTCCAGCGACCACTCGCAGTTCTA  -622

-621  CAGCGAAAGTGTTGATTTGGATTTCTAGTTTTTCTTCGTCTAACGGTTAGTATACTCCACATCCACCAATTCCGTCTGTCTGGTTGACTT  -532

-531  TTACCCAATCCGATGCTGGATCCAGTGTACAGTGCCCAACTTTCTGAAAAGAAAGATACTTTGGAAAGATAGAGATCTCAACAAACAACA  -442

-441  TATTTGACAAGGAGCAGAAAAAGTTCAATCAACGATCCTTAAATGTTTGGTTTTTAATAGTGACTAACTTTTGTTTAAAAAAACATTCTT  -352

-351  TAAAATGTTAAAACTGAAAAATATTAGTTGTTTGATCTTAAATCAAAAATTATAATTAATTAAAACTTTTATTAAAGTATGTAATATCGT  -262

-261  CAAAAGTTGAAGACAGCCCTTTGTTAATTATCCACGTTTCGATTAATTTTAAGGATTGCTCCTCTGCAAAGATACTCTTTCTTTTAAAGT  -172

-171  CATACATGTTCTGAGGCAACACCTACACGTATTTCATAATTTCACACTTACACACAAGATTACAATTAAAATCCATACCCAATCCGATTC  -82

-81   CCGAAAGCCCACTTCTCACTTCTCCTTCTAAAAACCGCCTCCGTTCTCGTTGTTGTTGCAGTGAAAACAGCCAGTAGCCAAGATGTGTGA  8
                                                                                      MetCysAs  (3)

9     CGATGAGGTTGCCGCATTGGTCGTGGACAATGGTTCCGGAATGTGCAAAGCAGGATTCGCCGGCGATGATGCGCCTCGCGCCGTCTTCCC  98
      pAspGluValAlaAlaLeuValValAspAsnGlySerGlyMetCysLysAlaGlyPheAlaGlyAspAspAlaProArgAlaValPhePr  (33)

99    CTCGATTGTGGGTCGTCCCCGTCATCAGGGCGTAATGGTGGGCATGGGACAGAAGGACTCCTATGTTGGTGATGAGGCCCAGAGCAAGCG  188
      oSerIleValGlyArgProArgHisGlnGlyValMetValGlyMetGlyGlnLysAspSerTyrValGlyAspGluAlaGlnSerLysAr  (63)

189   TGGTATCCTCACCCTGAAATACCCCATCGAGCACGGCATCATCACCAACTGGGACGATATGGAGAAGATCTGGCACCACACTTTCTATAA  278
      gGlyIleLeuThrLeuLysTyrProIleGluHisGlyIleIleThrAsnTrpAspAspMetGluLysIleTrpHisHisThrPheTyrAs  (93)

279   CGAGCTGCGCGTCGCCCCCGAGGAACACCCCGTCCTGCTGACCGAGGCCCCCCTGAACCCCAAGGCCAATCGCGAGAAGATGACCCAGAT  368
      nGluLeuArgValAlaProGluGluHisProValLeuLeuThrGluAlaProLeuAsnProLysAlaAsnArgGluLysMetThrGlnIl  (123)

369   CATGTTCGAGACCTTCAACGCACCCGCCATGTATGTGGCCATCCAGGCTGTGCTCTCGCTGTACGCCTCCGGTCGTACCACCGGTATTGT  458
      eMetPheGluThrPheAsnAlaProAlaMetTyrValAlaIleGlnAlaValLeuSerLeuTyrAlaSerGlyArgThrThrGlyIleVa  (153)

459   CCTCGACTCCGGTGACGGTGTCTCCCACACCGTGCCCATCTACGAGGGTTACGCCCTGCCCCACGCCATCCTGCGTCTGGATCTGGCTGG  548
      lLeuAspSerGlyAspGlyValSerHisThrValProIleTyrGluGlyTyrAlaLeuProHisAlaIleLeuArgLeuAspLeuAlaGl  (183)

549   TCGCGATTTGACCGACTACCTGATGAAGATCCTGACCGAGCGCGGTTACTCATTCACCACCACCGCTGAGCGTGAAATCGTTCGCGACAT  638
      yArgAspLeuThrAspTyrLeuMetLysIleLeuThrGluArgGlyTyrSerPheThrThrThrAlaGluArgGluIleValArgAspIl  (213)

639   CAAGGAGAAGCTGTGCTATGTTGCCCTGGACTTTGAGCAGGAGATGGCCACCGCCGCCGCCTCCACATCCCTGGAGAAGTCATACGAGCT  728
      eLysGluLysLeuCysTyrValAlaLeuAspPheGluGlnGluMetAlaThrAlaAlaAlaSerThrSerLeuGluLysSerTyrGluLe  (243)

729   TCCCGACGGACAGGTGATCACCATCGGCAACGAACGTTTCCGCTGCCCAGAGTCGCTGTTCCAGCCCTCTTTCCTGGGAATGGAATCGTG  818
      uProAspGlyGlnValIleThrIleGlyAsnGluArgPheArgCysProGluSerLeuPheGlnProSerPheLeuGlyMetGluSerCy  (273)
```

```
819  CGGCATCCACGAGACCGTGTACAACTCGATCATGAAGTGCGATGTGGACATCCGTAAGGATCTGTATGCTAACATCGTCATGTCGGGTGG  908
     sGlyIleHisGluThrValTyrAsnSerIleMetLysCysAspValAspIleArgLysAspLeuTyrAlaAsnIleValMetSerGlyGl  (303)

909  TACCACCATGTACCCTGGTATTGCCGATCGTATGCAGAAGGAGATCACCGCCCTGGCCCCGTCCACCATCAAGATCAAGATCATTGCCCC  998
     yThrThrMetTyrProGlyIleAlaAspArgMetGlnLysGluIleThrAlaLeuAlaProSerThrIleLysIleLysIleIleAlaPr  (333)

999  ACCGGAGCGCAAGTACTCCGTCTGGATCGGTGGCTCCATCCTGGCCTCCCTGTCCACCTTCCAGCAGATGTGGATCTCCAAGCAGGAGTA  1088
     oProGluArgLysTyrSerValTrpIleGlyGlySerIleLeuAlaSerLeuSerThrPheGlnGlnMetTrpIleSerLysGlnGluTy  (363)

1089 CGACGAGTCCGGCCCAGGAATCGTCCACCGCAAGTGCTTCTAAGCGATCTAAACACCACAGACACTGCAAACCACACGGGCATTGAGACC  1178
     rAspGluSerGlyProGlyIleValHisArgLysCysPheEnd                                                 (376)

1179 CAACCACACCACGCCACAGAACACCACACAACAACAACAAGAACAACATGAACAGCAACAACCAAATACCAAATCAAGATCTATAGCCTA  1268

1269 GTGCTATTGATGATTAATCTTAAGTTAAAACCTCTTGCTGCCCTGCCATCCAAAGAAAACCGAAGGAACCGCGATTGTAACAGCATGTAT  1358

1359 TATACTTATATTAATATTTATTGGAGAGCCGCTTGATGGCGCTGAAGGAGGAGTTGAGGAGACACAAGAATGCAAAATTTTACAGTTTTA  1448

1449 AAAATAAATTATACTAGCATCCTCTATAAATTAAATCTAAATTTTAAACGAAACGTATCTTTTATTCGCTGCAAGCGGCATGCTATGCGA  1538
       -----                                   |(A)n  |(A)n|(A)n

1539 TTATTTTTAGCGACGCACAGGAAATTACGAAATTTTGCACGCCCACTGCAAAGAGCGAAATCTGGAGGTGGATCTCCTCGACTGGGGTGC  1628

1629 ACATACATATGTACATATGTGGCTGGGGATGAGCACGGTAATCCCAGCATAGACGCCTCCAAGACAGTCCATTTTTGCCCATTGCCAGTC  1718

1719 GGTGCAGGAGCTGCCCCCCTCGTCGTGGATCTAAAAATACAGGCCAAAGGAAACAACAAAAGCGGCAAATCAACATGCCGAAGTATTAAC  1808

1809 AAATGTCTTCTAAGACTACAGTCAACCCACAGTAGATTGAACAAATATGTGACTTTGAATGTCAGAATGTCAACTTTAAAGGGATTCGAA  1898

1899 AATATATATTTTTAAAACTAAACTAATTAGGAATACAAGAGCTC  1942
```

Act87E SEQUENCE. Strain, *Oregon R.* Accession, X12452 (DROACT87EA), K00674 (DROACT87E).

in young adults. *Act79B* RNA is present in the various tubular-type muscles of the thorax: direct flight muscles, leg muscles and muscles that support the head and abdomen. *Act79B* transcripts are also present in muscles surrounding the male genitalia, but not in indirect flight muscles (Courchesne-Smith and Tobin 1989).

Act87E

Gene Organization and Expression

Expected mRNA sizes range between 1,568 and 1,580 bases. The 5′ end was determined by S1 mapping, by primer extension and by sequencing of several cDNA clones. Three poly(A) sites have been identified in five cDNA sequences. There is a leader intron with a donor site at −577 and an acceptor site at −20.

Act88F

-2066 TCTAGAATGCACAATAGGCAAATTTAGTTAAGATATGAATTTTTAAATAAATGGTGAGCCCAATCAATTCAGTGGTTGAATGACTTTTCA -1977

-1976 TAAATTAAAAAATAAAGATAAGAATGGTGAACAATTCTGTTCGCAGCCAATAACCTCTTGCTCAATACACGTGTCAATCAAGGCAATCCA -1887

-1886 AATAAAACGCTTTGGGAATGCCACCAATTCACTTCCGAGCATCAGTTCCTATCTTTAGCCAACCGATTCGATTATTTCATGTGGGCAAGC -1797

-1796 AATAAAAACGTAAATAGAAGAAGTAAAAAATAATTAAATCTACATAAAGGAATAAATACAGTTCGATTGAGAAAATACATTTTCGCTCGG -1707

-1706 TCTGGCTGGCAATGGTTGGTTAATTGCACTGATAAATGGTCGGCACGGTGATTTCGCAACTTCGGGATTGCATCGGCGCCGCAATGCAAA -1617

-1616 GTGCAGCAGCATTCTGTAGAATGCGATTGCAAATGTGGATGCAGCTTCCTCGAGCACCGCGCGGAGATCTGATCAACCTTGCGTGTTG -1527

-1526 ATTTATCGGTGCCGCTCTGCTTGGCGCGTCTATTTTAGATTCGCCTCGCTGCGTGCCCGTTGAAATGTCCCATTCTCCCAGTCCCTGCCG -1437

-1436 CGGATGCCAATTGTCTTGCGTCGGTCCTTCTAAGGTCCGTTTCTATTTTCCGAAGCTCTCAGCACCGAATGAGTCGTCCGCCGCAGCAGT -1347

-1346 CGCCCATTGGCAGCAGGATTGGGACAGAGATGGGGACGGAGATGGGGCTAATTGGCCGCTCGAGAGTGCTGATTGCCGTTTAGGTGGCCC -1257

-1256 ATACACCGCTATCACGCACCTCTGCTAATCACTCGGCTATGGCGTTCTCTTATCTTTCGAGAGCTTTCTCTCTCCTGGCACTCCCTACAA -1167

-1166 ATAATGAATAGGGTCCTAAGATTGATAGCTTACTTCCATCATATATTGTCAATTAATTAAATATTTCAGGATTAAAATATGAAACGAATT -1077

-1076 GAACATAAAGTTTCTACTACATAGTTATTTAAGCTGTTATATGTTATGAGACCATTTTCTCAGGATTTGTACCTACTAACAATGTGAAAA -987

-986 AAATATAAAATTGTCATATTTTCGCAGTTTGGAAATTCCCTCGTTTATTGAATTTATTGGTAATCTTAATAAATGATTCTATGCTTTATT -897

-896 AAGTATTTAATTGTGTGGCTTCCTTTTTTTTTGTTGAAAGCGCATTAATGAGTCGTCTTCGTGCAATGAGGCATCCAAACTTCTGACATG -807

-806 CTCGGCCAGAAGTCTGAAAACTGCTTATATGGATCGGTTCGAGTTGATTGTTCCGCAGCACTTTCGCTCAATCTTTTTCTCAGTGCCGCA -717

-->-646

-716 CTGGCATCCAATCAAATCGCTTCGAGGGAGAGCCGAGATATAAAAGGCAGGACAGACCGATCGGCGTGCCATTTGTTGTTGAATCTAGTT -627

-626 GTCAACAGGAATCGAACGTGCGACTCTATCCAATTTTTCTCCTTTCGTTGACCTAAAAGGTGTGTGAGTGCGACCTCAATGTCGAAGGAT -537

-536 CCAAGGATTATTACAGAAAAAGCCAAGAGGACTAAGGATATTAAAACTCTTTTTAATAAGTTCGGATTGTTTGATGGATTTTTCTACAAG -447

-446 TCACTAATCGGTCTTCGAAAGTTCAATATCTAAATATAAAGTGAAGAGTAATTGCAACGAAACGTATTTTCAATTAATTTGATACGTTTA -357

-356 AATTAAGTTCTATGAACTATTCTTTTCCGATATTTTTAGAGCACTGATTTAGTTTCAAGTGAATAACCAATTAGCATGACTCAAAAGGAA -267

-266 ATGGAATATACCAATTTTGGCAATTTTTCATGGTTTTATTTACTGAAATGTGCTCAAATGGACAATAGAGTTTCACTTCACTTCTTCAAT -177

-176 ATCTTAAAAAGTTAAATATTTTCTTGAGACACAAATTAGTTTTCTATGTTGTCATTAAAGTAGTAGAATTTAAAGAATTGAGATGTAGGT -87

-86 GGGAGCTATAAAACTTTACATATATAATCGACAGATCGAGCTAACCGAGTGCACTTCCATCTCCCTTCCAGATAAACAACTGCCAAGATG 3
Met (1)

4 TGTGACGATGATGCGGGTGCATTAGTTATCGACAACGGATCGGGCATGTGCAAAGCCGGCTTCGCCGGTGATGACGCTCCCCGTGCTGTC 93
CysAspAspAspAlaGlyAlaLeuValIleAspAsnGlySerGlyMetCysLysAlaGlyPheAlaGlyAspAspAlaProArgAlaVal (31)

```
  94  TTCCCCTCAATTGTGGGTCGTCCCCGACACCAGGGTGTGATGGTGGGTATGGGTCAGAAGGACTCGTACGTGGGCGACGAGGCGCAAAGC   183
      PheProSerIleValGlyArgProArgHisGlnGlyValMetValGlyMetGlyGlnLysAspSerTyrValGlyAspGluAlaGlnSer  (61)

                                                    A=KM88
 184  AAGCGCGGTATCCTGACGCTGAAGTACCCCATCGAGCACGGCATCATCACGAACTGGGACGACATGGAGAAGATCTGGCATCACACCTTC   273
      LysArgGlyIleLeuThrLeuLysTyrProIleGluHisGlyIleIleThrAsnTrpAspAspMetGluLysIleTrpHisHisThrPhe  (91)
                                                     End

 274  TACAACGAGCTGCGCGTGGCCCCCGAGGAGCATCCAGTATTATTGACCGAGGCTCCACTGAACCCCAAGGCCAATCGCGAGAAGATGACC   363
      TyrAsnGluLeuArgValAlaProGluGluHisProValLeuLeuThrGluAlaProLeuAsnProLysAlaAsnArgGluLysMetThr  (121)

 364  CAGATCATGTTCGAGACCTTCAACTCGCCGGCCATGTACGTGGCCATCCAGGCCGTGCTCTCCCTGTACGCCTCCGGTCGTACCACCGGT   453
      GlnIleMetPheGluThrPheAsnSerProAlaMetTyrValAlaIleGlnAlaValLeuSerLeuTyrAlaSerGlyArgThrThrGly  (151)

 454  ATTGTGCTGGACTCCGGCGATGGTGTCTCCCACACCGTGCCCATCTATGAGGGCTTCGCCCTGCCCCACGCCATTCTGCGTCTGGATCTG   543
      IleValLeuAspSerGlyAspGlyValSerHisThrValProIleTyrGluGlyPheAlaLeuProHisAlaIleLeuArgLeuAspLeu  (181)

 544  GCTGGTCGCGATCTGACCGATTACCTGATGAAGATCCTGACGGAGCGCGGCTACAGCTTCACCACCACCGCCGAGCGTGAGATCGTGCGC   633
      AlaGlyArgAspLeuThrAspTyrLeuMetLysIleLeuThrGluArgGlyTyrSerPheThrThrThrAlaGluArgGluIleValArg  (211)

 634  GACATCAAGGAGAAGCTGTGCTACGTGGCTCTGGACTTCGAGCAGGAGATGGCCACCGCTGCCGCCTCCACCTCGCTGGAGAAGTCGTAC   723
      AspIleLysGluLysLeuCysTyrValAlaLeuAspPheGluGlnGluMetAlaThrAlaAlaAlaSerThrSerLeuGluLysSerTyr  (241)

 724  GAGTTGCCTGACGGCCAGGTGATCACCATTGGCAACGAGCGCTTCCGCTGCCCCGAGGCCCTGTTCCAGCCCTCGTTCCTGGGCATGGAG   813
      GluLeuProAspGlyGlnValIleThrIleGlyAsnGluArgPheArgCysProGluAlaLeuPheGlnProSerPheLeuGlyMetGlu  (271)

 814  TCGTGCGGCATCCACGAGACCGTCTACAACTCGATCATGAAGTGCGACGTGGACATCCGCAAGGATCTGTATGCCAACTCCGTGCTGTCC   903
      SerCysGlyIleHisGluThrValTyrAsnSerIleMetLysCysAspValAspIleArgLysAspLeuTyrAlaAsnSerValLeuSer  (301)

                                        |>=DefKM129
 904  GGCGGTACCACCATGTACCCTGGTACACGGATCGTTCGCTTCAGCAGTTGCACTTGTGCTTAATCCTTTGGTGCACTTTCAGGTATTGCC   993
      GlyGlyThrThrMetTyrProG                                                            lyIleAla  (311)

 994  GATCGTATGCAGAAGGAGATCACTGCCCTGGCCCCATCGACCATCAAGATCAAGATCATTGCGCCACCCGAGAGGAAGTACTCCGTCTGG  1083
      AspArgMetGlnLysGluIleThrAlaLeuAlaProSerThrIleLysIleLysIleIleAlaProProGluArgLysTyrSerValTrp  (341)

                                                A=KM75                  A=HH5
1084  ATCGGTGGCTCCATCCTGGCCTCGCTGTCCACCTTCCAGCAGATGTGGATCTCGAAGCAGGAGTACGACGAGTCCGGCCCCGGAATCGTT  1173
      IleGlyGlySerIleLeuAlaSerLeuSerThrPheGlnGlnMetTrpIleSerLysGlnGluTyrAspGluSerGlyProGlyIleVal  (371)
                                                 End                       Ser

1174  CACCGCAAATGCTTTTAAGTCTTCGCCCGCCGCGAAAGCTCTTCAAAGGCAGCAACCAGCAGCGACCAACAAGCATCCATCTCGACCTTA  1263
      HisArgLysCysPheEnd                                                                          (376)

1264  CCCAACAACCTCGGCTCGGACAGTGATAGACAAAAGCAGCGAACCCATCGCGACAACAATTATCATCCAACTCAGATTCATAGCAGATAA  1353

1354  TCAGAGGCAACCTCGGTTGTCGGTGGTTATCTTATGGCATTTCATCGGCAGCGGTATAGCGGATTTTTATTTTGAAGAACTAATCGTAAT  1443

1444  CGTAAGAGTCGTGGTCTGCTCAGG  1467
```

Act88F Sequence. Strain, *Canton S.* Accession, M18830 (DROACT88F), and M13925 (DROACT88H). There are several discrepancies among published sequences, even within the coding regions; these could be due either to natural polymorphisms

(*continued*)

Transcription is directed toward the telomere (*Act87E* Sequence) (Fyrberg et al. 1981; Manseau et al. 1988).

Developmental Pattern

In embryos, transcripts are detectable in the developing musculature of the future larval body wall; the level of *Act87E* transcript is 5–10 times lower than for *Act57B* (Tobin et al. 1990).

Act88F

Gene Organization and Expression

The 5′ end was determined by primer extension and by cDNA sequencing (Geyer and Fyrberg 1986; Okamoto et al. 1986). The 3′ end has not been mapped. There is a leader intron with a donor site at −568 and an acceptor site at −15; there is another intron in the Gly-309 codon (*Act88F* Sequence) (Fyrberg et al. 1981; Sanchez et al. 1983).

Developmental Pattern

Transcription is undetectable in embryos (Tobin et al. 1990); it increases during the first larval instar, peaks during the second instar and diminishes during the third instar and in prepupae. There is another larger peak of expression during pupation (Sanchez et al. 1983); at this stage, transcription is most prominent in the indirect flight muscles (Geyer and Fyrberg 1986).

Promoter

Approximately 1,000 bp of 5′ flanking DNA are sufficient for normal levels of RNA production and for complementation of the *raised* mutation (*rsd*). A putative enhancer element was identified between −1,565 and −1,286 (Geyer and Fyrberg 1986).

References

Biggin, M. D. and Tjian, R. (1988). Transcription factors that activate the *Ultrabithorax* promoter in developmentally staged extracts. *Cell* **53**:699–711.

(*continued*) or to sequencing errors; I report the results of Geyer and Fyrberg (1986) with the modifications of Mahaffey et al. (1985) and Okamoto et al. (1986). These seem to correspond to the more common allele in *Canton S*. The nature of several mutations are shown (Karlik et al. 1984; Okamotot et al. 1986).

Bond, B. J. and Davidson, N. (1986). The *Drosophila melanogaster Actin 5C* gene uses two transcription initiation sites and three polyadenylation sites to express multiple mRNA species. *Mol. Cell. Biol.* **6**:2080–2088.

Bond-Matthews, B. and Davidson, N. (1988). Transcription from each of the *Drosophila Act5C* leader exons is driven by a separate functional promoter. *Gene* **62**:289–300.

Burn, T. C., Vigoreaux, J. O. and Tobin, S. L. (1989). Alternative 5C actin transcripts are localized in different patterns during *Drosophila* embryogenesis. *Dev. Biol.* **131**:345–355.

Chung, Y.-T. and Keller, E. B. (1990a). Regulatory elements mediating transcription from the *Drosophila melanogaster Actin 5C* proximal promoter. *Mol. Cell. Biol.* **10**:206–216.

Chung, Y.-T. and Keller, E. B. (1990b). Positive and negative regulatory elements mediating transcription from the *Drosophila melanogaster Actin 5C* distal promoter. *Mol. Cell. Biol.* **10**:6172–6180.

Couderc, J. L., Hilal, L., Sobrier, M. L. and Dastugue, B. (1987). 20-Hydroxyecdysone regulates cytoplasmic actin gene expression in *Drosophila* cultured cells. *Nucleic Acid Res.* **15**:2549–2561.

Courchesne-Smith, C. L. and Tobin, S. L. (1989). Tissue-specific expression of the 79B actin gene during *Drosophila* development. *Dev. Biol.* **133**:313–321.

Drummond, D. R., Hennessey, E. S. and Sparrow, J. C. (1991). Characterisation of missense mutations in the *Act88F* gene of *Drosophila melanogaster. Mol. Gen. Genet.* **226**:70–80.

Fyrberg, E. A., Bond, B. J., Hershey, N. D., Mixter, K. S. and Davidson, N. (1981). The actin genes of *Drosophila*: protein coding regions are highly conserved but intron positions are not. *Cell* **24**:107–116.

Fyrberg, E. A., Mahaffey, J. W., Bond, B. J. and Davidson, N. (1983). Transcripts of the six *Drosophila* actin genes accumulate in a stage- and tissue-specific manner. *Cell* **33**:115–123.

Fyrberg, E., Beall, C. and Fyrberg, C. C. (1991). From genes to tensile forces: genetic dissection of contractile protein assembly and function in *Drosophila melanogaster. J. Cell Sci. Suppl.* **14**:27–29.

Geyer, P. K. and Fyrberg, E. (1986). 5′-Flanking sequence required for regulated expression of a muscle-specific *Drosophila melanogaster* actin gene. *Mol. Cell. Biol.* **6**:3388–3396.

Karlik, C. C., Coutu, M. D. and Fyrberg, E. (1984). A nonsense mutation within the *Act88F* actin gene disrupts myofibril formation in *Drosophila* indirect flight muscles. *Cell* **38**:711–719.

Mahaffey, J. W., Coutu, M. D., Fyrberg, E. A. and Inwood, W. (1985). The flightless *Drosophila* mutant *raised* has two distinct genetic lesions affecting accumulation of myofibrillar proteins in flight muscles. *Cell* **40**:101–110.

Manseau, L. J., Ganetzky, B. and Craig, E. A. (1988). Molecular and genetic characterization of the *Drosophila melanogaster* 87E actin gene region. *Genetics* **119**:407–420.

Okamoto, H., Hiromi, Y., Ishikawa, E., Yamada, T., Isoda, K., Maekawa, H. and Hotta, Y. (1986). Molecular characterization of mutant actin genes which induce heat-stock proteins in *Drosophila* flight muscles. *EMBO J.* **5**:589–596.

Rao, J. P., Zafar, R. S. and Sodja, A. (1988). Transcriptional activity at the 3′ end of the actin gene at 5C on the X chromosome of *Drosophila melanogaster. Bioch. Biophys. Acta* **950**:30–44.

Sanchez, F., Tobin, S. L., Rdest, U., Zulauf, E. and McCarthy, B. J. (1983). Two

Drosophila actin genes in detail. Gene structure, protein structure and transcription during development. *J. Mol. Biol.* **163**:533–551.

Sparrow, J., Drummond, D., Peckham, M., Hennessey, E. and White, D. (1991). Protein engineering and the study of muscle contraction in *Drosophila* flight muscles. *J. Cell. Sci.* Suppl. **14**:73–78.

Tobin, S. L., Cook, P. J. and Burn, T. C. (1990). Transcriptions of individual *Drosophila* actin genes are differentially distributed during embryogenesis. *Dev. Genet.* **11**:15–26.

Vigoreaux, J. O. and Tobin, S. L. (1987). Stage-specific selection of alternative transcription initiation sites from the 5C actin gene of *Drosophila melanogaster*. *Genes Dev.* **1**:1161–1171.

3

Alcohol dehydrogenase: Adh, Adh-dup

Chromosomal Location:
2L, 35B2-3

Map Position:
2-50.1

Product

Alcohol dehydrogenase (ADH; alcohol:NAD$^+$ oxidoreductase, EC 1.1.1.1) (Grell et al. 1965).

Structure

ADH is a homodimer with subunits of 27.4 kD; the polypeptide is 255 amino acids long with Acetyl-Ser at the amino terminus. There are two common allozymes, Slow (S) and Fast (F), that differ in electrophoretic mobility due to a threonine/lysine substitution at position 192.

Unlike the ADH of other species, *Drosophila* ADH does not use Zn^{++} as a cofactor. Amino acid sequence comparisons reveal significant differences between *Drosophila* ADH on one hand and ADH from yeast or horse liver on the other (the latter two being quite similar); these observations suggest that the *Drosophila* protein is not homologous to other ADHs (Thatcher 1980; Benyajati et al. 1981). Rather, sequence comparisons show similarities between *Drosophila* ADH and *Klebsiella* ribitol dehydrogenase (Jörnvall et al. 1981). The evolution of ADH in the genus *Drosophila* has been discussed by Sullivan et al. (1990).

Function

ADH is more active on alcohols of 3–5 carbons than on ethanol and more active on secondary than on primary alcohols (Sofer and Ursprung 1968).

Tissue Distribution

ADH activity increases very rapidly from the second larval instar to immediately before pupariation; it declines during the pupal stages and increases again for the first 4–5 days after emergence of the adult. In larvae, the enzyme is distributed approximately equally between fat bodies and midgut (although it is absent from the middle midgut). In adults, most of the activity is in the fat tissues, with much lower levels in the Malpighian tubules and the male reproductive system (Ursprung et al. 1970; Maroni and Stamey 1983).

Mutant Phenotype

Null mutants are quite sensitive to a 5% ethanol solution. Even without an ethanol supplement, such mutants sometimes die as first instar larvae in cultures with very active yeast. *Adh* mutants, however, are more tolerant than wild-type flies to unsaturated secondary alcohols (O'Donnell et al. 1975).

Gene Organization and Expression

Open reading frame, 256 amino acids; expected mRNA length, 1,071 bases (distal promoter) and 1,010 bases (proximal promoter). The different-sized transcripts carry the same open reading frame but different 5′ untranslated regions (Benyajati et al. 1983). S1 mapping and primer extension sequencing of mRNA were used to determine 5′ ends while S1 mapping and cDNA sequences defined the 3′ end. Much of the extra length of the distal promoter transcript is in an intron with donor site at −690 and acceptor site at −35 (*Adh* Sequence and Fig. 3.1). *Adh* also has two small introns in the coding region. The first is after the codon corresponding to Lys-33 in the middle of the presumptive NAD$^+$-binding domain and the second after the codon corresponding to Ala-168 near the boundary between the presumptive NAD$^+$-binding and catalytic domains (Benyajati et al. 1981).

Developmental Pattern and Promoter

The upstream, distal promoter is expressed primarily in adults while the proximal promoter is used during larval stages (Savakis et al. 1986). Two

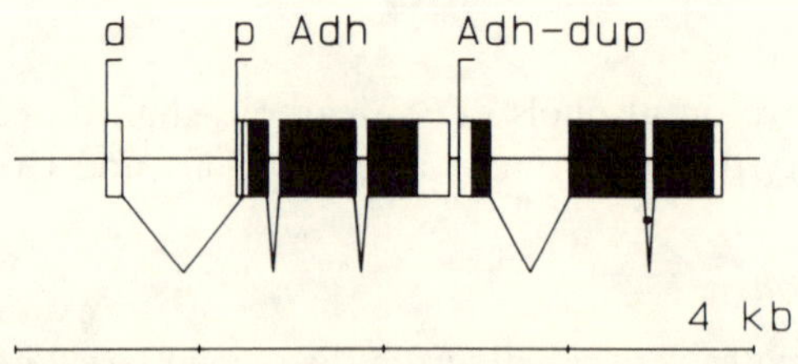

FIG. 3.1. Diagram of the organization and expression of *Adh* and *Adh-dup*

Adh

```
-1559  AGCTGCATTCGAAACCGCTACTCTGGCTCGGCCACAAAGTGGGCTTGGTCGCTGTTGCGGACAAGTGAGATTGCTAATGAGCTGCTTTTA  -1470

-1469  GGGGGCGTGTTGTGCTTGCTTTCCAACTTTTCTAGATTGATTCTACGCTGCCTCCAGCAGCCACCCCTCCCATCCCCATCCCCATCACCA  -1380

-1379  TCCAGTCCCGTTGGCTCCCAGTCACAGTATTACACGTATGCAAATTAAGCCGAAGTTCAATTGCGACCGCAGCAACAACACGATCTTTCT  -1290
                                                                           -----------------dep4=aef-1
                                                                                       ------

-1289  ACACTTCTCCTTGCTATGCTTGACATTCACAAGGTCAAAGCTCTTAATATTCTGGCTTGTGGCCCTACACTGTAAGAAATTACTATAGAA  -1200
       ---c/ebp            ------------dep1-2                        ---------------------------dep3

-1199  ATAACGGTACACGGAATAAGATATTTTTTTTAGTCCATATGCTTTTAACAAATGTGTTTTGAGTTTATGTTATATTATTGTTAGAAAACA  -1110

-1109  GGTGTTTTTTTTTAAATCGGTTAAAAAATTACTACGAGAGAAAAATACAAATTTTGTAAATAAGATTGACTCTTTTTCGATTTTGGAATA  -1020

-1019  TTTTCATTCATTTTATGTTTTTACGTTTTCACTTATTTGTTTCTCAGTGCACTTTCTGGTGTTCCATTTTCTATTGGGCTCTTTACCCCG  -930

 -929  CATTTGTTTGCAGATCACTTGCTTGCGCATTTTTATTGCATTTTACATATTACACATTATTTGAACGCCGCTGCTGCTGCATCCGTCGAC  -840
                                                                 -----    -----------------------
                                                                      -->-776
 -839  GTCGACTGCACTCGCCCCCACGAGAGAACAGTATTTAAGGAGCTGCGAAGGTCCAAGTCACCGATTATTGTCTCAGTGCAGTTGTCAGTT  -750
       -----------------d1           -------                   _
                                                                |
 -749  GCAGTTCAGCAGACGGGCTAACGAGTACTTGCATCTCTTCAAATTTACTTAATTGATCAAGTAAGTAGCAAAAGGGCACCCAATTAAAGG  -660
                                                               _|

 -659  AAATTCTTGTTTAATTGAATTTATTATGCAAGTGCGGAAATAAAATGACAGTATTAATTAGTAAATATTTTGTAAAATCATATATAATCA  -570

 -569  AATTTATTCAATCAGAACTAATTCAAGCTGTCACAAGTAGTGCGAACTCAATTAATTGGCATCGAATTAAAATTTGGAGGCCTGTGCCGC  -480

 -479  ATATTCGTCTTGGAAAATCACCTGTTAGTTAACTTCTAAAAATAGGAATTTTAACATAACTCGTCCCTGTTAATCGGCGCCGTGCCTTCG  -390

 -389  TTAGCTATCTCAAAAGCGAGCGCGTGCAGACGAGCAGTAATTTTCCAAGCATCAGGCATAGTTGGGCATAAATTATAAACATACAAACCG  -300
                                                         ----------------------------------------p2

 -299  AATACTAATATAGAAAAAGCTTTGCCGGTACAAAATCCCAAACAAAAACAAACCGTGTGTGCCGAAAAATAAAAATAAACCATAAACTAG  -210
                                                                                       ---------

 -209  GCAGCGCTGCCGTCGCCGGCTGAGCAGCCTGCGTACATAGCCGAGATCGCGTAACGGTAGATAATGAAAAGCTCTACGTAACCGAAGCTT  -120
       -------------------------------------p1    ----------------------p0
                                                                                            _
                .         .         .         .         .-->-69   .         .         .  |      .
 -119  CTGCTGTACGGATCTTCCTATAAATACGGGGCCGACACGAACTGGAAACCAACAACTAACGGAGCCCTCTTCCAATTGAAACAGATCGAA  -30
                         ------                                                              |_
                .         .         .         .         .         .         .    A=n11 .         .
  -29  AGAGCCTGCTAAAGCAAAAAAGAAGTCACCATGTCGTTTACTTTGACCAACAAGAACGTGATTTTCGTTGCCGGTCTGGGAGGCATTGGT  60
                                     MetSerPheThrLeuThrAsnLysAsnValIlePheValAlaGlyLeuGlyGlyIleGly  (20)
                                                                               Asp
                .         .         .      .CGATC|-def|=fn6   .         .         .  |-      .
   61  CTGGACACCAGCAAGGAGCTGCTCAAGCGCGATCTGAAGGTAACTATGCGATGCCCACAGGCTCCATGCAGCGATGGAGGTTAATCTCGT  150
       LeuAspThrSerLysGluLeuLeuLysArgAspLeuLys                                                     (33)
```

(continued)

```
         def   .-|G=fn4  .         .         .         . A=UF    .         A=F'      .         .
 151  GTATTCAATCCTAGAACCTGGTGATCCTCGACCGCATTGAGAACCCGGCTGCCATTGCCGAGCTGAAGGCAATCAATCCAAAGGTGACCG   240
                    AsnLeuValIleLeuAspArgIleGluAsnProAlaAlaIleAlaGluLeuLysAlaIleAsnProLysValThrV  (59)
                                                        Asp               Glu
               .        |-  def  -|=fn24     .         .         .         . T=n4    .         .
 241  TCACCTTCTACCCCTATGATGTGACCGTGCCCATTGCCGAGACCACCAAGCTGCTGAAGACCATCTTCGCCCAGCTGAAGACCGTCGATG   330
      alThrPheTyrProTyrAspValThrValProIleAlaGluThrThrLysLeuLeuLysThrIlePheAlaGlnLeuLysThrValAspV  (89)
                                                                             Ter

               .         .         .         .         .         .         .         .         .
 331  TCCTGATCAACGGAGCTGGTATCCTGGACGATCACCAGATCGAGCGCACCATTGCCGTCAACTACACTGGCCTGGTCAACACCACGACGG   420
      alLeuIleAsnGlyAlaGlyIleLeuAspAspHisGlnIleGluArgThrIleAlaValAsnTyrThrGlyLeuValAsnThrThrThrA  (119)

               .         .         .         .         .         .         .         .         .
 421  CCATTCTGGACTTCTGGGACAAGCGCAAGGGCGGTCCCGGTGGTATCATCTGCAACATTGGATCCGTCACTGGATTCAATGCCATCTACC   510
      laIleLeuAspPheTrpAspLysArgLysGlyGlyProGlyGlyIleIleCysAsnIleGlySerValThrGlyPheAsnAlaIleTyrG  (149)

               .         .         .         .         .         .         .         .         .
 511  AGGTGCCCGTCTACTCCGGCACCAAGGCCGCCGTGGTCAACTTCACCAGCTCCCTGGCGGTAAGTTGATCAAAGGAAACGCAAAGTTTTC   600
      lnValProValTyrSerGlyThrLysAlaAlaValValAsnPheThrSerSerLeuAla                                (168)

               .         .         .         .         .         .         .         .         .
 601  AAGAAAAAACAAAACTAATTTGATTTATAACACCTTTAGAAACTGGCCCCCATTACCGGCGTGACCGCTTACACCGTGAACCCCGGCATC   690
                                             LysLeuAlaProIleThrGlyValThrAlaTyrThrValAsnProGlyIle  (185)

               .         . C=F     .         .         .         .         .         .       T=F
 691  ACCCGCACCACCCTGGTGCACAAGTTCAACTCCTGGTTGGATGTTGAGCCCAGGTTGCTGAGAAGCTCCTGGCTCATCCCACCCAGCCA   780
      ThrArgThrThrLeuValHisLysPheAsnSerTrpLeuAspValGluProGlnValAlaGluLysLeuLeuAlaHisProThrGlnPro  (215)
                           Thr                                                               Ser
               .         .         .         .         . A=D     . A=nB    .         .   |-    .
 781  TCGTTGGCCTGCGCCGAGAACTTCGTCAAGGCTATCGAACTGAACCAGAACGGAGCCATCTGGAAACTGGACTTGGGCACCCTGGAGGCC   870
      SerLeuAlaCysAlaGluAsnPheValLysAlaIleGluLeuAsnGlnAsnGlyAlaIleTrpLysLeuAspLeuGlyThrLeuGluAla  (245)
                                                         Glu      Ter
               . def     .     -|=fn23       .         .         .         .         .         .
 871  ATCCAGTGGACCAAGCACTGGGACTCCGGCATCTAAGAAGTGATAATCCCAAAAAAAAAACATAACATTAGTTCATAGGGTTCTGCGAAC   960
      IleGlnTrpThrLysHisTrpAspSerGlyIleEnd                                                       (256)

               .         .         .         .         .         .         .         .         .
 961  CACAAGATATTCACGCAAGGCAATAAGGCTGATTCGATGCACACTCACATTCTTCTCCTAATACGATAATAAAACTTTCCATGAAAAATA   1050
                                                                           -------
               .         .         .         .         .         .         .         . -->1132 (Adh-dup)
1051  TGGAAAAATATATGAAAATTGAGAAATCCAAAAAACTGATAAACGCTCTACTTAATTAAAATAGATAAATGGGAGCGGCAGGAATGGCGG   1140
                                  |(A)n                    -------

               .         .         .         .         .         .         .         .         .
1141  AGCATGGCCAAGTTCCTCCGCCAATCAGTCGTAAAACAGAAGTCGTGGAAAGCGGATAGAAAGAATGTTCGATTTGACGGGCAAGCATGT   1230
                                                                      MetPheAspLeuThrGlyLysHisVa

               .         .         .         .         .         .         .         .         .
1231  CTGCTATGTGGCGGATTGCGGAGGAATTGCACTGGAGACCAGCAAGGTTCTCATGACCAAGAATATAGCGGTGAGTGAGCGGGAAGCTCG   1320
      lCysTyrValAlaAspCysGlyGlyIleAlaLeuGluThrSerLysValLeuMetThrLysAsnIleAla

               .         .         .         .         .         .         .         .         .
1321  GTTTCTGTCCAGATCGAACTCAAAACTAGTCCAGCCAGTCGCTGTCGAAACTAATTAAGTAAATGAGTTTTTCATGTTAGTTTCGCGCTG   1410

               .         .         .
1411  AGCAACAATTAAGTTTATGTTTCAGTTCGG  1440
```

Adh SEQUENCE. Slow allele from *Canton S*. Accession M14802 (DROADHA). Several other alleles have been sequenced and are listed under DROADH* in GenBank. Several mutations are indicated (Benyajati et al. 1982; Martin et al. 1985; Place et al. 1987; Thatcher 1980). Indicated under the sequence in the promoter regions are binding sites for various regulatory proteins. For the *Adh-dup*, initiation of transcription and translation, at 1,132 and 1,205, respectively, are suggested by sequence comparison to *Adh* (Schaeffer and Aquadro 1987).

enhancers that control expression of the two promoters were identified (Posakony et al. 1985):

Larval Enhancer and Proximal Promoter The larval enhancer is located between 5,000 and 1,845 bp upstream of the distal transcription initiation site; it can stimulate transcription from the proximal (but not the distal) promoter at all developmental stages (Corbin and Maniatis 1989a).

In the proximal promoter, three protein-binding regions were identified (p_0, p_1 and p_2 between −340 and −140 in the *Adh* Sequence) (Heberlein et al. 1985). Functional assays of promoter deletions demonstrated that those are the only regions in the neighborhood of the proximal promoter necessary for expression (Shen et al. 1989, 1991).

Adult Enhancer and Distal Promoter The adult enhancer is located between 600 and 450 bp upstream of the distal transcription initiation site (approximately −1,375 and −1,225 in the *Adh* Sequence); it stimulates transcription from both promoters but only during the late third larval instar and in adults (Corbin and Maniatis 1989a).

DNA-binding assays and *in vitro* transcription experiments defined a *cis*-acting region that extends from −860 to −820 as necessary for transcription from the distal promoter; a specific factor, ADF-1 (*Adh* distal factor 1), binds to this region (d_1 in the *Adh* Sequence) (Heberlein et al. 1985; England et al. 1990). In addition, a general transcription factor similar to human transcription factor SP2 is required (Heberlein et al. 1985).

Four distal enhancer binding proteins were obtained from cultured-cell nuclear extracts (DEP1–4) (*Adh* Sequence). DEP1 and DEP2 have partly overlapping binding sites (dep1 and dep2) in a segment that is required for full expression. DEP1 is FTZ-F1, a member of the steroid hormone receptor superfamily also involved in the control of the *fushi tarazu* (*ftz*) "zebra element" (Ayer and Benyajati 1992). The site dep4, also called aef-1, was identified as the binding site of a repressor (Falb and Maniatis 1992). Partly overlapping aef-1 is a binding site for mammalian C/EBP, and the authors suggest that the *Drosophila* homolog of C/EBP acts to stimulate transcription in fat body; competition between C/EBP and AEF-1 (=DEP4?) would determine the level of transcriptional activity. Overlapping C/EBP and QEF-1 binding sites were found in the regulatory sequences of another gene expressed in fat body, *Yp1*, one of the yolk protein genes (Falb and Maniatis 1992).

Down-regulation of the proximal promoter in adults is dependent on expression of the distal promoter, an apparent instance of transcriptional interference (Corbin and Maniatis 1989b). Transcriptional interference and the stage and promoter specificity of the two enhancers could explain the major promoter switch that occurs between larval and adult stages (Corbin and Maniatis 1989b).

Adh-dup

The putative 5′ end of this gene is positioned very near the 3′ end of *Adh* and probably originated as a duplication (*Adh* Sequence; Fig. 3.1). It is present in other *Drosophila* species (including those of the *pseudoobscura* group). The amino acid sequence of the two genes is approximately 38% identical, and the coding region introns are similarly positioned. The nature or function of the product is not known (Schaeffer and Aquadro 1987; Kreitman and Hudson 1991).

References

Ayer, S. and Benyajati, C. (1992). The binding site of a harmone receptor-like protein within the *Drosophila Adh*. *Mol. Cell. Biol.* **12**:661–673.

Benyajati, C., Place, A. R., Powers, D. A. and Sofer, W. (1981). *Alcohol dehydrogenase* gene of *Drosophila melanogaster*: Relationship of intervening sequences to functional domains in the protein. *Proc. Natl Acad. Sci.* (*USA*) **78**:2717–2721.

Benyajati, C., Place, A. R., Wang, N., Pentz, E. and Sofer, W. (1982). Deletions at intervening sequence splice sites in the alcohol dehydrogenase gene of *Drosophila*. *Nucl. Acids Res.* **10**:7261–7272.

Benyajati, C., Spoerel, N., Haymerle, H. and Ashburner, M. (1983). The messenger RNA for alcohol dehydrogenase in *Drosophila melanogaster* differs in its 5′ end in different developmental stages. *Cell* **33**:125–133.

Corbin, V. and Maniatis, T. (1989a). The role of specific enhancer-promoter interactions in the *Drosophila Adh* promoter switch. *Genes Dev.* **3**:2191–2200.

Corbin, V. and Maniatis, T. (1989b). Role of transcription interference in the *Drosophila melanogaster Adh* promoter switch. *Nature* **337**:279–282.

England, B. P., Heberlein, U. and Tjian, R. (1990). Purified *Drosophila* transcription factor, *Adh distal factor-1* (*Adf-1*) binds to sites in several *Drosophila* promoters and activates transcription. *J. Biol. Chem.* **265**:5086–5094.

Falb, D. and Maniatis, T. (1992). A conserved regulatory unit implicated in tissue-specific gene expression in *Drosophila* and man. *Genes Dev.* **6**:454–465.

Grell, E. H., Jacobson, K. B. and Murphy, J. B. (1965). Alcohol dehydrogenase in *Drosophila*: Isozymes and genetic variants. *Science* **149**:80–82.

Heberlein, U., England, B. and Tjian, R. (1985). Characterization of *Drosophila* transcription factors that activate the tandem promoters of the alcohol dehydrogenase gene. *Cell* **41**:965–977.

Jörnvall, H., Persson, M. and Jeffery, J. (1981). Alcohol and polyol dehydrogenases are both divided into two protein types, and structural properties cross-relate the different enzyme activities within each type. *Proc. Natl Acad. Sci.* (*USA*) **78**:4226–4230.

Kreitman, M. and Hudson, R. R. (1991). Inferring the evolutionary histories of the *Adh* and *Adh-dup* loci in *Drosophila melanogaster* from patterns of polymorphism and divergence. *Genetics* **127**:565–582.

Maroni, G. and Stamey, S. C. (1983). Developmental profile and tissue distribution of alcohol dehydrogenase. *Drosophila Inf. Ser.* **59**:77–79.

Martin, P. F., Place, A. R., Pentz, E. and Sofer, W. (1985). UGA nonsense mutation in the alcohol dehydrogenase gene of *Drosophila melanogaster*. *J. Mol. Biol.* **184**:221–230.

O'Donnell, J., Gerace, L., Leister, F. and Sofer, W. (1975). Chemical selection of mutants that affect alcohol dehydrogenase in Drosophila. II. Use of 1-pentyne-3-ol. *Genetics* **79**:73–83.

Place, A. R., Benyajati, C. and Sofer, W. (1987). Molecular consequences of two formaldehyde-induced mutations in the alcohol dehydrogenase gene of *Drosophila melanogaster. Biochem. Genet.* **25**:621–638.

Posakony, J. W., Fischer, J. A. and Maniatis, T. (1985). Identification of DNA sequences required for the regulation of Drosophila *Alcohol dehydrogenase* expression. *Cold Spring Harbor Symp. Quant. Biol.* **50**:515–520.

Savakis, C., Ashburner, M. and Willis, J. H. (1986). The expression of the gene coding for alcohol dehydrogenase during the development of *Drosophila melanogaster. Dev. Biol.* **114**:194–207.

Schaeffer, S. W. and Aquadro, C. F. (1987). Nucleotide sequence of the *Adh* gene region of *Drosophila pseudoobscura*: evolutionary change and evidence for an ancient gene duplication. *Genetics* **117**:61–73.

Shen, N. L. L., Subrahmanyam, G., Clark, W., Martin, P. F. and Sofer, W. (1989). Analysis of *Adh* gene regulation in *Drosophila*: studies using somatic transformation *Dev. Genet.* **10**:210–219.

Shen, N. L. L., Hotaling, E. C., Subrahmanyam, G., Martin, P. F. and Sofer, W. (1991). Analysis of sequences regulating larval expression of the *Adh* gene of *Drosophila melanogaster. Genetics* **129**:763–771.

Sofer, W. and Ursprung, H. (1968). *Drosophila* alcohol dehydrogenease. Purification and partial characterization. *J. Biol. Chem.* **243**:3110–3115.

Sullivan, D. T., Atkinson, P. W. and Starmer, W. T. (1990). Molecular evolution of the alcohol dehydrogenase genes in the genus *Drosophila. Evol. Biol.* **24**:107–147.

Thatcher, D. R. (1980). Complete amino acid sequence of three alcohol dehydrogenase alleloenzymes from the fruitfly *Drosophila melanogaster. Biochem. J.* **187**:875–886.

Ursprung, H., Sofer, W. H. and Burroughs, N. (1970). Ontogeny and tissue distribution of alcohol dehydrogenase in *Drosophila melanogaster. Wilhelm Roux' Arch.* **164**:201–208.

4

The α-Amylase Genes: *AmyA, AmyB*

Chromosomal Location:
2R, 54A

Map Position:
2-77.7

Product
α-Amylase (EC 3.2.1.1)

Structure and Function

α-Amylase is a monomeric enzyme of M_r 54.5 kD, which acts in the hydrolysis of starch. The mature protein is thought to be 476 amino acids long, with its N terminus, a derivatized Gln, being the 19th amino acid of the translation product. The first 18 amino acids of the translation product are thought to constitute the transport signal peptide. There is 55% identity between *Drosophila* α-amylase and α-amylase of the mouse pancreas (Fig. 4.1) (Boer and Hickey 1986).

Tissue Distribution

α-Amylase is most abundant in the midgut where it occurs in characteristic patterns under the genetic control of the *map* gene (Doane et al. 1975, 1983).

Organization of the Cluster

There are two divergently transcribed *Amy* genes separated by approximately 3.7 kb (Fig. 4.2). *AmyA* is the centromere proximal gene and *AmyB* the centromere distal one (Levy et al. 1985). The duplicated segments extend from approximately 130 bp upstream of the translation initiation site to the polyadenylation site. Within this region, divergence between the two genes is low in the coding region (the frequency of silent substitutions is ca. 1%) but it is considerable upstream and downstream of the coding region (frequency of substitutions, 30%). This observation led to the suggestion that gene conversions

```
      1                                                50                                                   100
   Dm MFLAKSIVCL ALLAVANAQF DTNYASGRSG MVHLFEWKWD DIAAECENFL GPNGYAGVQV SPVNENAV.. KDSRPWWERY QPISYKLETR SGNEEQFASM
Mouse ...MKFVLLL SLIGFCWAQY DPHTSDGRTA IVHLFEWRWV DIAKECERYL APKGFGGVQV SPPNENVVVH NPSRPWWERY QPISYKICTR SGNEDEFRDM
  CON ----K----L -L-----AQ- D-----GR-- -VHLFEW-W- DIA-ECE--L -P-G--GVQV SP-NEN-V-- --SRPWWERY QPISYK--TR SGNE--F--M

      101                                              150                                                  200
   Dm VKRCNAVGVR TYVDVVFNHM AADG...GTY GTGGSTASPS SKSYPGVPYS SLDFN...PT CAISNYNDAN EVRNCELVGL RDLNQGNSYV QDKVVEFLDH
Mouse VTRCNNVGVR IYVDAVINHM CGAGNPAGTS STCGSYLNPN NREFPAVPYS AWDFNDNKCN GEIDNYNDAY QVRNCRLTGL LDLALEKDYV RTKVADYMNH
  CON V-RCN-VGVR -YVD-V-NHM ---G---GT- -T-GS---P- ----P-VPYS --DFN----- --I-NYNDA- -VRNC-L-GL -DL-----YV --KV-----H

      201                                              250                                                  300
   Dm LIDLGVAGFR VDAAKHMWPA DLAVIYGRLK NLNTDHGFAS GSKAYIVQEV IDMGGEAISK SEYTGLGAIT EFRHSDSIGK VFR..GKDQL QYLTNWGTAW
Mouse LIDIGVAGFR LDAAKHMWPR DIKAVLDKLH NLNTKW.FSQ GSRPFIFQEV IDLGGEAIKG SEYFGNGRVT EFKYGAKLGT VIRKWNGEKM SYLKNWGEGW
  CON LID-GVAGFR -DAAKHMWP- D-------L- NLNT---F-- GS---I-QEV ID-GGEAI-- SEY-G-G--T EF------G- V-R------- -YL-NWG--W

      301                                              350                                                  400
   Dm GFAASDRSLV FVDNHDNQRG HGAGGADVLT YKVPKQYKMA SAFMLAHPFG TPRVMSSFSF .....TDTDQ ....GPPTTD GHNIASPIFN SDNSCSGGWV
Mouse GLVPSDRALV FVDNHDNQRG HGAGGSSILT FWDARMYKMA VGFMLAHPYG FTRVMSSYRW NRNFQNGKDQ NDWIGPPNNN GVTKEVTI.N ADTTCGNDWV
  CON G---SDR-LV FVDNHDNQRG HGAGG---LT ------YKMA --FMLAHP-G --RVMSS--- --------DQ ----GPP--- G------I-N -D--C---WV

      401                                              450                                                  500
   Dm CEHRWRQIYN MVAFRNTVGS DEIQNWWDNG SNQISFSRGS RGFVAFNNDN YDLNSSLQTG LPAGTYCDVI SGSKSGSSCT GKTVTVGSDG RASINIGSSE
Mouse CEHRWRQIRN MVAFRNVVNG QPFSNWWDNN SNQVAFSRGN RGFIVFNNDD WALSATLQTG LPAGTYCDVI SGDKVDGNCT GLRVNVGSDG KAHFSISNSA
  CON CEHRWRQI-N MVAFRN-V-- ----NWWDN- SNQ--FSRG- RGF--FNND- --L---LQTG LPAGTYCDVI SG-K----CT G--V-VGSDG -A---I--S-

      501        514
   Dm DDGVLAIHVN AKL*
Mouse EDPFIAIHAD SKL*
  CON -D---AIH-- -KL-
```

FIG. 4.1. Comparison of the mouse (Accession, V00718) and *Drosophila* (Dm) *AmyA* sequences. There is 55% overall identity between the two proteins. Sequences aligned with the GCG *Pileup* program.

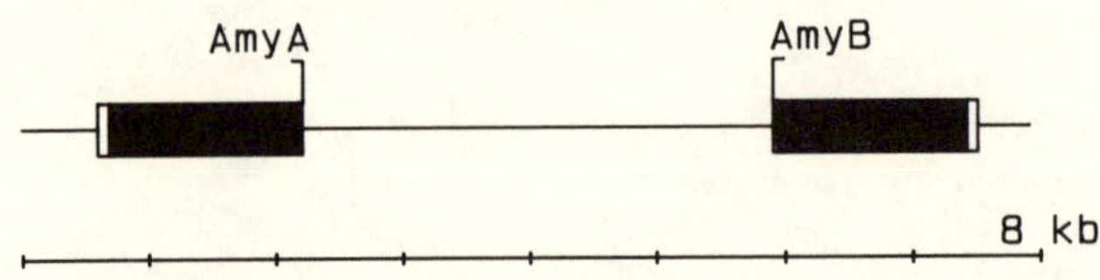

FIG. 4.2. The two *Amy* genes.

in the coding regions maintain a high degree of conservation (Hickey et al. 1991).

AmyA

Gene Organization and Expression

Open reading frame, 494 amino acids; predicted mRNA length, 1,601 bases. The 5′ end was determined by primer extension and the 3′ end from the sequence of one cDNA clone. There are no introns (Boer and Hickey 1986) (*AmyA* and *AmyB* Sequence).

Developmental Pattern

The methods used do not distinguish between *AmyA* and *AmyB* RNA. *Amy* transcription is subject to glucose repression: larvae grown in 10% glucose accumulate only 1% as much *Amy* mRNA as larvae grown in the absence of glucose (Benkel and Hickey 1987).

Promoter

An *AmyA* segment that extends from −142 to −50 in the *Amy* Sequence is sufficient to drive the glucose suppressible expression of *Adh* as a reporter gene. Deletion analysis showed that elements between −142 and −125 are required for full gene expression and that the sequences necessary for glucose repression are between −125 and −50 (Magoulas et al. 1992). Upstream *Amy* sequences have similarities with *cis*-acting elements that mediate glucose repression in yeast (Boer and Hickey 1986), and the *Drosophila AmyA* promoter is subject to glucose repression when introduced into yeast cells (D. A. Hickey, personal communication). Linker scanning mutations were used to identify functional CAAT and TATA boxes (Magoulas et al. 1992).

Amy

```
           .         .         .         .         .         .         .         .         .
      A T  -------  AGCG GT A T AAAA TC TTGC TA A  T GCAA TC AG  GTG TACATG      TTAC G TGGT  T
-565  CACTTCAGAACCCAGAGATCAAGTGGCCGCCAGTCAAGGCCAGAAGTCACGTATTCCAGAGAACGGCGCAGCCAAAGCTTCAAACCAAAA  -476

           .         .         .         .         .         .         .         .         .
         A ATA TGAA   A   TT    AC  TATAA C CCA   T AC TGCA TA  GTG A ATTAG  C   AT    T  TCCTTG
-475  TCGCTTGCTACCTTTATTTTCAACATTTTTAGGCGATATTGCATGATTTCAATGCTTTCAAATACGCTAAAAAATCCAAATAAC------  -386

           .         .         .         .         .         .         .         .         .
      TAGGCCAA G    G T G T A    -    TG     C A  --C  G G C        CT    TATC    A  A  TG
-385  --------AATTCACAGTAAACCCGCTCCTAGGAGCGTGAACGTAATAAATAGTCAATAAATTCCCAACTGAAACCGATTTCAAAGGAAT  -296

           .         .         .         .         .         .         .         .         .
           G  TT    -T   C T  GTCG AC   TT T   T    AC AGT A TAGG  T   -------------C G  CT  G T
-295  GCATTTTCCCGATGAGTTATTGATACAAATATAACGAAAATAAGCCGACTCACTAATCATCAGCGAAAAATTGCGATCTCCAGTCAATAC  -206

           .         .         .         .         .         .         .         .         .
      AA ACGC TT   C C     A TC   ----------   ATA AT CAA A   CGT G GAC TA  G           -----A T
-205  GTCTGCTCGGAATTGTGATTTGACAAACTAATCGCCAGTCAGACCCCATGCGTGAAAAAACCCCTTAGGGAGCGATAAGATCCCATGCAG  -116

           .         .         .         .         .         .         .         .         .
                                                                                         -->-32
      C G          G    A     ---------GAATAGGT T  TCATC         C   T  A  GACAC C  TTA T
-115  TCACAAATCACTCCCCGCGAAGCCCTCAGATAAAGTAGCAGTGGGGTCCACTATATAAGGAGCGGC-TCTGAGTAGTTCCGACCAGAGTG  -26
         -----                                             ------

           .         .         .         .         .         .         .         .         .
         T T G      C  AA         ||G=null-d
 -25  AAACTGAACTTCCATCTGGAATCATCATGTTTCTGGCCAAGAGCATAGTGTGCCTCGCCCTCCTGGCGGTGGCCAACGCCCAATTCGACA  64
                                MetPheLeuAlaLysSerIleValCysLeuAlaLeuLeuAlaValAlaAsnAlaGlnPheAspT  (22)
                                                                                     |

           .         .         .         .         .         .         .         .         .
                                                                                  G        G
  65  CCAACTACGCATCCGGTCGTAGTGGAATGGTCCACCTCTTCGAGTGGAAGTGGGACGACATCGCTGCCGAGTGCGAAAACTTCCTTGGAC  154
      hrAsnTyrAlaSerGlyArgSerGlyMetValHisLeuPheGluTrpLysTrpAspAspIleAlaAlaGluCysGluAsnPheLeuGlyP  (52)

           .         .         .         .         .         .         .         .         .
                                                                G                       A
 155  CCAATGGCTACGCCGGTGTTCAGGTCTCCCCTGTGAACGAGAACGCCGTCAAGGACAGCCGCCCCTGGTGGGAACGTTACCAGCCCATCT  244
      roAsnGlyTyrAlaGlyValGlnValSerProValAsnGluAsnAlaValLysAspSerArgProTrpTrpGluArgTyrGlnProIleS  (82)
                                                              Arg

           .         .         .         .         .         .         .         .         .
                                     G
 245  CCTACAAGCTGGAGACCCGCTCCGGAAACGAAGAGCAGTTCGCCAGCATGGTCAAGCGCTGCAACGCCGTCGGAGTGCGCACCTACGTGG  334
      erTyrLysLeuGluThrArgSerGlyAsnGluGluGlnPheAlaSerMetValLysArgCysAsnAlaValGlyValArgThrTyrValA  (112)

           .         .         .         .         .         .         .         .         .
                                 G            A=null-d
 335  ACGTGGTCTTCAACCACATGGCCGCCGACGGAGGCACCTACGGCACTGGCGGCAGCACCGCCAGCCCCAGCAGCAAGAGCTATCCCGGAG  424
      spValValPheAsnHisMetAlaAlaAspGlyGlyThrTyrGlyThrGlyGlySerThrAlaSerProSerSerLysSerTyrProGlyV  (142)
                                Gly         End

           .         .         .         .         .         .         .         .         .
           C=Canton S                                G
 425  TGCCCTACTCCTCGCTGGACTTCAACCCGACCTGCGCCATCAGCAACTACAACGACGCCAACGAGGTGCGCAACTGCGAGCTGGTCGGTC  514
      alProTyrSerSerLeuAspPheAsnProThrCysAlaIleSerAsnTyrAsnAspAlaAsnGluValArgAsnCysGluLeuValGlyL  (172)
           His                                 Arg

           .         .         .         .         .         .         .         .         .
                C               A=Canton S                                 C
 515  TGCGCGACCTTAACCAGGGCAACTCCTACGTGCAGGACAAGGTGGTCGAGTTCCTGGACCATCTGATTGATCTCGGCGTGGCCGGATTCC  604
      euArgAspLeuAsnGlnGlyAsnSerTyrValGlnAspLysValValGluPheLeuAspHisLeuIleAspLeuGlyValAlaGlyPheA  (202)
                                Asn
```

(continued)

```
           .         .         .         .         .         .         .         .         .
                                                null-p=A     T           G
 605  GCGTGGACGCCGCCAAGCACATGTGGCCCGCCGACCTGGCCGTCATCTATGGCCGCCTCAAGAACCTAAACACCGACCACGGCTTCGCCT   694
      rgValAspAlaAlaLysHisMetTrpProAlaAspLeuAlaValIleTyrGlyArgLeuLysAsnLeuAsnThrAspHisGlyPheAlaS   (232)
                                                     End

           .         .         .         .         .         .         .         .         .
                                                                                        A
 695  CGGGATCCAAGGCGTACATCGTCCAGGAGGTCATCGACATGGGCGGCGAGGCCATCAGCAAGTCCGAGTACACCGGACTGGGCGCCATCA   784
      erGlySerLysAlaTyrIleValGlnGluValIleAspMetGlyGlyGluAlaIleSerLysSerGluTyrThrGlyLeuGlyAlaIleT   (262)

           .         .         .         .         .         .         .         .         .
                                                     A             T
 785  CCGAGTTCCGCCACTCCGACTCCATCGGCAAGGTCTTCCGCGGCAAGGACCAGCTGCAGTACCTGACCAACTGGGGCACCGCCTGGGGCT   874
      hrGluPheArgHisSerAspSerIleGlyLysValPheArgGlyLysAspGlnLeuGlnTyrLeuThrAsnTrpGlyThrAlaTrpGlyP   (292)
                                                     Asn

           .         .         .         .         .         .         .         .         .
                                                    C                                G
 875  TCGCTGCCTCCGACCGCTCCCTGGTATTCGTCGACAACCACGACAATCAGCGCGGACATGGAGCAGGAGGCGCCGACGTTCTGACCTACA   964
      heAlaAlaSerAspArgSerLeuValPheValAspAsnHisAspAsnGlnArgGlyHisGlyAlaGlyGlyAlaAspValLeuThrTyrL   (322)

           .         .         .         .         .         .         .         .         .
                               A        T         C T
 965  AGGTGCCCAAGCAGTACAAGATGGCCTCCGCCTTCATGCTGGCGCACCCCTTCGGCACTCCCCGCGTGATGTCCTCCTTCTCCTTCACGG   1054
      ysValProLysGlnTyrLysMetAlaSerAlaPheMetLeuAlaHisProPheGlyThrProArgValMetSerSerPheSerPheThrA   (352)

           .         .         .         .         .         .         .         .         .

1055  ACACCGATCAGGGCCCGCCCACCACCGACGGCCACAACATCGCCTCGCCCATCTTCAATAGCGACAACTCCTGCAGCGGCGGCTGGGTGT   1144
      spThrAspGlnGlyProProThrThrAspGlyHisAsnIleAlaSerProIlePheAsnSerAspAsnSerCysSerGlyGlyTrpValC   (382)

           .         .         .         .         .         .         .         .         .
       C                                         G   G               C
1145  GTGAGCACCGCTGGCGCCAGATCTACAACATGGTGGCCTTCCGAAACACCGTGGGCTCGGACGAGATCCAGAACTGGTGGGACAACGGCA   1234
      ysGluHisArgTrpArgGlnIleTyrAsnMetValAlaPheArgAsnThrValGlySerAspGluIleGlnAsnTrpTrpAspAsnGlyS   (412)
                                                     Ala            Ala

           .         .         .         .         .         .         .         .         .

1235  GCAACCAGATCTCCTTCAGCCGAGGCAGCCGCGGCTTCGTGGCCTTCAACAACGACAACTACGACCTGAACAGCTCCCTGCAGACGGGCC   1324
      erAsnGlnIleSerPheSerArgGlySerArgGlyPheValAlaPheAsnAsnAspAsnTyrAspLeuAsnSerSerLeuGlnThrGlyL   (442)

           .         .         .         .         .         .         .         .         .
                                                    C
1325  TGCCCGCCGGCACCTACTGCGACGTCATCTCCGGCTCCAAGAGCGGTTCCTCCTGCACGGGCAAGACCGTCACCGTCGGATCCGACGGAC   1414
      euProAlaGlyThrTyrCysAspValIleSerGlySerLysSerGlySerSerCysThrGlyLysThrValThrValGlySerAspGlyA   (472)

           .         .         .         .         .         .         .         .         .
                                                                                 A     CAAAGACCA
1415  GGGCTTCCATCAACATTGGCAGCTCCGAGGACGACGGAGTGCTGGCCATTCACGTCAACGCCAAGTTGTAAACAGCTGGGG......AGC   1504
      rgAlaSerIleAsnIleGlySerSerGluAspAspGlyValLeuAlaIleHisValAsnAlaLysLeuEnd                      (494)

           .         .         .         .         .         .         .         .         .
       G C   GA    GA        T           C   - TTA          T   C G         A A AGGAAGA G GC
1505  ATGGCGAACAGCCAGGCAATTAATTGAGATTATTAATTGTACGAAATATATATGATGAGATTATAAACACACAACACTTTTATTCGCAAG   1594
                                                  ------                   |(A)n

           .         .         .         .         .         .         .         .         .
      TA  C   GT CA  T TATGGA AATG  AAAT TTAT  TACTTAAAATTGACCACAAATAACTGTTACGCATAATATGGCAAAAAAC
1595  GGATGATAAGATCTAATATATATATTATCTGGGCTAAGCTGA------------------------------------------------   1684

           .         .         .         .         .
      AACTTATGCGTGACCTTAAAAGCGCTGCCTTTTCATCTCGGTATTCAGCGTGATT
1685  -------------------------------------------------------  1739
```

AmyA AND *B* SEQUENCE. The sequence on the numbered line corresponds to the proximal gene (*A*) of *Oregon R* (allele *Amy*[1]). This sequence combines the nonoverlapping regions of two GenBank entries: Accession X04569 (DROAMYAG1)

(*continued*)

AmyB

Gene Organization and Expression

Open reading frame, 494 amino acids; predicted mRNA length, 1,606 bases. The 5′ and 3′ ends of *AmyB* were deduced from sequence similarity to *AmyA* (Okuyama and Yamazaki 1988; D. A. Hickey, personal communication) (*AmyA* and *AmyB* Sequence).

References

Benkel, B. F. and Hickey, D. A. (1987). A *Drosophila* gene is subject to glucose repression. *Proc. Natl Acad. Sci.* (*USA*) **84**:1337–1339.

Boer, P. H. and Hickey, D. A. (1986). The α-amylase gene in *Drosophila melanogaster*: nucleotide sequence, gene structure and expression motifs. *Nucl. Acids Res.* **14**:8399–8411.

Doane, W. W., Abraham, I., Kolar, M. M., Martenson, R. E. and Deibler, G. E. (1975). Purified *Drosophila* alpha-amylase isozymes: genetical, biochemical and molecular characterization. *In Isozymes: Current Topics in Biological and Medical Research*, ed. L. Markert (New York, NY: Alan R. Liss), Volume 4, 585–607.

Doane, W. W., Treat-Clemons, L. G., Gemmill, R. M., Levy, J. N., Hawley, S. A., Buchberg, A. M. and Paigen, K. (1983). Genetic mechanism for tissue-specific control of alpha-amylase expression in *Drosophila melanogaster*. In *Isozymes: Current Topics in Biological and Medical Research*, eds M. C. Rattazzi, J. C. Scandalios and G. S. White (New York, NY: Alan R. Liss), Volume 9, 63–90.

Hickey, D. A., Bally-Cuif, L., Abukashawa, S., Payant, V. and Benkel, B. F. (1991). Concerted evolution of duplicated protein-coding genes in *Drosophila*. *Proc. Natl Acad. Sci.* (*USA*) **88**:1611–1615.

Levy, J. N., Gemmill, R. M. and Doane, W. W. (1985). Molecular cloning of alpha-amylase genes from *Drosophila melanogaster*. II. Clone verification and organization. *Genetics* **110**:313–324.

(*continued*) and Accession Y00438 (DROAMYAR). On the line immediately above is the sequence of the distal gene of strain *Makokou*; only in those positions where there is a difference from the proximal gene is the base indicated. There are differences in six amino acid residues between these two sequences. In four of those six positions (Gly-121, Arg-156, Asn-278 and Ala-398), the *Makokou* proximal gene (not shown) has the same residue as the *Makokou* distal gene, reinforcing the idea that there is intergenic correction between these genes (Hickey et al. 1991). The Makokou sequences were kindly provided by Donal A. Hickey. A *Canton S* allele with two amino acid substitutions (Tyr-144 and Tyr-181) has the same electrophoretic mobility as $AmyA^1$. An *Amy*-null strain has two mutations in the distal gene, the addition of a G between positions 3 and 4, and a nonsense mutation at position 375 and one mutation in the proximal gene, with a nonsense mutation at position 654. This null strain apparently also has an inversion within the intergenic segment (Okuyama and Yamazaki 1988). The vertical bar marks the end of the signal peptide.

Magoulas, C., Bally-Ciuf, L., Loverre-Chyurlia, A., Benkel, B. and Hickey, D. (1992). A short flanking region mediates glucose repression of amylase gene expression in *Drosophila melanogaster*. *Genetics* (In press).

Okuyama, E. and Yamazaki, T. (1988). Nucleotide sequence of the duplicated *Amylase* structural genes in *Drosophila melanogaster*. *Proc. Japan Acad. Ser. B* **64**:274–277.

5

The Andropin and Cecropins Gene Cluster: *Anp, CecA1, CecA2, CecB, CecC*

Chromosomal Location:
3R, 99E

Map Position:
3-[101]

Products

Antibacterial peptides.

Structure

Sequence analysis suggests that each polypeptide may fold into two amphipathic α-helices separated by a four-amino-acid loop (Samakovlis et al. 1991).

By analogy to the better-characterized cecropins of the moth *Hyalophora cecropia*, processing is predicted to include the removal of the signal peptide and of an additional dipeptide at the N-terminus, and cleavage of the terminal Gly plus amidation at the C-terminus. These changes would give rise to mature cecropins 39 amino acids long (Kylsten et al. 1990, see Sequences).

Cecropins A1 and A2 are identical to each other and to the main cecropin from the flesh fly *Sarcophaga peregrina*. Cecropin B differs from A1 and A2 by four conservative substitutions in the mature protein (Arg-27, Ile-36, Ser-44 and Val-47) and four others in the signal peptide (Kylsten et al. 1990). Cecropin C is intermediate in sequence between A and B (Fig. 5.1) (Tryselius et al. 1992). The sequence similarities between andropin and the cecropins is restricted to the signal peptide (Samakovlis et al. 1991).

Tissue Distribution and Function

For the most part, cecropins are synthesized in response to bacterial infection and released in the hemolymph. Cecropins disrupt the cell membrane of Gram-positive and Gram-negative bacteria (Dunn 1986; Boman and Hultmark 1987). The related andropin is synthesized constitutively in the ejaculatory duct of males (Samakovlis et al. 1991).

```
      1                                                          50          63
 Anp  MKYFVVLVVL ALILAISVGP SDAVFIDILD KVENAIHNAA QVGIGFAKPF EKLINPK*.. ....
CecB  MNFNKIFVFV ALILAISLGN SEAGWLRKLG KKIERIGQHT RDASIQVLGI AQQAANVAAT ARG*
CecC  MNFYKIFVFV ALILAISIGQ SEAGWLKKLG KRIERIGQHT RDATIQGLGI AQQAANVAAT ARG*
CecA  MNFYNIFVFV ALILAITIGQ SEAGWLKKIG KKIERVGQHT RDATIQGLGI AQQAANVAAT ARG*
 CON  MNF--IFVFV ALILAI--G- SEAGWL-K-G K-IER-GQHT RDA-IQ-LGI AQQAANVAAT ARG-
                          | |        ^^
```

FIG. 5.1. Aligned cecropin and andropin peptide sequences. The vertical line under Ser-21 marks the last amino acid of the signal peptide, and, under Ala-23, the dipeptidase cleavage site. A caret marks the intron positions. The CON(sensus) line indicates positions in which three of the four sequences agree.

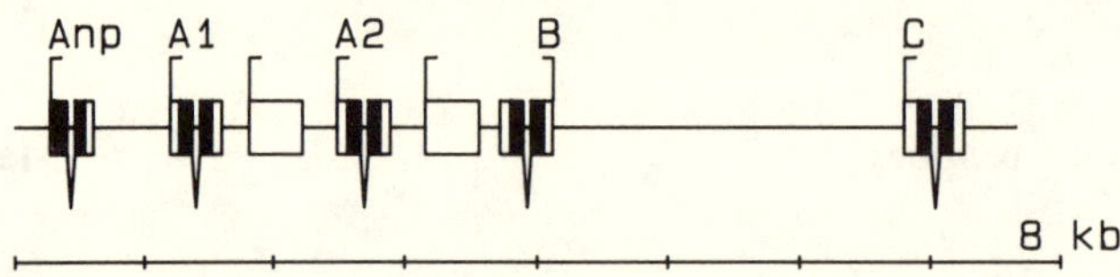

FIG. 5.2. The *Cecropin* cluster. Open boxes indicate the two pseudogenes.

Organization and Expression of the Cluster

Five genes and two pseudogenes are clustered in approximately 8.0 kb of DNA (Fig. 5.2). The pseudogenes contain vestiges of exons, introns and TATA boxes; but they also include numerous nonsense mutations, and they have lost the splicing signals.

Developmental Pattern

Transcription of *CecA*, *CecB*, and *CecC* is induced by injection or feeding of bacterial pathogens. Cecropin mRNAs, undetectable before infection, begin to accumulate 1 h after injection of bacteria, reach a maximum 2–6 h after injection, and soon thereafter they begin to decline. Twenty-four hours after injection, the RNAs return to their basal levels. *A1* and *A2* are expressed at high level in larval, pupal and adult stages. Transcription occurs primarily in the fat tissue, and the proteins accumulate in the hemolymph. *B* and *C*, by contrast, are inducible to a much lower extent than *A1* and *A2* in larvae and adults. They are active mainly during the early pupal stages in localized regions of tissues undergoing lysis, and this activation of *B* and *C* occurs in the absence of external agents (Kylsten et al. 1990; Tryselius et al. 1992).

AT-rich segments in the 3′ untranslated region of the mRNAs may play a role in their selective degradation (Kylsten et al. 1990).

Anp

```
          .         .         .         .         .         .         .         .         .
-287 TAACCTACAGAATTGTAGAACTTAATTACTATAGAACACTATTGAATGAAAACTTAGTAACTTGTTGAGGTTTTTAGTAATTCCAAGAAA  -198
          .         .         .         .         .         .         .         .         .
-197 TATGCTCTTGAATAAAAAACCTTTTTAAGTCTCTTTCAATGCAAAAACACGAGTTCTTTTTTTTTACATATTGTAATTAATATGTTTAAG  -108
                                                                            ----
          .         .         .         .         .         .         .-36--> -->-32        .
-107 GTCTAATTATTATTGTAAACGTTTTTCGGTGGGTTGATTGCCTATAAAGCCACTTGTTTTTCAGTCTAAATCATCAGTGTAAAATTCGGA  -18
                                               ------
          .         .         .         .         .         .         .         .         .
 -17 AAACCCAGCGATCTAGTTATGAAATACTTTGTGGTCCTTGTCGTCCTGGCCCTCATTTTGGCCATCAGCGTGGGTCCTTCGGATGCAGTA   72
                       MetLysTyrPheValValLeuValValLeuAlaLeuIleLeuAlaIleSerValGlyProSerAspAlaVal  (24)
          .         .         .         .         .         .         .         .         .
  73 TTTATTGATATTCTTGACAAAGTGGTTTGTTTCTTCTTTAAACAATTGTAGTTTACAATGAAGCTTAAACATTTGTATTTCTACAGGAAA   162
     PheIleAspIleLeuAspLysVal                                                               GluA  (34)
          .         .         .         .         .         .         .         .         .
 163 ACGCAATACACAATGCTGCTCAAGTGGGAATTGGCTTTGCTAAGCCCTTTGAAAAATTGATCAATCCGAAGTAATTCTGCACTGCAATTT   252
     snAlaIleHisAsnAlaAlaGlnValGlyIleGlyPheAlaLysProPheGluLysLeuIleAsnProLysEnd                  (57)
          .         .         .         .         .         .         .         .         .
 253 AATTAATGTATCGTTTAACGAAAATAAACACAAATTTTAAAATCTGAAAAACAACTAAGTTACTAACGCAAGACTTTTAGTTAAGTTAGT   342
          .         .         .         .         .         .         .         .         .
 343 TAATATAGACCGAGATGTATGTACATACATACCGCTTTCGCTTACAATAAAATGTTAAATAAGTTTTCAGATTCGTACGTGCTCAGTAAA   432
```

Anp SEQUENCE. Strain, *Canton S*. Accession, X16972 (DROCECPN).

Anp

Gene Organization and Expression

Open reading frame, 57 amino acids; expected mRNA length, 278 bases. Primer extension and sequence features were used to identify two 5′ ends, the upstream site being the major one. The 3′ end was obtained from a cDNA sequence. There is an intron after the Val-32 codon (*Anp* Sequence) (Samakovlis et al. 1991).

Developmental Pattern

Transcription is restricted to the ejaculatory ducts. mRNA level reaches a plateau 24 h after eclosion of the adult male and remains stable in virgin males; mating, however, causes a rise in the steady-state level of *Anp* mRNA (Samakovlis et al. 1991).

CecA1

Gene Organization and Expression

Open reading frame, 63 amino acids; expected mRNA length, 346 bases, in agreement with the 0.4 kb RNA detected in northern analysis of all cecropin

CecA1

```
 433 CAATTATTTTTTATTGTCATTTAATGCCTATTGAATTTTTCAAACTTAATTTAGTGCCTTTAGTAAAATATTGTAGTGATTCCCCTCGAA  522

 523 AAATACCACAAATTGGATGCGTTTATGTAAATAAATTGCCCTTGAGTGATAGAGTAAATTTGAATTTGACTGTCTTAGAAAGATAGAAAG  612

 613 AGATCAATTCAAAATGCCAAAAGGATAGAGTTATTAAAGCTCTAATTCAAATTGGCCCAGAACCGTTTAAAGGATATTACAATTTGTAAT  702

 703 TTACATATTTGGATTATAGCATTGAAATCCCCGATTGTTCCCTAGATGTGCAGATGTGTGCTTGGAATCAGATCGGTTACCTTCAGTGTA  792

 793 CTTTTCTCTGCAAAAATCCCCGTGCATGCCTTATCTGTCATTTTGTTTTTCAAGCTGCTGTTCGCCTATAAAAGCTCTCGCCTTTTGTAT  882
                                                                           ------
        A1 -->890                                                                    |963
 883 CGCAGTCATCAGTCGCTCAGACCTCACTGCAATATCAATATCTTTAGCTTCTCCTAAGAAAAAATCAAGAAAATATCACCATGAACTTCT  972
                                                                                    MetAsnPheT  (4)

 973 ACAACATCTTCGTTTTCGTCGCTCTCATTCTGGCCATCACCATTGGACAATCGGAAGCTGGGTGGCTGAAGAAAATTGGCAAGAAAATCG  1062
     yrAsnIlePheValPheValAlaLeuIleLeuAlaIleThrIleGlyGlnSerGluAlaGlyTrpLeuLysLysIleGlyLysLysIle  (33)

1063 TAAGTTCTTCCATTTGAAATCTGTTAAGACGGAAACTAACTGACTAACTTCTTTTCGAAGGAACGCGTTGGTCAGCACACTCGGGATGCC  1152
                                                                GluArgValGlyGlnHisThrArgAspAla  (43)

1153 ACAATCCAGGGACTGGGAATCGCTCAACAAGCCGCCAATGTCGCCGCAACTGCCCGAGGTTGACCACGATGATTATTTATAATTATTTAT  1242
     ThrIleGlnGlyLeuGlyIleAlaGlnGlnAlaAlaAsnValAlaAlaThrAlaArgGlyEnd                            (63)

1243 TTAAAGATCTATTTATTCTGTTGCTCCCTGTAAATAAAACAATTTTAAAAATTTAAAGAATTCTATTCAAACTTTGTTTTTTAAAGAGTT  1332
                                ------                  |(A)n

1333 GGAGAAAAGCGAACTCTTGAATTTATACACACATTTTAAATACACTTAAGAGGCATTATTTATACAGGATATTACAAATCGCTTCTTTTC  1422

1423 CGATTTGGAAAGGCCGAGATTATGTCTTATCTGTTGAAATATAATTCGTTTCACCTATAAAAGGACCAGTCTTTTAGTTTAAATTATCAG  1512
                                                             ------Psi1
```

CecA1 SEQUENCE. Strain, *Canton S*. Accession, X16972 (DROCECPN). The numbering system continues from *Anp* Sequence. Psi1 downstream of *A1* marks the TATA box of a pseudogene.

genes. Primer extension and sequence features were used to define the 5′ end. The 3′ end was obtained from cDNA sequences. There is an intron after the Ile-33 codon (*CecA1* Sequence).

Sequence similarity between *A1* and *A2* occurs in an interval that extends between 40 bp upstream of the 5′ end and 50 bp downstream of the 3′ end (Kylsten et al. 1990).

CecA2

Gene Organization and Expression

Open reading frame, 63 amino acids; expected mRNA length, 354 bases. Primer extension and sequence features were used to define the 5′ end. The 3′ end was

CecA2

```
1513  TCGCTTGTCAAATACTGAAACAATTAGATTAATTTGTGGATTTTATTTGTCCTCATCCTGACCACTTATTGGCCACAATTGGAAGCTGGC  1602

1603  TTCGACGGGACATTAGTAAGCTTAGTCATTTTAAAAGATTTCTTTGCATCTAACTATGATTCTAAATCCTCAGAAGGACGTTGGTCTATA  1692

1693  CACCCTAAATGCTACCCTGCAAGTTGCTGAAGTCGCTTCGAAAGCAGCCAATGTGGCAATCACTGCCAGGGGATAAACTTAAGTTAGGGT  1782

1783  ATTATTTATAAGAAATTAAATTAATAGATTTTATTTTATATATTTTTTGTATATTGTTATTCAAACTGATAATGTAATATACGCTTTTCA  1872

1873  AACGATCATTCCAAATCAGTTGTGGGCTTATCGCAAATGATTTCGTAGTGTTTTTATTTTGATTGATTCAAAGAAGGGGTTTCCTCTCTG  1962

1963  ATTCTTAGTCTCCCGCATTGACGAGGTAAAAAATCCCTATGCATATGAAATATGCAAATTTAAAAATCCCCCAATCCGACAGGTTGGTTT  2052

2053  TGATCGGTTTGGATTCCTCTCGTGTACTTTTCAGCCATAAAAATCCCCTTTCGAGCCTTATCAGGCGCTGAACTTAAGCTGATTCGCCTA  2142
                                                                                                  --
      .         .         .  -->2172 .         .         .         .         .         .
2143  TAAAAGCTCTCGGCGTTCCTGGTGCAATCAACAGTCGATCACTTTCCATTGCAACAGCAACATCAGAGCTATAGCTACTCTTGCAAAATC  2232
      ----
      .         .         |2253 .         .         .         .         .         .
2233  TAAAGTCAAATAAAACCACCATGAACTTCTACAACATCTTCGTTTTCGTCGCTCTCATTCTGGCCATCACCATTGGACAATCGGAAGCTG  2322
                          MetAsnPheTyrAsnIlePheValPheValAlaLeuIleLeuAlaIleThrIleGlyGlnSerGluAlaG  (24)

2323  GTTGGCTAAAGAAAATTGGCAAGAAAATCGTAAGTCCATTCTATTTGAAATTTGTTAAACCGGAAACTAACTAACTCCTTTTCATAGGAA  2412
      lyTrpLeuLysLysIleGlyLysLysIle                                                        Glu  (34)

2413  CGTGTTGGTCAGCACACTCGCGACGCCACAATCCAGGGACTGGGAATCGCTCAACAGGCCGCCAATGTTGCAGCCACTGCTCGAGGTTAA  2502
      ArgValGlyGlnHisThrArgAspAlaThrIleGlnGlyLeuGlyIleAlaGlnGlnAlaAlaAsnValAlaAlaThrAlaArgGlyEnd  (63)

2503  CCACGATGACTATCTAATAAATATTTATACAAAATCTTATTTATTTTTTTTGATCTAAGTAAATAAAACATTGGGAAAATCAATCTTTTG  2592
                                                                   ------             |(A)n

2593  TCTTCTCTCTAAAGATCTATTCAGCGAATAGTTGTGAATGAAAAGTGTATTATAAATCCTATCTATAGTTTTAGGAGCGCACGTGCGAAA  2682

2683  AATATATATACAACTAATAATCCACTAATTAATTTTGTTGTATTGTATAGATTGAAATTCTAATGATAATATTTTCGACTGGGAAAATCC  2772

2773  ACAAAAATATGCGTTATCTCCAAAAGTAGAAGATAGTTCGCCTATAAAAAGATCTAAGTCTAAGCTGTGAGCTTCAGTCCAAAAAATAAC  2862
                                                ------Psi2

2863  ATTAGCAAACAAACATTTGCTGCTTTTTCCAGTCTGTAATTATATATTACTTAATATGAACTTTAGCCATATTTTTTTGTTTGCTTTCAT  2952

2953  CATCCTGACAATTAACTTGCAACACTCGCATGCCGGTTGGCTGACGGATATAGTAATCTAAGACCGATCTAACTTAACTTCCCCTTCACA  3042

3043  GAAGAAGAAATCTGAGGAGACTTTTAAATACTTAAAAAACGCAGCATTGGAGGTCATTGACGTCGGCCAAAAAGCCGCGGATTTTGCTGC  3132

3133  CATTGCCAGGGGACAGAAAAAGTAGATCTCTACCAGATTTTTCTTGATGAGCTACAATTGCTGCAAATATTTAATAAAATCAAAAAGTAT  3222
```

CecA2 SEQUENCE. Strain, *Canton S*. Accession, X16972 (DROCECPN). The numbering system continues from *CecA1* Sequence. Psi2 downstream of *A2* marks the TATA box of a pseudogene.

CecB

```
-809  GAATTCATTATGCTGGGAGTGGATAAATGGGATAAATGAGTGTACAATAAATGGATAATGCCATGTTGATTGAGGGGATTTCTTATGTCC   -720

-719  AGGAAATATCATATTTCTACTGATGCTGTGTAAAGTTGTTGTTACCTTTTATTTCTGGGCTATAGAAAATAAATATATTAGTGTATAAAA   -630

-629  TAACATTTTTCTTGGAGTATTTATTTGCATTTGCTTCAATCTCCGACTTATTAACTCTGCTGATAATTCAGTTCCATTGCGAACTAAGTG   -540

-539  ACTGATAGTCTTATAAATTCTAAAAAAAAGAATACAGCATCTGTGACTGTAAAACGATGACAAATGGGATTTTGTCTGTAAAAAAAATAA   -450

-449  TAAAAATTAAAATAATAATAAAAATTACGGGAGGCTTGTCTTACGGGAATACTATATAGGGAAAAACACACTACACTTTAGTGTATGTTC   -360

-359  CCCTAAAAGTTTAAAAAGTAATGTTTCATTATAATTACTTTGTTTTTAATTGTAGTTTTACGTTATTTTTAAGCTAGTTTAAATCATCAT   -270

-269  AATTCAATAGATTAATCAAATCATAGCTTGCAACCAACCAGTTACTCTGAAATATCACTTGAGTAAGTCACTTTCATGGCGGTTCCGAAC   -180

-179  TGAGTCCATCTGCTGGTGAACTTTTGTCCCGCAGCAAAAAATTCCCGTCTGTGCAGCCGTAGCATCTGTTGGTATCGCTATATAAGCTCA   -90
                                                                                  -----

                     -->-70
 -89  ATCTCTTCGATGTCCAATCATCAGTCGCACAGTTCTCACTGCAACAGCTTAAGCTTTCTTTCAATCCGATCGTAAGCCAACAATCTCGTC   0

   1  ATGAACTTCAACAAGATCTTCGTCTTTGTGGCACTCATCCTGGCCATCAGCCTGGGAAACTCAGAGGCTGGTTGGCTTAGGAAGCTGGGA   90
      MetAsnPheAsnLysIlePheValPheValAlaLeuIleLeuAlaIleSerLeuGlyAsnSerGluAlaGlyTrpLeuArgLysLeuGly   (30)
                                                                         |     |

  91  AAAAAAATCGTATGGATTCCCTTCAAAACTAAACAAAATGAATTATTAATTTCGATTTTCCTTTTAGGAACGCATTGGTCAGCATACCAG   180
      LysLysIle                                                             GluArgIleGlyGlnHisThrAr   (41)

 181  GGATGCCTCAATCCAGGTCCTCGGAATCGCCCAACAGGCCGCCAATGTTGCAGCCACCGCTCGAGGTTGAAATCAAGTCTCGAAGATCCT   270
      gAspAlaSerIleGlnValLeuGlyIleAlaGlnGlnAlaAlaAsnValAlaAlaThrAlaArgGlyEnd                       (63)

 271  CGACCCGCTCATTTCTCTTATTTATTATTAATGCATTAGGAAGATTAACATAATGAAAATAGATACTCAATGCCAATGTCAAATTATTAA   360

 361  AATATAAGCAAGCAGATATTAATAAAAACAAATTAAGACACTATATACAACAATAAGAAATGGTGAAAATATATTCCCCTGTAGGCTTAT   450

 451  CAAGATGTAATCGCACAAGCTGGTTACTGGTTAAATTAAAATAGAATTTTGGAGGTTCTTATTATTTTATACTTTTTGATTTTATTAAAT   540

 541  ATTTGCAGCAATTGTAGCTCATCAAGAAAAATCTGGTAGAGATCTACTTTTTCTGTCCCCTGGCAATG   608
```

CecB SEQUENCE. Strain, *Canton S.* Accession, X16972 (DROCECPN).

obtain from cDNA sequences. There is an intron after the Ile-33 codon (*CecA2* Sequence; See *CecA1*) (Kylsten et al. 1990).

CecB

Gene Organization and Expression

Open reading frame, 63 amino acids; expected mRNA length, ca. 400 bases. Primer extension and sequence features were used to define the 5′ end. The 3′

```
                                             CecC
          .          .          .          .          .          .          .          .          .
-324  GAAAATATTGTTTAGAAGAAGTTAGCTATTGCTTTTTGCACACATGAGAGCTAAGCGAAGAACGCTCCATTTTTACTAGCAGCTGCTCAA   -235
          .          .          .          .          .          .          .          .          .
-234  ACAGATTACCGAAGACAGTCTTCGTCTAACAAAGAAGGGGATCCACTGCAGTCTTTCTCTTCTCGCTGCGAAAAGTTCCCCGTCGTCGCC   -145

                                                            !-91
          .          .          .          .          .          .          .          .          .
-144  TTATCGGCATCGCATTCTTCGCTATAAAAGCCGCCTGTGCCAGAAGTCCAGTCATCAGTCGCTCAGTTTCCACAGCAGCTAAACAGCTAA   -55
                          ------

          .          .          .          .          .          .          .          .          .
 -54  ATCGCAATCTATATATATATATATATACTAAGGAATTAAACCTAGAAAATTCACCATGAACTTCTACAAGATCTTCGTTTTCGTCGCCCT   35
                                                            MetAsnPheTyrLysIlePheValPheValAlaLe   (12)

          .          .          .          .          .          .          .          .          .
  36  CATCCTGGCCATCAGCATTGGACAATCGGAAGCCGGTTGGCTGAAGAAACTTGGCAAGAGAATCGTAAGTTCAGCAACAAAATATATTAA   125
      uIleLeuAlaIleSerIleGlyGlnSerGluAlaGlyTrpLeuLysLysLeuGlyLysArgIle                              (33)

          .          .          .          .          .          .          .          .          .
 126  ATACTTGCAAATTTACTAATTTGTTTTATATTTACTTGCAAAGGAGCGCATTGGCCAGCACACCCGGGATGCAACCATTCAAGGACTGGG   215
                                             GluArgIleGlyGlnHisThrArgAspAlaThrIleGlnGlyLeuGl   (49)

          .          .          .          .          .          .          .          .          .
 216  AATTGCGCAACAGGCCGCCAATGTGGCAGCCACCGCCAGAGGATGAGCCTTTAATGTCCATCAAAGGACTCTACCAGGATAACGCGCGTT   305
      yIleAlaGlnGlnAlaAlaAsnValAlaAlaThrAlaArgGlyEnd                                                (63)

          .          .          .          .          .          .
 306  TAATTATACACACTTATTTATTTACCAGCCATAGAAATAAACTAGCTTACATCCCCGTAATTT   368
                                              ------
```

CecC SEQUENCE. Strain, *Canton S*. Accession, Z11167 (DROCECCG).

end was not determined. There is an intron after the Ile-33 codon (*CecB* Sequence) (Kylsten et al. 1990).

CecC

Gene Organization and Expression

Open reading frame, 63 amino acids; expected mRNA length, ca. 380 bases. Sequence features were used to define the 5′ end, The 3′ end was not determined. There is an intron after the Ile-33 codon (*CecC* Sequence) (Tryselius et al. 1992).

References

Boman, H. G. and Hultmark, D. (1987). Cell-free immunity in insects. *Ann. Rev. Microbiol.* **41**:103–126.

Dunn, P. E. (1986). Biochemical aspects of insect immunology. *Ann. Rev. Entomol.* **31**:321–339.

Kylsten, P., Samakovlis, C. and Hultmark, D. (1990). The cecropin locus in Drosophila; a compact gene cluster involved in the response to infection. *EMBO J.* **9**:217–224.

Samakovlis, C., Kylsten, P., Kimbrell, D., Engström, Å. and Hultmark, D. (1991). The *Andropin* gene and its product, a male-specific antibacterial peptide in *Drosophila melanogaster*. *EMBO J*. **10**:163–169.

Tryselius, Y., Samakovlis, C., Kimbrell, D. and Hultmark, D. (1992). *CecC*, a cecropin gene expressed during metamorphosis in *Drosophila* pupae. *Eur. J. Biochem.* **204**:395–399.

6

bicoid: bcd

Chromosomal Location:
3R, 84A

Map Position:
3-[47.5]

Product

The following discussion refers to the 489 amino acid product of the major transcript, BCD. It is a DNA-binding regulatory protein of the homeodomain type. BCD controls the expression of early developmental genes in the anterior half of the embryo (Gehring 1987; Driever and Nüsslein-Volhard 1989; Hayashi and Scott 1990; Harrison 1991). For a review see Driever (1992).

Structure

The *bicoid* protein is a 55–58 kD protein, rich in Pro (10%) and probably phosphorylated. It has several sequence features of potential functional significance (Berleth et al. 1988):

1. The codons in the first exon include the PRD-repeat, alternating Pro and His, a pattern also found in the *paired* protein and other embryogenesis genes (Frigerio et al. 1986).

2. The amino-terminal region of the third exon (Pro-97 to Ser-156) encodes a homeodomain having weak (ca. 40%) similarity to other homeodomains.

3. There are several PEST sequences (rich in Pro, Ser and Thr), the most significant between amino acids 170 and 203. Such sequences are found in proteins of short half-life and are thought to be degradation signals (Rogers et al. 1986); although in this particular case, their deletion does not affect BCD stability (Driever 1992).

4. The carboxy half of the third exon is a Gln-rich region that results from the presence of repeated CAG (the M- or opa-repeat).

5. Further downstream, between positions 347 and 414 there is an acidic region.

Experiments with chimeric and mutant proteins in transgenic organisms established that the homeodomain is responsible for DNA binding and sequence recognition and that the carboxy-terminal two thirds of the protein are necessary to effect transcriptional activation. However, no single localized region of BCD seems unequivocally responsible for the latter function (*bcd* Sequence) (Struhl et al. 1989; Driever 1992 and references therein).

The ten residues from 138 to 147 constitute the *recognition alpha helix* of the homeodomain (helix 3, which corresponds to the second helix of the prokaryotic helix-turn-helix repressor proteins). The Lys at position 9 of the recognition helix provides the specificity that distinguishes the *bcd* homeodomain from the *Antp* class homeodomain in which a Gln occurs in that position (Hanes and Brent 1989; Treisman et al. 1989).

Function

The concentration of *bicoid* product determines "position" in the anterior embryo via regulatory action on other genes; that is, BCD is the "anterior morphogen" (Driever and Nüsslein-Volhard 1988b; Struhl et al. 1989).

BCD binds to the *hunchback* (*hb*) proximal promoter where it acts as a positive transcriptional regulator (Tautz 1988; Driever and Nüsslein-Volhard 1989). The BCD binding sites that occur in the *hb* promoter have the consensus TCTAAT*C*CC; in this segment, the central TAAT is the core necessary for homeodomain protein binding, and the C in position 7 ensures that BCD, but not ANTP, binds (Driever and Nüsslein-Volhard 1989; Hanes and Brent 1991).

BCD is also involved in the regulation of *Krüppel* (Hoch et al. 1990, 1991, 1992), *even-skipped* (Small et al. 1991; Stanojevic et al. 1991) and probably other early genes. A less-well-understood function of *bcd* is its role in the formation of the *caudal* RNA and protein gradients, since this is a post-transcriptional process (Mlodzik and Gehring 1987; Driever 1992).

Tissue Distribution

Production of BCD starts at the anterior tip of the egg shortly after oviposition (regardless of whether the egg is fertilized or not) and involves translation of a localized, pre-existing maternal message. By the syncytial blastoderm stage, the protein is localized in nuclei and distributed in a steep exponential gradient with the highest concentration at the anterior tip of the embryo and undetectable levels in the posterior 30% (Appendix, Fig. A.2). BCD reaches a maximum 2–4 h after oviposition; it begins to decline during blastoderm cellularization; and it is practically undetectable after gastrulation (Driever and Nüsslein-Volhard 1988a).

Mutant Phenotype

This is a maternal-effect gene: offspring of homozygous bcd^- females are inviable. In the absence of BCD, structures in the anterior half fail to differentiate; neither head nor thorax develops, and the terminal acron is transformed into a second telson (Frohnhöfer and Nüsslein-Volhard 1986).

bcd

```
-1414  GTCGACTGGAGTGTCTGTGAATTGACTTTTGTTGCCAGTTGGCAGCGGCAGAAGCAGCAAAGCCCGGCCAACAGCAACAAGCTCCTGCCA  -1325

-1324  GATCCCAAAAGCAAACACGACAATTATTTGGCAAATGTCATTAAAAAATATTTCACTTAAGGCCTTGCGACACTTGCTTAAAGGTCAACT  -1235

-1234  GGCTCGTTGGGTGTGTTTTAAAATGTTAAAGCTTGGGCCAATGCACTGAGCAACTTAATGCTTGTAGATATTTACACAATATTCTTCAAC  -1145

-1144  GCTAAACATATCGAATTTTCCAAATATGGAGCCTGAAAATAATAATTGCCAATCCTAGCTTAAAATCAGAAATGAGTAGAACAACTTAAA  -1055

-1054  AAAATTAACAAAAGAATCGAACGCTACAGCTAATTAACTCGACAACTGGTTACCTTTTATTCTTCTAATACATTTTATAATGCACTGCCT  -965

 -964  AACAGGTACAGATAGCAAGCACTATATGCTGTCTTACAAAACGATTATATGATATTTTCTTTCGTACGTAGCCGTTTGAGATCATTTGGA  -875

 -874  AAAACAAACTCGATCTCCACCATCCTTATTCTTTGTCCCAAGTCCTTATATATCTCGCGATACTAAGATTGAATAATGTAGTTATTAATA  -785

 -784  GCGGAAGTATGTAACAGAATAAACTACAAAGTGCACATTTTGTTCAATTCAGGCTGGACTGGACTGGAGCATATTAATATTATAATATTA  -695

 -694  ACAAAAATTCAAATTAAACATTCGACACTTGTCTAATTGATTCCTAAATTTGGGGTGCCTGTTTGTTAATTAAATGTTAATATTATGAAG  -605

 -604  TTCCAAACAGAGCAAAGAGTTTAAGTTTAATTGGTTCTACTTATTTGTTACAATATTCAAGCTTTTTTTATTATTATTCTCAAATGCAAA  -515

 -514  TCTCTACAAATAAATAAACCTCCGACGTTTTAGAACATTCACCTTTTGTCAGTGAGCACAACCTTTCAATACAGCCCGACAGGGGGCTCT  -425

 -424  CTACTGCTGTCTCTTCACGCCCCCTGGTGAAAACGCTGTGCACTCAATCGGTTTGCAGCTTTGCCGTACTGTTCGATTAAAAACTTTTAA  -335

 -334  ATTAGAGGCAAACATTTAAAAATAAAATGTCCAAATATTTGTCTAAAATGTATTGTAGACGCTTATTGATTTTTAAATTACTCAAAAGAA  -245

                                                                                     !-168
 -244  TGTTCATCGAGGGAGGGCCGCCAATTGTGCCATCTCTACATCTCTTCGCTCATCCCTAAATAACGGCACTCTGCAGATGCGAAGCAGTGG  -155

 -154  ATCGCAAAAACGCAAAATGTGGGCGAAATAAGTTCGCGAGCGTCTCGAAAGTAACCGGTTACTGAAAATACAAGAAAGTTTCCACACTCC  -65

  -64  TTTGCCATTTTTCCGCGCGGCGCTTGGAAATTCGTAAAGATAACGCGGCGGAGTGTTTGGGGAAAATGGCGCAACCGCCGCCAGATCAAA  25
                                                                     MetAlaGlnProProProAspGlnA  (9)

   26  ACTTTTACCATCATCCGCTGCCCCACACGCACACACATCCGCATCCGCACTCCCATCCGCATCCGCACTCGCATCCGCACCCACATCACC  115
       snPheTyrHisHisProLeuProHisThrHisThrHisProHisProHisSerHisProHisProHisSerHisProHisProHisHisG
             -----------------------------------------------------------------------------------  (39)

  116  AACATCCGCAGCTTCAGTTGCCGCCACAATTCCGAAATCCCTTCGATTTGGTGAGTTCCCATCGCAGCAGAGAAGGGCTCTTGTCCCAGG  205
       lnHisProGlnLeuGlnLeuProProGlnPheArgAsnProPheAspLeu                                         (55)
       -------- PRD REPEAT

  206  AAAGCTACAGTACAGATTCCCTATGGTGAACAAACAACCAGTGCGATCACTGATGACCATAAACATTTATTGAGCCGCAGCAAATGTGTT  295

  296  TCTAGAACATAGGGCGAAATCTTCTATTATCTTGTTTGTGACTTTTAAAGTATCGTAGCAGAATCTAAATAACAATTGATATTATTAATC  385

  386  GTTACAGTTAGTATAGTATATAATTGTATATGAATTGTGGGGCAACATGTTATTAGTGATTTGCCGAAATGTTCTAAAAGATGTTTCATT  475

  476  GAAATGGACGAATGTTAAACCTGTTGCACTCACACCGAATATCAGTAATGTCTATTTTTCAAAAGCCACATCTATGGCCACTGGGTATAC  565

  566  ATTATTGACTTAATACACTTCATACAACATATTTTCAAAAACAAGCATTGTTGTCCTGCATGATGATTAGTGAAAGTAATATTGCAAGAT  655
```

(continued)

```
          .         .         .         .         .         .         .         .         .
  656 TCGGTCCCCGAAGCGAATCGTCCTTTCACGTTTTTATATAAAGACAGTGTACCCCTTGATTCTTTGAAGCTTTTCGATGAGCGAACGGGA  745
                                                                           LeuPheAspGluArgThrGly  (62)

          .         .         .         .         .         .         .         .         .
  746 GCGATAAACTACAACTACATACGTCCGTATCTGCCCAACCAGATGCCCAAGCCAGGTGAGCTCAAAGCCAACAAAGTCAGCCATCGTCTT  835
      AlaIleAsnTyrAsnTyrIleArgProTyrLeuProAsnGlnMetProLysProA                                     (81)

                          _
          .         .    |alternate acceptor      .         .         .         .         .
  836 ATCAGATGTCTTTCCCTCAGAGGAGCTGCCCGACTCTCTGGTGATGCGGCGACCACGTCGCACCCGCACCACTTTTACCAGCTCTCAAAT  925
        spValPheProSerGluGluLeuProAspSerLeuValMetArgArgProArgArgThrArgThrThrPheThrSerSerGlnIl
                         |_                                |     *     *          * ------*----   (109)

          .         .         .         .      DefE6=|-   T=E4       .T=E3     .   |=DefE6
  926 AGCAGAGCTGGAGCAGCACTTTCTGCAGGGACGATACCTCACAGCCCCCGACTTGCGGATCTGTCAGCGAAACTAGCCCTGGGCACAGC  1015
      eAlaGluLeuGluGlnHisPheLeuGlnGlyArgTyrLeuThrAlaProArgLeuAlaAspLeuSerAlaLysLeuAlaLeuGlyThrAl  (139)
                                                          Phe         Leu
      -------*------------------H1      *        ------------------*--------------H2 *      ----

          .         .         .         .        .DefE1=|- .T=GB    .         .         .
 1016 CCAGGTGAAGATATGGTTTAAGAACCGTCGGCGTCGTCACAAGATCCAATCGGATCAGCACAAGGACCAGTCCTACGAGGGGATGCCTCT  1105
      aGlnValLysIleTrpPheLysAsnArgArgArgArgHisLysIleGlnSerAspGlnHisLysAspGlnSerTyrGluGlyMetProLe  (169)
                                                             End
      ----*-----*--*--*-----*--*H3*     *     *          | HOMEODOMAIN

                                                 T=085
          .         .         .         .                   .         .         .         .
 1106 CTCGCCGGGTATGAAACAGAGCGATGGCGATCCCCCCAGCTTGCAGACTCTTAGCTTGGGTGGAGGAGCCACGCCCAACGCTTTGACTCC  1195
      uSerProGlyMetLysGlnSerAspGlyAspProProSerLeuGlnThrLeuSerLeuGlyGlyGlyAlaThrProAsnAlaLeuThrPr  (199)
                                                 End
          .         .         .         . AA-|=DefE1          .         .         .         .
 1196 GTCACCCACGCCCTCAACGCCCACTGCACACATGACGGAGCACTACAGCGAGTCATTCAACGCCTACTACAACTACAATGGAGGCCACAA  1285
      oSerProThrProSerThrProThrAlaHisMetThrGluHisTyrSerGluSerPheAsnAlaTyrTyrAsnTyrAsnGlyGlyHisAs  (229)
          .         .         .         .         .         .         .         .         .
 1286 TCACGCCCAGGCCAATCGTCACATGCACATGCAGTATCCTTCCGGAGGGGGGCCAGGACCTGGGTCGACCAATGTCAATGGCGGCCAGTT  1375
      nHisAlaGlnAlaAsnArgHisMetHisMetGlnTyrProSerGlyGlyGlyProGlyProGlySerThrAsnValAsnGlyGlyGlnPh  (259)

          .  T=111 T=E5         .         .         .         .         .         .         .
 1376 CTTCCAGCAGCAGCAGGTCCATAATCACCAGCAGCAACTGCACCACCAGGGCAACCACGTGCCGCACCAGATGCAGCAGCAGCAACAGCA  1465
      ePheGlnGlnGlnGlnValHisAsnHisGlnGlnGlnLeuHisHisGlnGlyAsnHisValProHisGlnMetGlnGlnGlnGlnGlnGl  (289)
             End   End
          ------------            ---------         ---                  ---   -----------------

          .         .         .         .         .         .         .         .         .
 1466 GGCTCAGCAGCAGCAATACCATCACTTTGACTTCCAGCAAAAGCAAGCCAGCGCCTGTCGCGTCCTGGTCAAGGACGAACCGGAGGCCGA  1555
      nAlaGlnGlnGlnGlnTyrHisHisPheAspPheGlnGlnLysGlnAlaSerAlaCysArgValLeuValLysAspGluProGluAlaAs  (319)
       -   ------------                  ------   --- OPA REPEATS

          .         .         .         .         .         .         .         .         .
 1556 CTACAACTTCAACAGCTCGTACTACATGCGATCGGGAATGTCTGGCGCCACTGCATCGGCATCCGCTGTGGCCCGAGGCGCTGCCTCGCC  1645
      pTyrAsnPheAsnSerSerTyrTyrMetArgSerGlyMetSerGlyAlaThrAlaSerAlaSerAlaValAlaArgGlyAlaAlaSerPr  (349)

          .         .         .         .         .         .         .         .         .
 1646 GGGCTCCGAGGTCTACGAGCCATTAACACCCAAGAATGACGAAAGTCCGAGTCTGTGTGGCATCGGCATCGGCGGACCTTGCGCCATCGC  1735
      oGlySerGluValTyrGluProLeuThrProLysAsnAspGluSerProSerLeuCysGlyIleGlyIleGlyGlyProCysAlaIleAl  (379)
          .         .         .         .         .         .         .         .         .
 1736 CGTTGGCGAGACGGAGGCGGCCGACGACATGGACGACGGAACGAGCAAGAAGACGACGCTACAGGTCAGGCATGAGTCCACAACCTTTTT  1825
      aValGlyGluThrGluAlaAlaAspAspMetAspAspGlyThrSerLysLysThrThrLeuGln                            (399)

          .         .         .         .         .         .         .         .         .
 1826 TGATCTCTTGATTCTGAGTGTGGCGTTTATAAATTGAAGCTTTAAGCTTTGTAACTTTCAAACTGTCTGGTTTGAGATGTTATTCTGAAA  1915
```

```
1916 GTACTTCTATTTCCGATCGATGAGATTTGGGAGTTCTTCAATATTTAACATTTAACTTATTAAGTTTTTGTTTTCTAAATTAGACATGGC 2005

2006 ATTTCTGAAAGGGAAGTACAAGTGTTAAAGATGTATTTTAATATAGAATTTGTATCAAAGGTTAAGATTTCAACCGTTTGAAAGCCCTTA 2095

2096 GTTTTCAGGGTTTTTTACTTTTTTATTCATGTAATCACTCTTAATACACTGCAAGTTAAAATAGCATTTCTTTGACCAGAAAAATAAGAA 2185

2186 TCTATGCATTTTAAAAGTGAAAACAGACTCATATGCTGATGAACATTTTTAGCTATAAATTGTAACAATAATTTAGCAATTTCAATTGAA 2275

2276 TTTATTTATGTTCTAAATGCGTTCGCTCTCTCCCTAGATCTTGGAGCCTTTGAAGGGTCTGGACAAGAGCTGCGACGATGGCAGTAGCGA 2365
                                          IleLeuGluProLeuLysGlyLeuAspLysSerCysAspAspGlySerSerAs (418)

2366 CGACATGAGCACCGGAATAAGAGCCTTAGCAGGAACCGGAAATCGTGGAGCGGCATTTGCCAAATTTGGCAAGCCTTCGCCCCCACAAGG 2455
     pAspMetSerThrGlyIleArgAlaLeuAlaGlyThrGlyAsnArgGlyAlaAlaPheAlaLysPheGlyLysProSerProProGlnGl (448)

                   A=2-13
2456 CCCTCAGCCGCCCCTCGGGATGGGGGGCGTGGCCCTGGGCGAATCGAACCAATATCAATGCACGATGGATACGATAATGCAAGCGTATAA 2545
     yProGlnProProLeuGlyMetGlyGlyValAlaLeuGlyGluSerAsnGlnTyrGlnCysThrMetAspThrIleMetGlnAlaTyrAs (478)
                  His

2546 TCCCCATCGGAACGCCGCGGGCAACTCGCAGTTTGCCTACTGCTTCAATTAGCCTGGACGAGAGGCGTGTTAGAGAGTTTCATTAGCTTT 2635
     nProHisArgAsnAlaAlaGlyAsnSerGlnPheAlaTyrCysPheAsnEnd                                       (494)

2636 AGGTTAACCACTGTTGTTCCTGATTGTACAAATACCAAGTGATTGTAGATATCTACGCGTAGAAAGTTAGGTCTAGTCCTAAGATCCGTG 2725
                                                            |---

2726 TAAATGGTTCCCAGGGAAGTTTTATGTACTAGCCTAGTCAGCAGGCCGCACGGATTCCAGTGCATATCTTAGTGATACTCCAGTTAACTC 2815

2816 TATACTTTCCCTGCAATACGCTATTCGCCTTAGATGTATCTGGGTGGCTGCTCCACTAAAGCCCGGGAATATGCAACCAGTTACATTTGA 2905

2906 GGCCATTTGGGCTTAAGCGTATTCCATGGAAAGTTATCGTCCCACATTTCGGAAATTATATTCCGAGCCAGCAAGAAAATCTTCTCTGTT 2995

2996 ACAATTTGACATAGCTAAAAACTGTACTAATCAAAATGAAAAATGTTTCTCTTGGGCGTAATCTCATACAATGATTACCCTTAAAGATCG 3085

3086 AACATTTAAACAATAATATTTGATATGATATTTTCAATTTCTATGCTATGCCAAAGTGTCTGACATAATCAAACATTTGCGCATTCTTTG 3175

3176 ACCAAGAATAGTCAGCAAATTGTATTTTCAATCAATGCAGACCATTTGTTTCAGATTCTGAGATTTTTTGCTGCCAAACGGAATAACTAT 3265

3266 CATAGCTCACATTCTATTTACATCACTAAGAAGAGCATTGCAATCTGTTAGGCCTCAAGTTTAATTTTAAAATGCTGCACCTTTGATGTT 3355
                                                     ---| LOCALIZATION ELEMENT

3356 GTCTCTTTAAGCTTTGTATTTTTAATTACGAAAATATATAAGAACTACTCTACTCGGGTAAATTGTGACTAACTACACATAACTACATAC 3445
                                   ------                       |(A)n

3446 TTAGCCCATATTTCCGTCCCTTTCTAGAATGAACGAAAACAGTATCTGGTTTTCCCGAAAATCTTATGAATTTAAAAATGCACTTTATTG 3535

3536 CACATACTCACACATGCCTGCCATAAAATATGATTCGCGATTTTTCCGCGAACACCCGCGGATCATAAAACATTTGCACCAGCTGCCTGT 3625

3626 GTTTATTCACCTACCTGAAACCCATACTCTTATCGCCTGATCCTCGCGCGGTCGCACTATTTAGGTAGACACTGTACAGGCAGCACTAGC 3715
```

bcd SEQUENCE. Strain, *Oregon R*. Accession, X07870 (DROBCDG). An exclamation mark at −168 indicates the 5′ end of the longest cDNA. Dashes underline the region of PRD and OPA repeats. The boundaries of the RNA localization element and the homeodomain are indicated with vertical bars below the sequence. Within the homeodomain (Pro-97 to Ser-156), asterisks indicate conserved amino acids and dashes underline the presumptive helices. Mutations bcd^{E3}, bcd^{E4} and bcd^{E6} (which

(continued)

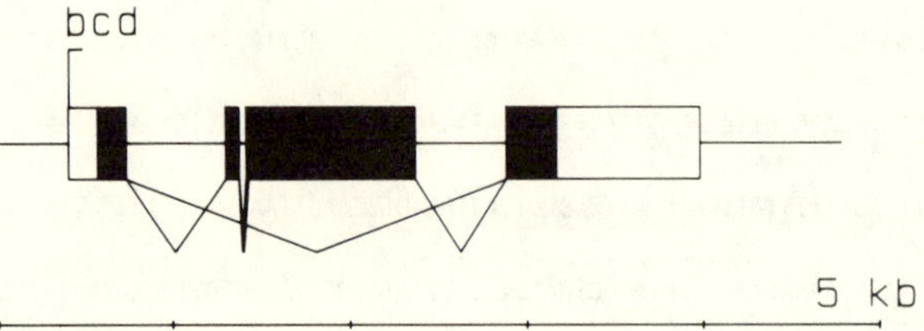

FIG. 6.1. Organization of *bcd*.

Gene Organization and Expression, Major Transcript

Open reading frame, 489 (the most abundant) or 494 amino acids depending on the acceptor site of the second intron (*bcd* Sequence); expected mRNA length, ca. 2,453 bases, in agreement with the prevalent 2.6 kb RNA band (Berleth et al. 1988). Minor Transcript: Open reading frame, 149 amino acids; expected mRNA length, ca. 1,436 bases in agreement with a 1.6 kb RNA band (Berleth et al. 1988).

The 5′ end of the longest cDNA is indicated in the Sequence at −168. No canonical TATA box is found in the appropriate position. The 3′ end was determined from the sequence of two cDNAs. There are three introns: after the Leu-55 codon, within the codon for Asp-81 (or Glu-81), and after the Gln-399 codon. There are three alternative splicing forms. Two of them represent the major 2.6 kb transcript that carries four exons. They differ with respect to the acceptor site of the second intron; the two sites are in frame and the difference is a five amino acid segment (*bcd* Sequence). In the minor transcript, the second and third exons are spliced out (Fig. 6.1) (Berleth et al. 1988). The mRNA that codes for the 489-amino-acid protein is sufficient for all the *bcd* functions and is probably the functional form (Driever 1992).

This gene is 35–40 kb closer to the centromere than *Deformed* (*Dfd*) in the *Antennapedia* complex, and it is transcribed toward the centromere (Berleth et al. 1988).

Developmental Pattern

Transcription of *bcd* begins early in oogenesis and seems to be restricted to the nurse cells. The RNA is transferred to the anterior region of the oocyte, together

---

(*continued*) affect the homeodomain) encode proteins unable to bind to *hb* sequences in yeast cells. Mutation bcd^{GB} (which truncates the polypeptide immediately downstream of the homeodomain) binds *hb* sequences, but it is unable to stimulate transcription in yeast cells (it is a strong allele *in vivo*). Mutations bcd^{085} and bcd^{E5} (which truncate further downstream) have some activating function left and are weaker alleles, specially bcd^{E5} (Struhl et al. 1989).

with other maternal RNAs, by passage through the ring canals. A special feature of *bcd* RNA is its ability to remain strictly localized or "anchored" in the anterior 20% of the oocyte, in the cortical zone. A discrete *cis*-acting segment necessary for this localization is present in the 3′ untranslated region of the *bcd* message. A 627-base segment (from 2,691 to 3,317) is sufficient to anchor mRNA to the anterior egg cap and includes sequences with the potential for extensive secondary structure (Macdonald and Struhl 1988). The *bcd* RNA remains highly localized until after the last embryonic cleavage division; then it is degraded, disappearing completely by blastoderm cellularization (Berleth et al. 1988). Microtubules and the products of maternal effect genes *swallow* (*swa*), *exuperantia* (*exu*) and *staufen* appear to be involved in the anchoring process (Schüpbach and Wieschaus 1986; Pokrywka and Stephenson 1991).

Promoter

A 4.0 kb segment in front of the gene is sufficient for normal expression (Berleth et al. 1988).

References

Berleth, T., Burri, M., Thoma, G., Boop, D., Richstein, S., Frigerio, G., Noll, M. and Nüsslein-Volhard, C. (1988). The role of localization of *bicoid* RNA in organizing the anterior pattern of the *Drosophila* embryo. *EMBO J.* **6**:1749:1756.

Driever, W. (1992). The bicoid morphogen: concentration dependent transcriptional activation of zygotic target genes during early *Drosophila* development. In *Transcriptional Regulation*, eds K. R. Yamamoto and S. L. McKnight (Cold Spring Harbor, NY: CSH Press).

Driever, W. and Nüsslein-Volhard, C. (1988a). A gradient of *bicoid* protein in *Drosophila* embryos. *Cell* **54**:83–93.

Driever, W. and Nüsslein-Volhard, C. (1988b). The *bicoid* protein determines position in the *Drosophila* embryo in a concentration-dependent manner. *Cell* **54**:95–104.

Driever, W. and Nüsslein-Volhard, C. (1989). The *bicoid* protein is a positive regulator of *hunchback* transcription in the early *Drosophila* embryo. *Nature* **337**:138–143.

Frigerio, G., Burri, M., Boop, D., Baumgartner, S. and Noll, M. (1986). Structure of the segmentation gene *paired* and the *Drosophila* PRD gene set as part of a gene network. *Cell* **47**:735–746.

Frohnhöfer, H. G. and Nüsslein-Volhard, C. (1986). Organization of anterior pattern in the *Drosophila* embryo by the maternal gene *bicoid*. *Nature* **324**:120–125.

Gehring, W. J. (1987). Homeo boxes in the study of development. *Science* **236**:1245–1252.

Hanes, S. D. and Brent, R. (1989). DNA specificity of the *bicoid* activator protein is determined by homeodomain recognition helix residue 9. *Cell* **57**:1275–1283.

Hanes, S. D. and Brent, R. (1991). A genetic model for interaction of the homeodomain recognition helix with DNA. *Science* **251**:426–430.

Harrison, S. C. (1991). A structural taxonomy of DNA-binding domains. *Nature* **353**:715–719.

Hayashi, S. and Scott, M. (1990). What determines the specificity of action of *Drosophila* homeodomain proteins? *Cell* **63**:883–894.

Hoch, M., Schröder, C., Seifert, E. and Jäckle, H. (1990). Cis-acting control elements for *Krüppel* expression in the *Drosophila* embryo. *EMBO J.* **9**:2587–2595.

Hoch, M., Seifert, E. and Jäckle, H. (1991). Gene expression mediated by *cis*-acting sequences of the *Krüppel* gene in response to the *Drosophila* morphogens *bicoid* and *hunchback*. *EMBO J.* **10**:2267–2278.

Hoch, M., Gerwin, N., Taubert, H. and Jäckle, H. (1992). Competition for overlapping sites in the regulatory region of the *Drosophila* gene *Krüppel*. *Science* **256**:94–97.

Macdonald, P. M. and Struhl, G. (1988). Cis-acting sequences responsible for anterior localization of *bicoid* mRNA in *Drosophila* embryos. *Nature* **336**:595–598.

Mlodzik, M. and Gehring, W. (1987). Hierarchy of the genetic interactions that specify the anteroposterior segmentation pattern of the *Drosophila* embryo as monitored by caudal protein expression. *Development* **101**:421–435.

Pokrywka, N. J. and Stephenson, E. C. (1991). Microtubules mediate the localization of *bicoid* RNA during *Drosophila* oogenesis. *Development* **113**:55–66.

Rogers, S., Wells, R. and Rechsteiner, M. (1986). Amino acid sequences common to rapidly degraded proteins: The PEST hypothesis. *Science* **234**:364–368.

Schüpbach, T. and Wieschaus, E. (1986). Maternal-effect mutations altering anterior posterior pattern in the *Drosophila* embryo. *Wilhelm Roux's Arch. Dev. Biol.* **195**:302–317.

Small, S., Kraut, R., Warrior, R. and Levine, M. (1991). Transcriptional regulation of a pair-rule stripe in *Drosophila. Genes Dev.* **5**:827–839.

Stanojevic, D., Small, S. and Levine, M. (1991). Regulation of a segmentation stripe by overlapping activators and repressors in the *Drosophila* embryo. *Science* **254**:1385–1387.

Struhl, G., Struhl, K. and Macdonald, P. M. (1989). The gradient morphogen bicoid is a concentration-dependent transcriptional activator. *Cell* **57**:1259–1273.

Tautz, D. (1988). Regulation of the *Drosophila* segmentation gene hunchback by two maternal morphogenetic centers. *Nature* **332**:281–284.

Treisman, J., Gonczy, P., Vashishtha, M., Harris, E. and Desplan, C. (1989). A single amino acid can determine the DNA binding specificity of homeodomain proteins. *Cell* **59**:553–562.

7

Blastoderm-specific gene at 25 D: *Bsg25D*

Chromosomal Location:
2L, 25D3

Map Position:
2-[16]

Product

Unidentified. Codon translation yields a 741-amino-acid protein with two regions of similarity to known products. A 95-amino-acid stretch (positions 250–344) shows significant similarity (22%) to a portion of the product of the *fos* oncogene. The other segment is 21 amino acids long (509–529) and shows similarity to the repeating segments of rabbit tropomyosin that are thought to bind F actin molecules (Boyer et al. 1987).

Gene Organization and Expression

Open reading frame, 741 amino acids; expected mRNA length, 2,645–2,774 bases when one of the proximal poly-A sites is used and approximately 4,749 bases when a distal site is used; this is in agreement with poly(A) + RNA bands of 2.7 and 4.5 kb. There are two introns in the coding region, one in the codon for Asp-78 and another after the codon for Gln-159. The 5′ end was determined by S1 mapping and primer extension; 3′ ends were determined by S1 mapping. Three of the proximal poly(A) sites scored have no corresponding poly(A) signals upstream; it is not clear whether these represent technical artifacts or whether they are true termini. A third RNA of 3.0 kb hybridizes to a *Bsg25D* cDNA, but it is very rare and could not be mapped (*Bsg25D* Sequence) (Boyer et al. 1987).

Developmental Pattern

The 3.0 and 4.5 kb RNAs are expressed during the first 8 h of embryogenesis, and the 2.7 kb RNA is blastoderm-stage specific (Roark et al. 1985).

Bsg25D

```
-420  ATCAATCTAACGATAGTGTATAACGATAGGAACAATGGTCCACGATATGGCCACCTCCGTGCAAGTTTGCTTAATGCCCTCCAGAGCGCG   -331
                                         ----
                                                -->-295
-330  CCACCGTGCTCGCTATACTGCATTAATTGTTTTTTATCAACTCGCTAGAAATACGCTATCCCAAAAAACCGCAAACCCGCGATGTTTATG   -241
                ----

-240  TTGCGTTCCGAAGTGCATATCATAGATTAGTAGTAGTAGTAACCCCTCAAACAGCCTGCTGTCCAAAAAACACGCGTGATTCCCCCGCCA   -151

-150  CCCACGCACATAGACCCCGATATTTCACTTTTCTTGTTTTCGACCCCTGACTGCGTTTGTGGATTTTCCCCCCAAGAAAAAAAAAGCGAA   -61

 -60  GTGAAAACGCAATTGAGCAGCCGATCGATTGGAACGGCAGGAATTCCCCGGGTTACGGATAATGGAGGTATCCGCCGATCCGTACGAGCA   29
                                                                      MetGluValSerAlaAspProTyrGluGl   (10)

  30  GAAGCTCTACCAAATGTTCCGCAGCTGCGAGACGCAGTGTGGACTTCTGGACGAGAAGTCCCTGCTGAAGCTCTGCTCACTGCTGGAGCT   119
      nLysLeuTyrGlnMetPheArgSerCysGluThrGlnCysGlyLeuLeuAspGluLysSerLeuLeuLysLeuCysSerLeuLeuGluLe   (40)

 120  CCGGGATCAGGGATCCGCACTGATCGCCAGCCTGGGCGGCAGCCATCAGCTGGGCGTGTCCTTTGGCCAGTTCAAGGAGGCGCTACTCAA   209
      uArgAspGlnGlySerAlaLeuIleAlaSerLeuGlyGlySerHisGlnLeuGlyValSerPheGlyGlnPheLysGluAlaLeuLeuAs   (70)

 210  CTTCCTGGGCTCCGAGTTCGATGGTAATACGTCATCGGGTTTCATTGGTGAGATAGCACAAAGAATCGATCACGCTATAGATTAACTTAT   299
      nPheLeuGlySerGluPheAspA                                                                     (78)

 300  ATAGTATAAAGATAATATTTGCTATAAGCTAACGCGACAGGTTCGCATAAAACAACATACGTTTTATCTGTAATTGCGCTTTAATTACCC   389

 390  ATCAAGCAACATCAGATAATTACGGAATGTTTGCCAGCCACTTATTAGAGATAGTAATTCAATTTTGACACGGATTTGGAACCGTGTGGG   479

 480  TTTCCCTATTAATAAAACACTGATCTAATGAACACATTTCTAGCAGTCTATAGATGAACAAAGCCATTACTTAATACTCAAAGAAGTGCT   569

 570  ACCATCTACGTGCTAATTTGCAAGGATTATGCACATTTACTTCAAACCTCCGCTTATCTGATTTGGAAACTTCTGGGCAAATTTAGGACA   659

 660  CCTTAGGGTACGAATATCATAATCAGCACGCGGATTAGCACGCGGCAGCTGGCGATCATAAAATCATAGATGCAATTGACACTTTTTTAC   749

 750  GACTCCCAACTGTTCTCGACTACCTGATCCTGCATGATCCTTATCAGGTAGATGGTTACAATGTCCTGTATAAATACGCGACACATTCAC   839

 840  CTGGGCAGTTTAGTCTAAATCAAAATGGGAACACGATTGTATTACCGCCGATCCGGCGGTCAGTTAACAGATCCGATAATTGAGAAGCTA   929

 930  GCCGCTCGTTTTGGTAGCCACCTAAGATCCATACAACTCTTCCAGTTCTCTGCTAACTTATATCTATTGAATCTTCCAGAGCGTTCACTG   1019
                                                                                        spArgSerLeu   (81)

1020  GTGATTACGGATGAGCCGCTAAACAACACATACATCGAGAGTCCGCCGGAGTCTTCCGATCGCGAGGTTTCACCCAAACTCGTCGTGGGC   1109
      ValIleThrAspGluProLeuAsnAsnThrTyrIleGluSerProProGluSerSerAspArgGluValSerProLysLeuValValGly   (111)

1110  ACCAAGAAATACGGTCGCCGGTCTAGGCCACAGCAGGGAATCTACGAGTTATCCGTCACGGACTCGGACAATACGGACGAGGACCAGTTG   1199
      ThrLysLysTyrGlyArgArgSerArgProGlnGlnGlyIleTyrGluLeuSerValThrAspSerAspAsnThrAspGluAspGlnLeu   (141)

1200  CAGCAGCAGCAAAATCAGCGAAGCCTCAACGGATGCGATGAGCTGGGAGTTCAGGTGAGTGTCGTTTGTCAAGTCACGTACGAAGTGGCG   1289
      GlnGlnGlnGlnAsnGlnArgSerLeuAsnGlyCysAspGluLeuGlyValGln                                      (151)

1290  ATACAACTTCTGGTATGTATGCAAAATTGCATAGTAAACAGATTTTGTTTAATCGTTATTATTGCTGATACAGTAGAGCATGCCTAAGTA   1379

1380  GCACTACCAAAGCAAACAAATTATCTTAAATATACATCATGATCATCATAAGCATCTTATTTTTCCAAACCACACAGGTGCAACGTTCCT   1469

1470  CGTCCCAGAGCGATCTTCCTGGCAGCCGGCGTCTGCGGTCCGTCCACACCAGCGGGAGCAAACTGAAGCGTTGTGCTTCACTGCCAGCCC   1559
```

```
      .         .         .         .         .         .         .         .         .
1560  GCCGGAAGATGAACAGCAACACCACGGAGCCACTACATCACCGACGGCAGCGGCCAAGTTGAAACAGCTTTCCATCCAGAGCCAGGCGCA  1649
      .         .         .         .         .         .         .         .         .
1650  GCACAGCAGCAGCGTGGAATCACTGGGTAAGTTTCCTCTGGCCAGACCAGCTTTGGCTAGCCGATCCCCCTTGTCCCTGCCACCCTCTGT  1739
      .         .         .         .         .         .         .         .         .
1740  TGTTGTTAGCCCAAAATGCCAAAATTACGTTTGAAGCAATGTTAAAAGCAAAACACTTGTTTGTCGGTACACACCGTAGCCATCGCCTGG  1829
      .         .         .         .         .         .         .         .         .
1830  CCACCAATCCCGCACCGTCGTCCGAGCACTGGAGATGCTACCACGGCGGCCGTTGGTCATGCTGCAAAGGTTTGTGCGCTCTGAAGCAAT  1919
      .         .         .         .         .         .         .         .         .
1920  TGTCAACACCCTCACACCCACCGAATCCCCAACCCAGTCATTCGGTATCTAATCGCACCCTATGTAGCCGCACATTTGATTCGTTTTTTT  2009
      .         .         .         .         .         .         .         .         .
2010  TACTCGTATAATAACATATCCTACATTTTCAACCCTTAGTAATGCTGTAATGCATTGACAATCAATTTAATTAAGGATTTCATATAAATC  2099
      .         .         .         .         .         .         .         .         .
2100  AATTTCAGTTAGAAAGGATATTTACTTATAATTTGTTCTATTTTCTTGATTTATTAGTTTCTACCTCTTTAAATAACACGGCAAAAATTT  2189
      .         .         .         .         .         .         .         .         .
2190  CTCATTTCTAAAAGCCATTTGATATAGAGAAATAACAAACTTTCGGCGCTTTTGCTTACACCATCGACACACACACACACCCTTCCCCAC  2279
      .         .         .         .         .         .         .         .         .
2280  TCCCAATCCCAATCCAATCCCACACCCACCTGGTATCTTGGGCTATATGTATAAAAATGTGTATATACAACAGCGAAGCCAATCTCATTC  2369
      .         .         .         .         .         .         .         .         .
2370  GTCCCACGCTAATTGTTAATTGCCATGATTTACAGACACCGTGACGCCGCAGCAATTGGAGACGATCTCAGTGCATAGCATTATGGAAGC  2459
                                                          GlnLeuGluThrIleSerValHisSerIleMetGluAl  (172)
      .         .         .         .         .         .         .         .         .
2460  CTGGGAGCTGGCCAGCATTCCCAACACTCGCAACCTACTTCACGTCCTGGGATTCGATGAGGAGGAGGAGGTGAACCTGCAGCAGCTAAC  2549
      aTrpGluLeuAlaSerIleProAsnThrArgAsnLeuLeuHisValLeuGlyPheAspGluGluGluGluValAsnLeuGlnGlnLeuTh  (202)
      .         .         .         .         .         .         .         .         .
2550  TAAGGCATTGGAGGAGGAGCTGCGGGGCATCGATGGGGATCACGAGCAATCGAATATGTTGCGCGCTCTGGCTGCTCTGCAGGCCACCGA  2639
      rLysAlaLeuGluGluGluLeuArgGlyIleAspGlyAspHisGluGlnSerAsnMetLeuArgAlaLeuAlaAlaLeuGlnAlaThrGl  (232)
      .         .         .         .         .         .         .         .         .
2640  GTTGGGCAACTACAGACTTGCCTATAGGCAGCAGCATGAGGAGAACCTCAAGCTGAGGGCCGATAATAAGGCGGCCAACCAAAGGGTGGC  2729
      uLeuGlyAsnTyrArgLeuAlaTyrArgGlnGlnHisGluGluAsnLeuLysLeuArgAlaAspAsnLysAlaAlaAsnGlnArgValAl  (262)
      .         .         .         .         .         .         .         .         .
2730  TTTGCTTGCCGTGGAAGTGGATGAGCGGCATGCGTCGCTGGAGGATAACTCCAAGAAGCAGGTGCAGCAGCTGGAGCAAAGACACGCCAG  2819
      aLeuLeuAlaValGluValAspGluArgHisAlaSerLeuGluAspAsnSerLysLysGlnValGlnGlnLeuGluGlnArgHisAlaSe  (292)
      .         .         .         .         .         .         .         .         .
2820  CATGGTGCGTGAAATAACGCTGCGGATGACTAATGACCGCGATCACTGGACCAGCATGACGGGAAAGCTGGAGGCACAGCTTAAATCGCT  2909
      rMetValArgGluIleThrLeuArgMetThrAsnAspArgAspHisTrpThrSerMetThrGlyLysLeuGluAlaGlnLeuLysSerLe  (322)
      .         .         .         .         .         .         .         .         .
2910  TGAGCAGGAGGAGATCCGTCTGAGAACGGAACTTGAACTGGTGCGCACTGAGAACACGGAGCTTGAGTCGGAGCAGCAAAAGGCTCACAT  2999
      uGluGlnGluGluIleArgLeuArgThrGluLeuGluLeuValArgThrGluAsnThrGluLeuGluSerGluGlnGlnLysAlaHisIl  (352)
      .         .         .         .         .         .         .         .         .
3000  CCAAATCACAGAGCTTCTCGAACAGAACATTAAGCTCAACCAGGAACTGGCCCAAAGGTCGAGCAGCATTGGTGGCACCCCGGAGCACAG  3089
      eGlnIleThrGluLeuLeuGluGlnAsnIleLysLeuAsnGlnGluLeuAlaGlnArgSerSerSerIleGlyGlyThrProGluHisSe  (382)
      .         .         .         .         .         .         .         .         .
3090  TCCATTGCGACCGAGAAGGCATAGCGAGGACAAGGAGGAGGAGATGCTCCAGCTAATGGAGAAGCTGGCTGCTCTTCAAATGGAGAACGC  3179
      rProLeuArgProArgArgHisSerGluAspLysGluGluGluMetLeuGlnLeuMetGluLysLeuAlaAlaLeuGlnMetGluAsnAl  (412)
      .         .         .         .         .         .         .         .         .
3180  CCAGCTGCGTGACAAGACTGACGAACTGACCATCGAAATCGAGAGCTTAAATGTGGAACTAATTCGCTCGAAAACCAAGGCTAAAAAGCA  3269
      aGlnLeuArgAspLysThrAspGluLeuThrIleGluIleGluSerLeuAsnValGluLeuIleArgSerLysThrLysAlaLysLysGl  (442)
      .         .         .         .         .         .         .         .         .
3270  AGAAAAACAGGAGAAACAAGAGGACCAGGAGTCGGCGGCCACGGCTACCAAAAGGCGTGGGGATTCGCCGAGCAAAACACATCTAACAGA  3359
      nGluLysGlnGluLysGlnGluAspGlnGluSerAlaAlaThrAlaThrLysArgArgGlyAspSerProSerLysThrHisLeuThrGl  (472)
      .         .         .         .         .         .         .         .         .
3360  GGAGAGCCCTCGCTTGGGGAAACAGCGCAAGTGCACCGAAGGAGAGCAGAGCGATGCCAGCAACAGCGGAGATTGGTTGGCTCTAAACTC  3449
      uGluSerProArgLeuGlyLysGlnArgLysCysThrGluGlyGluGlnSerAspAlaSerAsnSerGlyAspTrpLeuAlaLeuAsnSe  (502)
```

(*continued*)

```
3450 CGAGCTGCAAAGAAGTCAAAGCCAGGATGAGGAGCTAACAAGCCTTAGACAGCGGGTTGCTGAGCTAGAGGAGGAACTCAAGGCTGCAAA 3539
     rGluLeuGlnArgSerGlnSerGlnAspGluGluLeuThrSerLeuArgGlnArgValAlaGluLeuGluGluGluLeuLysAlaAlaLy (532)

3540 GGAAGGCAGATCTCTCACCCCGGAAAGCCGTTCGAAGGAACTGGAGACCAGTCTAGAGCAAATGCAGCGTGCCTATGAGGATTGCGAGGA 3629
     sGluGlyArgSerLeuThrProGluSerArgSerLysGluLeuGluThrSerLeuGluGlnMetGlnArgAlaTyrGluAspCysGluAs (562)

3630 CTACTGGCAAACGAAACTTAGCGAGGAGCGGCAGCTGTTTGAGAAGGAGCGACAGATCTACGAAGATGAGCAGCACGAGAGCGACAAGAA 3719
     pTyrTrpGlnThrLysLeuSerGluGluArgGlnLeuPheGluLysGluArgGlnIleTyrGluAspGluGlnHisGluSerAspLysLy (592)

3720 GTTCACCGAGCTGATGGAAAAGGTGCGCGAGTACGAGGAGCAGTTCAGCAAGGATGGCCGCCTCTCGCCCATTGATGAGCGCGATATGCT 3809
     sPheThrGluLeuMetGluLysValArgGluTyrGluGluGlnPheSerLysAspGlyArgLeuSerProIleAspGluArgAspMetLe (622)

3810 GGAACAGCAGTACTCGGAATTGGAGGCAGAGGCAGCCCAGCTGCGCTCGAGTTCCATTCAAATGCTCGAGGAGAAGGCTCAGGAAATCAG 3899
     uGluGlnGlnTyrSerGluLeuGluAlaGluAlaAlaGlnLeuArgSerSerSerIleGlnMetLeuGluGluLysAlaGlnGluIleSe (652)

3900 CTCACTGCAATCGGAGATCGAGGATTTGCGACAGAGATTGGGTGAGAGCGTTGAGATCCTTACAGGCGCCTGTGAACTCACCTCGGAGTC 3989
     rSerLeuGlnSerGluIleGluAspLeuArgGlnArgLeuGlyGluSerValGluIleLeuThrGlyAlaCysGluLeuThrSerGluSe (682)

3990 GGTAGCCCAACTGAGTGCCGAGGCGGGAAAAAGTCCAGCCAGCTCACCCATCAGCTACCTCTGGCTGCAGAGCACCATCCAAGAGCCAGC 4079
     rValAlaGlnLeuSerAlaGluAlaGlyLysSerProAlaSerSerProIleSerTyrLeuTrpLeuGlnSerThrIleGlnGluProAl (712)

4080 GAAATCGCTTGCCGATTCCAAGGATGAAGCCACCGCCAGTGCCATCGAATTGCTCGGAGGCTCACCATCGCACAAGACAGCCAGCCGGTG 4169
     aLysSerLeuAlaAspSerLysAspGluAlaThrAlaSerAlaIleGluLeuLeuGlyGlySerProSerHisLysThrAlaSerArgEn (741)

4170 AGTATGAGAAGCCTCTCGGTGTGTCCTTGGTGTGAGCATCCCTGTGTCTTCCTCATAATTTGCACTGTATGTCCTGTATATATGTTTCAG 4259
     d

4260 TTTGTCCCTCACATCTAACCATGTCTAATATAAGCTAATTTAATCCTTTTAATTGTATGTTTGTGCTTGTTTAATAAATATAATTTATAT 4349
                                         |(A)n          |(A)n                 |(A)-----

4350 TCATATAGAAATTCATCACATTATCGAAATTCATTGATTTATGATTTCAATAAATATACATTTAATATTTTACAAAAAAATTACTCTTTT 4439
                     |(A)n                        ------       |(A)n      |(A)n

4440 TCGGTTATGAAATTGGCTGCTGGAAATGGTTTTGTTTGCTTATTTTTCACATTTGTATCATTACACGTTTTGCATCTTTATGTTACATCT 4529

4530 TCAATCGTTTTTATTTTGTAAATCATGCCATTTAATGGTCCCTTAAACAGCAATAACCTCACCACTTCGGAAACATCCATCTTTAGCACT 4619

4620 ACACCCTTCGAAAGCTCTCAGTCGGGTCCTTCGCCCACGAACAGTGGCAACAGCAACGCCTACGGCCAATCCCGGCCCAGCTCCGATCAG 4709

4710 CAAGCCCAAGCGGTCCCAGAGTCCCCAACAGGCGGCTGCATCGGAGGGAGAGATAGCCGATTGCGAGACGTCGTCGACGGCGTCCGGCAA 4799

4800 AAGCTTCGAATCCAACAGTAAAACGTCTTGCCTTAGCCACGAGAAGTGCAGCAGTCCGTCGGCACTGAAGGAGGAACTGAAGCGCCTTAA 4889

4890 GTTCTTCGAGCTCTCCCTCAAGGAGCAAATCAAGGATCTGAGTCTGCAGCGGGACGGTCTGGTCATGGAACTGCAGCAGTTGCAGGAGGC 4979

4980 GCGACCCGTGCTCGAGAAGGCCTATGCGGTGAGCCATAGACTTTGTTGATCAGGGAACATATTCTAATCGTATCTGTGGACTCTTCTTTA 5069

5070 GCGAACAACGCATCCAACGCTTCAGCAGCGACTGAACCAATTGGAGCTGCGAAATCGCCATCTGCAGAATGTCATCAAGCAGCAGCAGCA 5159

5160 TTACACGGAGTCCCTGATGCAGCGTAAGTTGAAAAACTACCTACTGACCAAGAATGATAATGTAATATTTATTCATTAGAATCCTGGCGG 5249

5250 CAGCATCAAGTGGAGCTCAACGATTTGCATAGCCGAATCGAGACCAGGGTGTTTTACTGGCCGATCAGACACAGCGATTGCAGAGTGCCG 5339
```

```
5340  ACATCCTGGTGAAGGATCTATATGTGGAGAACTCCCATCTGACGGCCACGGTGCAGCGGTTGGAGCAGCAACGAGCTAGGGTGAACCTCA  5429
5430  TTCACCAGCAGCAGCAACAGCAGCGCCTTGTGGGCGGTGGACTGCCTGGCATGCCTTAGTTTGCCCCACCGGCAAACGTATATAGTTTAT  5519
5520  AGATAATTATGAAAAAGACAAACCTGAGGAGGGAGTGGTGCTCAGCATCGGCAGACATCGAACATGCACCTAACCATAGATCCTTATGAA  5609
5610  TGTTTAGACATATACAATTCTCGGTAGATTAAGTTTGCATACCCGTCGTATTCGTATTCGTACGTTGCGTTTTTTTTGTGAATGAATGTG  5699
5700  AATCCATGTTGTTCGACACGAGAGCACAGCAGCAATAACTAAAGTGACTTTAAACTAAACTTAAACTCACCCACGCGCAAATGAGGAACA  5789
5790  ATCCACACTAGTGTACCAATTTGTAACACATCTAGTAATCGAATCGACTAAACTATTTACACGAGCTACAGGACATATACGATGAAGTAC  5879
5880  CCACGTAGTATATGTTCGTGCAATGTTGACCTTACTAATTGACTACTGAAACAGTTATCGTATATTAATTATATTAGAAGAAACAGTATT  5969
5970  TTAAATTTGTTATGCGTCTGAGTAGGCGAGCACGTTTATCAATGTTTATCACGTGCCCAATCAAATGCATCGGAATTGTTGTTAATTTTA  6059
6060  TTGATAGAGAAAATGGAAATGAGCGTAAAAAATGATCTATGATATTGATATTGATGTAATATTTAACGACAAAAGACCTGTAAAGCTGTA  6149
6150  ACCATACACACGAATCTATGTATTTAAATTGCGATCTAAGTTAGCCAATACTCTTCAATATTGCTTTTGCGAACGCGACTTTTTGTTATA  6239
6240  TCTTCATTCGTCCCAATAACTCACTCGATTTATATGTAAAGAAAAAAAAACTCAACCTCAATCACACGATATCGTGTAATCAGTGCTTAA  6329
6330  ATCAATACTTTCGATCAAAATAGAAGTTTACTTTTTAAAAGTATAAAAAATAATACAACAAAAAAACCAAATATACAATTATTTATAAAA  6419
                                              ------                        |(A)n
6420  CCAAATTGTGATAACTCGTCTTTATTCTAAATAGTTATTAAAATGTTGCGGGAATATAAACTTATTGTTCATAAATACAACTTGCTTATC  6509
6510  AGTTTTTTGGAAATGTTAAGATTTTGTTTCTTATTAAATTAATTGTTATTAAAATTAAAAGATTTATGAAGTTTAAATTATATATTGATA  6599
6600  CGATAAACAATTTATTTTATTGCTTCAAAATATACATTACTTTTTTTTAGGAATTTAAATATCCGTTTTAAGTCTTTTAATTTTTAATTA  6689
6690  GGTTTTAAATATAATATCGATTAAATAGTTGACTCCATTGGAATATCGATACCGCGTCGATGTTTCTTCCAGCTCTATCGGGCACGCGCT  6779
6780  GTTAAAGTTTATTTGTACTGTTAACGCGAATTGGATTAAAATGTTTTGTTTTTTGTTAGTTTTCGGTGTAAATGTGGTTTTAGTGCACTT  6869
6870  AACAACTGGTGAGTGTGCGTTATAGTAAATGAACTTAAATGCAATTACCGAATTC  6924
```

Bsg25D Sequence. Accession, X04896 (DROBSG25D).

References

Boyer, P. D., Mahoney, P. A. and Lengyel, J. A. (1987). Molecular characterization of *bsq25D*: a blastoderm-specific locus of *Drosophila melanogaster*. *Nucl. Acids Res.* **15**:2309–2325.

Roark, M., Mahoney, P., Graham, M. and Lengyel, J. (1985). Blastoderm differential and blastoderm specific genes of *Drosophila melanogaster*. *Dev. Biol.* **109**:476–488.

8

Chorion Protein Genes: *Cp36, Cp38, Cp15, Cp16, Cp18, Cp19*

X-chromosome Cluster
*Cp*36 and *Cp*38

Chromosomal Location:
X, 7F1-2
Synonyms: *S36* and *S38*

Map Position:
1-[23]

Products

CP36 and CP38 (chorion proteins of 36 kD and 38 kD) are two of the six major protein components of the egg chorion; the other four are the product of genes on chromosome 3 (Petri et al. 1976). CP36 and CP38 are probably the main components of the innermost chorionic layer and the internal region of the thick endochorion (Parks and Spradling 1987; Orr-Weaver 1991). An N-terminal segment of approximately 20 amino acids is probably a signal peptide that is cleaved upon protein secretion (Waring and Mahowald 1979). As is true for all major chorion proteins, CP36 and CP38 are rich in Gly, Ala, Pro and Ser (in CP36, these amino acids constitute 40% of the residues and in CP38, 50%) and in Tyr, an amino acid that is extensively cross-linked in the mature chorion (Petri et al. 1976). Both proteins have runs of Ala and Gly-His, but overall sequence similarity is not striking.

Organization and Expression of the Cluster

Cp36 and *Cp38* lie within a 13 kb segment of DNA and are part of a cluster that includes six tandem transcription units. *Cp36* is positioned centromere distal, upstream of *Cp38*; both genes are transcribed toward the centromere. Downstream of *Cp38*, approximately 1.4 kb away and transcribed in the opposite orientation is *ovarian tumor* (see *otu* Fig. 23.1). The function of the

Cp36

```
-802  TTTCACATTGAGACGAAACAATCCACCGAAAATCCATAAAATATAAGAATGTTGCATTTTATTTTTAAAAATAAAGATGCCTTTTAAGAG  -713

-712  GAATAACTTAAATGTCTTTAATACCTTTGAATTTAATTATATGGCTAATAAACACAAACTTAAAGCTTAAAACTGCATCGAATTGAATGC  -623

-622  GGTTATAAATGTACTTATATATCTAATATAATCTGCTAATATGGTTTACATGGTATATCTTTCTCGGAAATTTTTACAAAAATTATCTAT  -533

-532  TCATATATCTCGAGCGTAAGATATTTATCAGTTTATAGATAACATCTTTAAATTTGGGTGATTAAAAAAAAACATTGCTGCAGGGCATGT  -443

-442  TTATGTACACATTTCAGTATAAGTCCCAAGTTAAAATGCAATGTAAAAACATATAAAGGATATTAACTTCAAACCCAAAGGATTGCAGAG  -353

-352  AGATTGCAGCACAGCTGTAATCATCGCAACAAGGCAACCAAAACGAGACTCTCGTAGCGTTGGCATCATATTCGATCTTTGGAAGAGCTA  -263

-262  TGATTCAAGCCAAGGGAAACAACTGCCAAAAAATAGAAGATTGCGACGAGCGGAAAGCAGAGTGGTGCACCACGGTGCATAGGTGCATAG  -173

-172  GAGGTTGGTTGTCTAGAGATCTGGGCACGATGGCGAGACAAAGATGCGGCGCAAAATCGGAAATGGAGATGGATCACGTAGCCGGCCATG  -83
                                                                       ------              ------
                                                                       ======
                                                                       -->-30
 -82  GCGGGCAGCGATTATAGGCGCTATAAAGAGCCGGAGTCATCGCCAGACAGCGAGCAGTAGACGATCCACACGTAAACGGCAACATGCAAC  7
                          ------                                                            MetGlnL  (3)

   8  TCGGTCTCTGGTTTGGCATTTTGGCCATCGCCGCCGCGCCGGTTAGTAGCTTTTATCCATGGCATTCGCATTGGGCATCCCGCTAATCTC  97
      euGlyLeuTrpPheGlyIleLeuAlaIleAlaAlaAlaPro                                                  (16)

  98  GCAAACTCTCTCTCTCTCTCTCTTACCTCCTCCTCATTAGCTGGTGAGCGCTAACTATGGTCCCGCTGGCGGACACGGACACGGACAT  187
                                                 LeuValSerAlaAsnTyrGlyProAlaGlyGlyHisGlyHisGlyHis  (32)

 188  GGACATGGACATGGACAGTACCTGTCCGGTCCCAATGCCGGACTCGAGGAGTACGTGAATGTGGCGTCTGGTGGCAACCAGCAGGCTGCC  277
      GlyHisGlyHisGlyGlnTyrLeuSerGlyProAsnAlaGlyLeuGluGluTyrValAsnValAlaSerGlyGlyAsnGlnGlnAlaAla  (62)

 278  AATCAGATCGCCTCACAGGCCGAGATCCAGCCCACGCCGGAGGAGGCCCGTCGTTTGGGTCGCGTCCAGGCCCAACTTCAGGCCCTCAAC  367
      AsnGlnIleAlaSerGlnAlaGluIleGlnProThrProGluGluAlaArgArgLeuGlyArgValGlnAlaGlnLeuGlnAlaLeuAsn  (92)

 368  GCCGATCCCAACTACCAGAAGCTGAAGAACTCCGAGGATATTGCCGAATCTCTGGCCGAGACCAATCTGGCCAGCAATATCCGTCAGGGC  457
      AlaAspProAsnTyrGlnLysLeuLysAsnSerGluAspIleAlaGluSerLeuAlaGluThrAsnLeuAlaSerAsnIleArgGlnGly  (122)

 458  AAGATTAAGGTGGTGTCGCCACAGTTCGTTGACCAGCATCTGTTCCGCTCCCTGTTGGTGCCATCGGGCCACAACAACCACCAGGTGATC  547
      LysIleLysValValSerProGlnPheValAspGlnHisLeuPheArgSerLeuLeuValProSerGlyHisAsnAsnHisGlnValIle  (152)

 548  GCCACCCAGCCCCTGCCACCAATCATTGTCCACCAGCCTGGTGCACCACCAGCCCATGTGAACAGCGGCCCACCGACTGTGGTGCGCGGC  637
      AlaThrGlnProLeuProProIleIleValHisGlnProGlyAlaProProAlaHisValAsnSerGlyProProThrValValArgGly  (182)

 638  AATCCGGTGATCTACAAGATCAAGCCCTCGGTCATCTACCAACAGGAGGTGATCAACAAGGTGCCCACTCCGCTGAGCCTCAACCCCGTC  727
      AsnProValIleTyrLysIleLysProSerValIleTyrGlnGlnGluValIleAsnLysValProThrProLeuSerLeuAsnProVal  (212)

 728  TACGTGAAGGTCTACAAGCCCGGCAAGAAGATCGAGGCTCCACTGGCCCCGGTGGTTGCACCCGTCTACAGCCAGCCCAGGGAGTACAGC  817
      TyrValLysValTyrLysProGlyLysLysIleGluAlaProLeuAlaProValValAlaProValTyrSerGlnProArgGluTyrSer  (242)

 818  CAGCCCCAGGGTTATGGTAGTGCCGGAGCTGCTTCCTCCGCCGCCGGTGCCGCCTCCTCTGCCGATGGCAATGCCTACGGCAACGAGGCT  907
      GlnProGlnGlyTyrGlySerAlaGlyAlaAlaSerSerAlaAlaGlyAlaAlaSerSerAlaAspGlyAsnAlaTyrGlyAsnGluAla  (272)

 908  CCACTGTACAACAGCCCCGCGCCCTATGGCCAGCCCAACTACTAAGGTGCTCATCCTGGGCATGGGTTGTTCCTCAGCTGCGACAGCTGG  997
      ProLeuTyrAsnSerProAlaProTyrGlyGlnProAsnTyrEnd                                              (286)
```

```
 998  TTTAATTTAAATTTTTGTTTTTTTTTTTCTTTGCCGAACTACTGAGCGCAAATAAATGAAAAAAAAAATACTTGAAACGTAAACGTTTTGT  1087
                                                            ------            |(A)n
1088  CATCAATTATTTTCGACCGGAAGGGGCTACCTGAGTGGCAATGAAGCACACAGATTAAGCACATATTTATGAATATATAAATATATACGA  1177
1178  ATGCATGAGAGAACAAAAAATTATATCTAGTTTTCTTCAAAAATAAATAATAAAGGGGAAAATGGTATAAAGTATGAACATAAAATTATG  1267
1268  AACGAGTATAAATAAGTTTGTAATTGAAATCTCTACGGTCATACAAGTATTTTAACTATTCTATAAATATGCATAAACATGGTACGCATT  1357
1358  TTATGAGATACAATTCGAAAGATATTGGATAGCATTATCATGCATGAAATTATTTCACACTCTTGTGATGGTAGCTATCTTATTCCAGTT  1447
1448  ATCGTTTAATTGCAAAGAATGGGATCAAAAGGTCATCTTTATCAACATATTGTTGATTCCGGAATGAATAATAAATAATAACATAACTAT  1537
1538  GAATTAATGGAGCAACTATAATTTTACGGCCTCTTTTCTTTTAAACAAAGAATATAGCACTTTTAATGCATTAAATACGTATTTAAACCT  1627
1628  TTTCTTTTGAAACGCCAAATTCATATTAGAGTTTCATAAGATTGTTTTAAAACATAACAACATAATAATTGAAGAATTGGAAATCTTTTT  1717
                           __EcoRI
1718  AGGTGTTTGTAAGCCTTTGA  1737
```

Cp36 SEQUENCE. Accession, X05245 (DROCHORS3). The *Cp36* sequence ends at the *Eco*RI site at which the *Cp38* sequence begins. Underlined are the regulatory chorion hexanucleotides, approximately 60 bp upstream of the transcription initiation site (see *Cp15 Promoter*).

three other transcription units in this cluster is unknown (Spradling et al. 1980).

The 13-kb segment is at the core of an 80–100-kb region that undergoes DNA amplification in the polyploid follicle cells prior to the time of programmed expression of *Cp36* and *Cp38*. This amplification results in a 15-fold increase in copy number (Spradling 1981). The amplification control element (ACE1), a *cis*-acting element, resides within a 3-kb segment that includes *Cp38*; a necessary portion of ACE1, at its upstream end, extends from −580 to −80 in the *Cp38* Sequence. In this region of *Cp38* are found the repeating pentanucleotide AATAC and related sequences (similar sequences are found in ACE3, the amplification control element of the third chromosome chorion–gene cluster). Whether other sequences within *Cp38* are also necessary for amplification is not known (Spradling et al. 1987). The mutation *ocelliless* (*In*(*1*)*oc*), is an inversion with one breakpoint 5 kb upstream of ACE1. Although homozygotes for this mutation amplify *Cp36* and *Cp38* in the new location, they do so to a reduced extent. The genes upstream of the breakpoint, which are left in place, fail to amplify but are correctly regulated (Spradling et al. 1979; Parks et al. 1986).

Developmental Pattern

All of the genes in the X-chromosome cluster are expressed exclusively in ovarian egg chambers during the last 6 h of oogenesis, a time when these cells are actively involved in the synthesis and deposition of the egg shell. *Cp36* and

Cp38 are transcribed during stages 10–13 of oogenesis (the chorion genes of the third chromosome cluster are expressed mainly during stages 13 and 14). Individual genes, however, have distinct temporal and spatial patterns of expression within the stages and cells mentioned, suggesting that each gene is independently regulated. With respect to *Cp36* and *Cp38*, in particular, *Cp38* transcripts accumulate in stages 11 and 12 while *Cp36* RNA is highest a little later, during stages 12 and 13 (Spradling and Mahowald 1979, Mahowald and Kambysellis 1980; Parks et al. 1986; Parks and Spradling 1987; Fenerjian et al. 1989).

A precise series of bursts of protein synthesis ensures that the different chorionic proteins are secreted in quick succession; this is accomplished by very fast mRNA turnover rates. Massive synthesis of each protein, on the other hand, depends on high levels of the corresponding mRNA. Because the mRNAs are short-lived, their accumulation depends on differential gene amplification in follicle cells, as described above (Mahowald and Kambysellis 1980; Parks et al. 1986; Parks and Spradling 1987).

Promoters

Approximately 60 bp upstream of the start of transcription, both *Cp36* and *Cp38* carry the sequence TCACGT, the chorion hexanucleotide, which is thought to be involved in the regulation of all major chorion genes in *Drosophila* as well as other insects (Kalfayan et al. 1985; Kafatos et al. 1985).

Cp36

Gene Organization and Expression

Open reading frame, 286 amino acids; expected mRNA length, 1,004 bases. The approximate position of the 5′ end was defined by primer extension; the exact position was suggested on the basis of sequence elements. The 3′ end was determined from a cDNA sequence. There is one 91-base intron after the Pro-16 codon (Spradling et al. 1987). There is a well-defined region of transcription termination between 0 and 210 bp downstream of the poly-A addition site (*Cp36* Sequence) (Osheim et al. 1986).

Promoter

An 84-bp segment (−162 to −79), sufficient for correct temporal expression, was defined by studies of germ line transformants carrying a reporter gene and fragments of the 5′ regulatory region. The reporter gene consisted of *lacZ* associated with the *Hsp70* basal promoter. These studies also suggest that the 84-bp segment may contain two or more regulatory elements: while the upstream half of this segment controls expression at the posterior pole of the

Cp38

```
      ____EcoRI    .         .         .         .         .         .         .         .
-822  ATTCCTAATTGGAATAGCTAAAGATCCATATTTCATCTTCAAATCTCTTTGCAACTAGAGATTTATTTTATCTGGCAATTATTAAGTATA  -733
        .         .         .         .         .         .         .         .         .
-732  CATTTTTATATGGTACTTTAAACTGATGGTTTAATCAGTTACATGGATTTTCTAAATTAAAAATGGTCATGTGAAGATAGCCACTCTTCT  -643
        .         .         .         .         .         .         .         .         .
-642  AACAATCTAATCACATTTATAGTAAGAAATACAATACAATACAATACAATACAATACAATACAATACAATAGAAAGACAATCGAATCTGC  -553
                                 --------------------------------------------
        .         .         .         .         .         .         .         .         .
-552  GCAATCCGTGTGAAATTCAAGGACTACAGCTGGGTGGCTAATCATTTCCCCCTATCCACTTACACCTCGGATTACCTCTTATTCCGACTC  -463
        .         .         .         .         .         .         .         .         .
-462  CCGGAGTCTTGTGTCTGCCAATGCGGAACTATTTTCGCTATCTGAACAGACGTTCGGACCTCGATATGCGGCAAAGATTCACAGCCCGGC  -373
        .         .         .         .         .         .         .         .         .
-372  TGTTGATTCCGATTCGGTGGCAATGTGTTCGTTGTTATTGTAAAACGGGCAATGGCAACTGGGCAGTGGGCAGTGGGGTTTTCGGGTTGT  -283
        .         .         .         .         .         .         .         .         .
-282  GGCTTCTACGTAAGTGGAAGAGACGCCGTGATATGCGCTGGCAGCGATGCGTGCGATCTATCAACATTTGGCCATGTTTCGTTCGCATCG  -193
        .         .         .         .         .         .         .         .         .
-192  CGTGGCCCGGAGCGGAACAGCCGGCACCGGAGTTGGCATCAATCCAAATGTCACGTACCCGGAGCCGGAGACGCGTCCGGAGCATATTT  -103
                                                ---- ----- ------                           -----
        .         .         .   -->-76.   .         .         .         .         .         .
-102  AAAGTAGTCGGCCACCAATGGAGGGCAGCAGAAGACAGCAGACAGTCCAAGCGGGAGCACACCAGAAGCCGAAGAGCAACTGGAACTGCA  -13
      ---
        .         .         .         .         .         .         .         .         .
 -12  ACTGGGAGACAAGATGACGAGATCGACCTACATTTGGGCGCTGGCCGCCTGCCTGATCGTAAGTGTTCAAGCTCGAAATCTTAGGATTAA  77
                   MetThrArgSerThrTyrIleTrpAlaLeuAlaAlaCysLeuIle                                (15)
        .         .         .         .         .         .         .         .         .
  78  TCCCAAGAAAACCAAGTCTATCAATTCTGACTGCTTTCGTTTGTGCCATGTAAATCGTACATGAAAAGCAAATTGACTTTCCTTTAAATT  167
        .         .         .         .         .         .         .         .         .
 168  ACTTGAAACGGAATCAAGCTATCTATCGATGCTAGACTTATTTTAAGTATATGTATATGTCGATCCAATTCTAATCCACCCCCCCCCCCT  257
        .         .         .         .         .         .         .         .         .
 258  CAATTTACTTTTAGGCCTGTGCAAGCGCCAACTACGGCAGTTCCCAGGGCTATGGACCCGAGTCCGGAAGCGGTGCCTCCGATGGCGGTG  347
                    AlaCysAlaSerAlaAsnTyrGlySerSerGlnGlyTyrGlyProGluSerGlySerGlyAlaSerAspGlyGlyA  (41)
        .         .         .         .         .         .         .         .         .
 348  CTGATGCCGCTTCAGCGGCCGCAGCAGCTGCCGGCGGTGCCGGTGGAGCTGGTGGCGAGTACGGTGGTGCTAACGCCGGTGCTGGTGCTC  437
      laAspAlaAlaSerAlaAlaAlaAlaAlaAlaGlyGlyAlaGlyGlyAlaGlyGlyGluTyrGlyGlyAlaAsnAlaGlyAlaGlyAlaL  (71)
        .         .         .         .         .         .         .         .         .
 438  TCGAATCCGGAGCCGATGCCGCCGGTGTGGCACAGGCTGGCCAGAGCAGCTACGGATCCGACCAGAACATTCCGTACAAGCCGGTGAACA  527
      euGluSerGlyAlaAspAlaAlaGlyValAlaGlnAlaGlyGlnSerSerTyrGlySerAspGlnAsnIleProTyrLysProValAsnT  (101)
        .         .         .         .         .         .         .         .         .
 528  CCAAGGGTAACACCCTGACCTCATCGATCACCTACCCGCAGAACAAGGGCGAGATCCTCATCCATCGTCCCGCTCCCATCATTGTCAAGC  617
      hrLysGlyAsnThrLeuThrSerSerIleThrTyrProGlnAsnLysGlyGluIleLeuIleHisArgProAlaProIleIleValLysA  (131)
        .         .         .         .         .         .         .         .         .
 618  GTCCGCCCACCAAGGTGCTGGTGAACCATCCACCATTGGTGGTTAAGCCCGCTCCCGTGGTGCTCCACAAGCCCCCAGCAATCGTTCTCC  707
      rgProProThrLysValLeuValAsnHisProProLeuValValLysProAlaProValValLeuHisLysProProAlaIleValLeuA  (161)
        .         .         .         .         .         .         .         .         .
 708  GCAAGGTCTACGTCAAGCACCACCCACGTCGCGTCAAGGTTGAGCCCGTGTTCGTCAATGTGGTCAAGCCCCCAGCAGAGAAGTACTTTG  797
      rgLysValTyrValLysHisHisProArgArgValLysValGluProValPheValAsnValValLysProProAlaGluLysTyrPheV  (191)
        .         .         .         .         .         .         .         .         .
 798  TCAACGAGAACAAGCAGGGCTACGGACAGGGCTCGCAGTCCCACGGACACGGCCATGGACACGGTGGCCATGGACACGGACACAGCGGAC  887
      alAsnGluAsnLysGlnGlyTyrGlyGlnGlySerGlnSerHisGlyHisGlyHisGlyHisGlyGlyHisGlyHisGlyHisSerGlyH  (221)
        .         .         .         .         .         .         .         .         .
 888  ACGGACACGGTGGACACGGTGCTGGACCCCATGGTCCTGGACCCCATGACGGTGGCCGTGCTCTGCCCGCCTACGCTTCGGGAGCTGATT  977
      isGlyHisGlyGlyHisGlyAlaGlyProHisGlyProGlyProHisAspGlyGlyArgAlaLeuProAlaTyrAlaSerGlyAlaAspS  (251)
```

```
 978  CCGCTGCCGCCAGCGCTGGCTATCAGCTGCTCCAGAGCGGCAACCAGGGTCTGTCCGCTCTTGCCAACATCGCCGGCGAGCGTGAGGGTC  1067
      erAlaAlaAlaSerAlaGlyTyrGlnLeuLeuGlnSerGlyAsnGlnGlyLeuSerAlaLeuAlaAsnIleAlaGlyGluArgGluGlyP  (281)

1068  CCTATGGTCCCGCTCCAAGCCATCAGCACTATAGCGCCGGTCCAGCCGGACATGGCGGCTATGCTGCTCCCGCCTATTAGGTAACAGATG  1157
      roTyrGlyProAlaProSerHisGlnHisTyrSerAlaGlyProAlaGlyHisGlyGlyTyrAlaAlaProAlaTyrEnd           (306)

1158  CGGAGGAGTTACGGATTGGATGACTGCTGCGGCTCCGGAATCAACTGAAGCGGCTGGTTTAGTCATTCGCTTATCCGGCTGATTAGTTAC  1247

1248  TATGTTTTTTTTTACAAAAAAAAAAAAAAAAACACAGCCTGATCGACCAACGCCCATGCCTACGCCCACGCCCACTCATGCACACCCCAA  1337

.338  TACCACCCACTCACCCATTCAACGGCCCAGGAGGGGCGTGGCACTCAGGTTTCTTTGCAAAACAAATAAAAAATTTGAACAAAAAAAAAA  1427
                                                                          ------
 428  AACAATTATACCCAAGCTGACTGTTGTTTTCGATGAAGGGTGAAATCTAGA  1478
                |(A)n
```

Cp38 Sequence. Accession, X05245 (DROCHORS3). The *Cp38* sequence begins at the *Eco*RI site at which the *Cp36* sequence ends. The bases underlined between −615 and −572 are part of ACE1. Also underlined are the regulatory chorion hexanucleotides, approximately 60 bp upstream of the transcription initiation site (see *Cp15 Promoter*).

egg chamber and the proximal half controls expression at the anterior pole, expression over the entire egg chamber requires the intact segment. A more distal element (−1,243 to −457), even though apparently not required, was found to allow weak expression (Tolias and Kafatos, 1990).

Cp38

Gene Organization and Expression

Open reading frame, 306 amino acids; expected mRNA length, 1,290 bases. The position of the 5′ end was determined by primer extension and S1 nuclease mapping. The 3′ end was obtained from a cDNA sequence. There is one 226-base intron after the Ile-15 codon (Spradling et al. 1987). There is a well-defined region of transcription termination between 220 and 585 bp downstream of the poly-A addition site (*Cp38* Sequence) (Osheim et al. 1986).

Chromosome 3 Cluster
Cp15, *Cp16*, *Cp18* and *Cp19*

Synonyms: *S15*, *S16*, *S18* and S*19*

```
     1                                                     50                                                      100
Cp15 .MKYLIVCVT LALFAYINAS PAYGNRGGYG .......... .GGYGGGYG. ......PVQR VVYEEVPAYG PSRG.....Y NSYP...RSL RSEGNGG...
Cp18 MMKFM..CIC LCAISAVSAN SYGRPRGGYG .......... .GAPVGGYAY QVQPALTVKA IVPSYGGGYG GNHGGYGGAY ESVPVPVSSV YSGANVGSQY
Cp16 ....MSATLR LLCLMACCVA LAVANRPHYG .......... .G........ .......... ......SGYG ASYGDVVKAA ETAEAQASAL TNAA......
Cp19 MNKFATLAVI FCACIVGSCY ANYGGQQSYG QRSYGQDSSA ASAASSAAAA GAEGQQRYER PVEIIAGGYR GSYAPEILRP IQVSGGYGGE RRGYNGGNYR
 CON --K------- L--------- -----R--YG ---------- -G-------- ---------- -V-----GYG -S-G------ ---------- ----N-G---

     101                                                   150                                            193
Cp15 SAA....... AAAAASAAAV NPGTYKQYAI PSYELDGARG YEIGHGYGQR AY*....... .......... .......... .......... ...
Cp18 SGS....... GYGGAPPVDA QAIALAKLAL AAPSAGAPLV WKEAPRYAQP VYPPTSYVNQ EYGHSEKVKG GSAAAAASSV AAGKKGYKRP SY*
Cp16 .GA....... AASAAKLDGA DWYALNRYGW EQGRPLLAKP YGPLDPLYAA ALPPRSFVAE VDPVFKKSQY GGSYGENAYL KTDAKLGVVA I*.
Cp19 RAGYGPRWTV QPAGATLLYP GQNNYKAYVS PPEYSKVILP IRPAAPVAKL FVPENQYGNQ YVSQYSAPRS SGY*...... .......... ...
 CON ---------- ----A----- -------Y-- ---------- ---------- --P------- ---------- ---------- ---------- ---
```

FIG. 8.1. Comparison of amino acid sequences for the chorion proteins in the chromosome 3 cluster. The sequences were aligned using the GCG *Pileup* program. The CON(sensus) line indicates positions at which three or more of the sequences agree.

Chromosomal Location:
3L, 66D11-15

Map Position:
3-[26.5]

Products

CP15, CP16, CP18 and CP19 (chorion proteins of 15, 16, 18 and 19 kD) are four of the six major chorion proteins; the other two are products of *Cp36* and *Cp38*, which occur on the X-chromosome (Petri et al. 1976). These proteins are localized mainly in the exochorion and in the outer portion of the endochorion (Parks and Spradling 1987). The 20 or so N-terminal amino acids in each protein probably represent signal peptides (Waring and Mahowald 1979). These basic proteins are rich in Gly, Ala, Pro and Ser (residues that represent approximately 50% of the total) and Tyr (Petri et al. 1976). As in chorion proteins CP38 and CP36, there are Ala-rich stretches but no pattern of strong sequence similarity (Fig. 8.1).

Organization and Expression of the Cluster

This cluster comprises four transcription units arranged in tandem (Fig. 8.2). In size, developmental expression and differential amplification, it is quite comparable to the X-chromosome chorion-gene cluster (Spradling et al. 1980).

The conserved position of introns (in all chorion genes but *Cp16*) and the presence of certain sequence elements in the 5′ regions of the major chorion protein genes are suggestive of a common phylogenetic origin for all chorion genes in this cluster (Levine and Spradling 1985; Spradling et al. 1987; Wong et al. 1985). Although various *Drosophila* species show considerable divergence with respect to specific chorion-gene sequences, the disposition of the genes and general organization of the two clusters are remarkably conserved (Fenerjian et al. 1989).

Amplification (see Chorion-Gene Cluster on the X-chromosome) reaches 60-fold in the third choromosome cluster, and the amplification control element, ACE3, resides in a 3.8 kb fragment that includes the genes *Cp15* and *Cp18* (Levine and Spradling 1985). Within this segment, ACE3 sequences essential for amplification have been localized to the interval −673 to −163 of *Cp18* (*Cp18* Sequence). A 440 bp segment is capable of autonomous amplification,

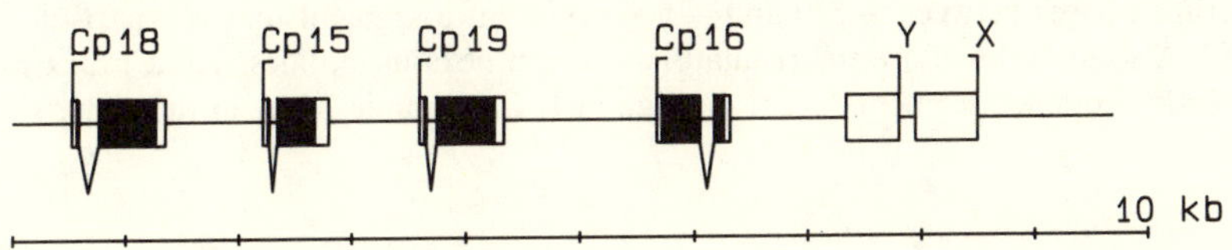

FIGURE 8.2. Chromosome-3 cluster organization. X and Y are two nonchorionic transcription units.

Cp18

```
-563  AAGCTTAGTGCGGCAGTTTGGAAAGTGGAACGGTTGTGTTTATAATTTTATTGTAATTTTATCTCAATTTTTTTTGCTTTTGTATATAAA  -474
                                       ---- --  -- -- ------- -- ---

-473  TTCTACCAACGCAGCAGAATTTTCAGGCCACTGCCTTGACTTCACTGTGTCACTGAAAAATCGGTGTCAAGCTCTCGGCACCGTGGGGCA  -384

-383  AAGCAACTGCAATACTGATCGAAACTATGCGGATCCGGAGCACGAAGAGTCATGCGGTCGGAATCTTACGTAATGGGTCTCGTCTCTGGT  -294

-293  AGACGATGGCGTAAGCACAGACGCCTGCTATCTGGACCGGCCCGAATTGAGAGCCAGCATTTTGGCCAGTGCGGATTCGGCCTGGCTGCA  -204

-203  CGTCTCCGGCGGCGTCTCAAGATTGCTGGACAAAGAGGCGAGGCCTGGAACTGCGTCTCCGGGAACCCGGAGAGCCGAAACTTGCATCAT  -114

                                                                            -->-43
-113  ATTCGTCACGTAAGAGTTGGGCCTCTGCCTGGATCTGGTATAAAAACAAAACATTGCGCCAGAATAAGACATTAGTTACCTTCGCATCGA  -24
           ------                                --------

 -23  TCAACTAACCAACTCAGCCTCAGAATGATGAAGTTCATGGTAAGCTTAAGTTCCAATATTGTTTCACCTCAACACCTCAACTGCGTCCAG  66
                              MetMetLysPheMet                                                   (5)

  67  TATGATCCTTTTAATAAAATATAACTACATATTATAATAATTTGAAATAATATGATTGGATCTTTCTTTTCTAATGCACTTCAGGCTATA  156

 157  CCCAAATTAATTGAATTTTTTCTTGAATCCCTTAGTGCATCTGCCTCTGCGCCATCTCTGCCGTTTCGGCCAACTCCTACGGACGTCCCC  246
                                         CysIleCysLeuCysAlaIleSerAlaValSerAlaAsnSerTyrGlyArgProA  (24)

 247  GTGGTGGATACGGTGGTGCCCCAGTCGGTGGCTATGCCTACCAGGTGCAGCCTGCCCTGACCGTTAAGGCGATCGTTCCCTCATACGGTG  336
      rgGlyGlyTyrGlyGlyAlaProValGlyGlyTyrAlaTyrGlnValGlnProAlaLeuThrValLysAlaIleValProSerTyrGlyG  (54)

 337  GTGGATACGGCGGAAACCATGGAGGATATGGCGGTGCCTACGAGTCGGTGCCTGTGCCCGTGTCCTCTGTCTACAGCGGTGCCAATGTGG  426
      lyGlyTyrGlyGlyAsnHisGlyGlyTyrGlyGlyAlaTyrGluSerValProValProValSerSerValTyrSerGlyAlaAsnValG  (84)

 427  GATCTCAGTACTCCGGTTCCGGCTACGGCGGTGCCCCACCAGTCGATGCCCAGGCCATTGCCCTCGCCAAGCTCGCCCTGGCCGCTCCCA  516
      lySerGlnTyrSerGlySerGlyTyrGlyGlyAlaProProValAspAlaGlnAlaIleAlaLeuAlaLysLeuAlaLeuAlaAlaProS  (114)

 517  GCGCTGGAGCTCCTCTGGTCTGGAAGGAGGCTCCCCGCTACGCCCAGCCCGTCTATCCCCCCACCAGCTACGTGAACCAGGAGTACGGAC  606
      erAlaGlyAlaProLeuValTrpLysGluAlaProArgTyrAlaGlnProValTyrProProThrSerTyrValAsnGlnGluTyrGlyH  (144)

 607  ACAGCGAGAAGGTGAAGGGAGGCTCCGCAGCCGCTGCTGCCAGCTCCGTGGCCGCCGGAAAGAAGGGCTACAAGAGGCCCAGCTACTAAG  696
      isSerGluLysValLysGlyGlySerAlaAlaAlaAlaAlaSerSerValAlaAlaGlyLysLysGlyTyrLysArgProSerTyrEnd  (172)

 697  TGGCAAAACGTTGAACAGTGAACCAAAAACTTACCTGCCAATAAGGAACTAGGTCATAATAATAAAAGCCAAAACATCAAGACTTAAAAT  786
                                                                 -------                 |(A)n

 787  TTTGAGTACTGTATTCTTGCTGGGTTTTTAGTTTCGGGCCAAGAGTTGAG  836
```

Cp18 SEQUENCE. Strain, *Oregon R.* Accession, X02497 (DROCHORSG). The underlined bases between −530 and −500 represent a segment that is a part of ACE3. Also underlined are the regulatory chorion hexanucleotides. The *Cp15*, *Cp18* and *Cp19* sequence segments occur contiguously in genomic DNA in the order shown in Fig. 8.2.

but sequences outside of it also seem to influence the process (Orr-Weaver and Spradling 1986; Carminati et al. 1992).

Developmental Pattern

Transcription of these genes occurs during oogenesis, a little later than transcription of *Cp36* and *Cp38*: *Cp16*, *Cp18* and *Cp19* are expressed mainly during stage 13 and to a lesser extent during stage 14; *Cp15* is expressed almost exclusively during stage 14 (Mahowald and Kambysellis 1980; Parks and Spradling 1987; Fenerjian et al. 1989).

Promoters

As in *Cp36* and *Cp38*, the sequence TCACGT is found approximately 60 bp upstream of the transcription initiation site (except for *Cp16*, in which it is found 80 bp from the 5′ end). Other sequence elements in the neighborhood of this hexanucleotide are also present in *Cp18*, *Cp15* and *Cp19* (Levine and Spradling 1985; Wong et al. 1985).

Cp15

Gene Organization and Expression

Open reading frame, 115 amino acids; expected mRNA length, 519 bases. One 71-base intron is present after the Leu-4 codon. The approximate position of the 5′ end was determined by S1 mapping, and the first nucleotide transcribed was assigned on the basis of similarities to canonical *Drosophila* sequences. The 3′ end was determined from the sequence of a cDNA clone (*Cp15* Sequence) (Levine and Spradling 1985; Wong et al. 1985).

Promoter

The 73-bp segment of DNA from −162 to −90 seems to be necessary and sufficient for correct tissue and temporal specificity in the expression of this gene; sequences between −858 and −162 may contribute to an elevation of the transcription rate. The TCACGT chorion hexanucleotide from −104 to −99 is indispensable for transcription as well as for follicular specificity. Another positive *cis*-acting element, between −116 and −107, activates expression late in oogenesis (stages 13 and 14). Element(s) between −162 and −124 act negatively to suppress early transcription (stages 11, 12 and early 13) (Mariani et al. 1988; Shea et al. 1990). By gel retardation assays, two protein–DNA complexes were detected that involve the −116/−107 and −104/−99 sites; there is partial overlap of the binding sites. Two cDNAs produce proteins that bind specifically to these sites: chorion factor I (CFI) binds to the chorion hexanucleotide while CFII binds to the late activator site (−116/−107). Both

Cp15

```
-857  ATCTGCATATCTTAGCTGAATTGGCAAAGACTTGCGGTTCATTGCAATGCCAAGCGATACTTTGAGCCAGCAAAAATTTCTTGGTTTCGT  -768

-767  AGTTAAATGAAAATGCTGCTTAAAGTGCTAAAGAATAATTGTCATGGCGAATGAAGCTGCAAAGCTAAAACTAAATTAATTTGTGGGGCC  -678

-677  AATTTAAAACTATAGTTTGTCAAAAGAGCCTTGACTTTTTTAAGTCACCATAAGTAAAGAATCTATTACATAAAACGCGATTAGATAGAA  -588

-587  TATAGTTTTGCTTGAAATTATGTTTTTGTAAAATTTCAAAATGATTGAAATACTTTAAAATGTTTTAGTTATAATTTTAAGTTTTGTATG  -498

-497  TGACTAGTAATCACTTTAAAGGAATGACTCTATATAGGTTTTATCAGAAAAACCGGCTGGAACCAGTTCTAGAAGAATCCTCACTTAGAC  -408

-407  AAGCCAAGTTCCGGACACAACCGATCTGGAAACCATTACCCCCGAGAATGTGGATAATATAAAGTTCAATTCAACAAATTTTGGAGTGTA  -318

-317  TTCGAAAATAAACGCGTTCGTGGTTCCCATTTGGAAGAGTCGCGTGTTCGTAGTGCTATCACCACCCAACACCCGGTAGAATAGCACATC  -228

-227  GCGTAACCAAGCGATTTTATAATGGCTTGACAACAAGTACATAAATCAAATGTGAGTATATTCCAGCCGGGCAATTATGAAATGCCATTT  -138

-137  CTGGGCTGAAACAGAACAATTAGTGTATATAGGTCACGTAAATGTCCAGGCTAAAATTTGCGTATAAAAGCGAGCGTTTCTGGTCGGTAA  -48
                          ---- ---------cf2------cf1                  -------
       -->-44
 -47  ATCATAGTTTGATTGATTACCCCAAACCAACAAAACTAAGCACTCACCATGAAGTACCTGGTAAGTTGTGGTAGTCCCCGTGAAGGAGTG  42
                                                       MetLysTyrLeu                                 (4)

  43  GCAGCCAACTGATCCTCCGGATTTCCCCTTTTCACCTTCAGATTGTCTGTGTTACCCTGGCCCTTTTCGCCTACATCAACGCCAGCCCAG  132
                                              IleValCysValThrLeuAlaLeuPheAlaTyrIleAsnAlaSerProA  (21)

 133  CGTACGGCAACCGTGGAGGTTATGGTGGTGGCTACGGTGGTGGCTACGGTCCTGTTCAGCGCGTCGTCTACGAGGAGGTGCCCGCCTACG  222
      laTyrGlyAsnArgGlyGlyTyrGlyGlyGlyTyrGlyGlyGlyTyrGlyProValGlnArgValValTyrGluGluValProAlaTyrG  (51)

 223  GACCATCCCGTGGCTACAACAGCTATCCCCGCAGCCTGCGATCGGAGGGTAATGGAGGAAGTGCCGCTGCCGCTGCCGCCGCTTCCGCCG  312
      lyProSerArgGlyTyrAsnSerTyrProArgSerLeuArgSerGluGlyAsnGlyGlySerAlaAlaAlaAlaAlaAlaSerAlaA  (81)

 313  CTGCCGTGAATCCCGGAACCTACAAGCAGTACGCCATTCCCTCCTACGAGTTGGATGGCGCTCGTGGCTACGAGATCGGACACGGCTACG  402
      laAlaValAsnProGlyThrTyrLysGlnTyrAlaIleProSerTyrGluLeuAspGlyAlaArgGlyTyrGluIleGlyHisGlyTyrG  (111)

 403  GCCAACGTGCTTACTAATTCTCGCTTCATCGGCAGTGAATTGAACTATCGACTCCTTGCTAAAATCCTCGAGTGGCTGTCATGGCGAAAC  492
      lyGlnArgAlaTyrEnd                                                                          (115)

 493  TCTGAGAATCAGTGAATAAAAGCAGCTTGAACGCAATGGAAAATACCGAAAAGAAATACGTATTGTGTGTTTTGCATTGTGACATACTTT  582
               -------                                    |(A)n
 583  TCAGCGCATT  592
```

Cp15 SEQUENCE. Strain, *Oregon R*. Accession, X02497 (DROCHORSG). Underlined is the regulatory chorion hexanucleotide. cf1 (which overlaps the chorion hexanucleotide) and cf2 indicate the binding sites of chorion factors I and II respectively. The *Cp15*, *Cp18* and *Cp19* sequence segments occur continuously in genomic DNA in the order shown in Fig. 8.2.

CFI and CFII RNAs are more abundant in follicle extracts than in extracts from other tissues, and CFII protein is more abundant in nuclear extracts from late follicles than in extracts from early follicles. CFI corresponds to the product

Cp16

```
-922  TGGCATCGAGTGCGGCACAATTCTTGGGAAAACTTGTCGTTGAAATTAAAACCATGTGTGTAGAGGTTTTGTCTGTTTGAATAATTTTAA  -833

-832  TTTTTCGTAAAAGTGAATTTATGTTTTGTGTTAAGCCGAAATATAAATAAAGTTTAATATTATTACTAACTAACCGTACGATCGTTTTTC  -743

-742  ATAAAACAGCTTAAAATTTGGTATTTAGCAACATTGTAATATTACATTAAAATAAATATAAGAATGAAATTCTAATAAAAAGGCATAACT  -653

-652  TAAATGCCAATGTATTTGAAACATAACTTAGAATATTGACGTAATAATCCACTTTGTTGCTATGCACATTTTTGTCCATTTTTAAATAAA  -563

-562  TTCATAGAACTGAGTTTACGATCCACAAACTTTTCAAAAACATTGTTCAGCTTTAAAACTAGAGTTTGCCCGACGTCCAAAGACTTTCGT  -473

-472  ACGTCGCTGGTTCCGGTTAGTGTTCATTGATCGGTGGCTTAACCCCAGTTGGCCCGATTTCCGATGCGTGCATGGCCGGATCGCCGTGG  -383

-382  AGTACGCCAAAGCCCCGATACCGCACACACCAGAAGCGAACAGAGCGTGCCGAGCAGGGGGAAGTCGCTATCAATGGAGCAGCTTCTCGG  -293

-292  GGTTCCCGGGGTTGTGGCAGTGCCGCAATTTGGCGCGCAATTTCAATGAGCATAGAAATTGGAGACGATCCCGTGGCCTTATCGCCCTGG  -203

-202  GCCGGAGGGGACGGAGGGGGCTGGAGATGCTGCCAGTGGCGGCCCCCCGAAAGTGACTGGTCATCGAGGTGGTTTGGTCACGTCTGGTGA  -113
                                                                                        ------
                                                                          -->-45
-112  GCTCACAAATCGCGGAGCAGCTCAAATGGTGTTGCTATAAAAGCAATTTGGACACACGCTCTGGTTAATTAGTTTTCGAAACAGTCCGTT  -23
         -----                       -----

 -22  CCTCGCACCACCACCAAAAAAAAATGTCCGCCACCCTACGCCTTCTCTGCCTGATGGCCTGCTGCGTCGCCCTGGCTGTGGCCAATCGCC  67
                                   MetSerAlaThrLeuArgLeuLeuCysLeuMetAlaCysCysValAlaLeuAlaValAlaAsnArgP  (23)

  68  CCCACTACGGCGGATCCGGATACGGAGCCAGCTACGGCGATGTGGTTAAGGCCGCTGAGACCGCCGAGGCTCAGGCTTCTGCCCTGACCA  157
      roHisTyrGlyGlySerGlyTyrGlyAlaSerTyrGlyAspValValLysAlaAlaGluThrAlaGluAlaGlnAlaSerAlaLeuThrA  (53)

 158  ACGCCGCCGGAGCAGCTGCCTCCGCCGCCAAGCTGGACGGTGCTGACTGGTATGCCCTCAACCGTTACGGATGGGAGCAGGGTCGCCCAC  247
      snAlaAlaGlyAlaAlaAlaSerAlaAlaLysLeuAspGlyAlaAspTrpTyrAlaLeuAsnArgTyrGlyTrpGluGlnGlyArgProL  (83)

 248  TTCTGGCCAAGCCCTACGGTCCTCTGGACCCGCTATACGCTGCTGCTCTGCCACCACGCTCCTTCGTGGCTGAGGTCGATCCAGGTGGGT  337
      euLeuAlaLysProTyrGlyProLeuAspProLeuTyrAlaAlaAlaLeuProProArgSerPheValAlaGluValAspProV     (111

 338  TCCTAAGCTAAGCTACAACATGGATAATATTGTTTATCCTATGATTTTGGATTGACTTCATAGCACCGCTTTGCCACCCATACTTACCTT  427

 428  CTTTTGTATCGTCTCTACCTTTCAGTCTTCAAGAAGAGCCAATACGGCGGATCTTACGGCGAGAATGCGTACCTGAAGACCGACGCCAAA  517
                                  alPheLysLysSerGlnTyrGlyGlySerTyrGlyGluAsnAlaTyrLeuLysThrAspAlaLys  (132)

 518  CTGGGTGTTGTGGCCATCTAAGAGCTTGGATTGTATAGCTCCAAAAGTGTTAATAAATAGTGATAGCTTAAAGCAATATAAATCAATGGA  607
      LeuGlyValValAlaIleEnd                    ------                                            (138)

 608  AATTCATTTATTGGGCTGGGAAACCAAACTTGAGCGAATCTTTATTTGCAAATGAGAATGTTTGTTTACTCCGACAACTTCTGCCTATTT  697

 698  TTGAATGCCATGAAACTTTAGATGGTTAAAAAAAAACTTCAAAAACTTTGAGTTGGCTATGCCAAACTTCATTACTTGAAGTCCACTAAG  787

 788  GTTCGCAGCTACACCATTTCTTGAAATCTTGAAGACCCCCCCAATTAGTAAAACCGAATTTCACTTACAATTTCTTATTGTTATTATTAT  877

 878  GAATAGAATTTCGTTTTTATTGCAAGATACAAAGTAAAAAATGTGAAAATTGCTCAGTTTTGTTGATGCTGATTTAATGTAAAATTCAAA  967

 968  TTCGTTACGAGCACACAGAAATTTACCTACTAAACATAAAGTGAACTAAAAACAAATAGTAGAAGCGGTTGTAACTCGGTTAACTCGATG  1057

1058  CTGCGGTGGCGTGCTTAGTGGGATATTTCGGTGACGATTATCATTTCCATTTCAAGTTATTAAGTTTTGTGCTTTTCGTTCAAATGGGCT  1147
```

(*continued*)

Cp19

```
-790  CGGGTTAAGATTTAGCGGTGGGTCATTATTATTATTCCACACACAAGATGGGTTTCAAAGTGGGGCAGCTAGAATATTCACTGCGGCAGA  -701

-700  TTGTACAATACTATATAGAAGTACTATTGCACTTTAAGCTACAAAGTCGACAGGTTAAGCTTCAGTGACTCAAGAATTTAGTCACCTATG  -611

-610  AAACCCTTAGTTTCACTAATAGATTCTTAGACGAACATCTTAAATTGTATAATCAAACAAATGGCTATGTATATATTACAATAACATATT  -521

-520  TGCCAATGTGCAAAAAGGCATAGACTTTGAAGTTATGTTTTATCGTTAAAATTTGGTTTGTTCTGTTTACTTGAAGGTATAGATAATATT  -431

-430  ATAGAATCCATATCCAATAACCATTGGTCAGTTGTGGGCCCGTTATCCCATTAACCCGCTTGGCTTCCCGCACGCACCCAAATGCAACCA  -341

-340  TTGATTTTGGGCCTCAGTTGGGAGCATCTGCATCTGCCACCCCCAACGAAGGTCAACCGGCGAATGGAGGCGATACGATACGCTGCGGTG  -251

-250  AGCAACCTGCTCGAGCCGAAACGAGCTCAACGTGGAGCCCCGATATCTGGCTAGGAAAAGCTAGAAATCCACAGAAAGTTCCCCAACAAA  -161

-160  CTGGCCGAGAAGAGACGGCGAAGCCAGCTCTTGAGCCGTGATAAATTTCTGGGCGAGATCACGTTTCGAGTGCAACAATAAATTTGCTTA  -71
                                                                                   ------

                                -->-44
 -70  TATAAAGAAGTGTGCTTGGCCATTTAATATGTTAATTCAGCCAACTGTGCCAAAACCCATACATCATAGCCATGAACAAGTTCGCTGTAA  19
      ------                                                                      MetAsnLysPheAla      (5)

  20  GTGTCCCTGAGAACCGCTTCCGTATTCCCTGCCGCTTTTTCATTTTCCGGACTTATGCTAACTGAAAGTTTTCCTGATTTTCCAGACTCT  109
                                                                                               ThrLe  (7)

 110  GGCAGTCATCTTCTGCGCCTGCATCGTGGGCAGCTGCTACGCCAACTACGGTGGCCAGCAGAGCTACGGACAGCGATCTTACGGTCAGGA  199
      uAlaValIlePheCysAlaCysIleValGlySerCysTyrAlaAsnTyrGlyGlyGlnGlnSerTyrGlyGlnArgSerTyrGlyGlnAs  (37)

 200  TAGCTCCGCCGCCTCCGCCGCCAGCTCAGCAGCTGCTGCTGGAGCCGAGGGTCAGCAGCGTTATGAGCGCCCCGTGGAGATCATCGCCGG  289
      pSerSerAlaAlaSerAlaAlaSerSerAlaAlaAlaAlaGlyAlaGluGlyGlnGlnArgTyrGluArgProValGluIleIleAlaGl  (67)

 290  CGGTTACCGCGGCAGCTATGCCCCCGAGATCCTGCGTCCCATCCAGGTCAGCGGTGGATATGGCGGTGAGCGACGTGGCTACAACGGTGG  379
      yGlyTyrArgGlySerTyrAlaProGluIleLeuArgProIleGlnValSerGlyGlyTyrGlyGlyGluArgArgGlyTyrAsnGlyGl  (97)

 380  CAACTACCGTCGTGCCGGCTACGGACCCCGTTGGACTGTCCAGCCCGCCGGTGCCACCCTCCTGTACCCCGGCCAGAACAACTACAAGGC  469
      yAsnTyrArgArgAlaGlyTyrGlyProArgTrpThrValGlnProAlaGlyAlaThrLeuLeuTyrProGlyGlnAsnAsnTyrLysAl  (127)

 470  TTACGTCTCGCCCCCGGAGTACAGCAAGGTGATCCTGCCCATCCGCCCCGCTGCTCCAGTGGCCAAGCTTTTCGTCCCAGAGAACCAGTA  559
      aTyrValSerProProGluTyrSerLysValIleLeuProIleArgProAlaAlaProValAlaLysLeuPheValProGluAsnGlnTy  (157)

 560  TGGCAACCAGTACGTTAGCCAGTACTCTGCACCCCGCAGCAGCGGCTACTAAGCGCATACATGATTATCCCCAGCCAACCTGGCGGATAC  649
      rGlyAsnGlnTyrValSerGlnTyrSerAlaProArgSerSerGlyTyrEnd                                         (173)

 650  TTGATCTCAGCCTGATCGTGTACATAATAAACAACAAGAAAAAATCATAATCATATTTGGAATATATATTTTTCGGGGCTTTTAGGTATT  739
                                ------                  |(A)n

 740  TTTTTATATCTATGAGAAAACAAATTTTCGGGTCTTTCGAGCTCAAATGCAGCTGCAGCAGCTGTTCAGAGTGGGTGGAGCATTGTTCAT  829

 830  TTGATTGCAGTCGCCACCGGGAATGTCTTTGAGTGGCTCGGCGGAAACGTGCTCCGGATTTGCTTGCTCCGGTTCGCTCGCAAATCACTC  919

 920  AGCAAGCCATAAACATTCAATTATTTATTGTGTCAGTCAGTCAATAATCTTGGGGCCAGAAAGCGCCACAGTTCGCCGAGTTTGCCGATT  1009

1010  GCCGCGCTCATTTTCATATTTTCTGTATTCTGGCTGGTAAGCAATCGCATCTGCTGACTTGTTTGGGGCCAAACTCTTGGCCAAGAGCTT  1099

1100  CAATGCTGCTGGCCATCGCTTGACATTCGAGTCGAGCGTGAATCACGGCAAGAATTC  1156
```

of the gene *ultraspiracle*, a steroid hormone receptor protein; and CFII contains C_2H_2 zinc finger motifs (Shea et al. 1990).

Cp16

Gene Organization and Expression

Open reading frame, 138 amino acids. One intron is present within the Val-111 codon. The position of the 5′ end was assigned on the basis of similarities to canonical *Drosophila* sequences (*Cp16* Sequence) (Fenerjian et al. 1989).

Cp18

Gene Organization and Expression

Open reading frame, 172 amino acids; expected mRNA length, 649 bases. One 176-base intron is present after the Met-5 codon. The approximate position of the 5′ end was determined by S1 mapping, and the first nucleotide transcribed was assigned on the basis of similarities to canonical *Drosophila* sequences. The 3′ end was determined from the sequence of a cDNA clone (*Cp18* Sequence) (Levine and Spradling 1985; Wong et al. 1985).

Cp19

Gene Organization and Expression

Open reading frame, 173 amino acids; expected mRNA length, 653 bases. One 89-base intron is present after the Ala-5 codon. The approximate position of the 5′ end was determined by S1 mapping, and the first nucleotide transcribed was assigned on the basis of similarities to canonical *Drosophila* sequences. The 3′ end was determined from the sequence of a cDNA clone (*Cp19* Sequence) (Wong et al. 1985).

Cp16 SEQUENCE (*page 83*). Accession, X16715 (DROCHORS16). Underlined is the regulatory chorion hexanucleotide.

Cp19 SEQUENCE (*opposite*). Strain, *Oregon R*. Accession, X02497 (DROCHORSG). Underlined is the regulatory chorion hexanucleotide. The *Cp15*, *Cp18* and *Cp19* sequence segments occur continuously in genomic DNA in the order shown in Fig. 8.2.

References

Carminati, J. L., Johnston, C. G. and Orr-Weaver, T. L. (1992). The *Drosophila* ACE3 chorion element autonomously induces amplification. *Mol. Cell. Biol.* **12**:2444–2453.

Fenerjian, M. G., Martínez-Cruzado, J. C., Swimmer, C., King, D. and Kafatos, F. C. (1989). Evolution of the autosomal chorion cluster in *Drosophila*. II. Chorion gene expression and sequence comparisons of the *S16* and *S19* genes in evolutionarily distant species. *J. Mol. Evol.* **29**:108–125.

Kafatos, F. C., Mitsialis, S. A., Spoerel, N., Mariani, B., Lingappa, J. R. and Delidakis, C. (1985). Studies on the developmentally regulated expression and amplification of insect chorion genes. *Cold Spring Harbor Symp. Quant. Biol.* **50**:537–547.

Kalfayan, L., Levine, J., Orr-Weaver, T., Parks, S., Wakimoto, B., de Cicco, D. and Spradling, A. C. (1985). Localization of sequences regulating *Drosophila* chorion gene amplification and expression. *Cold Spring Harbor Symp. Quant. Biol.* **50**:527–535.

Levine, J. L. and Spradling, A. (1985). DNA sequence of a 3.8 kilobase pair region controlling *Drosophila* chorion gene amplification. *Chromosoma* **92**:136–142.

Mahowald, A. P. and Kambysellis, M. P. (1980). Oogenesis. In *The Genetics and Biology of* Drosophila, eds. M. Ashburner and T. R. F. Wright (London: Academic Press), Volume 2d, pp. 141–224.

Mariani, B. D., Lingappa, J. R. and Kafatos, F. C. (1988). Temporal regulation in development: Negative and positive cis regulators dictate the precise timing of expression of a *Drosophila* chorion gene. *Proc. Natl Acad. Sci.* (*USA*) **85**:3029–3033.

Orr-Weaver, T. L. (1991). *Drosophila* chorion genes: cracking the eggshell's secrets. *Bioessays* **13**:97–105.

Orr-Weaver, T. L. and Spradling, A. C. (1986). *Drosophila* chorion gene amplification requires an upstream region regulating S18 transcription. *Mol. Cell. Biol.* **6**:4624–4633.

Osheim, Y. N., Miller, O. L. and Beyer, A. L. (1986). Two *Drosophila* chorion genes terminate transcription in discrete regions near their poly(A) sites. *EMBO J.* **5**:3591–3596.

Parks, S. and Spradling, A. (1987). Spatially regulated expression of chorion genes during *Drosophila* oogenesis. *Genes Dev.* **1**:497–509.

Parks, S., Wakimoto, B. and Spradling, A. (1986). Replication and expression of an X-linked cluster of *Drosophila* chorion genes. *Dev. Biol.* **117**:294–305.

Petri, W. H., Wyman, A. R. and Kafatos, F. C. (1976). Specific protein synthesis in cellular differentiation. III. The eggshell proteins of *Drosophila melanogaster* and their program of synthesis. *Dev. Biol.* **49**:185–199.

Shea, M. J., King, D. L., Conboy, M. J. and Kafatos, F. C. (1990). Proteins that bind to *Drosophila* chorion *cis*-regulatory elements: A new C2H2 zinc finger protein and a C2C2 steroid receptor-like component. *Genes Dev.* **4**:1128–1140.

Spradling, A. C. and Mahowald, A. C. (1979). Identification and genetic localization of mRNAs from ovarian follicle cells of *Drosophila melanogaster*. *Cell* **16**:589–598.

Spradling, A. C., Wang, G. L. and Mahowald, A. P. (1979). *Drosophila* bearing the ocelliless mutation underproduce two major chorion proteins both of which map near this gene. *Cell* **16**:609–616.

Spradling, A. C., Digan, M. E. and Mahowald, A. P. (1980). Two clusters of genes for major chorion proteins of *Drosophila melanogaster*. *Cell* **19**:905–914.

Spradling, A. C., de Cicco, D. V., Wakimoto, B. T., Levine, J. F., Kalfayan, L. J. and Cooley, L. (1987). Amplification of the X-linked *Drosophila* chorion gene cluster requires a region upstream from the S38 chorion gene. *EMBO J.* **6**:1045–1053.

Spradling, A. C. (1981). The organization and amplification of two chromosomal domains containing *Drosophila* chorion genes. *Cell* **27**:193–201.

Tolias, P. P. and Kafatos, F. C. (1990). Functional dissection of an early *Drosophila* chorion gene promoter. Expression throughout the follicular epithelium is under spatially composite regulation. *EMBO J.* **9**:1457–1464.

Waring, G. L. and Mahowald, A. C. (1979). Identification and time of synthesis of chorion proteins in *Drosophila melanogaster*. *Cell* **16**:599–607.

Wong, Y.-C., Pustell, J., Spoerel, N. and Kafatos, F. C. (1985). Coding and potential regulatory sequences of a cluster of chorion genes in *Drosophila melanogaster*. *Chromosoma* **92**:124–135.

9

Cuticle Protein Genes: *Lcp1, Lcp2, Lcp3, Lcp4, Pcp, Edg78E, Edg84A, Edg91A*

Larval Cuticle Protein Gene Cluster on Chromosome 2: *Lcp1-Lcp2-Lcp3-Lcp4*

Chromosomal Location:
2R, 44D

Map Position:
2-[58]

Products

Members of the cutin family. These are four of the five major protein components of the third-instar larval procuticle (the main layer of the cuticle) (Fristrom et al. 1978; Silvert et al. 1984).

Structure and Function

These proteins bind chitin and can be solubilized from untanned cuticles with 7 M urea; upon tanning of the cuticle, they become cross-linked and insoluble. The solubilized (untanned) proteins have an apparent M_r of 8–20 kD. The only detectable modification of these proteins is the excision from each of them of the first 16 amino acids, the signal peptide; the resulting N-terminus is unmodified. Direct amino acid sequencing of 50–75% of the residues confirmed the sequence predicted from nucleic acids data (Fristrom et al. 1978; Snyder et al. 1982; Silvert et al. 1984).

Tissue Distribution

Like the other components of the cuticle, LCPs are secreted by epithelial cells, the epidermis, probably in response to the steroid 20-hydroxyecdysone (20-HE). During its life cycle, *Drosophila* produces five different cuticles, three larval, one pupal and one adult. LCP1–4 contribute only to the third larval instar cuticle: LCP3 and LCP4 accumulate early in the third instar while LCP1 and LCP2 synthesis predominates late in the third instar (Chihara et al. 1982; Kimbrell et al. 1988).

Organization and Expression of the Cluster

The four genes are clustered in less than 8 kb of DNA, and they are best regarded as two pairs: *Lcp1* and *Lcp2* versus *Lcp3* and *Lcp4*. The two pairs are transcribed divergently (Fig. 9.1). In the coding regions, the similarity within the *Lcp1–2* gene pair is 91%, within the *Lcp3–4* pair it is 85%; the similarity between pairs is approximately 60% (Fig. 9.2). Considerable similarities also occur in the 5′ untranslated regions and in the 200 bp just upstream of the site of transcription initiation. The observed similarities suggest that the four-gene cluster evolved via an inverted duplication that gave rise to two ancestral genes followed by direct duplications of each of the two ancestral genes to give rise to the two pairs (Snyder et al. 1982).

A pseudogene carrying numerous disabling mutations lies between genes 1 and 2. It was probably generated by unequal crossing over between *Lcp1* and *Lcp2* (Snyder et al. 1982).

Developmental expression

Lcp1 and *Lcp2* are transcribed primarily late in the third larval instar while *Lcp3* and *Lcp4* are transcribed primarily earlier, as might be expected from the pattern of protein synthesis (Snyder et al. 1982).

Gene Organization and Expression

Transcription initiation sites were defined by primer extension and sequence features. The 3′ ends have not been determined (Snyder et al. 1982).

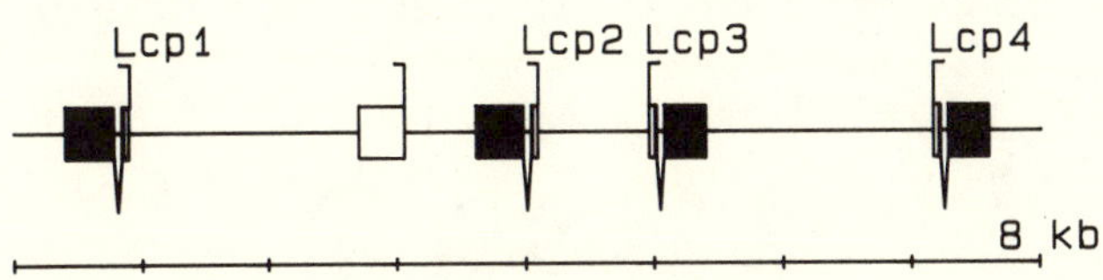

FIG. 9.1. *Lcp* cluster organization. Open box, pseudogene.

```
         1                                                    50                                                      100
Lcp1   MFKFVMICAV LGLAVANPPV PHSLGRSEDV HADVLSRSDD VRADG...FD SSLHTSNGIE QAASGDAHGN IHGNFGWISP EGEHVEVKYV ANENGYQPSG
Lcp2   MFKFVMILAV VGVATALAPV ....SRSDDV HADVLSRSDD VRADG...FD SSLHTSNGIE QAASGDAHGN IHGNFGWISP EGEHVEVKYV ANENGYQPSG
Lcp3   MFKILLVCSL AALVAANA.. .......... NVEVKELVND VQPDG...FV SKLVLDDGSA SSATGDIHGN IDGVFEWISP EGVHVRVSYK ADENGYQPQS
Lcp4   MFKILLVCAL VALVAANE.. .......... NPEVKELVND VQADG...FV SKLVLDNGSA ASATGDVHGN IDGVFEWVSP EGEHVRVSYK ADENGYQPQS
CON1   MFK----CA- --L--AN--- ---------- ---V-----D V-ADG---F- S-L---NG-- --A-GD-HGN I-G-F-WISP EGEHV-V-Y- A-ENGYQP--

Edg78  MYKYLFCLAL IGCACADNI. ........NK DAQIRSFQND .ATDAEGNYQ YAYETSNGI. QIQEAGNANG ARGAVAYVSP EGEHISLTYT ADEEGYHPVG
  Pcp  MYLLVNFIVA LAVLQVQAGS SYIP....DS DRNTRTLQND LQVERDGKYR YAYETSNGIS ASQEGLGGVA VQGGSSYTSP EGEVISVNYV ADEFGYHAHI
CON2   M-K-----AL ---A-A-A-- --------D- -A---S--ND -Q-D------ ----TSNGI- QS--G----- --G-----SP EGEH----YV ADE-GYQP-G
          ^             |


       101                                               150                                              193
Lcp1   AWIPTPPPIP EAIGRAVAWL ESHPPAPEHP RHH*...... .......... .......... .......... .......... .......... ...
Lcp2   AWIPTPPPIP EAIARAVAWL ESHPPAPEHP RHH*...... .......... .......... .......... .......... .......... ...
Lcp3   DLLPTPPPIP AAILKAIAYI EANPSKN*.. .......... .......... .......... .......... .......... .......... ...
Lcp4   DLLPTPPPIP EAILKAIAYI QAHPSKE*.. .......... .......... .......... .......... .......... .......... ...
CON1   ---PTPPPIP EAI--A-A-- E-HP------ ---------- ---------- ---------- ---------- ---------- ---------- ---

Edg78  DHLPTPPPVP AYVLRALEYI RTHP...... ..PAPAQKEQ Q*........ .......... .......... .......... .......... ...
  Pcp  ......PQVP DYILRSLEYI RTHPYQIKDY YTGELKTVEH DAAAFNVYTR NIQDHTIPQS RPSTTPKTIY LTHPPTTTSR PLRQRRALPT H**
CON2   ---PTPPP-P --ILR---YI --HP------ ---------- ---------- ---------- ---------- ---------- ---------- ---
```

FIG. 9.2. Comparison of the four larval (LCP1–4) and two pupal EDG78E and PCP) cutins. The sequences were aligned with the GCG program *Pileup*. CON1 indicates positions where at least three LCP proteins have the same residue. CON2 indicates positions where the pupal proteins agree with the larval ones. Ala-16 is the last amino acid of the signal peptide. A caret under residue 4, indicates the presence of an intron in all the genes discussed in this chapter with the exception of *Edg91A*.

Lcp1

```
          .         .         .         .         .         .         .         .         .
-707 AACATTAGGTTTTCTTAACAACTTTAATTGTCGCTAAAAAACTGTATTTATTTCGGTAGCCTCCTATATTGAACAGGTCTTATTATCTTA  -618

-617 TTTAATTTTACGAAATATAAAAAAATAATATAAGGCGCTATGCATATGAAATATAATGTAAAACACACTTGAATTTGTTTAAAAGCAAAC  -528

-527 TGCAACCTGTTCGTCGAGAGGAATTGATAAAAAAAAGAAGAAATGGTGCCAAGGTAGAGACACACGTTTATATATAACAAAACATCGAGT  -438

-437 CTAAGAAGTCGGCGATGCTTTGTAGTCCATGGAGTCTTGATGGGACTACAAAAGTGGTTCACGGCCTGGCAATGCCAAGTCAAGCTCAAA  -348

-347 GGAGGGGATTTAATGAAGGGGCGGGTCAAACTCGTTTCGATTTCGGGATGCCACCCGACCCGTTTGCCCCTTATTGATGCGATTGTTTCA  -258

-257 TTTTAGCATCTATTAAGCGATTATATATAGTACTTATCCCGTTGTTTGGCATTTGCTAAGCTGTCGCATGTGACGATGCTTTTTAATGGG  -168

-167 TGTGGGCGCATCCGCGAAGTCAACCCATAACTCAGCGAACCAATTGAATGCAAGATGTAGAGTTTTGATATGGGTTCACTTTGGGTGGCA  -78
                                             ----
                                        -->-41
 -77 ATCATATAAAAAGGCTCTGCCCGACCACAATCAGTTATCAGTCAACGTTCGTTCTCGACCAGACAGAAGTCAGCCAATATGTTCAAGTTT  12
       ------                                                                      MetPheLysPhe  (4)

  13 GTAAGTGTCCGCAGGATACGAACCAACATACTCGATCCCTAACGAATGCCTATTTCTCCTTCAGGTCATGATCTGCGCAGTTTTGGGCCT  102
                                                                     ValMetIleCysAlaValLeuGlyLe  (13)

 103 GGCGGTGGCCAACCCCCCGGTGCCCCATTCCCTAGGCCGTTCGGAGGATGTCCACGCCGATGTCCTTTCCCGATCCGATGATGTTCGTGC  192
     uAlaValAlaAsnProProValProHisSerLeuGlyArgSerGluAspValHisAlaAspValLeuSerArgSerAspAspValArgAl  (43)

 193 CGATGGATTCGATTCCAGCCTGCACACCTCCAACGGAATCGAGCAGGCCGCCAGCGGTGATGCCCATGGCAACATCCACGGCAACTTCGG  282
     aAspGlyPheAspSerSerLeuHisThrSerAsnGlyIleGluGlnAlaAlaSerGlyAspAlaHisGlyAsnIleHisGlyAsnPheGl  (73)

 283 CTGGATCTCACCCGAGGGCGAGCACGTCGAGGTTAAGTACGTCGCCAATGAGAACGGATACCAGCCCTCGGGAGCCTGGATCCCCACTCC  372
     yTrpIleSerProGluGlyGluHisValGluValLysTyrValAlaAsnGluAsnGlyTyrGlnProSerGlyAlaTrpIleProThrPr  (103)

 373 TCCTCCAATCCCAGAGGCCATCGGCCGCGCCGTCGCCTGGCTAGAGTCCCACCCACCAGCACCCGAGCACCCCCGTCATCACTAGAACCT  462
     oProProIleProGluAlaIleGlyArgAlaValAlaTrpLeuGluSerHisProProAlaProGluHisProArgHisHisEnd       (130)

 463 CTATGAAAGCGGATCGCACTACGGACTGTTCCCCGAAGACCTTTCGAACTATTAGCTTAAGTAATCGTACTGTTTGTAAAATACACGCAA  552

 553 TTGTTAACGGCAGAAACCAGTTTGCAACCTTGACTTTGAATTTGGCAAACAACTGTAACGGTTTCGAACCCGTCCTACCCGTTTACCACC  642

                                                                                 ______EcoRI
 643 TTCGATTTACTTAGTTGTTTAGCACGTTCAGTACAATATGGTAATGTGGTCTCTACCTGGACCGTAAACCGAATTC 718
```

Lcp1 SEQUENCE. *Canton S* strain. Accession, J01080 (DROCTCL1). The sequence of *Lcp1* extends to the first *Eco*RI site downstream of that gene.

Lcp1

Open reading frame, 130 amino acids; expected mRNA size, ca. 545 bases. There is one 64-base intron after the Phe-4 codon (*Lcp1* Sequence) (Snyder et al. 1982).

Lcp2

```
                    HindIII
-568 AACTCTGGCCAAAAGCTTTGCGGGTTTTTTAAATTAAACAGTGACATCCAAAATATTGAGATACAACAAAATGTCATAGGCAACTAGCA  -479

-478 CGTTAATATGCAGTATCACTTGCGAAATCGTTTATTCCGGTATATTGTTTATTACCACTTCGGAACCTTTTAAAATAGATGGGACTGCTA  -389

-388 TCAAGTGAAGTGTATTGGGTTTTTTGATTTTGTACAGGCATGATTGATAAAGACTTGGTCAACTCGAAACGTCATCGATGAGCACAGAAT  -299

-298 CCGAAAACCGTACTCCATCGCCCCTACAAAATTTCTACCGAAGCATGTTTCATTTCGGAATCTGTTCAGCAGCGCAAGACTTGTTTTTTG  -209

-208 ACATTTGTATCGCAGAGTCAAGTGGAGAATTTATGGGCCCTGCCTTTTGTTGGCATCATGGGCGTTTCGTGATAACTTAGATTTGGCCCA  -119
                                                                                      -----
                                                                                   -->-41
-118 AAAAGTAATAAGCAATTCGTTTGGAAAGCAACCAAATTGGGAATCATATAAAAAGACTCTGTCGACCAAAGTCAGTTATCAGTCAACGTT  -29
         ----                       ------

 -28 CGTTCTCGACCAGACAGAAATCAGCCAACATGTTCAAGTTTGTGAGTGGCTCACAGGACATTTATGAACTCGCCATCTAATTGGTATCAT   61
                                 MetPheLysPhe                                                  (4)

  62 TTCCTCTATCCAGGTGATGATTCTCGCCGTTGTGGGAGTGGCTACCGCCCTAGCCCCAGTTTCCCGCTCCGATGATGTACACGCTGATGT  151
                 ValMetIleLeuAlaValValGlyValAlaThrAlaLeuAlaProValSerArgSerAspAspValHisAlaAspVa (30)

 152 CCTTTCCCGATCGGACGACGTTCGTGCCGACGGATTCGACTCCAGCCTGCACACCTCAAACGGAATCGAGCAGGCCGCCAGCGGTGATGC  241
     lLeuSerArgSerAspAspValArgAlaAspGlyPheAspSerSerLeuHisThrSerAsnGlyIleGluGlnAlaAlaSerGlyAspAl (60)

 242 CCATGGCAACATCCACGGCAACTTCGGCTGGATCTCACCCGAGGGCGAGCACGTTGAGGTAAAGTACGTCGCGAATGAAAACGGATACCA  331
     aHisGlyAsnIleHisGlyAsnPheGlyTrpIleSerProGluGlyGluHisValGluValLysTyrValAlaAsnGluAsnGlyTyrGl (90)

 332 GCCCTCGGGAGCCTGGATCCCCACTCCTCCTCCAATCCCAGAGGCCATCGCCCGCGCCGTTGCCTGGCTGGAGTCTCACCCCCCAGCACC  421
     nProSerGlyAlaTrpIleProThrProProProIleProGluAlaIleAlaArgAlaValAlaTrpLeuGluSerHisProProAlaPr (120)

 422 CGAGCACCCCCGTCATCACTAGGACTCGTCACCCGGATCCCGGACCACTACACGGACTGTTCTCCCGAAACAAATCGCCCAAGTTGTTTA  511
     oGluHisProArgHisHisEnd                                                                    (126)

 512 GCTGTACTTCTTGACTTTCAAAAAAATACATGCACTTGCTTATAGCAGTAAAAATGTGTGTGTTCTTCATTGCACTTTTTAGGTAGTTCC  601

 602 TGTAATAATACGAGCTTTTATACCTCTACCTTCGCTGGGAATGCTTCCTTCTACCTTTTATATTCGATTCACTAAATCCATTTATCAAAA  691

 692 ATGAGTATATGTGTCCATAAAGAAAAGATGTGCTGAATTAA  732
```

Lcp2 SEQUENCE. *Canton S* strain. Accession, J01081 (DROCTCL2). The sequence of *Lcp2* starts in the neighborhood of a *Hind*III site between *Lcp2* and *Lcp3*.

Lcp2

Open reading frame, 126 amino acids; expected mRNA size, ca. 533 bases. There is one 62-base intron after the Phe-4 codon (*Lcp2* Sequence) (Snyder et al. 1982).

Promoter

Approximately 800 bp upstream of the *Lcp2* transcription initiation site is sufficient for correct developmental regulation, but other sequences still farther upstream may also be necessary for full expression. A 270-bp segment does not support any detectable transcription in transgenic animals (Kimbrell et al. 1989). It should be noted that the distance between the divergently transcribed genes *Lcp2* and *Lcp3* is approximately 870 bp.

Lcp3

Open reading frame, 112 amino acids; expected mRNA size, ca. 494 bases. There is one 56-base intron after the Ile-4 codon (*Lcp4* Sequence) (Snyder et al. 1982).

Lcp4

Open reading frame, 122 amino acids; expected mRNA size, ca. 494 bases. There is one 57-base intron after the Ile-4 codon (*Lcp4* Sequence) (Snyder et al. 1982).

Pupal Cuticle Proteins: Pcp, Edg78E, Edg84A *and* Edg91A

Pcp
(*Pcp* in the *ade3* gene intron 1)

Chromosomal Location:
2L, 27D1-3

Map Position:
2-20

Synonym: *Pcpgart*

Product

Probably a pupal cuticle protein. The amino acid sequence shows clear similarities to larval and pupal cuticular proteins (Fig. 9.2), including the presence of a putative signal peptide (Silvert et al. 1984; Henikoff et al. 1986).

Gene Organization and Expression

Open reading frame, 184 amino acids; expected mRNA length, 718 bases, in agreement with an RNA band of 0.9 kb. Primer extension, mRNA sequencing and the sequence of two cDNAs were used to define the 5′ end. The 3′ end was obtained from a cDNA sequence. There is an intron after the Leu-4 codon (*Pcp*

Lcp3 and *Lcp4*

```
-650  GTGCTCATCGATGACGTTTCGAGTTGACCAAGTCTTTATCAATCATGCCTGTACAAAATCAAAAAACCCAATACACTTCACTTGATAGCA  -561

-560  GTCCCATCTATTTTAAAAGGTTCCGAAGTGGTAATAAACAATATACCGGAATAAACGATTTCGCAAGTGATACTGCATATTAACGTGCTA  -471

                                                                                 ______HindIII.
-470  GTTGCCTATGACATTTTGTTGTATCTCAATATTTTGGATGTCACTGTTTAATTTAAAAAAACCCGCAAAGCTTTTGGCCAGAGTTGTCAA  -381

-380  CGTGCCACACACCAAATGAAACACCGAAAACTATGCTATGCTTAAGTTTAGTTCATATTGAAGTTGAATTTTAGAAAATTAAATATTGTA  -291

-290  CTGCTTAATAATTATTCTGGTTTCTGGTCCGGTTTGCTTTGCATTTCGGTTAGACTAGGGCGAATATTTCAGTTGAATAAATAACTAAGA  -201

-200  ATGCTCATCTCCTAATGAAAGTGGTTAAGCCATCTCAAGTCGACTAATTTGCATCCCAGACGGTTTTTATTATATGCATCACATTGACTT  -111
                               HMS Beagle insertion                                        ----
                                       ||=n1                              -->-44 Lcp3
-110  AATTATAATACGCACATTGCATCAGCTTTTGATGATATATAAACACCGATTTGAGCATAGATTGTCATCAGTCTTAGAAGATTTCTAGTC  -21
                                       ------

 -20  CGACAATCCACCCAAATCAAAATGTTCAAGATCGTAAGTATGCCTTGAGGAGCATAGTGACTTCGCAGTCTAATCCTGGATTATCCTAGC  69
                            MetPheLysIle                                                       L  (5)

  70  TGCTTGTCTGTTCTCTCGCCGCCCTGGTGGCCGCCAACGCTAATGTGGAGGTCAAGGAGCTGGTCAACGATGTCCAGCCCGATGGCTTTG  159
      euLeuValCysSerLeuAlaAlaLeuValAlaAlaAsnAlaAsnValGluValLysGluLeuValAsnAspValGlnProAspGlyPheV  (35)

 160  TCAGCAAGTTGGTCCTCGACGACGGATCTGCCTCCTCCGCCACCGGAGACATCCACGGCAACATCGACGGAGTCTTCGAGTGGATCTCCC  249
      alSerLysLeuValLeuAspAspGlySerAlaSerSerAlaThrGlyAspIleHisGlyAsnIleAspGlyValPheGluTrpIleSerP  (65)

 250  CCGAGGGTGTCCATGTGCGAGTGAGCTACAAGGCTGACGAGAACGGATACCAGCCCCAGAGTGACCTGCTGCCCACTCCTCCTCCGATCC  339
      roGluGlyValHisValArgValSerTyrLysAlaAspGluAsnGlyTyrGlnProGlnSerAspLeuLeuProThrProProProIleP  (95)

 340  CAGCTGCCATCCTGAAGGCTATCGCCTACATCGAGGCTAACCCCAGCAAGAACTAAGTGAACCCGCCGACTAGGAACATGAAAGATTGGA  429
      roAlaAlaIleLeuLysAlaIleAlaTyrIleGluAlaAsnProSerLysAsnEnd                                   (112)

 430  GACAGCTAGGTTGAGTTTGGATAATTTCTTACCAGTTGTTTTAAATTTAAGGAAAATGTTATCGAATTCGAAAATAAATTAAACCTTGCA  519
                                                                                    ------

 520  ATATAAACCAAGTGCATGTTTTACAAATCTGACAGTTCGATTTAAGAGAAGGCTCCCGGTATTATATGGTATAAGAAGGTACAATTAGAA  609

 610  GATTAAAAGTAATCAAAGACACTTTGGCCTTCATTAAATTACAATTGTGTTGTTATAGTATAGTACGAAATTAATTTAAATACAAAAATC  699

 700  TTTAAAGCATCTAAAATAAATGTAAACATTACAAAAACCTTACCTGGACAAGCCGATATCTCCTTGCATTAATTTCATATTTCCGAAAAC  789

 790  TGGGTTATAACTAGTTATTATTTTAAGTTAAGTTCATAGGCAGCCACAAGTAATTAAATGTTGCCAACCTGATGCATCCCAGATAAGATC  879

 880  GCAGTATGATGAAAACGACGAGGAACTTTTTTATATCTATTATTTGTAGAGGATAAGGGTACACTTGAATTGTTAGAACGCATGTCGGTA  969

 970  TTATGGGTATTAAGGGTATTATGAAGCGTTTTCGAACCTAAAAAGTATGTATATTCATCATCATTATCACTAGCCGAGTCGAATTTTTGT  1059

1060  TTTTATATTGGTCTTTTTATGAATATAACTGAAATTGGCATTATAAGCCTAGATGTAAAAATCAAATTCTTCATCTTTTTTTTAACCTTT  1149

1150  TTTAAATAGTCATACACGTAACAAAAAATAACACAGACTTCCCTGAGGTTACACGGTTATAAGATCTTGTAGTGATTTTTGGAGAAATAT  1239

1240  CAATCAAATTGCTGTGCTTTCGGATTTTTGATTATATTATGATATTGTAACTTAAGTGTTTAATAGTGATTGTATAAGTAAGAATCGTAT  1329
```

```
1330 ACATTGTATATGTCAAACTCCCCGGGAATGTTCATATTGACTTAACGGAAACTAGAGATAAAATATACACACAATGTTTTTTTTATTAA 1419

1420 ACGAAATTATTTACAATAATTTAATTGACTAGCAATAGTACGCTCTTCTTAGGCAACCCAATCTTATCGGTATCAATTTAAACTATTCTT 1509

1510 AATATCTATGTTATTTACAAAGGTTATATGAGTAAGAGTTTTTGAGGAATAGAATGTTTATGCAGATTTTAATTTAGTAGGATTTATGTC 1599

1600 AAGTCCCGGTCAGTCAAGTCTTGAGGGTGGTGAACACAGAATGTTAGATTCCATAAACCCGTTCCCAGTCATTTCGCAGATAGAAACCAA 1689

1690 ATGATGCTCCGAAAGGTATGCTGGATCTACAAGCGGTTCGCAAAAAAGTTTTGTTTTCTAGTTATTTTTCACCTCCTAATAATTAAACTT 1779

1780 CTACTATCAGCAGCTTAGACATTATTCAATCAAGTTATTTTTATATGATTTGTCTGGAGTAATTCAAAGTTATCTGACTAAATATTCCGG 1869

1870 AAGATGTTAAATTATTTCAATGAGAAGGTGGACTTACCCTTTTCCGAGTAACCCGATTCTTTTTAGAATAATTACGGTAGCGATTTGCAT 1959

1960 AGACAATAGAAATCAAAAAGAGTGCAGCAGACGATTTTTATCGCCACCAAGCATGTCACTTGAACCAGTCCGTAAAACCAAACGAGACCT 2049

2050 ATGCTGGCCGAAATGTTAATTAAAAACGGGTTGCATCAGCTTTTGATCAGCTTTAAGATTTCGTGGGGAGTGCGTATAAGCGTATAAAAG 2139
                                                                                       ------

                 -->2163 Lcp4
2140 CCGACGAGTGATCCCGAATTGGCATCAGTCTCACGAGTTCTTTAGTCTGACAATCTAACCAAGTCAAAATGTTCAAGATCGTAAGTATCT 2229
                                                                        MetPheLysIle       (4)

2230 GAAGTTTAAAGCCGGACAGTTCAATGAGTAATCCCGGAATATCCTAGCTGCTTGTCTGCGCCCTTGTCGCCCTGGTGGCCGCCAACGAGA 2319
                                               LeuLeuValCysAlaLeuValAlaLeuValAlaAlaAsnGluA (19)

2320 ATCCCGAGGTCAAGGAACTGGTCAACGATGTCCARGCCGATGGCTTCGTAAGCAAGTTAGTCCTGGACAACGGTTCCGCTGCTTCTGCTA 2409
     snProGluValLysGluLeuValAsnAspValGlnAlaAspGlyPheValSerLysLeuValLeuAspAsnGlySerAlaAlaSerAlaT (49)

2410 CCGGAGATGTCCACGGAAACATCGACGGAGTTTTCGAGTGGGTCTCCCCCGAGGGCGAACACGTCCGTGTGAGCTACAAGGCCGACGAGA 2499
     hrGlyAspValHisGlyAsnIleAspGlyValPheGluTrpValSerProGluGlyGluHisValArgValSerTyrLysAlaAspGluA (79)

2500 ACGGATACCAGCCCCAGAGCGACCTCCTGCCCACTCCTCCTCCAATCCCAGAGGCCATCCTGAAGGCCATCGCCTACATCCAGGCCCATC 2589
     snGlyTyrGlnProGlnSerAspLeuLeuProThrProProProIleProGluAlaIleLeuLysAlaIleAlaTyrIleGlnAlaHisP (109)

2590 CCAGCAAGGAATAAGCAATCGACACGACCAGGACCCACATTCGAATCGGAGGTGCAACTCCAAAGACCTTGCCCTCTAACCCTTAGAATT 2679
     roSerLysGluEnd
                                                                                           (112)

2680 TAAACAGCATGCAGACATTATAAATGATTATCGAGTTAGGAAATAAATTCGATACTCTTTGGCAACAAATCTATTTAGATATGGTCTTAA 2769
                                            ------

2770 TTTCCCTGACGGCAGCAGGAGTTACCTTGTTTATGGCTGATTTATTTTGGCCGAGGGAACAAACGCTGCTTCAGATTATCACGAACTGTC 2859

2860 TGGATGTTACGTGATTGATCTTAGCCAATAGTAACCTGTTTAATTAGCGATACATAAAGTGAAGACCATCAAACCAGATTTAGGTATAAA 2949

2950 TTCGGTCTGTTTATTACAGTTTTAAATGCAATAAAATATTTCATTAAACAAAAGTCATGGCTGAGCAAAATATAACCGGATTGGAATTGC 3039

3040 TTGCGTTACTCTTCATCTTCATATTGTTAAAAGAACAGTAAAGAACGGTATAGTGAAATTTTCGAATACTTATTATTATTATTACTCGGT 3129

3130 TTAAATGTTGGTGGTACACCGATAGAAATTTGCAAGAAAAAAGTTAAAATAACCATTTTTTTGAAAGAATTTCGGTGCCAAAATGAGACG 3219

3220 GTTTGAGAGCGTTACACTGGAAAAAAAAACCCGATGCAAACATGGCTTTAACGATCGACTACCTGTTATACAATACCCTTCACATTGTCAA 3309

3310 TCATCTAGTATAAACTTCAAATCTAGGAGTAGAGAGTTGGTAAAAACATCCTTGAAGATGTTAATGGACTAGCTGTTATCATGATTATAT 3399
```

Lcp3–4 SEQUENCES. *Canton S* strain. Accession, J01081 (DROCTCL2). The sequence of *Lcp3–4* (the opposite strand of the two previous sequences) starts near the same *Hind*III site. Indicated is a mutation of *Lcp3* caused by an insertion in its TATA box.

Pcp

```
                                               -->-32
 -64  GATATAAAAACGAAAGCGTCCGAGAAGCAGGCAAGCAGTTTAGAACCAAACTCGAACGCGACACCATGTATTTGCTTGTAAGCATCAGCT   25
-244  AAAATCATTTTATTTATGACTGACTAAGGCGACCAGCAGCGATGAGATGTTTGTAGATGGAGACGATCATGACGATGACGAGCGGAGATG  -155

-154  GAGATGGAGACGGCAACGGCAACGGCAACGGCAACTCGGAACTGGGTTTCCGAGGCGATGTATAGCCAAAAATCCGCTGGTGAGCGGATG   -65
```

```
                                               -->-32
 -64  GATATAAAAACGAAAGCGTCCGAGAAGCAGGCAAGCAGTTTAGAACCAAACTCGAACGCGACACCATGTATTTGCTTGTAAGCATCAGCT   25
      ------                                                                 MetTyrLeuLeu      (4)

  26  GGGAATTTCCCGAAAATGGATTATAATCGCCGACTCTCGTCTCGAATCCCGCCCACAGGTGAACTTCATCGTTGCGCTGGCCGTGCTGCA  115
                                                      ValAsnPheIleValAlaLeuAlaValLeuGl     (15)

 116  GGTGCAAGCCGGCTCATCCTACATTCCGGACTCGGATCGCAACACACGCACCCTGCAGAACGATCTGCAGGTGGAGCGGGATGGCAAGTA  205
      nValGlnAlaGlySerSerTyrIleProAspSerAspArgAsnThrArgThrLeuGlnAsnAspLeuGlnValGluArgAspGlyLysTy  (45)

 206  TCGGTATGCCTACGAGACCTCCAATGGCATTTCCGCATCGCAGGAGGGATTGGGTGGCGTGGCCGTACAGGGCGGCAGTAGTTACACATC  295
      rArgTyrAlaTyrGluThrSerAsnGlyIleSerAlaSerGlnGluGlyLeuGlyGlyValAlaValGlnGlyGlySerSerTyrThrSe  (75)

 296  ACCCGAGGGCGAAGTAATTAGTGTGAACTATGTGGCCGATGAGTTTGGCTATCATCCCGTGGGCGCACATATACCCCAGGTGCCGGACTA  385
      rProGluGlyGluValIleSerValAsnTyrValAlaAspGluPheGlyTyrHisProValGlyAlaHisIleProGlnValProAspTy  (105

 386  CATACTGCGCTCCCTGGAGTACATTAGGACGCATCCCTACCAGATCAAGGACTACTACACCGGGGAGCTGAAGACCGTGGAGCACGATGC  475
      rIleLeuArgSerLeuGluTyrIleArgThrHisProTyrGlnIleLysAspTyrTyrThrGlyGluLeuLysThrValGluHisAspAl  (135

 476  AGCCGCCTTCAATGTGTACACACGCAACATTCAGGATCATACGATCCCCCAATCCCGACCGAGCACCACGCCCAAGACCATATACCTCAC  565
      aAlaAlaPheAsnValTyrThrArgAsnIleGlnAspHisThrIleProGlnSerArgProSerThrThrProLysThrIleTyrLeuTh  (165

 566  CCATCCGCCCACGACCACGTCGCGACCTCTGCGCCAGAGACGAGCTCTTCCGACGCACTGATGATCGATGGACGTGACTCTATGGCGGGG  655
      rHisProProThrThrThrSerArgProLeuArgGlnArgArgAlaLeuProThrHisEndEnd                            (184

 656  CAAGGGGCTGGTCTCTTCGGCGGCCAGCGGGCGAATCTGTGAATTTTGATCTAAACAATTAATTAAGCCACGAACAATAAATAGAAGTGC  745
                                                                                ------

 746  TAAGCAAACATAAGCTAAAGTGTAATCGATCTGTCGAGTTGTCTGCTGGGGATCATGGATCACATCATGGAGCGACATAAACAATTTTGG  835
                |(A)n

 836  GTATTCGATTCTGTTTATGGC  856
```

Pcp SEQUENCE. Accession, J02527 (DROGART).

Sequence). The *Pcp* gene is completely within the long first intron of *ade3* (*Gart*), a gene that encodes two polypeptide chains involved in purine biosynthesis. *Pcp* and *ade3* are transcribed from opposite strands (Henikoff et al. 1986).

Developmental Pattern

Pcp RNA is present in prepupae and possibly in larvae and pupae as well. *In situ* hybridization in 11 h prepupae, shows *Pcp* RNA to be present in the larval epidermal cells that secrete abdominal cuticle, and to a lesser extent in the imaginal cells that secrete cephalic and thoracic cuticle (see *Edg78E*) (Henikoff et al. 1986).

Edg
(*Ecdysone dependent genes*)

These genes were identified because their transcripts accumulate in imaginal discs in response to a pulse of the steroid 20-HE (Fetchel et al. 1988).

Edg78E

Chromosomal Location:
3L, 78E

Map Position:
3-[47]

Product

Pupal cuticle protein (Fetchel et al. 1988, 1989; Apple and Fristrom 1991).

Structure and Function

Sequence features indicate a signal peptide at the N-terminus. Other sequence features characterize *Edg78E* as a member of the cutin family of cuticle proteins (Fig. 9.2) (Apple and Fristrom 1991). It is immunoprecipitated by antibodies against low molecular weight pupal cuticle proteins (L-PCP) (Fetchel et al. 1988).

Tissue Distribution

The pupal procuticle is produced in the prepupal stage. It is subdivided into the exocuticle, secreted between 8 and 12 h after puparium formation, and the endocuticle, secreted between 12 and 20 h. The main protein components of the exocuticle are of low molecular weight (L-PCP; M_r, 8–25 kD). Six L-PCPs have been identified by gel electrophoresis, but it is not known which one of them corresponds to EDG78. Because the endocuticle is characterized by high molecular weight proteins (H-PCP; M_r, 40–82 kD), it is inferred that EDG78 is localized in the exocuticle (Fetchel et al. 1988, 1989 and references therein).

Edg78E

```
-1000 CTACCTGGGCTGGGAAAAATATACCATTTTATGTACGTTTATTTCCTGGGTCGTTTGGCGATTTCTTGAATCGAAGTCTACACATATGTA  -911

 -910 GAGAGATAAGTGTGAACTACATTTAATTACTAGCTTACTTCGGATTTTGCACACTTCCTTGTTTACCGAAACGATCTCAGCAATTAACAG  -821

 -820 CAGTTTGCAATGGTTGATGGTGTTTGCCAAGCTAATTTCGAAACAAAAAAATATCTTCGTTCGAACCTATCGCTGCGTCCATTGCAACCA  -731

 -730 ATCGATTCGCCGAAGATCAAAGTGAACAATTAATTAAAGTCATAAATGTAGGGTATCAGAAGATCACACGTAACATCGCACTGCATGGCT  -641

 -640 GGATCATCTTCGGCGGCGCTCCGGGTGTCATGCTGATGCTGCCCGATGACCTTGTCCATGTTTCAACAGCTTTCCAGGGGCACAAGGTAT  -551

 -550 ACTCGCACCATACTAGACCATCGCACCTGCCACTCCATTTGGAAGCCTCGAGCCCAGGGCGCAACTCCAATTGAAAGTTGTAACAAGAAA  -461

 -460 TCTTCAGCTCGTGTTGGGAATTTCCAACGCTGTTTTGAATGGGCCGAAAGCGTCACATTAAACAGCAATATTTATTCGATTGTAATTCAA  -371

 -370 CAGTTAATGGTATCTCGTGCGAAACCGAAACCGAAATCGAATCTGAAACTGAAACCGAATTAAAGCATACAATATAAATTGTTGGCAAAT  -281

 -280 GACTCATGTATTTTAACTATAGGCCAGCCGAGCACTGTTGTTATTGTTGCCCAATATTGGTGGTATGATAAGAAGACATTTTGGCAATTA  -191

 -190 TACTCACCTCTCGGGTCCTTGTGATTCCACGAGAAAAAACTTGTTTAGCTGCTAAACTAAAGAGACTAAAGACAAGGGTCTCATCCATAT  -101
                                                                                           ---
                             -->>>>-75/-72
 -100 AAAAGACGCACTTGAGCTGATCAAATTAAACAGTTGCACTGCAAGCACCATCATCACAGCATCACCGCTTTAAGAGAAGAAAATTCCCAA  -11
      ---

  -10 TTCCCATCATCATGTACAAATATGTAAGTTCGGTTGGACTTGGCACGCCTCATACCCCAGAGTACCAATACTGATCATTGTACTTTGATC  79
                  MetTyrLysTyr                                                                (4)

   80 CCAAAAGCTGTTCTGTCTTGCTCTCATCGGCTGCGCCTGCGCCGACAACATCAACAAGGATGCCCAGATCCGCAGCTTCCAGAACGACGC  169
             LeuPheCysLeuAlaLeuIleGlyCysAlaCysAlaAspAsnIleAsnLysAspAlaGlnIleArgSerPheGlnAsnAspAl  (32)

  170 TACCGATGCTGAGGGCAACTACCAGTACGCCTACGAGACCAGCAATGGCATCCAGATCCAAGAGGCGGGCAACGCCAACGGAGCACGTGG  259
      aThrAspAlaGluGlyAsnTyrGlnTyrAlaTyrGluThrSerAsnGlyIleGlnIleGlnGluAlaGlyAsnAlaAsnGlyAlaArgGl  (62)

  260 TGCCGTGGCTTACGTGTCGCCCGAGGGCGAGCACATCTCGCTGACATACACCGCCGACGAGGAGGGCTACCATCCAGTGGGTGACCACCT  349
      yAlaValAlaTyrValSerProGluGlyGluHisIleSerLeuThrTyrThrAlaAspGluGluGlyTyrHisProValGlyAspHisLe  (92)

  350 GCCCACCCCGCCCCCAGTTCCGGCTTACGTTCTCCGTGCCCTGGAATATATCCGCACCCATCCCCCGGCGCCCGCCCAGAAGGAGCAGCA  439
      uProThrProProProValProAlaTyrValLeuArgAlaLeuGluTyrIleArgThrHisProProAlaProAlaGlnLysGluGlnGl  (122

  440 GTAATCTGGAGTAGCACCAGCACTCCAAAGCAGCAACCCCACATCTAAACTGCGGCCAGTCATTGTTATTTAGGTAGTTATCGTTAATAA  529
      nEnd

  530 AGGATTTCGATACAGATCATTTTCGTTTTTAGTAATGTAGTAAAGATGGAAAATAAATGTTTCATGTATATGTATTCATATGTAAATGAA  619

  620 CATATGTATAGTTCTTCGAAAAATATAGAAGCGTACACTATCTTCAATAGAAACAAATTTCAGGCGGATGGAGTTTACATTTTGAAACAT  709

  710 TTCTTTATCTTAACATTGCTCTTTTTTCTTTCAAATGAACAATTTGAAGAATGTATATGTTAGTTAATGATTTCGGCAGCCAGTAATTGT  799

  800 ATAAAACCATTTATCTATGTAATAGATTTTGATTTATGTCATTTATTTTTCCACTTTCATTTATACTCAACGCATTATGATTTCCGAACT  889

  890 ACAATAGTTAAATTTTTGAAAACCAATCCAGCGGTGATGCACAGATGAGATAAATTAAAAGAAACAAAATCTCGTAGATGAGATAAATTA  979
                                                         ------           |(A)n
```

Edg78E SEQUENCE. Strain, *Canton S.* Accession, M71247 (DROEDG78A).

Gene Organization and Expression

Open reading frame, 122 amino acids; predicted mRNA length, 962–966 bases, somewhat larger than the 0.6 kb band detected by northern analysis. Primer extension was used to define the 5′ ends (there seem to be four clustered transcription initiation sites). The 3′ end was obtained from a cDNA sequence that included a poly-A tail. There is an intron after the Tyr-4 codon (*Edg78E* Sequence) (Fetchel et al. 1988; Apple and Fristrom 1991).

Developmental Pattern

As would be expected for a secreted protein, the *Edg78E* mRNA is preferentially associated with the membrane-bound polysome fraction. Low levels of this RNA are detected only in prepupal stages (Fetchel et al. 1988). By *in situ* hybridization, *Edg78E* RNA can be detected both in the larval epidermal cells that secrete abdominal cuticle and in the imaginal cells that secrete cephalic and thoracic cuticle. The peak of accumulation is in 10 h prepupae (Fetchel et al. 1989).

In imaginal discs in culture, *Edg78E* transcription is stimulated by a pulse of 20-HE, 6 h in 1 μg/ml hormone and 8.5 h without hormone. Transcription, however, is inhibited if the hormone treatment is continuous or if hormone is re-added to the medium after an original pulse that stimulates transcription. This hormonal regimen mimics the endocrine status during the larva-to-pupa molt. Thus, a 20-HE peak would stimulate *Edg78E* expression, and its product would presumably contribute to the exocuticle being produced at that time. A second rise in hormone titer, which signals the transition from exo- to endocuticle production, would repress *Edg78E* and induce expression of other genes whose products are characteristic of the endocuticle (Fetchel et al. 1988; Apple and Fristrom 1991).

Edg84A

Chromosomal Location:
3R, 84A

Map Position:
3-[47]

Product

Probably a cuticular protein.

Structure and Function

It has sequence features that indicate a signal peptide and sequence similarities to cuticular proteins of *Hyalophora cecropia* and *Locusta migratoria* but not to cutins (Apple and Fristrom 1991).

Gene Organization and Expression

Open reading frame, 188 amino acids; in northern analysis, a 0.9 kb band is detected. Primer extension was used to define the 5′ ends (there seem to be three clustered transcription initiation sites). The 3′ end was not determined. There is an intron after the Lys-4 codon (*Edg84A* Sequence) (Fetchel et al. 1988; Apple and Fristrom 1991). *Edg84A* is part of a cluster of small genes with related sequences located within the Antennapedia Complex, between *labial* and *proboscipedia* (Pultz et al. 1988; Fetchel et al. 1988).

Developmental Pattern

As would be expected for a secreted protein, the *Edg84A* mRNA is preferentially associated with the membrane-bound polysome fraction. This RNA is detected only in prepupal stages (Fetchel et al. 1988). By *in situ* hybridization, *Edg84A* RNA can be detected only in the imaginal cells that secrete cephalic and thoracic cuticle but not in the larval epidermal cells that secrete abdominal cuticle. The peak of accumulation is in 10 h prepupae (Fetchel et al. 1989).

As for *Edg78E*, *Edg84A* transcription is stimulated by a pulse of 20-HE in imaginal discs in culture (Fetchel et al. 1988; Apple and Fristrom 1991).

Edg91A

Chromosomal Location:	**Map Position:**
3R, 91A	3-[64]

Product

Probably a cuticular protein.

Structure and Function

It has sequence features that indicate a signal peptide and sequence similarities to insect egg-shell and egg-casing structural proteins. It also has some similarities to vertebrate cytokeratins. EDG91 is a hydrophobic protein with very high (32%) Gly content (Apple and Fristrom 1991).

Edg84A

```
-818  GAATTCTTTTTTATTAATTTTAAAGTTACATTTTTTCTAAATAACACATATTTTTACGATGGAAATATAAAACATTTTTGTAAACCATTT  -729

-728  TGTTACCTGTATATATGTATTTGTTTGATTTATTTATAAGGAAAGCGAAATCAGGAAATTTAGCACCACCTGTTGGTCAGCAAGAAAAAA  -639

-638  TATTCTTGCATACTTTTGGGCTGACTATGAATATTCAAAAAATTGCTCCCAAATGGTAATGGTTTTTTATTTCGGTCTAATACTACAACT  -549

-548  AATGAGCCATAGCAGTACATTATAAATTCGAAGTATGTCTTTGCATTAGGGCTTATATTTTGGGCGACATATTTGAGCAGTCTGCAAACA  -459

-458  ATCGGCAAAATTTTATAAAAATGTTTCCTGTCTTAGTTACAATATCATCAATTTGAAATTGAGCAAGGCGATTATTATTATATTTGCAAG  -369

-368  TTGTCCTTAAATAAGGAAGTTAATAAAAAAACATACAAATTATCAAATTTTGGTGAGGAATGACTCCGCGAAATTATGGACGGAGCCCAT  -279

-278  ATCCCGGACAGCAAGTAAAAAACGGTCTGAAAACCTGCCGATTGCCCGATAAACTTGTTGGGGCATCTCAACGCCAATTAAGCGGTCTAC  -189

-188  AAAGTGACTGGGCTGGAGGTCCCCGCGATGACCTTGTTAAGATCCAGATGCAGAAACAGGCCACTGTGGCACTGGGTCGACGGCAAGGAA  -99

                                               -->> -->-60/-59,-55
 -98  GCCGCCTATAAAAGCCGATGTGAGTACCGTAGTGAACTTGTGTAAAATCAACTACCGACAGGAGCAAACCTAATTCATCAACCTAAAAAT  -9
             ------

  -8  TCGATCAGCATGTTGGTTAAGGTATATCATGTGTTATTTACAAGTTGGCTTGCCTTTATCCTAGTCCTTTAACCACGTACAGACTGCGCT  81
               MetLeuValLys                                                                ThrAlaLe  (7)

  82  ATTTGTGACCCTCATCGGCTTGGCTCAAGCTGGTCCACTGCCCGCGAAATCATCTGGAAGTGAGGACACCTATGATTCTCATCCGCAGTA  171
      uPheValThrLeuIleGlyLeuAlaGlnAlaGlyProLeuProAlaLysSerSerGlySerGluAspThrTyrAspSerHisProGlnTy  (37)

 172  CTCATTTAACTATGATGTTCAGGATCCAGAGACAGGAGATGTTAAGTCCCAGTCGGAGTCTCGGGATGGCGATGTAGTCCACGGTCAGTA  261
      rSerPheAsnTyrAspValGlnAspProGluThrGlyAspValLysSerGlnSerGluSerArgAspGlyAspValValHisGlyGlnTy  (67)

 262  CAGCGTGAATGATGCCGATGGTTACAGACGAACCGTGGACTACACGGCCGATGATGTCCGTGGATTCAACGCCGTGGTGCGTCGTGAACC  351
      rSerValAsnAspAlaAspGlyTyrArgArgThrValAspTyrThrAlaAspAspValArgGlyPheAsnAlaValValArgArgGluPr  (97)

 352  ACTTTCCAGTGCCGCGGTGGTTGTGAAGCCACAGGCTACAGCAGTCGTTCCAAAAGTTCAGTTAAAGCCTCTGAAGAAGTTGCCAGCCCT  441
      oLeuSerSerAlaAlaValValValLysProGlnAlaThrAlaValValProLysValGlnLeuLysProLeuLysLysLeuProAlaLe  (127)

 442  GAAGCCGCTTTCTCAGGCATCGGCTGTGGTGCACCGATCCTTTGCACCGGTGGTCCACCATGCCCCAGTGACCCATGTCGTGCACCACGC  531
      uLysProLeuSerGlnAlaSerAlaValValHisArgSerPheAlaProValValHisHisAlaProValThrHisValValHisHisAl  (157)

 532  AGCTCCGGCGCATTCTTTCGTCTCTCACCACGTTCCCGTGCTGAAGACTACCGTGCACCACGCCCATCATCCCCATGCCATTTCATATGT  621
      aAlaProAlaHisSerPheValSerHisHisValProValLeuLysThrThrValHisHisAlaHisHisProHisAlaIleSerTyrVa  (187)

 622  GTTCTAGA                                                                                    629
      lPheEnd                                                                                     (188)
```

Edg84A Sequence. Strain, *Canton S.* Accession, M71249 (DROEDG84A).

Gene Organization and Expression

Open reading frame, 159 amino acids; mRNA length, 581–591 bases. Primer extension was used to define the 5′ ends (there seem to be three clustered

Edg91A

```
-1116  CTGCAGGTCGATTAAAGGCTCGATTGACCAAATGTAAAATCCCAAATAAGAAAGACTTTACTCGTTGAGTTTTTGTAAGAAACTAATTTT  -1027

-1026  ATTTGGAAATATCTTCGGTTTAAATAGGTGACATGAGAATCGCATCTTAAAGTAAATGGCCTACGCAGAGGCTAAGTAAATAGTCCCCGC  -937

-936   CTTATCGAGGTCCCACGCTCGGGCACATCTGCCTATCTTGAGCGGCGAGGACCTTATCTGTGGTCTCCCACTAAGGGACTATTTTAGGAG  -847

-846   GCGGGGAACGATCTCAAGTGACTGACTCATGTAGTGTGCACTTAAATTACATTTTTGAGCAATGCACCCATGTCGCCTTGGATAACAAAA  -757

-756   TCCTAAATATAATTTATCGCTCTCGATTCATTTACATAAGATATGAACGGAGCCCAAAATTGTAAGTCTTTAAATATATTCGTGTTCATG  -667

-666   TGTGAACAACAAGCATTTGGGTTTAACCCTGCTATTGTAACCCATTAAAAGAAATATTTTATCAAAATTAATATTTATAAAATATTTATA  -577

-576   TAGCCTTTAAATACTCCTTTCATTCTGATTTGAAGTGGCTAAATTAATAGGTAAATTATTATTTATCAACTCATACTTTTAAGAAATTAG  -487

-486   TTTCTTTACATTGAAATTTTTTAAAGATATGCTTAGTTTAAAATTTTATATTTTTAAATTGCAGAGTCATCTATCGGTTACAGTGGAATA  -397

-396   TTATATTCGTATTTCAACATTTTTCTGGTTGGTCTTGAAATTACCGGGTGATTGTAGTATGCGATCGCTCAGTGATATTTTTATGGTTCA  -307

-306   CGATCTTGATGACCGGCAACTAAGACAACCTCAAAAATGATAATTAGTTGGGCCTGTGACTTCAAGAAATTAACGCGTTCTGGGGCCAAG  -217

-216   TGAAGCACTGGTAGGCAAAGTGTCTCTTGGGGGATTCCAAAGTTACGTCACAAACTGGTTTCGCTTTCGCCGTGTTTGTTCTGCAATTTG  -127

-126   CGTAGAATCACTTGGCAATGCGTAGCGCGTACTTGAGCTTCTTGGCCAGATTGAAGCGGCGGTATAAAAGCGGTGGGCACTTCACAACTT  -37
                       ----                                                    ------
         -->>-33/-34-->-23
-36    GCAATTTAGTTTCATCCAAGAAGCGCTCGTTATCGCAATGGCTCTGGTTCGCGTGAGTTGTGTAAGTCCGGCTGCTATTTCCGCTCCGAT  53
                                             MetAlaLeuValArgValSerCys                                  (8)

54     TGGGATGCACTGAATCGATTTGGTTACCTTGCAGATGCTGGCCCTTTTGCTGATTGCCGGTCAAGGTCAGGCGGCGCCGGTGAAGACCGA  143
                                          MetLeuAlaLeuLeuLeuIleAlaGlyGlnGlyGlnAlaAlaProValLysThrGl  (27)

144    AGGTCGCACCTTGGGCCTTCTGGGCGGTGGATTTGGTGGCAGTGTAGGACTTAGTGCCGGCATCGGAGTGGGTGGTGGCCTGTATAGCGG  233
       uGlyArgThrLeuGlyLeuLeuGlyGlyGlyPheGlyGlySerValGlyLeuSerAlaGlyIleGlyValGlyGlyGlyLeuTyrSerGl  (57)

234    TTTCGGAGGCGGTGGCTATCCTGGTGGCTATGCGAGTGGATACCCAGGTGGATATGGTGGTGGCTACTCAGGCTATAACGGCTACGGAGG  323
       yPheGlyGlyGlyGlyTyrProGlyGlyTyrAlaSerGlyTyrProGlyGlyTyrGlyGlyGlyTyrSerGlyTyrAsnGlyTyrGlyGl  (87)

324    CAGTGGATTCGGAGGTGGCTACTATCCAGGAGGAGGTTACTCCGGCTTTGGACACAGGCCGCATTACCACGGAGGATACTATCCGGGCGG  413
       ySerGlyPheGlyGlyGlyTyrTyrProGlyGlyGlyTyrSerGlyPheGlyHisArgProHisTyrHisGlyGlyTyrTyrProGlyGl  (117)

414    TGGATCGTACCACAATCAGGGCGGATCTTATGGCGGCCACTATAGTCAGTCACAGTACTCGAATGGATATTACGGAGGTGGTGGCTATGG  503
       yGlySerTyrHisAsnGlnGlyGlySerTyrGlyGlyHisTyrSerGlnSerGlnTyrSerAsnGlyTyrTyrGlyGlyGlyGlyTyrGl  (147)

504    AGGCGGTGGCTATGGAGGCAATGGCTTCTTTGGAAAGTAAAGATGCCAAATCTTGCCACCGGGATAGTTAAGTACTTGTGATTGACCCTT  593
       yGlyGlyGlyTyrGlyGlyAsnGlyPhePheGlyLysEnd                                                    (159)

594    TGTAGATTGTAAAATAAACGAAAAAACATAACCAGATTTAGTAAGCTCAATTCAAGGCACTTAAAAATCCGGTTTTCCTGTTGGAAATAT  683
                   ------        |(A)n

684    TGTCCTTGGCGCTGCCTTTGTGGTTATTCTCTCACTGATTTTTATGAAGCAGACGCGACGTGCATAAATTTAATGGCCAAAGATCCAAGA  773

774    TTTATGCGCAAGTCTGACTAATCCATTGCCTCGAAATTATCTGGGAATTC  823
```

Edg91A SEQUENCE. Strain, *Canton S.* Accession, M71250 (DROEDG91A).

transcription initiation sites). The 3′ end was obtained from a cDNA sequence. There is an intron after Cys-8 (*Edg91A* Sequence) (Apple and Fristrom 1991).

Developmental Pattern

As is true for *Edg78E*, *Edg91* is expressed during the time of pupal exocuticle synthesis (8–12 h after pupariation) in both larval and imaginal epidermal cells. Also as for *Edg78E*, a 20-HE pulse in imaginal discs *in vitro*, induces transcription of *Edg91A* (Apple and Fristrom 1991).

References

Apple, R. T. and Fristrom, J. W. (1991). 20-Hydroxyecdysone is required for, and negatively regulates, transcription of *Drosophila* pupal cuticle genes. *Dev. Biol.* **146**:569–582.

Chihara, C. J., Silvert, D. J. and Fristrom, J. W. (1982). The cuticle proteins of *Drosophila melanogaster*: Stage specificity. *Dev. Biol.* **89**:379–388.

Fetchel, K., Natzle, J. E., Brown, E. E. and Fristrom, J. W. (1988). Prepupal differentiation of Drosophila imaginal discs: identification of four genes whose transcripts accumulate in response to a pulse of 20-hydroxyecdysone. *Genetics* **120**:465–474.

Fetchel, K., Fristrom, D. K. and Fristrom, J. W. (1989). Prepupal differentiation in *Drosophila* imaginal discs: distinct cell types elaborate a shared structure, the pupal cuticle, but accumulate transcripts in unique patterns. *Development* **106**:649–656.

Fristrom, J. W., Hill, R. J. and Watt, F. (1978). The procuticle of *Drosophila*: Heterogeneity of urea-soluble proteins. *Biochemistry* **17**:3917–3924.

Henikoff, S., Keene, M. A., Fetchel, K. and Fristrom, J. W. (1986). Gene within a gene: nested *Drosophila* genes encode unrelated proteins on opposite DNA strands. *Cell* **44**:33–42.

Kimbrell, D. A., Tojo, S. J., Alexander, S., Brown, E. E., Tobin, S. L. and Fristrom, J. W. (1989). Regulation of larval cuticle protein gene expression in *Drosophila melanogaster*. *Dev. Genet.* **10**:198–209.

Kimbrell, D. A., Berger, E., King, D., Wolfgang, W. J. and Fristrom, J. W. (1988). Cuticle protein gene expression during the third instar of *Drosophila melanogaster*. *Insect Biochem.* **18**:229–235.

Pultz, M. A., Diederich, R. J., Cribbs, D. L. and Kaufman, T, C. (1988). The proboscipedia locus of the Antennapedia-Complex: a molecular and genetic analysis. *Genes Dev.* **2**:901–920.

Silvert, D. J., Doctor, J., Quesada, L. and Fristrom, J. W. (1984). Pupal and larval cuticle proteins of *Drosophila melanogaster*. *Biochemistry* **23**:5767–5774.

Snyder, M., Hunkapiller, M., Yuen, D., Silvert, D., Fristrom, J. and Davidson, N. (1982). Cuticle protein genes of *Drosophila*: Structure, organization and evolution of four clustered genes. *Cell* **29**:1027–1040.

10

The Cytochrome c Gene Cluster: *Cytc1, Cytc2*

Chromosomal Location: 2L, 36A10-11

Map Position: 2-[52]

Synonyms: DC4 and DC3

Product

Cytochromes c, small heme-binding proteins important in the mitochondrial electron-transport chain.

Structure and Function

Two Cys residues near the N-terminus bind the heme group. Another region near the N-terminus has a primary role in the import of cytochromes c into mitochondria *in vitro*; other portions of the molecule are also necessary for this transport (Sprinkle et al. 1990). These proteins are ubiquitous among eukaryotes and, judging from comparisons made among cytochromes c from 30 species, they are highly conserved. The CYTC1 sequence is very similar to the consensus sequence for other eukaryotic cytochromes c: at every position, the residue present in CYTC1 is found also in some other eukaryotic cytochrome c. CYTC2, on the other hand, is more divergent and has some unique characteristics: at 12 positions, the residues found in CYTC2 are not represented in any other eukaryotic cytochrome c (Fig. 10.1) (Limbach and Wu 1985). It is not known whether the two *Drosophila* proteins have specialized functions.

Organization of the Cluster

The two genes are arranged in tandem with approximately 2.5 kb between the 3′ end of *Cytc2* and the 5′ end of *Cytc1*. These are probably the only genes

```
       1                                                   50                                                                       100 101       111
Dm c1  .MGVPAGDVE KGKKLFVQRC AQCHTVEAGG KHKVGPNLHG LIGRKTGQAA GFAYTDANKA KGITWNEDTL FEYLENPKKY IPGTKMIFAG LKKPNERGDL IAYLKSATK* .
Human  .....MGDVE KGKKIFIMKC SQCHTVEKGG KHKTGPNLHG LFGRKTGQAP GYSYTAANKN KGIIWGEDTL MEYLENPKKY IPGTKMIFVG IKKKEERADL IAYLKKATNE *
Dm c2  ...MGSGDAE NGKKIFVQKC AQCHTYEVGG KHKVGPNLGG VVGRKCGTAA GYKYTDANIK KGVTWTEGNL DEYLKDPKKY IPGTKMVFAG LKKAEERADL IAFLKSNK*. .
Yeast  MTEFKAGSAK KGATLFKTRC LQCHTVEKGG PHKVGPNLHG IFGRHSGQAE GYSYTDANIK KNVLWDENNM SEYLTNPKKY IPGTKMAFGG LKKEKDRNDL ITYLKKACE* .
  CON  ------G--- -G---F---C -QCHT-E-GG -HK-GPNL-G --GR--G-A- G--YT-AN-- K---W-E--- -EYL--PKKY IPGTKM-F-G -KK---R-DL I--LK----- -
```

FIG. 10.1. Comparison of the human (Accession, M22877), yeast (Accession, V01298) and *Drosophila* (Dm) sequences. The CON(sensus) line displays all positions for which there is agreement among the four sequences. There are 77% and 67% overall identities between the human protein and CYTC1 and CYTC2, respectively. Sequences aligned with the GCG *Pileup* program

Cytc1

```
              .         .         .         .         .         .         .         .         .
-766 TCTGATGACGTTGCGACGCCCTCCACGCGCGTATTAGTGAGAGCAAAGTATGTGGGTTAAAAAGGGGGTGGCCGCAAATGGAAATGCAGA  -677
              .         .         .         .         .         .         .         .         .
-676 CTACGTTAGATAATAATTTCGGGCCTTATCAGAAACAACAGCCGACTAATGCACTTAGCATGAGCAATTTTAATAATTCCGTTTCCGCAG  -587
              .         .         .         .         .         .         .         .         .
-586 GAGCTTATCAATTGTTTACATAACGGGGCAAGGGGACAAATATTAATTCACGGTCCATAACTACCTACATTAACCCATTATTTCCCAAGC  -497
              .         .         .         .         .         .         .         .         .
-496 ATGGAAATTTTTGATGATATAAAGACGTTATTATTTTAATACCTTAAAAATATATAATATTATATAAGTAACGTTGGGAAATCAACTGGT  -407
              .         .         .         .         .         .         .         .         .
-406 TAATAAATTTTAAATTTCGGGTTTATTTATTCAATAATCTTTTGATAATGTATGGCTGAAAGTGAAGCTTTTATCAGTATCTACACAATG  -317
              .         .         .         .         .         .         .         .         .
-316 GTTCATTGTGGCTAATAATAAATGGTATCAAATATCGTATAACTATTTTTTGCAGTGAAACCAGAATTTCGGACTAAGTACATAAGCAAA  -227
              .         .         .         .         .         .         .         .         .
-226 TGATATAAAATATATATTGTAATCAATTTATCAGAATAGAACAAATTAATTGGTAGCAGTTATGAACTTTGAACCATTTAGTAGCCAGTT  -137

              .         .         .         .         .         .         .  -->-67?.        .
-136 TTTTAAGTTTTTCAAACCTAAGATGTAAGATAACAGATATATTGGTTACCCTTGTTTTATGAACCACTCATTAATAACAAACATTTCTTT  -47
                                               ------
              .         .         .         .         .         .         .         .         .
 -46 TTACAGTCGAGTCCGTGTTAACACATTAATTAACCACATAATCCATAATGGGCGTTCCTGCTGGTGATGTTGAGAAGGGAAAGAAGCTGT   43
                                                    MetGlyValProAlaGlyAspValGluLysGlyLysLysLeuP  (15)
              .         .         .         .         .         .         .         .         .
  44 TCGTGCAGCGCTGCGCCCAGTGCCACACCGTTGAGGCTGGTGGCAAGCACAAGGTTGGACCCAATCTGCATGGTCTGATCGGTCGCAAGA  133
     heValGlnArgCysAlaGlnCysHisThrValGluAlaGlyGlyLysHisLysValGlyProAsnLeuHisGlyLeuIleGlyArgLysT  (45)
                 ***       ***

              .         .         .         .         .         .         .         .         .
 134 CCGGACAGGCGGCCGGATTCGCGTACACGGACGCCAACAAGGCCAAGGGCATCACCTGGAACGAGGACACCCTGTTCGAGTACCTGGAGA  223
     hrGlyGlnAlaAlaGlyPheAlaTyrThrAspAlaAsnLysAlaLysGlyIleThrTrpAsnGluAspThrLeuPheGluTyrLeuGluA  (75)
              .         .         .         .         .         .         .         .         .
 224 ACCCCAAGAAGTACATCCCCGGCACCAAGATGATCTTCGCCGGTCTGAAGAAGCCCAACGAGCGCGGCGATCTGATCGCCTACCTGAAGT  313
     snProLysLysTyrIleProGlyThrLysMetIlePheAlaGlyLeuLysLysProAsnGluArgGlyAspLeuIleAlaTyrLeuLysS  (105)
              .         .         .         .         .         .         .         .         .
 314 CGGCGACCAAGTAATGGTGCTGTCCATCAACTTACCCACAACAACTGCAGGATGTCAAACTGTATTATTGTGTTCAGTCACAGTCCGGCA  403
     erAlaThrLysEnd
                                                                                                 (108)
              .         .         .         .         .         .         .         .         .
 404 CGCAAATGCAGCAGCAGCAACAACTACAACTACAAATCAACATAGTACAGACCTAAAGAACTACAATTATGTTAATTATAAAGTTTAAAT  493
              .         .         .         .         .         .         .         .         .
 494 AGGACAATTTATTATTTAATTTAAATAAAAAGTGGAATATTTAATTCAAACCCGATGAGAATTGTGACATCCACAAAAAAGTTAAATAAT  583
                           ------
              .         .         .         .         .         .         .         .         .
 584 AAAAAAAGAACTAAAAAATGATATAAAATCTGTTTTATGCGAGGACCTGGTTTTTGTAGCTCGCAGGTCAAAAAGAATAAAAAAAGCTTC  673
              .         .         .         .         .         .         .         .         .
 674 TTCAGATTTTTGACTCGGGCAACTCAAATTAAAAATAAGAGATACCAATCATATTTATAAAACAATTGTCCTGGCAATTTCTATCAATAG  763
              .         .         .
 764 GTATCTGTTAGTCGTCAAACTCGACTGCG  792
```

Cytc1 SEQUENCE. Strain, *Canton S.* Accession, X01760 (DROCYCDC4).

Cytc2

```
-916  GTAATATAAATATATAAATAATATCATTCTCTGAAAAATATCAAATGCACTCTTGTAAATTTAAAACAAATTTAATTTTAAGATAATTGG  -827

-826  TTGAGATAAACATAGTTAATATTTTCAATTGATCCTTTAAATTTTAAATTGCAGGTGAATATCATCCCTGTGTGACCGTTGTATGCGGCA  -737

-736  TGGTTCCATGTCTCTTTCCCGTTATTCATTTCCCTCTGCTTTGTTTTTTTTTTTTTTTTTGGTTTTGTTATGCGGTGGCATCTGTTTTTG  -647

-646  CCACAGGAAAAATGTTAAGAGAGGGGAAGGCAGGGGGCGAAAACGGAGAGTGCGTAAATTGCGTTTTAATTGGAAGGAAATGTTCAAAGC  -557

-556  ACATGTGCATCTGCTAGTCAACGAATTGGTTGGGAAAGGGGGTGGAAAAGGGGTTGCAAGCCGAATGTGTCTGCTAATTGAATTACTTTC  -467

-466  GGTTGCTTTTCCCATTAGAAGTGCCGCCAAGTTCTCGAGCTGCTTGTTTGCTTTTCATTTAATACCCATTTTGATTTAATTTTCGTTTTT  -377

-376  CCTATTTTTCTGACCCAATTTTGTTTTGCTTTCGTGCATTAGCAGCTGTCTCTGTCTATCGCTGTGCAGCCAAGAGAGTGACCAAGAGAA  -287

-286  ACGCTCTCTCTCTCTCTCAGTTGTCCAGGACTTGCACTTTTCAAACGGTTTTTTAGGACACTGAAACAAATTGAATCTGTTTTTCTTT  -197

-196  TCTATCAAATTTTTAGTTCTACACTTTTCTTTTTCTTTTCTTTTTTTTTTTTTTCGGAATCAACCAATTTCTATTTGATCCAAATTAACAA  -107
                                                                      ----
                                                                      -->-43?
-106  AAAACAAATAACAAAAAATTAAAAAATATAGAAATAAAAGCTGCATAAAAAGTTGAATTCTAAATCATAAATATCATTTTTCCCTATTTG  -17
                                ------

 -16  TCTTTCAGGCTTCCAAGATGGGTTCTGGTGATGCAGAGAACGGCAAGAAGATATTTGTGCAGAAGTGCGCCCAGTGCCACACCTACGAAG  73
                        MetGlySerGlyAspAlaGluAsnGlyLysLysIlePheValGlnLysCysAlaGlnCysHisThrTyrGluV  (25)
                                                                           ***      ***

  74  TGGGGGGCAAACACAAGGTGGGCCCAAATCTTGGCGGGGTCGTGGGTCGCAAGTGTGGCACAGCAGCGGGATACAAGTATACCGATGCCA  163
      alGlyGlyLysHisLysValGlyProAsnLeuGlyGlyValValGlyArgLysCysGlyThrAlaAlaGlyTyrLysTyrThrAspAlaA  (55)

 164  ATATAAAGAAGGGCGTTACCTGGACAGAGGGGAATTTGGACGAGTACCTCAAGGACCCGAAGAAATACATTCCCGGAACAAAGATGGTGT  253
      snIleLysLysGlyValThrTrpThrGluGlyAsnLeuAspGluTyrLeuLysAspProLysLysTyrIleProGlyThrLysMetValP  (85)

 254  TCGCAGGTCTTAAAAAGGCTGAGGAGCGGGCCGATTTGATTGCCTTCCTCAAGTCAAACAAGTAGAATCGCCTGCGAAACAACAAGATCG  343
      heAlaGlyLeuLysLysAlaGluGluArgAlaAspLeuIleAlaPheLeuLysSerAsnLysEnd                          (105)

 344  GCCACCATGCTATCCAGAAAACTGCGCTTAAAGACTACAAACATATTCAAAAGATGACGTATTTCACTTGGATTTCGAAACTTTGATTGG  433

 434  GAATGGTCGAGCTCAAATACATTTCAAAAAGGTTTACTTTCACTTTAGCCAATTAAAGTTGATAAACCAAAAACCCTCTTCTTAATTCAA  523

 524  GTTGTGTGCGACGCGGGTGGAGGAAAGTGTTGTACCAATCAGCTTTGGTCACAGTTGGTTTTATGGTCCTACTAGCAAAATGTAATAAAT  613
                                                                                          ------

 614  TGGAGAAGCTTGTTAAATAATGCAAATTTTCCAGAGGCTTTCCAATATAGTCCCCTTAATAGGGGAAAAAATTACTTATACGCCGTGTGG  703

 704  TGGATAAATACGGGTACAAAAGCTT  728
```

Cytc2 Sequence. Strain, *Canton S.* Accession, X01761 (DROCYCDC3).

responsible for cytochrome c production in *Drosophila* (based on Southern analysis) (Limbach and Wu 1985).

Cytc1

Gene Organization and Expression

Open reading frame, 108 amino acids. The 5′ and 3′ ends have not been identified, a tentative site of transcription initiation was indicated based on sequence elements. A putative TATA box at −99 and a polyadenylation signal at 517 suggest a mRNA of approximately 600 bases, in reasonable agreement with an observed RNA of 0.9 kb bases. There are no introns in the coding region (*Cytc1* Sequence) (Limbach and Wu 1985).

Developmental Pattern

Expression is highest in first instar larvae and adults and lowest in third instar larvae. In adults, expression is higher in the muscle-rich thorax than in the head or abdomen. Expression of *Cytc1* is 25–150 times higher than that of *Cytc2* (Limbach and Wu 1985).

Cytc2

Gene Organization and Expression

Open reading frame, 105 amino acids. The 5′ and 3′ ends have not been identified, a tentative site of transcription initiation was indicated based on sequence elements. A putative TATA box at −80 and a polyadenylation signal at 607 suggest that the mRNA is approximately 700 bases long, but the only transcript detected by northern analysis is 2.1 kb long; this indicates that the elements described here do not constitute the whole gene. There are no introns in the coding region (*Cytc2* Sequence) (Limbach and Wu 1985).

Developmental Pattern

Cytc2 is present uniformly in all postembryonic stages and in adult head, thorax and abdomen. Expression is at very low levels relative to that of *Cytc1* (Limbach and Wu 1985).

References

Limbach, K. J. and Wu, R. (1985). Characterization of two *Drosophila melanogaster* cytochrome c genes and their transcripts. *Nucl. Acids Res.* **13**:631–644.

Sprinkle, J. R., Hakvoort, T. B. M., Koshy, T. I., Miller, D. D. and Margoliash, E. (1990). Amino acids sequence requirements for the association of apocytochrome c with mitochondria. *Proc. Natl. Acad. Sci.* (*USA*) **87**:5729–5733.

11

The *Dopa decarboxylase* Cluster: *Ddc, l(2)amd, Cs, DoxA2*

Chromosomal Location:			**Map Position:**
Ddc	2L,	37C1-2	2-54
l(2)amd	2L,	37B13-C2	2-54
Cs	2L,	37B13-C2	2-54
DoxA2	2L,	37B10-13	2-53.9

Organization of the Cluster

The *Ddc* cluster is arbitrarily defined as those genes that fail to complement *Df(2L)TW130*, 37B9–C1 to 37D1–2, an 8–12-band deletion in the left arm of chromosome 2. The cluster contains 18 genetically identified genes plus three transcription units for which no mutations are known. Some of the genes in this cluster seem to be functionally related, most of them being involved in the formation, sclerotization and pigmentation of cuticle. Several genes in the cluster have mutant alleles that are female sterile. For three genes, *Ddc*, *l(2)amd* and *DoxA2*, some of the gene-product biochemistry is known; these genes are involved in catecholamine metabolism (Fig. 11.1) (Wright 1987).

Most of the genes are grouped in two very dense subclusters. The centromere-proximal sub-cluster contains nine elements in 25 kb of DNA, 70% of which is transcribed; the distal sub-cluster includes seven genes in 22 kb (Fig. 11.1) (Wright 1987).

The sequences of *Ddc* and *l(2)amd* are related, and it is probable that the genes originated by duplication. It appears unlikely, however, that all the genes in the cluster are members of a single family; the sequences of *l(2)37Cc* and *Cs*, for example, are not obviously related to *Ddc* or *l(2)amd.* Three genes in the proximal cluster and one in the distal cluster are presented here.

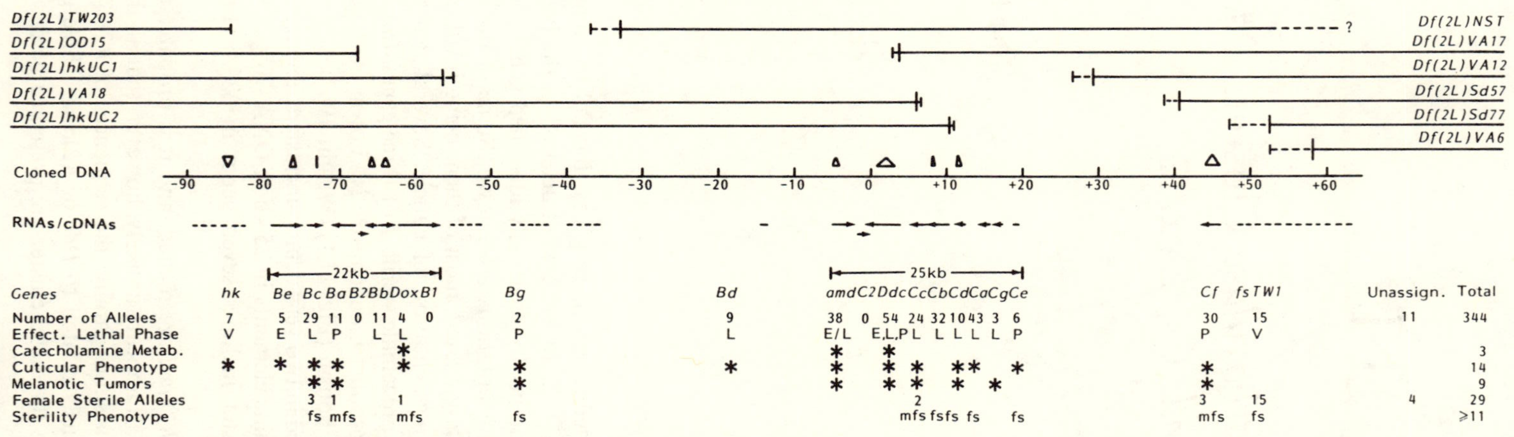

FIG. 11.1. *Ddc* cluster (centromere to the right), from Wright (1987), updated in 1992 by T. R. F. Wright: "The genetic and molecular organization of the *Ddc*-region. Deficiencies: *Solid lines* represent deleted DNA with *dashed lines* indicating uncertainty of the position of the breakpoint. Cloned DNA coordinates in kb from Gilbert et al. (1984). *Small triangles* above the cloned DNA line physically locate small deletion mutations and *short lines underneath* designate regions which hybridize to mRNAs or cDNAs with *arrowheads* representing direction of transcription. Transformed DNA lines indicate the segments of DNA that have been transformed by P elements. All the gene symbols except *hk*, *Dox* = *DoxA2*, *Bh*, *amd* = *1(2)amd*, *Cs*, *Ddc*, and *fsTWI* = *fs(2)TWI* should be preceded by '*1(2)37*', e.g., *1(2)37Ba*. Effective lethal phase designations: *E* embryonic; *L* larval; *P* pupal; *V* viable. *Asterisks* underneath a gene symbol indicate the mutant alleles of that gene alter catecholamine metabolism, express a mutant cuticular phenotype, or produce melanotic tumors. Sterility phenotype: Individuals hemizygous for female sterile, ts, or hypomorphic alleles or heterozygous for complementing heteroalleles are female sterile = *fs* or both male and female sterile = *mfs*. See text for the sources of the information included in this figure." The transcription unit *Cs* is designated *C2* in this figure. The transcription unit *Cf* is actually transcribed toward the centromere.

Ddc
(Dopa decarboxylase)

Product

Dopa decarboxylase (DDC, EC 4.1.1.26).

Structure

DDC is a homodimer of 54 kD subunits (Clark et al. 1978). Two forms of the enzyme, which are generated by alternative splicing, have been isolated; one is found in the central nervous system and the other in the epidermis (Morgan et al. 1986).

The amino acid sequence has considerable similarity with the DDC of mammals (Fig. 11.2) (Scherer et al. 1992), and prokaryotes (Jackson 1990). The heptapeptide consisting of residues 332 through 338 has similarities with the pyridoxal binding sites of porcine DDC and feline glutamate decarboxylase. Lys-337 is probably the pyridoxal-binding residue. See also *1(2)amd* below.

Function and Tissue Distribution

This enzyme catalyzes the decarboxylation of dopa (3,4-dihydroxy-L-phenylalanine) to dopamine, and of 5-hydroxytryptophan to serotonin. DDC is involved in tanning of the cuticle, and most of the enzyme is found in the epidermis where its activity peaks during the molting episodes. DDC is also involved in the synthesis of neurotransmitters and is present in a group of 150 serotonergic and dopaminergic neurons of the central and visceral nervous system (Wright et al. 1976a, 1976b; Konrad and Marsh 1987; Beall and Hirsh 1987; for reviews see Wright 1987 and Hirsh 1989).

Mutant Phenotype

Amorphic mutations are lethal; death occurs mostly in late-embryonic and larval stages. A few individuals survive to the pupal stage. Survivors have cuticular structures that are characteristically incompletely pigmented and sclerotized.

Gene Organization and Expression

Open reading frame, 475 or 510 amino acids; expected mRNA size, 2,067 or 1,923 bases, depending on splicing. The 5′ end was defined by S1 mapping and primer extension. The 3′ end was defined by cDNA sequencing. There are three introns, one in the leader, spanning −692 through −57, one after the Ser-33 codon and one in the Arg-62 codon. Two alternative splicing products have

```
     1                                                   50                                                  100
Dm   MSHIPISNTI PTKQTDGNGK ANISPDKLDP KVSIDMEAPE FKDFAKTMVD FIAEYLENIR ER.VLPEVKP GYLKPLIPDA APEKPEKWQD VMQDIERVIM
Rat  .......... .......... .......... .....MDSRE FRRRGKEMVD YIADYLDGIE GRPVYPDVEP GYLRALIPTT APQEPETYED IIRDIEKIIM
CON  ---------- ---------- ---------- -----M---E F----K-MVD -IA-YL--I- -R-V-P-V-P GYL--LIP-- AP--PE---D ---DIE--IM

     101                                                 150                                                 200
Dm   PGVTHWHSPK FHAYFPTANS YPAIVADMLS GAIACIGFTW IASPACTELE VVMMDWLGKM LELPAEFLAC SGGKGGGVIQ GTASESTLVA LLGAKAKKLK
Rat  PGVTHWHSPY FFAYFPTASS YPAMLADMLC GAIGCIGFSW AASPACTELE TVMMDWLGKM LELPEAFLAG RAGEGGGVIQ GSASEATLVA LLAARTKMIR
CON  PGVTHWHSP- F-AYFPTA-S YPA--ADML- GAI-CIGF-W -ASPACTELE -VMMDWLGKM LELP--FLA- --G-GGGVIQ G-ASE-TLVA LL-A--K---

     201                                                 250                                                 300
Dm   EVKELHPEWD EHTILGKLVG YCSDQAHSSV ERAGLLGGVK LRSVQSE.NH RMRGAALEKA IEQDVAEGLI PFYAVVTLGT TNSCAFDYLD ECGPVGNKHN
Rat  QLQAASPELT QAALMEKLVA YTSDQAHSSV ERAGLIGGVK IKAIPSDGNY SMRAAALREA LERDKAAGLI PFFVVVTLGT TSCCSFDNLL EVGPICNQEG
CON  ------PE-- ------KLV- Y-SDQAHSSV ERAGL-GGVK -----S--N- -MR-AAL--A -E-D-A-GLI PF--VVTLGT T--C-FD-L- E-GP--N---

     301                                                 350                                                 400
Dm   LWIHVDAAYA GSAFICPEYR HLMKGIESAD SFNFNPHKWM LVNFDCSAMW LKDPSWVVNA FNVDPLYLKH DMQGSA..PD YRHWQIPLGR RFRALKLWFV
Rat  VWLHIDAAYA GSAFICPEFR YLLNGVEFAD SFNFNPHKWL LVNFDCSAMW VKKRTDLTEA FNMDPVYLRH SHQDSGLITD YRHWQIPLGR RFRSLKMWFV
CON  -W-H-DAAYA GSAFICPE-R -L--G-E-AD SFNFNPHKW- LVNFDCSAMW -K-------A FN-DP-YL-H --Q-S----D YRHWQIPLGR RFR-LK-WFV

     401                                                 450                                                 500
Dm   LRLYGVENLQ AHIRRHCNFA KQFGDLCVAD SRFELAAEIN MGLVCFRLKG SNERNEALLK RINGRGHIHL VPAKIKDVYF LAMAICSRFT QSEDMEYSWK
Rat  FRMYGVKGLQ AYIRKHVKLS HEFESLVRQD PRFEICTEVI LGLVCFRLKG SNQLNETLLQ RINSAKKIHL VPCRLRDKFV LRFAVCSRTV ESAHVQLAWE
CON  -R-YGV--LQ A-IR-H---- --F--L---D -RFE---E-- -GLVCFRLKG SN--NE-LL- RIN----IHL VP----D--- L--A-CSR-- -S------W-

     501        516
Dm   EVSAAADEME QEQ*..
Rat  HIRDLASSVL RAEKE*
CON  -----A---- ------
```

FIG. 11.2. Comparison of the rat (Accession, M27716) and *Drosophila* (Dm) DDCs. There is 60% overall identity between the two proteins. Sequences aligned with the GCG *Pileup* program.

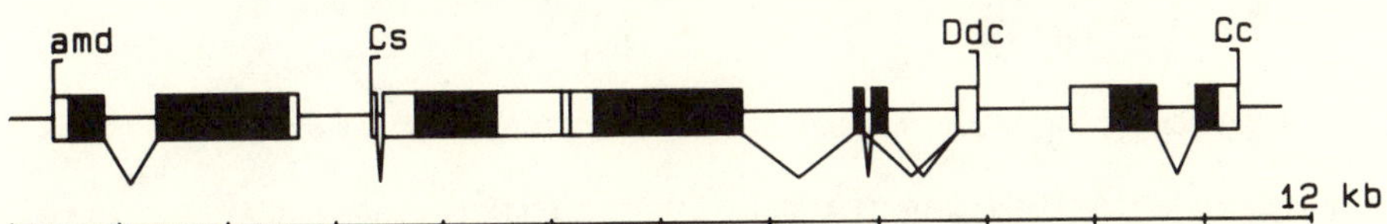

FIG. 11.3. Organization of the genes in the immediate vicinity of *Ddc*.

been detected. One is a 2.3 kb RNA in which all exons are present. The other, the most abundant, is a 2.1 kb RNA produced when the small second exon is spliced out together with the first two introns (*Ddc* Sequence). In the latter case the leader is spliced, in frame, onto the middle of the original open reading frame, and translation seems to start from an AUG six bases downstream of the splice site (Met-36) (Eveleth et al. 1986; Morgan et al. 1986). Transcription is toward the telomere (Fig. 11.3) (Spencer et al. 1986a).

Another gene in this cluster, *Cs*, is located immediately downstream of *Ddc*. The two genes are transcribed convergently and their untranslated 3′ ends overlap by 76 bp (*Ddc* Sequence) (Spencer et al. 1986a; Eveleth and Marsh 1987).

Developmental Pattern

The splicing reaction is tissue-specific with the 2.3 kb RNA occurring in embryos and in the nervous system and the 2.1 kb RNA involved in cuticular tanning. The 2.1 kb RNA is the predominant form during larval development; it is found in the integument fraction, and its level fluctuates according to the intensity of cuticle deposition (Eveleth et al. 1986; Morgan et al. 1986; Krieger et al. 1991).

Promoter

Proximal Elements P-element-mediated transformation of genes carrying 5′ deletions established that the 209 bp upstream of the transcription initiation site (up to position −1,093 in the *Ddc* Sequence) are sufficient for normally regulated full expression of *Ddc* in the epidermis. Deletions that leave only 25 bp of the 5′ region (up to position −909 in the *Ddc* Sequence) result in much lower levels of mRNA production, but transcription is started correctly despite the absence of the TATA box (Hirsh et al. 1986). Progressively lower levels of DDC are produced when deletions are introduced in the segment between −1,093 and −922. In that segment, five putative regulatory elements have been identified on the basis of sequence similarities between the distant species *D. melanogaster* and *D. virilis*. Each of the putative regulatory elements includes the consensus sequence C(A/T)GCG(G/A) (Scholnick et al. 1986). In addition, a dimer of this consensus sequence, designated element I and lying between positions −970 and −957 is necessary for central nervous system expression in both glial cells and neurons. Element I is totally conserved in the two species, and this is the only segment of the proximal promoter region that is protected

Ddc

```
-2521  CCAATTAATTACAGATCGATCCTAAAACGAATCTAATCACTTGCCCATATCATATAGATTCAGACTAAATACGTGACCTATTGAAGCTCA  -2432

-2431  GCGATGTGATGTGTACACCAAACACCCGCTCGTTTATCTCTGCCCTTGTTTACCCCATATGATGCCTGTTTATGCAATCCCCCTCTCAAA  -2342
                                                                -----------------Df6

-2341  GGCGCCATTCGACCCCTATAAGCGGAGAATACTTTCGCATTCATTCGCAATCTAAGATGGTCATAAATCAAATTGTAGTAAACTTCGCCT  -2252
                                                             ---------------------cf1

-2251  CAATCGACCCGAACTCCAGCCACCCGTAAAGCAGCATAATGTGGGTGGGTAGTTGGGCGACTGGTGGCTGGTGGCTGTTGGCTGGCGTGT  -2162
                       ------------------------bf2      -------------------uf3

-2161  GGGTGGAGCACCCAGCGCATTAAAATCGAAAGCAGAGCCGTTGGCATGGCCGTATAAATCTGTTGATTCAGCCAAGTGATTTGCCAAAGT  -2072

-2071  GGCTTCGTTGAAATGTCAGGCACACGCACTTTGCTCGGCACTCAGCAACAGTTGGACCACCCGCAGGATTCTTAGCAGCCCTACACTGAA  -1982

-1981  AGAAATTATTTTCTTTTGTCGTAGGCTAAAAATGTTTACTTGATTCTTTTAAATAGTAATTAAAGGAAGAGAATGATTTTCCTGTCGTAT  -1892
                                                         ---------------uf7          -------------

-1891  TCCAGGATCATTAGCCGAGCCGATATACCCATGTTTGTCTGTCCGTATAAACTTCGAGATTTTGGGAACTTTAAAAAAAAAAAACGGTCA  -1802
       -----------------uf8

-1801  CGAAAACAGTTTTGAAAAATATTTTGAATTTTTGTATTATATCTCTCGATATATTTGGCATAAACATTTAAGCCACATATTTATTGTTTC  -1712
                    --------------uf9                                          -------------uf10

-1711  TTGCCAATTTCTATTGATATTTCAACTGAATTTTGAAATTCCGGCCAAGTAACTGGCATCCAAAAGCTTTCTATAGTAATTTTGAATTTT  -1622

-1621  TCTCAGTGTATGCGGAACTGCCCGCTCAAAAGGCTCAACCTAGCCCACTTCCCCTAGCACAATGCGAAAGTGAGTGAGAGCATTGGATTA  -1532

-1531  TTTGACGTCACAATTCCATGAGCGGTTCAAAAAGCACGTCATATGTGGTGCTCTAATAACCGGTTTCCAAGATGCGCGTAAAGCTGCCAT  -1442

-1441  TCCACGGCTTAATCAATTTCTTGTCTTTCCTACGAATATAACTTTGTTTACATTTTTTGCGTGATTTTTTCTTCGGGGAGTCCAAGAAAA  -1352

-1351  ACCCTGTTTCGAGTGACTCATAATTGGGGGATTCCTGACGAGATCGCTCTCTTTCCACAAATTCGAGTTGGGAACGACGTGAGCAGAATT  -1262

-1261  CAAAATGTTTTGCTTGCTGTTTTAAATATCACTAGGTTCTCAAACTAATTTCAAAAATAATCAAATTAAGTTCACAGAGCTGGCAAATAA  -1172

-1171  AATGTAATAGCTTGCATGTATGTATATATATATATTTTTTTAAATTCTAAATAAATCCATGAAAAATAATGCCTTTGATATCCAGTTACT  -1082

-1081  GATTCAGCGCCCAATTAATGCATGTTCCAAAAAAGTGTCAAAAAACGTGCACAAATCAAACGAGAGCAGAATTTGTTTTTACGACAGCGG  -992

 -991  CTGCGATTCGAAGTTCAGCGGCTGCGGACTGCGATTGAACCGGTCCTGCGGAATTGGCAGCGCTGCTGGACGGGCTTTAAAAGCCATGGC  -902
                         I ------ ---------                                               ------
                        -->-883

 -901  CAAGAGCCGGGCAGCGCTCAGTTAAGAGGAGAACGCCAAGCGCACAGCAATCAGCACCGAAATATCAGCATCGAAATATCAGCAAATAAA  -812

 -811  TATTAGCTGTTCTAAACCAGGAGGGCAAACTGAACTTGGAGCAAAGATTTAGTTCGGAACGGAAGTAAAGCTCGGCAACAAGTGCAAACA  -722
                                   _
                                    |
 -721  ATTAAAAGCAGGTTAACTAAAGTGCAACGGTGAGAGACGAAAGTGTGGCTCCTCAACAGCCTCAGCTGCCTGAAGTGCTTGGCCAACATA  -632
                                   _|

 -631  ATGAGTGCATGGTGCATGCGAAAGATTCATTTCGGGGCTAACGCTGCGTATACGTAATGTGTATCTAAAACTGGGCATATACTATAGCCT  -542
```

```
-541  TGCTTCGGTTCAATTTGATAGTTCGGGCCCCGAATTCTATAGTGCTTAAGCTTTTCTCGGCTTTCGGTATCTGCATGCTTTTGTGTATCT  -452

-451  ATTAAAATAAGATTTTAGCTGGCAACAAGTCGTCGTCTCAATGCCAACTTGTTTACGTTGTTAAAATTGGAATTTAGAAAAAAAAAAAAT  -362

-361  ATAAAGCAGTCTTGATTAATGCAAGAATGCATTAAACATTCTAATTACCATACTAATTCACAGCCTATACTTAAGCAGCGCACTCGATGG  -272

-271  GAAAACGCTTTAAACTATTAATACCTTAATACCTTATTATTATAACTATTATCATCATCGTTTTGCCTATCAAGTAATTAGTATTCAATG  -182

-181  TCGTTCATTTGTCGTGTTTGCAGCGATACAGTTTTTTGTTTGAGCTGCTGCACTAATTAGCACTATCTTCAAAAACGCACTTCTATTAAT  -92

-91   AACACTTTCAATAATCGCACATTCTTTCATATTAGCTCTAACCATTCGAGTTCATATCATTGCAAAAGTCAAACGAAAAGTAAATCTCTG  -2
                                                                                             GATC
                                                                                              ||=B-ORF
-1    AAATGAGCCACATACCCATTAGTAACACAATTCCAACAAAACAAACTGATGGTAATGGTAAAGCTAACATTTCGCCGGATAAGCTGGATC  88
        MetSerHisIleProIleSerAsnThrIleProThrLysGlnThrAspGlyAsnGlyLysAlaAsnIleSerProAspLysLeuAspP  (30)

89    CCAAGGTTTCGGTATGTCTATTGGGTTTAGGTATAGAGCCAACAATTATGCACGTCTGATAACTAAATACTTTTGCATCCACATCAAGAT  178
      roLysValSer                                                                              Il  (34)

179   CGACATGGAGGCGCCGGAGTTCAAGGATTTTGCCAAGACAATGGTCGACTTTATAGCCGAATATCTGGAGAATATACGCGAAAGGTGAGC  268
      eAspMetGluAlaProGluPheLysAspPheAlaLysThrMetValAspPheIleAlaGluTyrLeuGluAsnIleArgGluAr        (62)

269   CAGATTTAGACTTCCTACTCAATTAGCTTGAATTAAACTTAATTTAGCGTATAAATTTCATTTATATGGTATCAGAATCAGTCGCTTGAC  358

359   CTCAGCATTTTACGTTCGAATCGAAAGTTCGTTCTGCTCGATTCGAATCCCCGGGCAAGTGAATGACATTTCGCACACGTTTTGAGATTA  448

449   GTCACGGGAAAGTCGCACCGATCGGACATTTCCATTGCTATATATATCTATATATATATATCATTTTGTTTAGGGGGTTGAGGCGACCTT  538

539   CCCATTAGCTCGAGGGCCAAGTACTTTCGCTGCTCTTGGGCCGAAAACTAATTAATTAAATGGCTTTGTTGAGTTGGCGTGTCAAGGTCG  628

629   TTTTTCATGTATACGAGTATAGATATAATTGCACTGCTAACGCCTTGGCCAAAAGCAATTCGGGTATTTCACTATTCTTGGGCAATTCTT  718

719   CTAACGGCTTCGTTTCCATTACCTTGAAAATCAAAGTCAGCTAAGTAAACAATTTTCTATACTACAGCTGCTGAGTTTGTTTGCCCATCG  808

809   ACAGTCGCTGAAATTAATGGTTAATTGAAAATCAAGCTTAAGTAGAGCGTAATATAATAATTCATTTTGCTTTATTAAAGTTCCTTCGAC  898

899   ATTGAAGTTTCAAAACTATTTTCTTAGTTAGATAACTTTTTAAACGAATCTTTGTTAATTGAAGATACATATATATAGAGAAATTATCTT  988

989   TTTATTTTCTTTTTTTCACCTCTTAGTAGTACTTCCTTTTAATTGAAAGGATAGAAAATCCCACCATCATTATCAGCATTGCCTCTCTAT  1078

1079  CTATATTCTGTTCCCATAGCAATTTGCTACATATTCGTATTGATTTTCATGGCAGTGGCAACAAGTTGGGGGTGGTTGGATGGGTGTCCT  1168

1169  TCAACCCCAATGATTCCTGATGCCTTTGTTGGCTAACTGAGTTTCGCAGCCAATTAGCAAGGAGCTTTTACTGAATGGGCGCCAAAATGC  1258

1259  AATCAGAACGTAACGCAATTTTCGCAATTACAGGCGCGTTCTGCCGGAAGTGAAGCCTGGCTACCTGAAGCCATTGATTCCGGATGCTGC  1348
                                      gArgValLeuProGluValLysProGlyTyrLeuLysProLeuIleProAspAlaAl  (81)

1349  GCCCGAGAAGCCGGAGAAGTGGCAGGATGTGATGCAGGACATCGAGCGAGTCATCATGCCGGGCGTGACACACTGGCACAGTCCCAAGTT  1438
      aProGluLysProGluLysTrpGlnAspValMetGlnAspIleGluArgValIleMetProGlyValThrHisTrpHisSerProLysPh  (111)

1439  TCATGCCTACTTCCCCACGGCCAACTCGTATCCAGCGATCGTTGCGGACATGCTGAGTGGAGCGATTGCCTGCATCGGATTCACGTGGAT  1528
      eHisAlaTyrPheProThrAlaAsnSerTyrProAlaIleValAlaAspMetLeuSerGlyAlaIleAlaCysIleGlyPheThrTrpIl  (141)
```

(*continued*)

```
1529 CGCCAGTCCCGCGTGCACGGAACTCGAGGTGGTCATGATGGATTGGCTGGGCAAGATGCTGGAGCTGCCGGCAGAGTTCCTGGCCTGTTC 1618
     eAlaSerProAlaCysThrGluLeuGluValValMetMetAspTrpLeuGlyLysMetLeuGluLeuProAlaGluPheLeuAlaCysSe (171

1619 GGGCGGCAAGGGTGGCGGTGTCATCCAGGGCACGGCCAGTGAGTCCACACTGGTGGCTCTGCTGGGAGCCAAGGCCAAGAAGTTGAAGGA 1708
     rGlyGlyLysGlyGlyGlyValIleGlnGlyThrAlaSerGluSerThrLeuValAlaLeuLeuGlyAlaLysAlaLysLysLeuLysGl (201

1709 GGTGAAGGAGCTCCATCCGGAGTGGGATGAGCACACCATCTTGGGCAAGTTGGTGGGCTACTGCTCGGACCAGGCTCACTCATCCGTGGA 1798
     uValLysGluLeuHisProGluTrpAspGluHisThrIleLeuGlyLysLeuValGlyTyrCysSerAspGlnAlaHisSerSerValGl (231

1799 GCGGGCTGGTCTTCTGGGCGGAGTAAAGCTCCGTTCCGTGCAGTCCGAGAATCACAGAATGCGTGGTGCTGCCCTGGAAAAGGCCATCGA 1888
     uArgAlaGlyLeuLeuGlyGlyValLysLeuArgSerValGlnSerGluAsnHisArgMetArgGlyAlaAlaLeuGluLysAlaIleGl (261

1889 ACAGGATGTGGCCGAGGGTTTGATTCCCTTCTACGCGGTGGTCACCCTGGGCACCACCAACTCCTGCGCCTTCGACTACTTGGATGAGTG 1978
     uGlnAspValAlaGluGlyLeuIleProPheTyrAlaValValThrLeuGlyThrThrAsnSerCysAlaPheAspTyrLeuAspGluCy (291

1979 TGGACCGGTGGGAAACAAGCACAATTTGTGGATCCATGTGGACGCTGCCTATGCCGGATCCGCTTTCATTTGCCCCGAGTATCGCCACCT 2068
     sGlyProValGlyAsnLysHisAsnLeuTrpIleHisValAspAlaAlaTyrAlaGlySerAlaPheIleCysProGluTyrArgHisLe (321)

2069 GATGAAGGGCATCGAATCAGCAGACTCTTTCAATTTCAATCCACACAAATGGATGCTGGTGAACTTTGACTGCTCGGCCATGTGGCTGAA 2158
     uMetLysGlyIleGluSerAlaAspSerPheAsnPheAsnProHisLysTrpMetLeuValAsnPheAspCysSerAlaMetTrpLeuLy (351)
                                                    ---------------------PYR

2159 GGATCCCAGTTGGGTGGTCAACGCGTTCAATGTGGACCCTCTTTACCTGAAGCACGACATGCAGGGATCAGCTCCGGACTATCGTCACTG 2248
     sAspProSerTrpValValAsnAlaPheAsnValAspProLeuTyrLeuLysHisAspMetGlnGlySerAlaProAspTyrArgHisTr (381)

2249 GCAAATCCCACTTGGACGGCGATTCAGGGCACTGAAGCTCTGGTTCGTCCTCCGGCTGTACGGTGTCGAGAATCTCCAGGCCCACATCCG 2338
     pGlnIleProLeuGlyArgArgPheArgAlaLeuLysLeuTrpPheValLeuArgLeuTyrGlyValGluAsnLeuGlnAlaHisIleAr (411)

2339 CAGACACTGCAACTTTGCCAAGCAGTTCGGGGATCTCTGCGTGGCGGACTCCAGATTTGAACTGGCCGCCGAGATCAATATGGGATTGGT 2428
     gArgHisCysAsnPheAlaLysGlnPheGlyAspLeuCysValAlaAspSerArgPheGluLeuAlaAlaGluIleAsnMetGlyLeuVa (441)

2429 CTGCTTCCGGCTGAAGGGCAGCAACGAGCGGAACGAAGCTCTTCTCAAGCGAATCAATGGACGCGGCCACATCCACTTGGTTCCCGCCAA 2518
     lCysPheArgLeuLysGlySerAsnGluArgAsnGluAlaLeuLeuLysArgIleAsnGlyArgGlyHisIleHisLeuValProAlaLy (471)

2519 GATCAAGGATGTCTACTTCCTCGCGATGGCCATTTGCTCGCGATTCACCCAGTCCGAGGACATGGAGTACTCGTGGAAGGAGGTCAGCGC 2608
     sIleLysAspValTyrPheLeuAlaMetAlaIleCysSerArgPheThrGlnSerGluAspMetGluTyrSerTrpLysGluValSerAl (501)

2609 CGCTGCCGACGAGATGGAACAGGAGCAGTAAAGTGGTTGTGCAGGTCTGTTCCGTGTTTAGTATATAAATTAATATAGTAAACTTAAATT 2698
     aAlaAlaAspGluMetGluGlnGluGlnEnd                                                            (510)

2699 GGACCAGTATGATATATAATGCATTGTGACTTGGAACCCGGAACAGACCATACACTTTCCACTTGCGACATGTTTAGGGAATTTACATCG 2788

2789 CAACAAAAGATGGTTCGTCCATCGCTACATTATATTTATAGTATCCTATCATTGTATCATTGATGTTGTTCATGATTTTTATTGTTAACG 2878
                                                                 Cs  n(A)|

2879 TTATGCGCCTAATTAAAACAAATGTATTCTGCTTAAAAATACAAACGAATTGTAACTATAAATTTTGACTAGTTTTCGTGTTGATATACA 2968
                                                ------            |(A)n  Ddc

2969 CTGTACATTTAGCAGCCCATTCGGATTTCCATTTCACT 3006
```

Ddc SEQUENCE. Accession, X04661 (DRODDC). The sequence was corrected by J. Hirsh by addition of a G at position −932. The acceptor site of the leader intron is 15 bases upstream from the position proposed by Morgan et al. (1986) (Shen and Hirsh, personal communication). Footprints in the promoter region are indicated by underlining; there are eight in the distal region and one in the proximal region. *B-ORF*,

(*continued*)

from nuclease digestion by an extract from embryonic nuclei (Bray et al. 1988, 1989). It has been reported that *Ddc* and *Ubx* may have a regulatory protein in common (Biggin and Tjian 1988).

Distal Elements In addition to the proximal elements, expression in the central nervous system also requires certain more distal *cis* sequences located in an 863-bp segment between −2,506 and −1,643 (Johnson et al. 1989). Eight protein-binding sites were detected within that segment (*Ddc* Sequence) by nuclease protection assays. Partial deletions of the distal promoter region, re-introduced into transgenic organisms, showed that uf8, uf9 and uf10 are not essential for neuronal expression. On the other hand, deletion of either uf7 or bf2 and uf3 leads to complete loss of neuronal activity. The element cf1 appears to be essential for expression in the medial dopaminergic neurons (Johnson et al. 1989). The gene for a POU/homeobox protein that binds to cf1 has been cloned (Johnson and Hirsh 1990). A 40-bp segment between −2,519 and −2,479 is necessary for expression in serotonergic neurons (Johnson et al. 1989).

l(2)amd
(α-methyl dopa sensitive)

Product

Unknown.

Structure

The sequence of the coding regions show 55% identity with the dopa decarboxylase sequence. The amino acid sequence similarity is particularly high near a putative pyridoxal-binding site (starting at position 298) (Eveleth and Marsh 1986; Marsh et al. 1986).

Function

AMD is thought to be involved in the metabolism of catecholamines judging by the α-methyl dopa sensitivity of mutants (for a review, see Wright 1987).

(*continued*) between positions 84 and 85, is a 4-bp insertional mutation that alters the reading frame of the second exon and leads to the absence of DDC in the central nervous system but not in the epidermis (Morgan et al. 1986). The putative pyridoxal-binding site starting at Asn-332 is underlined. The poly(A) site, $_n$(A)| (near 2,850) of the partly overlapping gene *Cs*, is indicated.

l(2)amd

```
-304  TCAAGCTAAATTAGTTAGATCAAAGAATAAACAAGTCAGTTGCGCCGTTTTAATGATTCTCAAAACTAGCCAGATTGGCTGAACCGACAG   -215
                                                                                  ----
                                                                                 .!-149
-214  CTCTGGAGGCTGTCCAGAGAAGTCGGAGTATAAAAGGCCAGTCACCGGCGATCGGTTTCAGAGTGAACCTCAGGCAACTTGGAGGAGCAT   -125
                                ------

-124  CAACGGATCGGGAACTGAAATCGAGTTGGGCAAACAAATCAAAAACGAAAACGGGGAAATAAAACCAAAACAAACAGAACGTAAAAAGTG   -35

 -34  CAAATAGAAAACGATATCGCAACATTGTCAGCGGTATGGATGCCAAGGAGTTTCGGGAATTCGGCAAGGCCGCCATTGACTACATAGCCG   55
                                        MetAspAlaLysGluPheArgGluPheGlyLysAlaAlaIleAspTyrIleAlaA   (19)

  56  ACTATCTGGAGAATATTCGGGATGACGACGTACTGCCCAATGTGGAGCCAGGCTATCTGTTGGACCTGCTGCCCACAGAGATGCCGGAAG   145
      spTyrLeuGluAsnIleArgAspAspAspValLeuProAsnValGluProGlyTyrLeuLeuAspLeuLeuProThrGluMetProGluG   (49)

 146  AGCCCGAAGCGTGGAAGGATGTCCTCGGCGACATTAGTCGCGTCATCAAGCCGGGACTGACCCACTCGGAGTCGCCTCACATGCATGCCT   235
      luProGluAlaTrpLysAspValLeuGlyAspIleSerArgValIleLysProGlyLeuThrHisSerGluSerProHisMetHisAlaT   (79)

 236  ACTACCCCACCAGCACCTCGTATCCCTCCATTGTGGGCGAGATGCTGGCCAGCGGGTTCGGCGTCATCGGATTCAGCTGGGTATGTTGGT   325
      yrTyrProThrSerThrSerTyrProSerIleValGlyGluMetLeuAlaSerGlyPheGlyValIleGlyPheSerTrp           (105

 326  TTATGGTGAAATCTGCTGCTGCTGCTGCTGCTGCTTCTGGCCCATTGTTTTGGCCGGCTGAATGGGCGCTCATTGTGCCGGGTGGGTGAA   415

 416  AGTGAATCCAAGAACTCGACAAACAGGTTGCCACTGCACCGGACCGAAGAGAGTTGTTCACACAAATCAATCGGCAATTGTCACCAAATA   505

 506  AAAGCAATAAAATTGGGCAGCAGACTCACCTTAAAGGCATACAAATAAATTTAAATGTATATAGAATTGATTAGCCAAACCCAAACTAGA   595

 596  CGTTTCGCAAACAATATTTGTCATTGCGAACAAAGAAGTTACCACCGAACAAAAACTTAGTGAAATAAACCCTAGTTTAAATTATAATAT   685

 686  ATTTGTAAAAATTACTATATGTATGTATTCCGGATTTAATAGTGTATTACAAACGGATGGAGTTATCTTCAATGCATAATTTCTTACATA   775

 776  ATAATTCGTATAATCCCCCACAGATCTGCAGTCCCGCCTGCACAGAACTGGAGGTGGTGGTCATGGACTGGCTGGCCAAGTTCCTGAAGC   865
                                IleCysSerProAlaCysThrGluLeuGluValValValMetAspTrpLeuAlaLysPheLeuLysL   (12

 866  TGCCCGCACACTTCCAGCACGCCAGCGATGGACCAGGAGGCGGGGTGATCCAGGGATCAGCTAGCGAGGCTGTGTTGGTGGCTGTCCTAG   955
      euProAlaHisPheGlnHisAlaSerAspGlyProGlyGlyGlyValIleGlnGlySerAlaSerGluAlaValLeuValAlaValLeuA   (15

 956  CTGCCAGGGAACAAGCTGTGGCCAACTACAGGGAATCGCATCCGGAGCTGAGCGAAAGTGAGGTGCGTGGCCGCTTGGTGGCCTACTCCT   104
      laAlaArgGluGlnAlaValAlaAsnTyrArgGluSerHisProGluLeuSerGluSerGluValArgGlyArgLeuValAlaTyrSerS   (18

1046  CGGACCAGAGTAACAGCTGCATTGAGAAGGCTGGAGTCCTGGCTGCCATGCCGATTCGATTGCTGCCGGCTGGAGAGGATTTCGTACTTA   113
      erAspGlnSerAsnSerCysIleGluLysAlaGlyValLeuAlaAlaMetProIleArgLeuLeuProAlaGlyGluAspPheValLeuA   (21

1136  GAGGCGATACACTGAGAGGAGCCATCGAGGAGGACGTGGCAGCGGGCAGGATTCCGGTGATCTGCGTTGCCACTCTGGGCACCACGGGCA   122
      rgGlyAspThrLeuArgGlyAlaIleGluGluAspValAlaAlaGlyArgIleProValIleCysValAlaThrLeuGlyThrThrGlyT   (24

1226  CTTGTGCCTATGACGATATTGAATCCCTGTCCGCTGTCTGCGAGGAATTCAAGGTGTGGCTCCATGTTGATGCCGCGTATGCCGGTGGAG   131
      hrCysAlaTyrAspAspIleGluSerLeuSerAlaValCysGluGluPheLysValTrpLeuHisValAspAlaAlaTyrAlaGlyGlyA   (27

1316  CCTTTGCTCTGGAGGAATGTTCGGATTTGCGAAAGGGATTGGATCGCGTGGACTCGCTAAACTTCAACCTGCACAAGTTCATGCTGGTCA   140
      laPheAlaLeuGluGluCysSerAspLeuArgLysGlyLeuAspArgValAspSerLeuAsnPheAsnLeuHisLysPheMetLeuValA   (30
                                                                       -------------------- PYR
```

```
1406 ACTTCGATTGCTCGGCCATGTGGCTAAGGGATGCCAACAAGGTGGTCGACAGCTTCAATGTGGATCGCATCTATCTGAAGCACAAGCACG  1495
     snPheAspCysSerAlaMetTrpLeuArgAspAlaAsnLysValValAspSerPheAsnValAspArgIleTyrLeuLysHisLysHisG  (338)

1496 AGGGTCAGTCGCAAATTCCTAGACTTCCGTCATTGGCAAATCCCCTGGGTCGCCGCTTCCGAGCTCTAAAAGTCTGGATCACATTCCGCA  1585
     luGlyGlnSerGlnIleProArgLeuProSerLeuAlaAsnProLeuGlyArgArgPheArgAlaLeuLysValTrpIleThrPheArgT  (368)

1586 CTCTGGAAGCCGAGGGATTGCGAAACCATGTCGCGAAGCACATCGAGTTGGCCAAACAGTTTGAGCAGCTTGTGCTCAAGGATTCGCGAT  1675
     hrLeuGluAlaGluGlyLeuArgAsnHisValAlaLysHisIleGluLeuAlaLysGlnPheGluGlnLeuValLeuLysAspSerArgP  (398)

1676 TCGAGCTGGTGGCTCCTCGTGCCCTGGGACTGGTTTGTTTCCGGCCCAAAGGTGACAATGAGATTACCACCCAGTTGCTGCAACGGCTTA  1765
     heGluLeuValAlaProArgAlaLeuGlyLeuValCysPheArgProLysGlyAspAsnGluIleThrThrGlnLeuLeuGlnArgLeuM  (428)

1766 TGGATCGAAAGAAGATCTACATGGTTAAGGCCGAGCATGCGGGTCGTCAGTTTCTGCGATTCGTCGTATGCGGCATGGACACCAAAGCCT  1855
     etAspArgLysLysIleTyrMetValLysAlaGluHisAlaGlyArgGlnPheLeuArgPheValValCysGlyMetAspThrLysAlaS  (458)

1856 CCGATATTGATTTCGCCTGGCAGGAGATCGAGTCTCAACTGACGGACCTGCAGGCGGACGAATCCTTGGTGGCCCGCAAATCGGGAAACG  1945
     erAspIleAspPheAlaTrpGlnGluIleGluSerGlnLeuThrAspLeuGlnAlaAspGluSerLeuValAlaArgLysSerGlyAsnV  (488)

1946 TCGGCGATCTTGCGCACGACTTCCAGATCCATCTGAGCACCGAAAATGCAACGCACGAGAAATCTCAGTGAGAAAAACGGATAAACTATT  2035
     alGlyAspLeuAlaHisAspPheGlnIleHisLeuSerThrGluAsnAlaThrHisGluLysSerGlnEnd                    (510)

2036 TATGTTTAGGGACAACCTAGTTAGTTTGCGATGTAGTTTTTAACTTTCCACATGTTTATAAATAAAGTGAAATTAAATGTACGATCATTT  2125
                                                                 ------            |(A)n

2126 GGCACGTTTTCTATAAAGGTAGAGTGGTTTTTCCCTGTCATTTTTTTTTGTCGAAAAAGTTCCCAACATTCTCTGTTAAACTTTCTGCCG  2215

2216 AGGCTTTAGTTTTTTAAGCATTACAAATATCGTCGACTTTTATTTTAAAATTTAAAACCAAAATTTTCGCGGCTTAGTGTGACTGCATTT  2305

2306 GGTTATGAATCGATACACTTCTTCATCGCCCTTCGATAAGTTCGCCAAGGTCTATCGTCATGTGCCGATCCGCAGGGCAAACAGCTGTTT  2395

2396 CTCCCAATTGGGACCACCTGATATCGGTTAAATAACAAAGTATAAACAAAACAAAAATATCTGTTTCCCTTAATTCAATATTTTGATTAG  2485
                                                                             ______EcoRI
2486 CTTTGAATAGCGTTTAGTGCTATTTCTCATAAATATAGAATAGAAGCAGCCGCGGCTCGCCTTTGTACGAATTC  2559
```

l(2)amd SEQUENCE. Strain, *Canton S*. Accession, X04695 (DROL2AMD). The sequence ends at the *Eco*RI site at which the *Cs* Sequence, begins. The exclamation mark indicates the 5′ end of the longest cDNA sequenced.

Gene Organization and Expression

Open reading frame, 510 amino acids; expected mRNA size, 1,782 bases. The 5′ end was tentatively identified on the basis of sequence features in the neighborhood of the 5′ end of a cDNA clone. The 3′ end was identified from the sequence of two cDNA clones. There is one intron after the Trp-105 codon. The distance from the polyadenylation site of *1(2)amd* to the transcription initiation site of *Cs* is 682 bp.

Although the length of the coding region is the same as that of *Ddc*, and the sequence is similar, the position of the introns in the two genes do not match. In the aligned sequences, the 5′ end of *amd* coincides approximately with the second *Ddc* intron; and the *amd* intron is approximately 250 bp away from the position of the third *Ddc* intron (Fig. 11.3 and *1(2)amd* Sequence) (Eveleth and Marsh 1986).

Cs

```
      ______EcoRI
-658  GAATTCTCAGATTTCGTGAGTAAATAATCATATATGTAACATACAAATACATCCGAATTACTATAACCCTTTCAGCCAAGTTTGGAATTA  -569

-568  GACACCCAAACTGCCAATTGGATAACCGGCGACCATTGTGGTGGATACTATGATTCTGCTTTTTAAAAACAACTTGGACGTGTGGCACTT  -479

-478  TAAAGGCCTATACGCCTCTCAGCCTGTCAACAAATATTAAAAAATCTGGCAAAATTCTAAAAATAACTTGATTTACTTTCGGAACTCCAG  -389

-388  GAAACTAGGCCCGCTTGCCATGCAATGGTAAGTTGAACAGCTCCAGCGGATTTGAATGTGCAAACTAAACCTTCTCTTGATCCCCGCAGT  -299

-298  TTTAAACTGGCCAGCAGGCGCAGCTTATACAATGCACGGGTTCTACAGGCGGATAACATCGGCGACAAGCAACGCAGTCCAGATCTGGAG  -209

-208  CGGCGCGCCAAAATACCCAGATAGTGGTCGTGGGCGCAGGACTCGCCGGTCTCTCGGCGGCCCAGCACCTCTTGTCGCACGGCTTTCGGC  -119

-118  GCACTGTGATCCTGGAGGCCACAGATCGTTATGGCGGCAGGATTAACACCCAGCGCTTTGGTGACACCTACTGTGAACTAGGCGCCAAGT  -29

 -28  GGGTAAAGATCGATGGATCGCAGGATTCGATGTATGAACTGCTACGCAACACGGAAGGCTTGGGGAAGCAGATAAAGCAGGCCGGATCGG  61
                                  MetTyrGluLeuLeuArgAsnThrGluGlyLeuGlyLysGlnIleLysGlnAlaGlySerG  (21)

  62  GCCACCTATCTTCAGGATGGAAGCCGCATCAATCCAGCCATGGTCGAGCTTATCGACACGCTATTTCGGCAGCTTTGCCAGGCTTCAAGG  151
      lyHisLeuSerSerGlyTrpLysProHisGlnSerSerHisGlyArgAlaTyrArgHisAlaIleSerAlaAlaLeuProGlyPheLysV  (51)

 152  TCTCCGAACGAGTTAAAACGGGTGGTGACCTGCACTCGCTGGACAATGTCATGAACTACTTTAGAACAGAAAGCGATCGCATCATTGGCG  241
      alSerGluArgValLysThrGlyGlyAspLeuHisSerLeuAspAsnValMetAsnTyrPheArgThrGluSerAspArgIleIleGlyV  (81)

 242  TCTCCTTCCAGCATCCTAAGGATCAACTGGCGGCACGCGAGATCTTCCAATCGCTGTTCAAGGAGTTCGGCAGCATCTTGGGATGCTGCC  331
      alSerPheGlnHisProLysAspGlnLeuAlaAlaArgGluIlePheGlnSerLeuPheLysGluPheGlySerIleLeuGlyCysCysL  (111

 332  TGGAGTACGTGAACATCGAACACATAACCAAGTGTCCAGTGCAGCAGGAACAGCGCCCGCGTTATGTGCCCACTGGTCTAGATAATGTAG  421
      euGluTyrValAsnIleGluHisIleThrLysCysProValGlnGlnGluGlnArgProArgTyrValProThrGlyLeuAspAsnValV  (141

 422  TGGACGATCTCATTCAGAACATGGACAAAGCGCAGCTGCAGACCGGAAAGCCTGTGGGCCAGATACAGTGGACACCAGCGCCGATGAAAA  511
      alAspAspLeuIleGlnAsnMetAspLysAlaGlnLeuGlnThrGlyLysProValGlyGlnIleGlnTrpThrProAlaProMetLysS  (171

 512  GTGTGGGTTGCCTGGATGGCAGTCTTTACAACGCCGATCACATAATATGCACCCTGCCGCTCGGGGTGCTCAAAAGCTTTGGCGCGTTCT  601
      erValGlyCysLeuAspGlySerLeuTyrAsnAlaAspHisIleIleCysThrLeuProLeuGlyValLeuLysSerPheGlyAlaPheC  (201

 602  GTTTCGACCCACGCTGCCGCTGGACAAGATGCTGGCTATCACGCAACCTCGGCCTTTGGCAATCCCCTCAAGATATATCTCTCCTACAAG  691
      ysPheAspProArgCysArgTrpThrArgCysTrpLeuSerArgAsnLeuGlyLeuTrpGlnSerProGlnAspIleSerLeuLeuGlnG  (231

 692  AAGCCATTCTGGTGGCTAAAGGGAAGCTGCGCCATGGAACGTTCTGAATCTTCGTAGAGCAGCAACCGAACGCAACTGGACGCAGCAGGT  781
      luAlaIleLeuValAlaLysGlyLysLeuArgHisGlyThrPheEnd                                             (245

 782  CGTGGAGATAGCCAGGTGCCCAGCAGTCAGCATGTGCTGGAGGTGCATGTGGTGGCGGATACTACGAGGAGATCGAGAAGCTGCCCGATG  871

 872  AGGAGCTGCTGGAGCAGATAACTGGTCTGCTAAGGCGCTGCGTGAGCAGTCACCTGGTGCCGTACCCACAGGAACTGCTGCGTTCCAACT  961

 962  GGAGCACCTCGGCCTGCTACCTCGGCCGGTCCGTCCTTACTTCTCCACCAACAGCAGTGCCCGGGATGTCCAGCGACTGGCCGCTCCGGC  1051

1052  TGGGCGAGAAGTCCGGGGTCTGCTCTTTGCTGGGGATGCAACCTCGCTGAAAGGCTTTGGAACCATTGATGCCGCCACGTCCAGTGGCAT  1141

1142  CCGAGAAGCCCAATGTATCATTGACTACTATCTGAAAAGCGTGCACTGCGGTTAAGTGAAATGGGAAATCCGAATGGGCTGCTAAATTGT  1231
```

```
              .         .         .         .         .         .         .         .         .
1232  ACAGTGTATATCAACACGAAAACTAGTCAAAATTTATAGTTACATTCGTTTGTATTTTTAAGCAGATACATTTGTTTTAATTAGGCGCAT  1321
                                Ddc  n(A)|
              .         .         .         .         .         .         .         .         .
1322  AACGTTAACAATAAAAATCATGAACAACATCAATGATACAATGATAGGATACTATAAATATAATGTAGCGATGGACGAACCATCTTTTGT  1411
                ======      |(A)n
              .         .         .         .         .         .         .         .         .
1412  TGCGATGTAAATTCCCTAAACATGTCGCAAGTGGAAAGTGTATGGTCTGTTCCGGGTTCCAAGTCACAATGCATTATATATCATACTGGT  1501
              .         .         .         .         .         .         .         .         .
1502  CCAATTTAAGTTTACTATATTAATTTATATACTAAACACGGAACAGACCTGCACAACCACTTTACTGCTCCTGTTCCATCTCGTCGGCAG  1591
              .         .         .
1592  CGGCGCTGACCTCCTTCCACGAGTACTCCA  1621
```

Cs SEQUENCE. Strain, *Canton S*. Accession, X05991 (DROCSG). The sequence starts with the *Eco*RI site at the end of the *1(2)amd* Sequence. The exclamation mark indicates the 5′ end of the longest cDNA sequenced. The poly(A) site, $_n$(A)| (at 1,267) of the partly overlapping gene *Ddc*, is indicated.

Developmental Pattern

A 2 kb RNA is detected in 8–16 h embryos and, at a much lower level, in adults.

Cs

Product

Unknown. It has been questioned whether this protein is ever synthesized, although the corresponding mRNA is found in association with polysomes. No mutations have been recovered in this transcription unit despite intensive screens involving the region (Eveleth and Marsh 1987 and references therein).

Gene Organization and Expression

The longest open reading frame is 245 amino acids; but several smaller open reading frames exist, some with the starting codon upstream of the longest one. The presence of those upstream AUGs and the very poor codon bias displayed by this mRNA suggest that translation may be very inefficient. The expected mRNA length is 1,696 bases, in agreement with a 1.9 kb band detected in gels. A cDNA sequence was used to define the 5′ end. The 3′ end was obtained from the sequence of two cDNAs that included poly(A) tails. There is a leader intron at −361/−300 (*Cs* Sequence) (Eveleth and Marsh 1987).

Developmental Pattern

Transcription of *Cs* occurs mainly in the first 8 h of embryonic development; the highest levels of transcript are detected in 3 h embryos (Spencer et al., 1986b).

DoxA2

```
          .          . l(2)37Bb.   <--   .         .         .         .         .         .
-399  GTGCGATGTTATCGGAGTATCGATATCGAAAAGGCTTAACGGAATTGTGGTAATGTTTTATTGCAATTTAAATAAAAATACGCTTTTACT   -310
          .         .         .         .         .         .         .         .         .
-309  GAGTGCGTAACTTAAGAAATTCCTAACCCAAATTAAAGCAATAGATACATTTACTGTAAAACATTAAAAATAAATTCCTATAACATATGA   -220
          .         .         .         .         .         .         .         .         .
-219  GACTGAAAGTTCGCTCAGTGTACCGTAAAACGTATCGATAAATTGAAACGTAACGGCTTAACAGCTCTGTTAACCAACTAAATTTACCAG   -130
                                                ----
          .         .         .         .-->-89 .         .         .         .         .
-129  CACTGCCTGTAGCCGAAAACGAATAAGAAGAAGAAGCGACATTACTAGGCATTTTTGATTGGGATTGAGAAAAACAAAAGAAAAGTCGGC   -40
          .         .         .         .         .         .         .         .         .
 -39  TATATTTGTGACCCCAGTAAATTGAGAGTTCCATTACAAAATGACCAACGCAACGGACATCGGTGCTAACGACGTGGAGATGGAGGTGGA   50
                                                MetThrAsnAlaThrAspIleGlyAlaAsnAspValGluMetGluValAs   (17)
          .         .         .         .         .         .         .         .         .
  51  TCCAACGGCGGAGACGCTGGCTGACGAGAAGAAGAACCAAGATGTGGCCGCCGTGCAGGAGATCCGCGAGCAGATTCGTCAGATTGAGAA   140
      pProThrAlaGluThrLeuAlaAspGluLysLysAsnGlnAspValAlaAlaValGlnGluIleArgGluGlnIleArgGlnIleGluLy   (47)
          .         .         .         .         .         .         .         .         .
 141  GGGGGTAGCCTCGAAAGAGTCGCGGTGAGTAGTGCAAGAATTAAAATCTTGTCCCTTCCTTATTATGGCTCATTTCCGCCAACAGCTTCA   230
      sGlyValAlaSerLysGluSerAr                                                           gPheI   (57)
          .         .         .         .         .         .         .         .         .
 231  TCCTGCGCGTCCTTCGCAATTTGCCCAACACTCGTCGCAAGCTGAACGGCGTCGTCTTCCGGAATCTTGCACAGAGTATTTACCCCGCTG   320
      leLeuArgValLeuArgAsnLeuProAsnThrArgArgLysLeuAsnGlyValValPheArgAsnLeuAlaGlnSerIleTyrProAlaG   (87)
          .         .         .         .         .         .         .         .         .
 321  GTGCAGATCGTGAGGCGGCCGTGGCTTTGATGCCCGCTGTGGAGAAAGACGCCACCGAGCTGCCCGATGTTCCCAAAAAACAAGTTGCCA   410
      lyAlaAspArgGluAlaAlaValAlaLeuMetProAlaValGluLysAspAlaThrGluLeuProAspValProLysLysGlnValAlaT   (117)
          .         .         .         .         .         .         .         .         .
 411  CCAAGGCTCCAATCGCCGAGGTCGATGCCTACTTCTACCTGCTCCTGCTGGTCAAGCTCATCGACGCCAGTGATTTAAAGCGGGCCGGAA   500
      hrLysAlaProIleAlaGluValAspAlaTyrPheTyrLeuLeuLeuLeuValLysLeuIleAspAlaSerAspLeuLysArgAlaGlyI   (147)
          .         .         .         .         .         .         .         .         .
 501  TTAGCGCCGACGCCCTAATGGCCAAAATCTCCATCCAAAACCGACGCACCCTTGATCTGATTGGTGCCAAGTCCTACTTCTATTTTTCAA   590
      leSerAlaAspAlaLeuMetAlaLysIleSerIleGlnAsnArgArgThrLeuAspLeuIleGlyAlaLysSerTyrPheTyrPheSerA   (177)
          .         .         .         .         .         .         .         .         .
 591  GAGTGGCGGAGCTAAAAAACTCACTGGAAGGCATACGCTCGTTCCTGCACGCTCGTCTGCGCACCGCTACGCTGCGTAATGATTTTGAAG   680
      rgValAlaGluLeuLysAsnSerLeuGluGlyIleArgSerPheLeuHisAlaArgLeuArgThrAlaThrLeuArgAsnAspPheGluG   (207)
          .         .         .         .         .         .         .         .         .
 681  GCCAGGCGGTGCTTATTAACTGTTTGCTCCGCAACTACTTGCACTATGCTTTGTACGACCAAGCCGACAAGCTGGTAAAGAAATCCGTCT   770
      lyGlnAlaValLeuIleAsnCysLeuLeuArgAsnTyrLeuHisTyrAlaLeuTyrAspGlnAlaAspLysLeuValLysLysSerValT   (237)
          .         .         .         .         .         .         .         .         .
 771  ACCCGGAATCGGCCAGCAACAATGAATGGGCGCGTTTCCTGTACTATCTAGGTCGGATTAAGGCCGCTAAGCTGGAGTACAGCGATGCCC   860
      yrProGluSerAlaSerAsnAsnGluTrpAlaArgPheLeuTyrTyrLeuGlyArgIleLysAlaAlaLysLeuGluTyrSerAspAlaH   (267)
          .         .         .         .         .         .         .         .         .
 861  ACAAGCATCTGGTCCAGGCCCTGCGTAAGTCGCCGCAGCACGCTGCCATCGGCTTTCGTCAGACGGTTCAAAAGCTAATTATCGTTGTGG   950
      isLysHisLeuValGlnAlaLeuArgLysSerProGlnHisAlaAlaIleGlyPheArgGlnThrValGlnLysLeuIleIleValValG   (297)
          .         .         .         .         .         .         .         .         .
 951  AGCTGCTTTTGGGCAACATCCCGGAGCGTGTGGTGTTCCGGCAAGCCGGTCTTCGCCAATCTCTTGGTGCCTACTTCCAGCTCACGCAGG   1040
      luLeuLeuLeuGlyAsnIleProGluArgValValPheArgGlnAlaGlyLeuArgGlnSerLeuGlyAlaTyrPheGlnLeuThrGlnA   (327)
          .         .         .         .         .         .         .         .         .
1041  CCGTGCGTCTGGGCAACTTGAAGCGCTTCGGCGACGTGGTATCCCAATACGGACCCAAGTTCCAACTGGACCACACATTCACCCTGATTA   1130
      laValArgLeuGlyAsnLeuLysArgPheGlyAspValValSerGlnTyrGlyProLysPheGlnLeuAspHisThrPheThrLeuIleI   (357)
          .         .         .         .         .         .         .         .         .
1131  TCCGGCTGCGCCACAATGTGATCAAGACGGCAATCCGCTCCATCGGACTATCGTACTCACGCATCTCGCCGCAAGACATTGCCAAGCGGC   1220
      leArgLeuArgHisAsnValIleLysThrAlaIleArgSerIleGlyLeuSerTyrSerArgIleSerProGlnAspIleAlaLysArgL   (387)
```

```
         .         .         .         .         .         .         .         .         .
1221 TAATGCTAGACTCCGCGGAGGATGCCGAGTTTATTGTATCGAAGGCTATACGGGACGGCGTGATTGAGGCTACGTTGGACCCAGCCCAGA  1310
     euMetLeuAspSerAlaGluAspAlaGluPheIleValSerLysAlaIleArgAspGlyValIleGluAlaThrLeuAspProAlaGlnA  (417)
         .         .         .         .         .         .         .         .         .
1311 ATTTCATGCGCAGCAAGGAAAGTACGGACATCTACAGCACCCGGGAACCGCAGCTGGCCTTTCACGAGCGCATCTCGTTCTGCCTGAACC  1400
     snPheMetArgSerLysGluSerThrAspIleTyrSerThrArgGluProGlnLeuAlaPheHisGluArgIleSerPheCysLeuAsnL  (447)
         .         .         .         .         .         .         .         .         .
1401 TGCACAACCAGAGCGTTAAGGCCATGCGCTATCCCCCAAAGTCCTACGGCAAGGATTTGGAGAGCGCCGAGGAGAGACGCGAGCGGGAGC  1490
     euHisAsnGlnSerValLysAlaMetArgTyrProProLysSerTyrGlyLysAspLeuGluSerAlaGluGluArgArgGluArgGluG  (477)
         .         .         .         .         .         .         .         .         .
1491 AGCAGGACCTTGAGCTGGCCAAGGAGATGGCCGAGGATGATGAGGATGGTTTCTAAGCGGCTGATTCTGCAAATTAATTTGTGCTTGCAT  1580
     lnGlnAspLeuGluLeuAlaLysGluMetAlaGluAspAspGluAspGlyPheEnd                                   (494)
         .         .         .         .         .         .         .         .         .
1581 TCATTTTTATAGAAATATAATCCGCAATTAAATAAGTTACAATAATTTCGGAACTTTTTAATTAGGTATTGGAATCAAATAGTTCAGAAC  1670
             ======    ======        |(A)n  |(A)n
         .         .         .         .         .         .         .         .
1671 TGATCTTCTTTATTCAAGCAAAGTTGTATGTTGTTGTTGGTAGACATCAAATTCATCGTAGAATGAACATTAAGTTCCATTCTG  1754
```

DoxA2 Sequence. Accession, M63010 (DRODOXA2). At −364 is indicated the 5′ end of the neighboring gene *1(2)37Bb*, which is transcribed in the opposite direction.

DoxA2 (*Diphenol oxidase component A2*)

Product

Component A2 of phenol oxidase (PO) (EC 1.10.3.1).

Structure

Sequence comparisons involving entire amino acid sequences show 57% identity between DOXA2 and the mouse *tum*$^{-}$ transplantation antigen P91A; the similarity is even greater in the C-terminal two-thirds of the protein (Pentz and Wright 1991).

Function

PO has three components: A1 acts on monophenols, A2 and A3 on diphenols, including dopa and its derivatives; it is involved in the oxidation of catecholamines to quinones, compounds that are subsequently utilized to produce melanin or to cross-link cuticular proteins. Thus, PO plays a central role in eggshell and cuticular sclerotization, in melanization and in defense against pathogens. A2 (like A1 and A3) is synthesized as a proenzyme and activated, probably by proteolysis, via an activation cascade.

Mutant Phenotypes

Homozygous *DoxA2* mutants die primarily during the larval stages; however, rare pharate adults can be recovered, and these are totally unpigmented (Pentz et al. 1986 and references therein).

Gene Organization and Expression

Open reading frame, 494 amino acids; expected mRNA length, 1,649 and 1,657 bases in agreement with a 1.7 kb band detected in gels. There are two alternate 3′ ends; the positions of these were obtained from two cDNA sequences terminating in poly(A) tails. S1 mapping and a cDNA sequence were used to define the 5′ end. There is no apparent TATA box. There is an intron in the Arg-55 codon (*DoxA2* Sequence) (Pentz and Wright 1991).

References

Beall, C. and Hirsh, J. (1987). Neuronal and glial expression of the *Drosophila* dopa decarboxylase gene. *Genes Dev.* **1**:510–520.

Biggin, M. D. and Tjian, R. (1988). Transcription factors that activate the *Ultrabithorax* promoter in developmentally staged extracts. *Cell* **53**:699–711.

Bray, S. J., Johnson, W. A., Hirsh, J., Heberlein, U. and Tjian, R. (1988). A *cis*-acting element and associated binding factors required for CNS expression of the *Drosophila melanogaster DOPA decarboxylase* gene. *EMBO J.* **7**:177–188.

Bray, S. J., Burke, B., Brown, N. and Hirsh, J. (1989). Embryonic expression pattern of a family of Drosophila proteins which interact with a CNS regulatory element. *Genes Dev.* **3**:1130–1145.

Clark, W. C., Pass, P. S., Vankataraman, B. and Hodgetts, R. B. (1978). Dopa decarboxylase from *Drosophila melanogaster*. Purification, characterization and analysis of mutants. *Molec. Gen. Genet.* **162**:287–297.

Eveleth, D. D. and Marsh, J. L. (1986). Evidence for evolutionary duplication of genes in the *DOPA decarboxylase* region of *Drosophila. Genetics* **114**:469–483.

Eveleth, D. D. and Marsh, J. L. (1987). Overlapping transcription units in *Drosophila*: Sequence and structure of the *Cs* gene. *Mol. Gen. Genet.* **209**:290–298.

Eveleth, D. D., Gietz, R. D., Spencer, C. A., Nargang, F. E., Hodgetts, R. B. and Marsh, J. L. (1986). Sequence and structure of the *DOPA decarboxylase* gene of *Drosophila*. Evidence for novel RNA splicing variants. *EMBO J.* **5**:2663–2672.

Gilbert, D., Hirsh, J. and Wright, T. R. F. (1984). Molecular mapping of a gene cluster flanking the Drosophila DOPA decarboxylase gene. *Genetics* **106**:679–694.

Hirsh, J. (1989). Molecular genetics of dopa decarboxylase and biogenic amines in *Drosophila. Dev. Genetics* **10**:232–238.

Hirsh, J., Morgan, B. A. and Scholnick, S. B. (1986). Delimiting regulatory sequences of the *Drosophila melanogaster Ddc* gene. *Mol. Cell Biol.* **6**:4548–4557.

Jackson, F. R. (1990). Prokaryotic and eukaryotic pyridoxal dependent decarboxylases are homologous. *J. Mol. Evol.* **31**:325–329.

Johnson, W. A. and Hirsh, J. (1990). A Drosophila "POU Protein" binds to a sequence element regulating gene expression in specific dopaminergic neurons. *Nature* **343**:467–470.

Johnson, W. A., McCormick, C. A., Bray, S. J. and Hirsh, J. (1989). A neuron-specific enhancer of the *Drosophila Dopa decarboxylase* gene. *Genes Dev.* **3**:676–686.

Konrad, K. D. and Marsh, J. L. (1987). Developmental expression and spatial distribution of DOPA decarboxylase in *Drosophila. Dev. Biol.* **122**:172–185.

Krieger, M., Coge, F., Gros, F. and Thibault, J. (1991). Different messenger RNAs code decarboxylase in tissues of neuronal and nonneuronal origin. *Proc. Natl Acad. Sci.* (*USA*) **88**:2161–2165.

Marsh, J. L., Erfle, M. P. and Leeds, C. A. (1986). Molecular localization, developmental expression and nucleotide sequence of the alpha-methyl dopa hypersensitive gene of *Drosophila. Genetics* **114**:453–467.

Morgan, B. A., Johnson, W. A. and Hirsh, J. (1986). Regulated splicing produces different forms of DOPA decarboxylase in the central nervous system and hypoderm of *Drosophila melanogaster. EMBO J.* **5**:3335–3342.

Pentz, E. S. and Wright, T. R. F. (1991). *Drosophila melanogaster* diphenol-oxidase A2: gene structure and homology with the mouse mast-cell tum^- transplantation antigen, P91A. *Gene* **103**:239–242.

Pentz, E. S., Black, B. C. and Wright, T. R. F. (1986). A diphenol oxidase gene is part of a cluster of genes involved in catecholamine metabolism and sclerotization in Drosophila. I. Identification of the biochemical defect in *Dox-A2* [*l(2)37Bf*] mutants. *Genetics* **112**:823–841.

Scherer, L. J., McPherson, J. D., Wasmuth, J. J. and Marsh, J. L. (1992). Human dopa decarboxylase: localization to human chromosome 7p11 and characterization of hepatic cDNAs. *Genomics* **13**:469–471.

Scholnick, S. B., Bray, S. J., Morgan, B. A., McCormick, C. A. and Hirsh, J. (1986). Central nervous sytem and hypoderm regulatory elements of the *Drosophila melanogaster* DOPA decarboxylase gene. *Science* **234**:998–1002.

Spencer, C. A., Gietz, R. D. and Hodgetts, R. B. (1986a). Overlapping transcription units in the dopa decarboxylase region of *Drosophila. Nature* **322**:279–281.

Spencer, C. A., Gietz, R. D. and Hodgetts, R. B. (1986b). Analysis of the transcription unit adjacent to the 3′ end of the dopa decarboxylase gene in *Drosophila. Dev. Biol.* **114**:260–264.

Wright, T. R. F., Bewley, G. C. and Sherald, A. F. (1976a). The genetics of dopa decarboxylase in *Drosophila melanogaster.* II. Isolation and characterization of dopa decarboxylase deficient mutants and their relationship to the alpha-methyl dopa hypersensitive mutants. *Genetics* **84**:287–310.

Wright, T. R. F. (1987). The genetic and molecular organization of the dense cluster of functionally related vital genes in the DOPA decarboxylase region of the *Drosophila melanogaster* genome, in *Results and Problems in Cell Differentiation*, ed. W. Hennig (New York, NY: Springer-Verlag), Volume 14, pp. 95–120.

Wright, T. R. F., Hodgetts, R. B. and Sherald, A. F. (1976b). The genetics of dopa decarboxylase in *Drosophila melanogaster.* I. Isolation and characterization of deficiencies that delete the dopa-decarboxylase-dosage-sensitive region and the alpha-methyl-dopa-hypersensitive locus. *Genetics* **84**:267–285.

12

Elongation Factor Genes: *Ef1α1*, *Ef1α2*

Chromosomal Location: **Map Position:**

Ef1α1	2R,	48D	2-[64]
Ef1α2	3R,	100E	3-[102]

Synonyms: *Ef-1αF1* and *Ef-1αF2*

Products

Translation elongation factor 1 alpha (EF1α), one of three components of elongation factor 1.

Structure

There is remarkable conservation of the amino acid sequence in very distant species (Fig. 12.1). The similarities are particularly noteworthy in a region near the N-terminus that is thought to be the GTP binding site and in the neighborhoods of Ala-92, Lys-244 and Lys-273, residues that are considered important for tRNA binding (Walldorf et al. 1985; Hovemann et al. 1988).

Function

EF-1α is involved in the GTP-dependent binding of charged tRNAs to the acceptor site of the ribosome. A decrease in EF-1α levels after emergence of adults seems to play a role in the aging process (Webster 1985). Conversely, increased expression of *Ef1α1* under the control of a heat-shock promoter leads to extended life spans (Shepherd et al. 1989).

Comparison Between *Ef1α1* and *Ef1α2*

There is 90.5% identity and 93.3% similarity between the *Drosophila* sequences. Differences between the amino acid sequences of EF1α1 and EF1α2 are comparable to the interspecific differences found between the fly and rat

```
Dm Ef1af1       I                                            Q                               A       A
Dm Ef1af2       I                                            Q                               A
 Rat Ef1a       T                                            A                               S
  Eh Ef1a   P   T                                Q          SA                N              S           F
      CON  MGKEK-HINI VVIGHVDSGK STTTGHLIYK CGGIDKRTIE KFEKEA-EMG KGSFKYAWVL DKLKAERERG ITIDI-LWKF ETSKYYVTII DAPGHRDFIK
           1                                                  50                                                     100

Dm Ef1af1                Q D                D          F                 S      EA           S     V    AA          H
Dm Ef1af2                  D                           F                 T      EA           S     I    AS          H
 Rat Ef1a                  V   V                       Y                 T      QK      V    T     I    DT          N
  Eh Ef1a             V I  V                      I  S Y      M          AIQ. . KQE         I AFL  T    DKIP       FQ
      CON  NMITGTSQAD CAVLI-AAGT GEFEAGISKN GQTREHALLA -TLGVKQLIV GVNKMDS-EP PYS--RYEEI KKEVS-YIKK -GYNP--VAF VPISGW-GDN
           101                                                150                                                    200

Dm Ef1af1       T         E G  E  N D K  VD    A    A      A                            V    T  V    A IT              Q
Dm Ef1af2       EK        S E  E  K E KC ID    A    Q                                   L    M  N    V LV              T
 Rat Ef1a       A         K T  D  S S T  LE    C    T                                   V    M  T    V VT              S
  Eh Ef1a   I   T     Y......... .... P  IG     SVT E   V               S               I    TI Q    SGVSS C   I    T  A
      CON  MLEPS-NMPW FKGW-V-RK- G-A-G-TL-- ALD-ILPP-R PTDKPLRLPL QDVYKIGGIG TVPVGRVETG -LKPG-VV-F AP-N--TEVK SVEMHHEAL-
           201                                                250                                                    300

Dm Ef1af1    V                EL    Y       AN  K                    AN                   A IL  V    S TT EN . F
Dm Ef1af2    M                EL    Y        N  R                    AN                   S IK  Y    T GTT D  . A
 Rat Ef1a    L                DV    N        D  ME  G       I        SA  A                A LK  I    S KL  D  . FL
  Eh Ef1a  Q I          R LT  DI K  N  S A   Q AV CE         M       RK           S       E LLS I    T SM  G  E EY  N  S
      CON  EA-PGDNVGF NVKNVSVK-- RRG-VAGDSK N-PP-GAADF TAQVIVLNHP GQI--GYTPV LDCHTAHIAC KF-E--EK-D RR-GK--E-G -PK-IKSGDA
           301                                                350                                                    400

Dm Ef1af1     NL  S       A QE                           NF DASG      A E  T G K*
Dm Ef1af2    IVL  S       S QE                         S NF ETTS      A E  Q  K *.
 Rat Ef1a     DM  G   M   S SDY                          DK AAGA      S Q  Q  A *.
  Eh Ef1a   L KI  T       E AK              K       V    TP*..... .......... ....
      CON  AIV--VP-KP LCVE-F--FP PLGRFAVRDM RQTVAVGVIK AV--K----G KVTK-A-KA- K-K--
           401                                                450              465
```

FIG. 12.1. Comparison of the two *Drosophila* Ef1α sequences (Dm) to the corresponding sequences of *Rattus norvegicus* (Rat) (Accession, X63561) and *Entamoeba histolytica* (Eh) (Accession, M92073). The CON(sensus) line indicates positions at which all four sequences agree. There is 86% overall identity between the rat and *Drosophila* proteins. Sequences aligned with the GCG *Pileup* program.

Ef1α1

```
-1881  CTCAAGCTTCCATTGTTATTTAAAGTTCTATTACGTTAGGGTTCACATACAAATTAAAGTGGCAGGTTCTATCTCAAAACATTCGTTCAA  -1792
                                                                                   ------------

-1791  AATGCGGACTACTAATGCAATTGTTATTGTTTTTACATATTAAAAGATATGTGTTCCAATATTACGTATAGAAATTATAGACATCGTTTT  -1702

-1701  GTAGAAAATACTTTTGGAATCACTGATTATTTAGTTTTTCATATAAAAACAATGTCGAGCAAACAAGGTTTTTAAATTCCTCAATCTTT  -1612

-1611  AGGTTATTGTATTTTGCCACTTTCAATCACTTAAATTTCAATAAAATGAAGTGCTTCATTCGCGCGTAGTGGAAACACCGCAGTGGGAAC  -1522

-1521  ACGGTTTCTGCTCTTTTGACAGTTGCGTAGCTTCGGTCACACCATGTGTCAAACGAGGCTTCCTGTGCTGAGCTCTGCCGAACGCTCGTT  -1432

                                                                _
       -->-1431                                                 |
-1431  CACTTTGTTCGAATCCGTCGCCGCTTAGACTTCGTGATTTCTCATTCAGCTTATTAGAGAGTAAGTTTTACCTGCGAGGCTATAATTAAG  -1342
                                                               _|

-1341  TGATTTCTGCAAAAAAACTGCAGGGGGGAAACAATTTATAAACAAATATGCAGCTGAGACGCCGAATTTGTGCATATTTCCAGTGTTTTT  -1252

-1251  CCTGTGTGTGTGTAATAAACCCGGAGATAACCTCTAACTGCGGTTTTCCAAAGTGAAAGGTGGCCATAGAAGCAAACACGTGGCAAGTCT  -1162

-1161  GCAAAGGCAAAAATTTTAACTGGCGTTCCCAGTTAAAGTTCCCAGCATTCTCAAAATAATTTTCCGGCTTTTCCGGCCGCATTTTCGCCC  -1072

-1071  TGCAATATGGTGCACTTAGCGTGTAATTACTTTGCCACGCCCACGCCGGACACAGAGGTCATCCACCAGATGTGCTCATTAACCGAGAAA  -982

 -981  AAAAAACGTGCTTTCTCTCTTGCCTTTGTCATGGCCTATAGATATTCCTTATTCTTTCTTTTTGCGGCATGGAATTCTAAAATGGCGACC  -892

 -891  CAGTGGCGTGAGTCAAGTGGGCGAAAAAATTCGCCTGGCAACAAGCGAAAAAATGTGCTTTTTTGGGTTTCCAGCCCATTAGCATATCTG  -802

 -801  GTGTAATGGCACTCGCATCAGCTATTTCGCCATTTCCAACCGACTCAATAATTGGTTTTGGTAAAATGGCTGCCGCTGCACTACGTTCTT  -712

 -711  GATTAATTCGTTGTGTGCCCCTCTCTTTTTCATTTCTTTCCAATTACCAATTGTGCCACCGCGGCGGAGACGCTTGCATTTGTACAAGTC  -622

 -621  ACACACGCACACTAATGCACATCCGCCATTTTGGTCTCTCTCTCTTCCTCTCTTACTTTTTCCGGCCGGCAACAGCGTCACACAAATACA  -532

 -531  CAGGCATAGATATACACACGCATAGGCAGATAAGCACATGTGTATTTGCGAATTAAATTTGCTGGAATTTTCCTTTGGACTCTTCGATTT  -442

 -441  AACATGATGATGATTTTTCAGTTCTGCTACTGAAGAGAGTTGACAGAAAGCAAAAATACCAAAATCACTGAAACAAAATCGAGTTTCCAT  -352

 -351  ATGGAATTTTATTTGCACGCTCTTTTCTGTAGTTGCGCCCCACTCGTTTTACCCACACCCCTACATGCGGGCACTGGTCCTAACCTCAAA  -262

 -261  AAACACGTTTTGTACGGCTGCAAGAGTTTGAGGTTAGGTTGTGCTCGCGCATGCAAACAAAAGTCGAACGTACGCTAGGGAAATGAGAAA  -172

 -171  GTGTTATACCCACTAATAATTGTAGTTGTAATCCCACCGAATTGTTTTACCCTTTGTTTATTCCAACCTCTCTTGCTCGCCAACCCGCCG  -82

                                                                   _
                                                                   |
  -81  AACCCTGCAACCTTCCAATGTTCCAACGTTCCGTTAATCCAACACTCGAATACACACAACAGCCATAGTGTAATCATCCAACATGGGCAA  8
                                                                   |_                   MetGlyLy  (3)

    9  GGAAAAGATTCACATTAACATTGTCGTGATCGGACACGTCGATTCCGGTAAGTCGACCACCACCGGACACTTGATCTACAAGTGCGGTGG  98
       sGluLysIleHisIleAsnIleValValIleGlyHisValAspSerGlyLysSerThrThrThrGlyHisLeuIleTyrLysCysGlyGl  (33)

   99  TATCGACAAGCGTACCATCGAGAAGTTCGAGAAGGAGGCCCAGGAGATGGGAAAGGGATCCTTCAAGTACGCCTGGGTTTTGGATAAGTT  188
       yIleAspLysArgThrIleGluLysPheGluLysGluAlaGlnGluMetGlyLysGlySerPheLysTyrAlaTrpValLeuAspLysLe  (63)
```

```
189   GAAGGCTGAGCGCGAGCGTGGTATCACCATCGATATCGCCCTGTGGAAGTTCGAAACTGCCAAGTACTACGTGACCATCATTGATGCCCC   278
      uLysAlaGluArgGluArgGlyIleThrIleAspIleAlaLeuTrpLysPheGluThrAlaLysTyrTyrValThrIleIleAspAlaPr   (93)

279   CGGACACAGGGATTTCATCAAGAACATGATCACTGGTACCTCGCAGGCCGATTGCGCCGTGCAGATTGACGCCGCCGGAACCGGAGAATT   368
      oGlyHisArgAspPheIleLysAsnMetIleThrGlyThrSerGlnAlaAspCysAlaValGlnIleAspAlaAlaGlyThrGlyGluPh   (123)

369   CGAGGCCGGTATCTCGAAGAACGACCAGACCCGCGAGCACGCCCTGCTCGCCTTCACCCTGGGTGTGAAGCAGCTGATCGTTGGTGTGAA   458
      eGluAlaGlyIleSerLysAsnAspGlnThrArgGluHisAlaLeuLeuAlaPheThrLeuGlyValLysGlnLeuIleValGlyValAs   (153)

459   CAAGATGGACTCCTCCGAGCCACCATACAGCGAGGCCCGTTATGAGGAAATCAAGAAGGAAGTGTCCTCTTACATCAAGAAGGTCGGCTA   548
      nLysMetAspSerSerGluProProTyrSerGluAlaArgTyrGluGluIleLysLysGluValSerSerTyrIleLysLysValGlyTy   (183)

549   CAACCCAGCCGCCGTTGCCTTCGTGCCCATTTCCGGATGGCACGGCGACAACATGTTGGAACCCTCTACCAACATGCCCTGGTTCAAGGG   638
      rAsnProAlaAlaValAlaPheValProIleSerGlyTrpHisGlyAspAsnMetLeuGluProSerThrAsnMetProTrpPheLysGl   (213)

639   ATGGGAAGTGGGACGCAAGGAGGGTAACGCTGACGGCAAGACCCTGGTCGATGCCCTCGATGCCATCCTTCCCCCAGCCCGTCCCACCGA   728
      yTrpGluValGlyArgLysGluGlyAsnAlaAspGlyLysThrLeuValAspAlaLeuAspAlaIleLeuProProAlaArgProThrAs   (243)

729   CAAGGCCCTGCGTCTGCCCCTGCAGGATGTGTACAAAATTGGCGGTATTGGAACAGTACCCGTGGGTCGTGTGGAGACTGGTGTGCTGAA   818
      pLysAlaLeuArgLeuProLeuGlnAspValTyrLysIleGlyGlyIleGlyThrValProValGlyArgValGluThrGlyValLeuLy   (273)

819   GCCCGGTACCGTTGTGGTCTTCGCCCCTGCTAACATCACCACTGAGGTCAAGTCCGTGGAGATGCACCACGAGGCCCTGCAGGAGGCCGT   908
      sProGlyThrValValValPheAlaProAlaAsnIleThrThrGluValLysSerValGluMetHisHisGluAlaLeuGlnGluAlaVa   (303)

909   TCCCGGAGACAACGTTGGCTTCAACGTCAAGAACGTGTCCGTGAAGGAGCTGCGTCGTGGCTACGTTGCCGGTGACTCCAAGGCTAACCC   998
      lProGlyAspAsnValGlyPheAsnValLysAsnValSerValLysGluLeuArgArgGlyTyrValAlaGlyAspSerLysAlaAsnPr   (333)

999   CCCCAAGGGAGCCGCCGACTTCACCGCCCAGGTCATCGTGCTGAACCACCCCGGTCAGATTGCCAACGGCTACACCCCAGTGTTGGATTG   1088
      oProLysGlyAlaAlaAspPheThrAlaGlnValIleValLeuAsnHisProGlyGlnIleAlaAsnGlyTyrThrProValLeuAspCy   (363)

1089  CCACACCGCTCACATTGCTTGCAAGTTCGCTGAGATCTTGGAGAAGGTCGACCGTCGTTCCGGCAAGACCACCGAGGAGAACCCCAAGTT   1178
      sHisThrAlaHisIleAlaCysLysPheAlaGluIleLeuGluLysValAspArgArgSerGlyLysThrThrGluGluAsnProLysPh   (393)

1179  CATCAAGTCTGGCGATGCTGCCATCGTCAACCTGGTGCCCTCTAAGCCCCTGTGCGTGGAGGCCTTCCAGGAGTTCCCCCCTCTGGGTCG   1268
      eIleLysSerGlyAspAlaAlaIleValAsnLeuValProSerLysProLeuCysValGluAlaPheGlnGluPheProProLeuGlyAr   (423)

1269  CTTCGCTGTGCGTGACATGAGGCAGACCGTGGCTGTCGGTGTCATTAAGGCTGTCAACTTCAAGGATGCCTCCGGTGGCAAGGTCACCAA   1358
      gPheAlaValArgAspMetArgGlnThrValAlaValGlyValIleLysAlaValAsnPheLysAspAlaSerGlyGlyLysValThrLy   (453)

1359  GGCCGCCGAGAAGGCCACCAAGGGCAAGAAGTAGCTGGTTTGCTTCCACTCAACAACAACAACAACACGCAGTAGTAGCAGCAACAACAA   1448
      sAlaAlaGluLysAlaThrLysGlyLysLysEnd                                                         (463)

1449  GCATATAACCAACATCATAATGCAGCCAACAACACCACTCAATAATACCAGCAACAGCAGCAGCGAACACAATAGTAGTATAACACCAAC   1538

1539  ACCTGTCCTGCGCAAGATGACCGATAAGATGATGTTTCAGCAGAAGCATAAGTTTAATTTCTTCCATCGAAAGGAGTTTCGACGGATACG   1628

1629  AATGCTAAATGCAGACGAGGCCGCCTTCACTGGGAAATCGGTGGATCCCAAGGATAAGAGTGCACACTGGGAAAACACTTGCATTTATGC   1718

1719  ATCCACTCCTCATCCACTTCCCCGTCGATCTTTAGTTTACTAAATATGGTATGATGCACGCAGTTGACTTCGTTTTATCATATCATATAT   1808

1809  AGGAATCCTCTGTAGCATTTATGATATCGTTTAAATTAACCTTTATACTTTGATATGTATCATTTATCTTACCCTACTTTTGCACACACT   1898

1899  ACTTTGTACACAAGAAAAGAACCAGAATAGAAGCGATAAACTATATTTACAAAAAAAATAAAAACCCTATTTTTGTATTTCTTTTGTTTT   1988
                                                                 ------            |(A)n
1989  TACCACCCAGCCCGTAAAAGAGCACTCTCTTTTTGGTTGTTGCCTCCCGATTT  2041
```

Ef1α1 SEQUENCE. Strain, *Canton S*. Accession, X06869 (DROEF1AF1).

sequences (Fig. 12.1); this suggests that the two genes in *Drosophila* originated as an ancient duplication. The sequence similarity between the two genes outside of the coding regions is very limited; and there is great discrepancy in the number of introns (Walldorf et al. 1985; Hovemann et al. 1988).

Ef1α1

Gene Organization and Expression

Open reading fame, 463 amino acids; expected mRNA size, 2,054 bases, in agreement with a single RNA band of 2 kb. The 5′ end was defined by primer extension, cDNA sequencing and RNA sequencing. There is no apparent TATA box. The 3′ end was obtained from S1 mapping and the sequence of a cDNA clone. There is one intron at −1,371/−20, in the leader (*Ef1α1* Sequence) (Walldorf et al. 1985; Hovemann et al. 1988).

Developmental Pattern

Expression is high throughout development, but it declines with age in adults (Webster 1985). It is also 5–10 times higher in adult females than in males (Walldorf et al. 1985; Hovemann et al. 1988).

Promoter

At −1,804 (373 bp upstream of the transcription initiation site) there is a sequence very similar to the HOMOL1 box of yeast. In yeast, this sequence occurs upstream of several genes for translation factors and ribosomal proteins (Walldorf et al. 1985; Hovemann et al. 1988).

Ef1α2

Gene Organization and Expression

Open reading frame, 462 amino acids; expected mRNA size, 2,555/2,558 bases, in agreement with a single RNA band of 2.5 kb. The 5′ end was defined by primer extension and by sequencing of a cDNA. There is no apparent TATA box. The 3′ end was obtained from a cDNA sequence. There are four introns: two in the leader, at −1,811/−567, and −479/−30, −27 (this intron has two acceptor sites, and both are used), one in the Gly-275 codon and one after the Gln-343 codon (*Ef1α2* Sequence) (Walldorf et al. 1985; Hovemann et al. 1988).

Ef1α2

```
-2156  TAAGCGAATAGTGTGCACAATGTCTTTTGCAATTAGTGGTGAATGTGCATACTTTAGTGACAGTCCGTGAAAGTACTATATTATTTTATC  -2067

-2066  TGCAAAAGACTCAGTTTAAGAGAATATAAAATATTCCATGAATGGTAGTAAAATTGTATTACTATTTTTATTTTGGTACGTTTTATACTT  -1977

-1976  AAGGGATGGAAACTTTATTTAAGTCAAGAAATCCGCATAATGCAATAGGAAACCCAAGGCCCTTGTCATACATGGAATCCTGTGCCATCT  -1887
                                          ----
                                                                -->-1833
-1886  CTAGGTCGGAATCAGTTCAGCTCCGTTCACCTCAGCATCGTTGCTTTTTCGGTCTTTCCGTTTTGTGATTTCGAGGTAAGTGCACGCAGA  -1797

-1796  GCTCCCGTTAAAATTGTGAAAATATTAATAGGCATTGATTAGTTGTGGAAATGTAAAAAGGGAAAGTCCCAGAATTCCCTACCCTGCATT  -1707

-1706  ATTAGGCGAATTTCGGTTCGATTTCCAACCTAAAGAAAGTTCTAAAGTAAAGAAAGTTCCGAAAAGTGAGAGAGTGTAAGTGATTTGCGC  -1617

-1616  TGCCGGCCGGTCTTCTCATTCCTTTTGCATAATAGCTGTGTAAATCGATTCGAATTGGAAATTGGTTTTCCAGCGACCTTAAATTGCAAG  -1527

-1526  TAAATTAATAAAGTTGCATAGACTTTCGAATTCCAACATGGCGACCGGCTGCATGTGTGTGCGCGTTCGATTTTGCCTGGATTGTACCCG  -1437

-1436  TTTCTCCTTCCCGTTCTCAAGCCGTTTATTCCCGAGTAGTTTCTATTGGAATTCGCAGGCAAAAAAAAAATATCCGCGGCATGATGGCAC  -1347

-1346  ATGGTTAGCAGATTATTTTCTTGCCCTGCATCTCTGACGAAGTATTTTGCATATTCTTTCCCCCTTCATTCCCATTGCTTCTTCCAATTT  -1257

-1256  GCACTTCGATGCAAATACAAAGATTTAAAAATGGCATGCAGGAAAATCGGCAAGTGAAACTGTCACTGGGGTAGAAAATAAATCACAACG  -1167

-1166  CCCTGCAGTTCTCGCCGTCTCTTTCCCTTCCTTCTCTGCATGACCAGCAAGTGCACTGCGCCCGTTCGCCGTCCCTTTCTCTCCCGCTCT  -1077

-1076  CTCCATCTCCCTCTACAGTTTTTCACCCTTTGGAATCGCGGGATTTTCGCCGCACGACCGCCACCGAATGCCGATGCTTTTGGCCATTTC  -987

 -986  CCTTTGGATTTTCTTCCACCGTGCTGCGAAAGTTGCCAAATTTCGGCATTTCGACATTTGGCTTAATTGAAATCCGTTTGGGTGTGCGAT  -897

 -896  TTTCATTGGTTTTCCCACTAAAAACGCCGGCCGGCACATTTTCGCCATGCACTGCCGCACTTCCCGGCTTTCCGACGAGGGTTTCTCTTC  -807

 -806  GGCTTAATCCTCTCCAGCCGAGGAGAGTGCATTTTCCCAGTACGCACACTTCGGCTCCATTCGTTTCTGTCTGGGGCTCGTTATTGATTT  -717

 -716  TTCGCCCGGTGCACTTCGGCAGAGGATATACACGGCAGTCTTTAACCAACAGACACTTGGCCCGGTCGTGGTCCGGCTGCAGAGTACGGA  -627

 -626  AGATCCGCATAGAGTTTAAAAACTGCCATTTTTATGACAACGATTTCCTTCTAATTCTAGGATATAGCGTCGCGTGGGTTTGTGATCAGT  -537

 -536  TTCTAAGTGCGCCAGTTGCCGAGTAATAAGAAACTCTAGAAAGTCTCGTGAAAACAGGTGAGTTTTTCTGCTTGTAAATTCTTGCTGCAT  -447

 -446  AGATTTGTGGGCAAAAATATTATGGGAATATGGGTGTATTTCTCAATCGTACACATTAGTGTCCATAAGAGTCCGTAAAAACATACATGT  -357

 -356  GTATTTATATTTTTCCTATTATTCAGTATAAGGCTTAATTTGAACTAATTGGTAAACTTTTCGCGTGATTTTCGTGTTTACTCTTGAATT  -267

 -266  GTTTAAAATTCGTATTTTCGAAATATAAAAGTTCAACGGTTTTCCCTGTGTACGTTTGTGCCGTCCGTATGAAGTGTGCTTTTGGTGTCG  -177

 -176  CCACCACGATGACACGACCCACAGCATACAGACGTCACTCGTCTGCACCACCCATTAAGTTCAGACCCACATTGGCATGCTACCTCCCCG  -87
```

(continued)

```
 -86  AGTACGGAAACCACCCACTTTGCTCATCCGAATACCTGCATCCCTTCTGTCTCCCAGCAGCTCTAAAAAATAGCTTAATCTGCAAGGATG     3
                                                                                           Met   (1)

   4  GGCAAGGAGAAGATCCATATTAACATTGTGGTCATTGGCCATGTGGACTCCGGCAAGTCGACGACCACCGGCCACTTGATCTACAAATGC    93
      GlyLysGluLysIleHisIleAsnIleValValIleGlyHisValAspSerGlyLysSerThrThrThrGlyHisLeuIleTyrLysCys  (31)

  94  GGCGGCATCGACAAGCGTACGATTGAGAAGTTCGAGAAGGAGGCCCAGGAAATGGGAAAAGGCTCCTTTAAGTACGCTTGGGTACTGGAC   183
      GlyGlyIleAspLysArgThrIleGluLysPheGluLysGluAlaGlnGluMetGlyLysGlySerPheLysTyrAlaTrpValLeuAsp  (61)

 184  AAGCTGAAGGCAGAGCGGGAGCGGGGCATCACCATCGACATTGCCCTATGGAAGTTCGAGACGTCCAAGTACTATGTGACCATCATCGAT   273
      LysLeuLysAlaGluArgGluArgGlyIleThrIleAspIleAlaLeuTrpLysPheGluThrSerLysTyrTyrValThrIleIleAsp  (91)

 274  GCCCCTGGTCACAGGGATTTCATCAAGAACATGATTACCGGTACCTCTCAGGCCGATTGTGCGGTGCTGATCGACGCCGCCGGAACTGGA   363
      AlaProGlyHisArgAspPheIleLysAsnMetIleThrGlyThrSerGlnAlaAspCysAlaValLeuIleAspAlaAlaGlyThrGly (121)

 364  GAGTTCGAGGCCGGGATCTCGAAGAACGGCCAGACCCGCGAGCACGCCCTTCTGGCATTCACGCTGGGCGTGAAGCAGCTTATTGTGGGC   453
      GluPheGluAlaGlyIleSerLysAsnGlyGlnThrArgGluHisAlaLeuLeuAlaPheThrLeuGlyValLysGlnLeuIleValGly (151)

 454  GTCAACAAGATGGACTCCACTGAGCCGCCGTACAGCGAGGCCCGCTACGAGGAGATCAAGAAGGAGGTGTCCTCGTACATCAAGAAGATC   543
      ValAsnLysMetAspSerThrGluProProTyrSerGluAlaArgTyrGluGluIleLysLysGluValSerSerTyrIleLysLysIle (181)

 544  GGCTACAATCCGGCCTCGGTGGCCTTCGTGCCCATCTCCGGATGGCACGGCGACAATATGCTGGAGCCGTCCGAGAAGATGCCCTGGTTC   633
      GlyTyrAsnProAlaSerValAlaPheValProIleSerGlyTrpHisGlyAspAsnMetLeuGluProSerGluLysMetProTrpPhe (211)

 634  AAGGGATGGTCCGTGGAGCGCAAGGAAGGCAAGGCAGAGGGCAAGTGCTTGATCGACGCGCTGGACGCGATCCTTCCACCCCAGCGTCCC   723
      LysGlyTrpSerValGluArgLysGluGlyLysAlaGluGlyLysCysLeuIleAspAlaLeuAspAlaIleLeuProProGlnArgPro (241)

 724  ACCGACAAGCCGCTGCGCCTGCCGCTCCAGGACGTCTACAAGATCGGAGGCATCGGAACCGTACCAGTAGGTCGTGTGGAGACTGGTCTC   813
      ThrAspLysProLeuArgLeuProLeuGlnAspValTyrLysIleGlyGlyIleGlyThrValProValGlyArgValGluThrGlyLeu (271)

 814  CTCAAGCCAGGTAAGGCTCCGGGTTGATGAGGTCGGGTGTGGGCCCTCTTTTCTCTTTGGGCACTTCATACATGTATTCTGCAAAATTTG   903
      LeuLysProG                                                                                 (275)

 904  GGTCGACAGTGGGCTGGCATCCAACAGCCACCGCCTCCAAAGCGGAGCCGCAACGAAGTCTTGCGCATGTATGCATTATTGAGCGAACGT   993

 994  CTTCGTCGAGAGCGAGACCCTCCACCTCATGCACTTGGTGAAATTCTCACTCCGAAGAGCTTCCATTTTCAACATGAAAGTGAAAGGCCA  1083

1084  TTAAAATAAAATAACCCTAGCTAACATATTAATATATGTAGAGCTATTGATTCAAATAAAAATAAATTGGAGTTAGTTCGAATAATATCG  1173

1174  CTCCACGTTTCTCTCTCTGTATGCACCCACCCCCATCCAAATGTCTACACATAACGTCCGGATATGTAACTTCGTTTCGGTCGCTTCGTT  1263

1264  TCCGGTTTCGTTTCAGGCATGGTCGTCAACTTTGCGCCGGTCAACCTGGTCACCGAAGTAAAGTCTGTGGAGATGCACCACGAGGCTCTC  1353
                      lyMetValValAsnPheAlaProValAsnLeuValThrGluValLysSerValGluMetHisHisGluAlaLeu (299)

1354  ACCGAAGCCATGCCCGGCGACAACGTTGGCTTCAACGTGAAGAACGTGTCCGTGAAGGAGCTCCGTCGTGGCTATGTGGCCGGCGATTCC  1443
      ThrGluAlaMetProGlyAspAsnValGlyPheAsnValLysAsnValSerValLysGluLeuArgArgGlyTyrValAlaGlyAspSer (329)

1444  AAGAACAATCCTCCTAGGGGAGCAGCCGACTTTACCGCTCAGGTAGGGTAACAAAGATGAGAAATCTTTGATAGTTGAACTCATCTTTGT  1533
      LysAsnAsnProProArgGlyAlaAlaAspPheThrAlaGln                                                 (343)

1534  TTGGTTTTTTTTTTTTCTTTTTGCCCACAGGTGATTGTGCTCAACCATCCGGGCCAGATCGCCAATGGGTACACTCCCGTCTTGGATTGC  1623
                                                ValIleValLeuAsnHisProGlyGlnIleAlaAsnGlyTyrThrProValLeuAspCys (363)
```

```
1624 CACACGGCGCACATTGCCTGCAAGTTTTCCGAGATCAAGGAGAAGTACGACCGCCGTACGGGCGGAACCACCGAAGACGGCCGAAGGCT  1713
     HisThrAlaHisIleAlaCysLysPheSerGluIleLysGluLysTyrAspArgArgThrGlyGlyThrThrGluAspGlyProLysAla (393)

1714 ATCAAGTCCGGGGATGCGGCCATCATTGTGCTGGTGCCCAGCAAGCCGTTGTGCGTAGAGAGCTTCCAGGAGTTCCCACCGCTGGGACGG  1803
     IleLysSerGlyAspAlaAlaIleIleValLeuValProSerLysProLeuCysValGluSerPheGlnGluPheProProLeuGlyArg (423)

1804 TTCGCTGTGCGCGACATGAGGCAGACCGTGGCCGTGGGCGTCATCAAGTCGGTGAACTTTAAAGAGACGACCTCGGGCAAGGTGACAAAA  1893
     PheAlaValArgAspMetArgGlnThrValAlaValGlyValIleLysSerValAsnPheLysGluThrThrSerGlyLysValThrLys (453)

1894 GCCGCTGAGAAGGCACAGAAGAAGAAATAACTAGGGTACCAGCAGAACAACGTCATCACTCGAACCCAACAACAACAAAAACAGACGGCT  1983
     AlaAlaGluLysAlaGlnLysLysLysEnd                                                             (462)

1984 AGAGCAACAGCAGCAACAACACACAACAACAATACACATGTCAAAATTATAATACCCACTCGACGATCAAATTCACACCTTGACTCCATG  2073

2074 GCAAGAGAGACACCAATTACTACTATTACTAGCTGCTGGGAGAAGCGGCAGATATTAACCGAAATCGAGCAGATTATACCCTATATAATA  2163

2164 ACCACACGTACGATTAGCGAGGAGAGGAGCATCAGGTGCAGCGAGGATGCGAAGGAGGAGCCCTTCCAGCCTCGCCGGGTCGGTTTTGGT  2253

2254 CGCCTTCGCCGTGGTGGTCTACTGCAGCTATCTGAACATGTATCGTCACCGCAAGTCCTTTCGTAGGAAACCACCCGCTAGCCACTCCGC  2343

2344 AGAGTGGATAGGGGCCTCCGGAGCACTGCTGTAGCCCGCCCCTTCGATATATACTCATCTCTAAAACTAACCTTACACTTGATTAGCAGC  2433

2434 CACACATCCGGTCGCATCCACCTGTTTCGAATGGATTTTAAACACTTTTTATACTTTTGATAAGTCAAGTCGGAGGCATTCGATTTAAAA  2523

2524 TCTATTGAAATATGTAATTTCCGAATTTAGTTTTAAACCACGTCCGCGCTCCCAAAAATCCCCCGAACCGAAAAGACTACATTCGCGATG  2613

2614 AATTCAAAATTTCTCTTGAAACCAAAAAAAACAAATGCTTAAGAAGTATTACAAAAAAGAAATCAACATTACACACATAATCATGCGGTT  2703

2704 TTTGAAAACATTATAAATGTTTAATCGAGCCTCATTTGCATTTGCATATTACATAATATACGTTAGCCACATGTCATCTCATTGCCCATA  2793

2794 ATAACCTGCATCCTGCATATTATACACGTTAATCTCACACTCTGAATTTATACAAACCGAAGACAATTGTAACCGACACCAGAACAATTC  2883

2884 TTGGATACAGAACATGTTGGCTTGATAAAAGATCTTTTAAATGATGAGAAAAATAAAGGAAGCTTAACCGTAAAATACCACACACGAACG  2973
                                                   ------                |(A)n
2974 CCTTTTAATTGAAAAATACTTGAATATCTATGAAGAAAATGAATTC  3019
```

Ef1α2 SEQUENCE. Strain, *Canton S*. Accession, X06870 (DROEF1AF2).

Developmental Pattern

The level of expression is lower than that of *Ef1α1*, and it peaks during the pupal stages (Walldorf et al. 1985; Hovemann et al. 1988).

Promoter

There are no obvious similarities with the promoter region of *Ef1α1* (Walldorf et al. 1985; Hovemann et al. 1988).

References

Hovemann, B., Richter, S., Walldorf, U. and Cziepluch, C. (1988). Two genes encode related cytoplasmic elongation factors 1-alpha (EF-1a) in *Drosophila melanogaster* with continuous and stage specific expression. *Nucl. Acids Res.* **16**:3175–3194.

Shepherd, J. C. W., Walldorf, U., Hug, P. and Gehring, W. J. (1989). Fruit flies with additional expression of the elongation factor EF-1α live longer. *Proc. Natl Acad. Sci. (USA)* **86**:7520–7521.

Walldorf, U., Hovemann, B. and Bautz, E. K. F. (1985). F1 and F2: Two similar genes regulated differently during development of *Drosophila melanogaster. Proc. Natl Acad. Sci. (USA)* **82**:5795–5799.

Webster, G. C. (1985). Protein synthesis in aging organisms. In *Molecular Biology of Aging, Gene Stability and Gene Expression*, eds R. S. Sohal and R. G. Culter (New York: Raven Press), pp. 263–289.

13

even-skipped: eve

Chromosomal Location:
2R, 46C3-11

Map Position:
2-58

Product

A DNA-binding regulatory protein of the homeodomain type important in establishing the segmentation pattern of the embryo.

Structure

The homeodomain occurs toward the N-terminus (Val-70 to Arg-129). The Gln residue in position 9 of the third homeodomain helix (*eve* Sequence, H3) makes EVE a homeoprotein of the *Antennapedia* (*Antp*) class (Hanes and Brent 1991). Another noteworthy sequence feature is the Ala-rich segment spanning Ala-146 to Ala-179. Similar Ala repeats have been found in the genes *caudal*, *engrailed* (*en*), *Ultrabithorax* (*Ubx*) and *Krüppel* (*Kr*); in the *Kr* product the Ala-rich region seems to be associated with the repressor function of that protein (Macdonald et al. 1986; Hoey et al. 1988; Biggin and Tjian 1989; Licht et al. 1990; Harrison 1991).

Function

Binding sites for EVE have been found in the region of the *eve* promoter proximal to the site of transcription initiation and in the *en* promoter. The sequences of the binding sites are quite different in the two promoters. The consensus for the EVE binding site of the *en* promoter is TCAATTAAAT; this is similar to binding sites of other homeodomain proteins of the *Antp* class and was designated as class I (Levine and Hoey 1988; Hanes and Brent 1991). In contrast, the EVE binding sites near *eve* have in common the sequence TCAGCACCG and were designated as class II (Hoey and Levine 1988). EVE binding sites with segments combining features of both class I and class II sequences also exist in the *eve* autoregulatory region, 5.4–5.2 kb upstream of

eve

```
        .         .         .         .         .         .         .         .         .
-5400   CCCGGGCAGTGAGGAATTCCTCCGAAAGTCGGGTCCTCCGTTCTCCAGCCGAAGATTTTTCGAGCAACCAAAATATTATGGTGTGCCCC   -53
                                                                              -------------->eveD

        .         .         .         .         .         .         .         .         .
-5310   GCTGTTCTCGCACAGTCAGCGCGAATTTGCTGCGGTGAGTCGATGCTGTTTCGCAGGACCTTCTTCCATTTTCGTCTCCCTCTTGCTCAG   -52
                  ------------------------eveP                     <==========denf2
          <//////////denf1

        .         .         .
-5220   CCTGTCCCTGTTCCTCTGCAG    -5200


        .         .         .         .         .         .         .         .         .
-1600   AATATAACCCAATAATTTGAAGTAACTGGCAGGAGCGAGGTATCCTTCCTGGTTACCCGGTACTGCATAACAATGGAACCCGAACCGTAA   -15
           -------kr6

        .         .         .         .         .         .         .         .         .
-1510   CTGGGACAGATCGAAAAGCTGGCCTGGTTTCTCGCTGTGTGTGCCGTGTTAATCCGTTTGCCATCAGCGAGATTATTAGTCAATTGCAGT   -14
                                                        ----------k5         -----------------------
                                                       ---------bcd5         --------bcd4

        .         .         .         .         .         .         .         .         .
-1420   TGCAGCGTTTCGCTTTCGTCCTCGTTTCACTTTCGAGTTAGACTTTATTGCAGCATCTTGAACAATCGTCGCAGTTTGGTAACACGCTGT   -13
        ---gt3                                  ---------------------------------------------gt2

        .         .         .         .         .         .         .         .         .
-1330   GCCATACTTCATTTAGACGGAATCGAGGGACCCTGGACTATAATCGCACAACGAGACCGGGTTGCGAAGTCAGGGCATTCCGCCGATCTA   -12
                                              ---------bcd3 ----------kr4

        .         .         .         .         .         .         .         .         .
-1240   GCCATCGCCATCTTCTGCGGGCGTTTGTTTGTTTGTTTGCTGGGATTAGCCAAGGGCTTGACTTGGAATCCAATCCCGATCCCTAGCCCG   -11
                                                 ---------bcd2

        .         .         .         .         .         .         .         .         .
-1150   ATCCCAATCCCAATCCCAATCCCTTGTCCTTTTCATTAGAAAGTCATAAAAACACATAATAATGATGTCGAAGGGATTAGGGGCGCGCAG   -10
                                              -----------------------gt1     -----------bcd1/kr3
                                                    ----------hb3

        .         .         .         .         .         .         .         .         .
-1060   GTCCAGGCAACGCAATTAACGGACTAGCGAACTGGGTTATTTTTTTGCGCCGACTTAGCCCTGATCCGCGAGCTTAACCCGTTTTGAGCC   -9
                                      ----------kr2                           -----------kr1
                                            ----------hb2

        .         .         .
-970    GGGCAGCAGGTAGTTGTGGGTGGACCCCACGATTTTTTTG  -931
                                     ----------hb1

                                                                          .         .
                                                                  -498 CCCGCCCGTCCCGCTCGCTCCTGCGG   -47
                                                                                        =======>11

        .         .         .         .         .         .         .         .         .
-472    AGCAAGCCTGCGGGCGGGCGAGACAAAAGATTCGTTCGCTCATCGCTATAATACCAAATCGAACTCTCTCTCTCTCCAGCTCGGGAGTGC   -38
                                            <////10                    <////////////9

        .         .         .         .         .         .         .         .         .
-382    CATGCCAGCATGGCCAGGACCTCCTCATGGTCCTGCCGAGCAGAGAACGCGGCTCCATCCCGCTGCTCCGGGTCCTGCTCCTCCGCTTTG   -29
                     <=======8b     ========>8a                                =======>7

        .         .         .         .         .         .         .         .         .
-292    TCCCGCCTCGTTATCGCCGCTCAGCACCGAGAGCACAGCAGCGCATCCACTCTCAGCACCGCACGATTAGCACCGTTCCGCTCAGGCTGT   -20
                         <////6b    ////>6a              <////5                      <////4c <///
                           -------->e4                       -------->e5b -------->e5a

        .         .         .         .         .         .         .         .         .
-202    CCCGCTCGCACCTGCCTGGGCCGCTGCGATTGGCCGCTCCCAGCGACGGCGGCCATTTGCCTGCAGAGCGCAGCGGTATAAAAGGGCGCG   -1
        /4b <////4a                 ----  <//////3                     ///////>2   ------
```

```
        .         .(-->  -->)-98/-93  .         .         .         .         .         .
 -112 GGGTGGCTGAGAGCAGCACACTCGAGCTGTGACCGCCGCACAGTCAACAACTAACTGCCTTCGTTAATATCCTTTGAATAAGCCAACTTT  -23
              ////>1

        .         .         .         .         .         .         .         .         .
  -22 GAATCACAAGACGCATACCAAACATGCACGGATACCGAACCTACAACATGGAGAGCCACCATGCCCATCACGACGCCAGTCCCGTGGACC  67
                             MetHisGlyTyrArgThrTyrAsnMetGluSerHisHisAlaHisHisAspAlaSerProValAspG  (23)

        .         .         .         .         .         .         .         .         .
   68 AGAAGCCCCTGGTTGTGGACCTCTTGGCCACCCAGTACGGCAAGCCCCAGACACCGCCTCCCTCGCCAAATGGTAAGTTTAAAGATAAAG  157
      lnLysProLeuValValAspLeuLeuAlaThrGlnTyrGlyLysProGlnThrProProProSerProAsnG                    (47)

        .         .         .         .         .         .         .         .         .
  158 CCGAGCAAACGTGACGAGTTACTTACACCCAATCTTTCCTCTGTCCAAAACAGAATGCCTATCCAGTCCGGATAACTCCTTGAACGGCAG  247
                                                           luCysLeuSerSerProAspAsnSerLeuAsnGlySe  (59)

        .         .         .         .         .         .         .         .         .
  248 CCGCGGCTCGGAGATTCCCGCCGACCCGTCGGTACGCCGCTATCGCACCGCCTTCACCCGTGACCAGCTGGGTCGCTTGGAGAAGGAGTT  337
      rArgGlySerGluIleProAlaAspProSerValArgArgTyrArgThrAlaPheThrArgAspGlnLeuGlyArgLeuGluLysGluPh  (89)
                                      |    *     *           *  ------*-----------*-------------

        .         .         .         .         .         .         .         .         .
  338 CTACAAGGAGAACTACGTGTCCCGTCCCCGTCGCTGCGAACTGGCCGCCCAGCTGAACCTCCCGGAGAGCACGATCAAGGTGTGGTTCCA  427
      eTyrLysGluAsnTyrValSerArgProArgArgCysGluLeuAlaAlaGlnLeuAsnLeuProGluSerThrIleLysValTrpPheGl  (119)
      ----H1       *        ------------------*--------------H2 *     ---------*-----*--*--*----

        .         .         .         .         .         .         .         .         .
  428 GAACCGCCGCATGAAGGACAAGCGTCAGAGGATCGCCGTCGCCTGGCCCTACGCAGCCGTCTACTCCGATCCCGCCTTCGCCGCCTCCAT  517
      nAsnArgArgMetLysAspLysArgGlnArgIleAlaValAlaTrpProTyrAlaAlaValTyrSerAspProAlaPheAlaAlaSerIl  (149)
      -*--*H3*     *     *          |HOMEODOMAIN

        .         .         .         .         .         .         .         .         .
  518 CCTCCAGGCCGCCGCCAACAGCGTGGGCATGCCCTATCCGCCCTACGCCCCCGCTGCTGCCGCCGCTGCTGCCGCCGCCGCTGCCGTGGC  607
      eLeuGlnAlaAlaAlaAsnSerValGlyMetProTyrProProTyrAlaProAlaAlaAlaAlaAlaAlaAlaAlaAlaAlaValAl  (179)

        .         .         .         .         .         .         .         .         .
  608 CACCAATCCGATGATGGCCACCGGAATGCCCCCGATGGGCATGCCCCAGATGCCCACAATGCAGATGCCCGGACACTCGGGACATGCCGG  697
      aThrAsnProMetMetAlaThrGlyMetProProMetGlyMetProGlnMetProThrMetGlnMetProGlyHisSerGlyHisAlaGl  (209)

        .         .         .         .         .         .         .         .         .
  698 CCATCCCTCGCCCTACGGACAGTACCGCTACACGCCCTACCACATCCCCGCCCGCCCGGCGCCGCCACATCCCGCTGGTCCTCATATGCA  787
      yHisProSerProTyrGlyGlnTyrArgTyrThrProTyrHisIleProAlaArgProAlaProProHisProAlaGlyProHisMetHi  (239)

        .         .         .         .         .         .         .         .         .
  788 TCATCCGCACATGATGGGATCCAGCGCCACGGGATCGTCGTACTCCGCCGGTGCCGCCGGCCTTTTGGGCGCTCTGCCCTCCGCCACCTG  877
      sHisProHisMetMetGlySerSerAlaThrGlySerSerTyrSerAlaGlyAlaAlaGlyLeuLeuGlyAlaLeuProSerAlaThrCy  (269)

        .         .         .         .         .         .         .         .         .
  878 CTATACCGGACTGGGTGTGGGTGTGCCCAAGACCCAGACGCCGCCGCTGGATCTGCAGTCGTCGTCATCGCCGCACTCCTCCACGCTGTC  967
      sTyrThrGlyLeuGlyValGlyValProLysThrGlnThrProProLeuAspLeuGlnSerSerSerSerProHisSerSerThrLeuSe  (299)

        .         .         .         .         .         .         .         .         .
  968 CGTCTCGCCAGTGGGATCCGATCACGCCAAGGTGTTCGACCGCAGTCCAGTGGCTCAATCCGCTCCATCAGTTCCTGCTCCCGCTCCACT  1057
      rValSerProValGlySerAspHisAlaLysValPheAspArgSerProValAlaGlnSerAlaProSerValProAlaProAlaProLe  (329)

        .         .         .         .         .         .         .         .         .
 1058 GACCACCACCAGCCCGCTGCCCGCTCCCGGCCTCCTGATGCCCAGTGCCAAGCGGCCTGCCTCCGACATGTCGCCGCCGCCGACGACAAC  1147
      uThrThrThrSerProLeuProAlaProGlyLeuLeuMetProSerAlaLysArgProAlaSerAspMetSerProProProThrThrTh  (359)

        .         .         .         .         .         .         .         .         .
 1148 TGTGATTGCGGAGCCCAAGCCGAAGCTCTTCAAGCCCTACAAGACTGAGGCGTAAGCCCGCGATCCACACACACTCTCGCCCCCCCCCCC  1237
      rValIleAlaGluProLysProLysLeuPheLysProTyrLysThrGluAlaEnd                                    (376)

        .         .         .         .         .         .         .         .         .
 1238 CTGCTCCCCCAAAGATTGTACAAACTAGTCTTAGTCAGCCTCATCTATTTATTCCCGAAGATTGTACAGATTGTAGAGTAGCTAATTGTA  1327

        .         .         .         .         .         .         .         .         .
 1328 GTCATAATTAAGGCGCAAAATCAAATTAAGAAATAAATGCGAAAATAACATTGAAAATTATACGACACACACTGTTTATTTGCACTACCT  1417
        .                            ======                           |(A)n
 1418 GGTACC  1423
```

the transcription initiation site (see *Promoter*). The binding of EVE to these sites is required for autoregulatory function as shown by germline transformation experiments (Jiang et al. 1991).

EVE is important in establishing segmentation in the early embryo. The anterior borders of EVE stripes define the anterior border of the corresponding *en* stripes at the anterior borders of odd-numbered parasegments. In the trailing edge of the stripes, EVE represses *fushi tarazu* (*ftz*) expression, thus defining the anterior border of FTZ stripes in even-numbered parasegments (Lawrence et al. 1987; Ish-Horowicz et al. 1989). EVE seems to act directly on *eve*, *ftz*, *en* and *wingless* (Macdonald et al. 1986; Harding et al. 1986; Frasch et al. 1988).

Tissue Distribution

As detected by antibody staining, EVE protein is localized in nuclei; and it peaks briefly during the cellular blastoderm and gastrulation stages of embryonic development. EVE is first detectable in division-cycle-12 nuclei throughout the embryo; by cycle 13, staining disappears from the poles and becomes restricted to a band that extends from 70% to 20% egg length. Soon afterwards, the striped pattern along the antero-posterior axis of the embryo develops (Appendix, Fig. A.3). After germ band elongation, EVE protein persists only in neurogenic cells. The developmental pattern of EVE protein follows closely the distribution of *eve* transcript (Frasch et al. 1987; see below).

Mutant Phenotypes

eve is one of the pair-rule genes, hypomorphic *eve* mutants are embryonic lethals having only half the correct number of segments. The missing elements correspond to the posterior region of T2 and the anterior of T3, the posterior of A1 and anterior of A2, etc.; i.e., every other segment boundary and neighboring areas (corresponding to odd-numbered parasegments) are missing. In amorphic mutants, the bands of ventral denticles are replaced with a uniform "lawn" of denticles, so that all obvious trace of segmentation is lost (Nüsslein-Volhard and Wieschaus 1980; Nüsslein-Volhard et al. 1985; Akam 1987).

(*previous pages*) *eve* SEQUENCE. The segment from −202 to 1,423 has accession number M14767 (DROEVE). The segment −498 to −203 is from Read et al. (1990). The segment −1,601 to −931, the stripe 2 element, is from Stanojevic et al. (1991). The segment −5,400 to −5,200, the autoregulatory region, is from Jiang et al. (1991). GAGA (////>) and TCCT (====>), cores of the GAGA and TKK regulatory protein-binding sites, are underlined and numbered; dashes (----) underline EVE, BCD, HB, GT and KR binding sites. The limits of the homeodomain are marked by vertical lines under the sequence, asterisks indicate conserved amino acids, and dashes underline the presumptive helices.

Gene Organization and Expression

Open reading frame 376 amino acids; expected mRNA length, 1,416/1,421 bases. There is a small uncertainty about the position of the 5′ end of the transcript: RNase protection experiments localized the 5′ end at position −93 (Macdonald et al. 1986) while S1 mapping and primer extension indicate it is at position −98 (Frasch et al. 1987). RNase protection was used to define the 3′ end. There is an intron within the Glu-47 codon (*eve* Sequence) (Macdonald et al. 1986; Frasch et al. 1987).

Developmental Pattern

This section was excerpted from Macdonald et al. (1986) and Frasch et al. (1988). The level of *eve* transcript is very low in 0–2 h embryos, it peaks in 2–4 h embryos and then persists in ever-decreasing amounts until the first larval instar. Early in nuclear cycle 13 (syncytial blastoderm), the transcript is localized in the peripheral cortical region of the embryo forming a broad band, as indicated by *in situ* hybridization. Over the next 30 min, this band intensifies and expands until it covers most of the future segmented portion of the embryo (20–70% egg length). As it expands, the band becomes subdivided into two, then four and then seven stripes to produce the "zebra" pattern of expression that is characteristic of pair-rule genes.

By the middle of nuclear cycle 14A (late syncytial blastoderm), expression is localized in seven stripes, six of them being five- or six-nuclei wide while the seventh posterior-most stripe is 6–8 nuclei wide; the stripes are separated by 2–3 nuclei wide spacers. Each *eve* stripe is asymmetric, with the anterior cells showing the highest level of expression; this is the first sign of segment polarity. Some transcript is also detectable in the yolk nuclei occupying the central region of the embryo. During blastoderm cellularization, the stripes narrow to a width of 2–3 nuclei.

At the beginning of gastrulation, the most anterior *eve* stripe is positioned immediately anterior to the cephalic fold. The *ftz* transcripts, which also display a seven-stripe pattern, are shifted posteriorly relative to *eve* such that the two genes are expressed in alternating parasegments. As gastrulation proceeds, seven minor *eve* stripes appear between the major ones. A similar pattern of alternating major and minor stripes is also exhibited by *en*; however *en* major stripes occur in even-numbered parasegments while *eve* major stripes are localized in odd-numbered parasegments. At this stage, *eve* expression seems to be localized to the anterior region of each of the 14 parasegments.

During germ band elongation, the segmented expression of *eve* disappears, and a new site of accumulation appears posterior to the last major stripe. This new site corresponds to cells of the proctodeal primordium, cells that also express *en*, *hairy* and *paired*.

After gastrulation, *eve* expression can be detected only in small clusters of neural ganglion mother cells in each parasegment; *eve* expression continues in the nerve cord until late in embryogenesis. The rapid disappearance of *eve* transcript and protein suggest that these molecules have very short half-lives. In the grasshopper, expression of the *eve* cognate gene occurs in neuroblasts that occupy equivalent positions to those that express *eve* in *Drosophila*; the "zebra" pattern of expression of early embryos, however, is absent. This suggests that the role of *eve* in short germ band embryos is restricted to neurogenesis, and the pair rule function was acquired secondarily during the evolution of higher insects (Patel et al. 1992).

In null mutations, the sites of major and minor stripes, i.e., odd-numbered parasegments and the anterior regions of even-numbered parasegments, are missing. Thus, only the posterior regions of even parasegments are left; they correspond to the denticle belts of T2, A1, A3, etc., which, becoming fused without any naked cuticle to separate them, form the denticle "lawn" mentioned above. In weaker alleles, only the sites of major stripes, i.e., the odd-numbered parasegments, are missing (Nüsslein-Volhard et al. 1985; Macdonald et al. 1986).

Promoter

Regulation of the seven major stripes was investigated in some detail. The production of the striped pattern seems to occur in two phases, an early phase when seven regions of expression are established and a late phase when these regions become narrower and expression intensifies such that stripes become more sharply defined. The early phase is regulated by the gap gene products and the maternal morphogen BCD (product of *bicoid*), all of which are expressed in broad, non-periodic and partly overlapping areas (Appendix, Fig. A.2). The late phase is controlled by the pair-rule gene products, EVE included, which are distributed periodically along the antero-posterior axis of the embryo (Goto et al. 1989; Harding et al. 1989 and references therein).

Early expression in stripes 1, 4, 5, and 6 seems to require unidentified *cis*-acting elements located more than 8.0 kb upstream of the transcription initiation site. An element located between 3.8 and 3.0 kb upstream of the transcription initiation site is required for early expression in stripe 3; and elements between 1.65 and 1.15 kb upstream of the transcription initiation site are required for expression in stripes 2 and 7 (Goto et al. 1989; Harding et al. 1989).

The late or autoregulatory function is controlled by a segment between 5.9 and 5.2 kb upstream of the transcription initiation site. A construction in which the 5.9–5.2 kb segment is linked to a reporter gene is expressed in all seven stripes only if the host organism is wild-type for all pair-rule genes. In the absence of the stripe-specific, early control elements, however, expression, is much weaker (Goto et al. 1989; Harding et al. 1989).

In the segment regulating transcription in stripe 2, there are the following protein binding sites: five for BCD, three for the *hunchback* (*hb*) product (HB), three for the *giant* (*gt*) product (GT) and six for the *Krüppel* (*Kr*) product (KR) (*eve* Sequence). In the stripe 3 promoter segment, there are 18 HB binding sites. The BCD binding sites have the consensus GGGATTAGA; KR binding sites are derivatives of the decamer AACGGGTTAA and the HB binding sites have the consensus G/CA/CATAAAAAA (Stanojevic et al. 1989; Small et al. 1991).

When KR binding sites are inserted into the promoter region of a reporter gene, the expression of the reporter is repressed by KR (Licht et al. 1990). Studies on cultured cells transfected with one or more of the putative regulatory genes (*bcd*, *hb*, *gt* or *Kr*) under the control of the *Actin 5C* promoter and co-transfected with a reporter gene under the control of the stripe 2 regulatory segment showed that BCD and HB are activators and that GT and KR are repressors of stripe 2 transcription. These results suggest that relatively high levels of BCD and HB in a region that includes the second stripe stimulate *eve* transcription; posterior to stripe 2, the band of KR accumulation represses transcription, thus defining the posterior boundary of stripe 2 (Appendix, Fig. A.2 and Fig. A.3); anteriorly, a region of GT accumulation defines the anterior border. The interactions between the regulatory factors seem to occur through direct competition for binding sites. Eight binding sites in the stripe 2 segment are sufficient for proper regulation, and these sites are arranged in two clusters: the proximal cluster includes a BCD and an HB site, overlapped respectively by KR and GT binding sites; the distal cluster includes two BCD sites also overlapped by a KR and a GT site (*eve* Sequence) (Small et al. 1991). This view of the regulation of stripe 2 expression is supported by studies on transgenic organisms carrying various binding-site mutations (Stanojevic et al. 1991).

The results described above are consistent with genetic studies indicating that KR is a repressor of *eve* and with the finding that establishment of the "zebra" pattern (early phase) requires the function of gap genes *hb*, *Kr*, *Knirps* and *tailless* (Frasch and Levine 1987).

Maintenance and refinement of the striped pattern (late phase) is dependent on the pair-rule genes *eve*, *hairy* and *runt* (Frasch and Levine 1987) and the autoregulatory region, 5.9–5.2 kb upstream of the transcription initiation site. EVE, in cooperation with the general transcription factor GAGA (and possibly with a zinc-finger protein coded by the gene *tramtrack* [*ttk*]), seems to interact directly with a 200-bp segment in the autoregulatory region (*eve* Sequence) (Jiang et al. 1991; Read and Manley 1992). The GAGA binding site has a sequence related to GAGAG (Biggin and Tjian 1988) while the putative binding site for the *ttk* product includes the octamer GGTCCTGC (see below) (Jiang et al. 1991).

Two other clusters of EVE binding sites in the *eve* promoter are necessary for transcription, one in a region 3.1–2.9 kb upstream of the transcription initiation site and the other in a proximal region, 295–44 bp upstream of the transcription initiation site (e4 and e5 in the *eve* Sequence). The proximal ones belong to the class II of EVE binding sites as already discussed (Hoey and Levine 1988). Sites e4 and e5 also bind the product of *prd*; e5 comprises two sections and can bind two PRD molecules, one through the homeodomain and the other through the paired domain (Treisman et al. 1991).

In vitro assays in the presence of proteins from embryonic nuclear extracts showed that sequences between 179 and 72 bp upstream of the transcription initiation site are required for transcription. Protein-binding assays identified

12 binding sites in the segment between 574 bp upstream of the transcription initiation site and 175 bp downstream of the transcription initiation site. Eleven of those sites are between 390 and 0 bp upstream of the transcription initiation site and one is 45 bp downstream. Eight of the eleven upstream sites probably bind the GAGA factor (1–6, 9, and 10 in the *eve* Sequence) while the remaining three (7, 8, and 11) seem to bind a different factor and share sequences related to GGTCCTGC. The GAGA-binding protein is relatively constant through development; but the TCCT-binding factor, the product of *ttk*, is apparently restricted to developmental stages when *eve* is active (Read et al. 1990; Read and Manley 1992).

References

Akam, M. E. (1987). The molecular basis for metameric development in the *Drosophila* embryo. *Development* **101**:1–22.

Biggin, M. D. and Tjian, R. (1988). Transcription factors that activate the *Ultrabithorax* promoter in developmentally staged extracts. *Cell* **53**:699–711.

Biggin, M. D. and Tjian, R. (1989). A purified *Drosophila* homeodomain protein represses transcription *in vitro*. *Cell* **58**:433–440.

Frasch, M. and Levine, M. (1987). Complementary patterns of *even-skipped* and *fushi tarazu* expression involve their differential regulation by a common set of segmentation genes in *Drosophila*. *Genes Dev.* **1**:981–995.

Frasch, M., Hoey, T., Rushlow, C., Doyle, H. and Levine, M. (1987). Characterization and localization of the *even-skipped* protein of *Drosophila*. *EMBO J.* **6**:749–759.

Frasch, M., Warrior, R., Tugwood, J. and Levine, M. (1988). Molecular analysis of *even-skipped* mutants in *Drosophila* development. *Genes Dev.* **2**:1824–1838.

Goto, T., Macdonald, P. and Maniatis, T. (1989). Early and late periodic patterns of *even skipped* expression are controlled by distinct regulatory elements that respond to different spatial cues. *Cell* **57**:413–422.

Hanes, S. D. and Brent, R. (1991). A genetic model for interaction of the homeodomain recognition helix with DNA. *Science* **251**:426–430.

Harding, K., Rushlow, C., Hoyle, H. and Levine, M. (1986). Cross-regulatory interactions among pair-rule genes in *Drosophila*. *Science* **233**:953–959.

Harding, K., Hoey, T., Warrior, R. and Levine, M. (1989). Autoregulatory and gap gene response elements of the *even-skipped* promoter of *Drosophila*. *EMBO J.* **8**:1205–1212.

Harrison, S. C. (1991). A structural taxonomy of DNA-binding domains. *Nature* **353**:715–719.

Hoey, T. and Levine, M. (1988). Divergent homeo box proteins recognize similar DNA sequences in *Drosophila*. *Nature* **332**:858–861.

Hoey, T., Warrior, R., Manak, J. and Levine, M. (1988). DNA-binding activities of the *Drosophila melanogaster even-skipped* protein are mediated by its homeo domain and influenced by protein context. *Mol. Cell. Biol.* **8**:4598–4607.

Ish-Horowicz, D., Pinchin, S. M., Ingham, P. W. and Gyurcovics, H. G. (1989). Autocatalytic *ftz* activation and metameric instability induced by ectopic *ftz* expression. *Cell* **57**:223–232.

Jiang, J., Hoey, T. and Levine, M. (1991). Autoregulation of a segmentation gene in *Drosophila*: combinatorial interaction of the *even-skipped* homeo box protein with a distal enhancer element. *Genes Dev.* **5**:265–277.

Lawrence, P. A., Johnston, P., Macdonald, P. and Struhl, G. (1987). The *fushi tarazu* and *even-skipped* genes delimit the borders of parasegments in *Drosophila* embryos. *Nature* **328**:440–442.

Levine, M. and Hoey, T. (1988). Homeobox proteins as sequence-specific transcription factors. *Cell* **55**:537–540.

Licht, J. D., Grossel, M. J., Figge, J. and Hansen, U. M. (1990). *Drosophila Krüppel* protein is a transcriptional repressor. *Nature* **346**:76–79.

Macdonald, P. M., Ingham, P. and Struhl, G. (1986). Isolation, structure, and expression of *even-skipped*: a second pair-rule gene of *Drosophila* containing a homeo box. *Cell* **47**:721–734.

Nüsslein-Volhard, C. and Wieschaus, E. (1980). Mutations affecting segment number and polarity in *Drosophila. Nature* **287**:795–801.

Nüsslein-Volhard, C., Kluding, H. and Jürgens, G. (1985). Genes affecting the segmental subdivision of the *Drosophila* embryo. *Cold Spring Harbor Symp. Quant. Biol.* **50**:145–154.

Patel, N. H., Ball, E. E. and Goodman, C. S. (1992). Changing role of *even-skipped* during the evolution of insect pattern formation. *Nature* **357**:339–342.

Read, D. and Manley, J. L. (1992). Alternatively spliced transcripts of the *Drosophila tramtrack* gene encode zinc finger proteins with distinct DNA binding specificities. *EMBO J.* **11**:1035–1044.

Read, D., Nishigaki, T. and Manley, J. L. (1990). The *Drosophila even-skipped* promoter is transcribed in a stage specific manner *in-vitro* and contains multiple, overlapping factor-binding sites. *Mol. Cell. Biol.* **10**:4334–4344.

Small, S., Kraut, R., Warrior, R. and Levine, M. (1991). Transcriptional regulation of a pair-rule stripe in *Drosophila. Genes Dev.* **5**:827–839.

Stanojevic, D., Hoey, T. and Levine, M. (1989). Sequence-specific DNA-binding activities of the gap proteins encoded by *hunchback* and *Krüppel* in *Drosophila. Nature* **341**:331–335.

Stanojevic, D., Small, S. and Levine, M. (1991). Regulation of a segmentation stripe by overlapping activators and repressors in the *Drosophila* embryo. *Science* **254**:1385–1387.

Treisman, J., Harris, E. and Desplan, C. (1991). The paired box encodes a second DNA-binding domain in the Paired homeo domain protein. *Genes Dev.* **5**:594–604.

14

fushi-tarazu: ftz

Chromosomal Location: 3R, 84B1-2

Map Position: 3-47.5

Product

A DNA-binding regulatory protein of the homeodomain type important in establishing the segmentation pattern of the embryo.

Structure

The homeodomain occurs between Ser-257 and Arg-316 (Laughon and Scott 1984). Like *Antennapedia*'s (*Antp*) homeodomain (ANTP-HD), the *ftz* homeodomain (FTZ-HD) has a Gln in position 9 of helix 3 (*ftz* Sequence, H3); this gives FTZ a binding specificity that distinguishes it from the *bicoid* (*bcd*) and *paired* (*prd*) products, in which Lys and Ser respectively occupy position 9 (Treisman et al. 1989). FTZ occurs as a family of phosphorylated isoforms; 19 differently charged forms were detected. Given the numerous Ser and Thr residues available for modification, the total number of specific isoforms could be much larger. Some isoforms are specific to certain embryonic stages (Krause and Gehring 1989).

Function

The binding of FTZ-HD to the *engrailed* (*en*) promoter binding site bs2 is similar to the binding of ANTP-HD to this site ($K_D = 6\text{–}8 \times 10^{-10}$ M); in particular, the Gln in position 9 of H3 interacts with CC in the bs2 sequence GCCATTAGA (Percival-Smith et al. 1990). *In vitro*, FTZ binds as a monomer to 10–12 bp binding sites. Six of those base pairs are critical, the optimal sequence being C/TAATTA with an equilibrium dissociation constant of 2.5×10^{-11} (Florence et al. 1991).

FTZ is required for embryonic expression of the *Antp* proximal promoter

ftz

```
-1020 AAGCTTTATATTCTCAACAATATTATGCTATTAAAATATTGCTGGTTTTCTGCTGTTATAGAATCATTTTTAAAAGTATAACGTAAAAAA  -931

-930  TAAAATAAAACTAGTATTCATTTGAAAAATCAGCGGGCATATAATTTATATCATATTTTTAAAATTTCGGCAAAGGATGTTTGCATAAAG  -841

-840  TTTTTACTGTTTACTAGTCATTTTGGAAGTGCGTTTGTTGGTTTTTAGGCAAATACCGGGCACAGGAGTGAGTTTGGGAATCGGGAGTTG  -751

-750  CGCACTTGCTTGGCCACGAGGGCAAACAAAAAGCGCAAACACGCGACCCTCGGCCACGCGTATTCCTGATCCCAGGGATCGGACGTAATG  -661

-660  TTATCCTTTGGCCGCCCAGTGCCACGAAATAAATTCGGAGGGAAAGGGCATCGGGTTCCGGAACAACTGGCAGCCAGTCTTCGGTGTTTT  -571

-570  GCGCGCTGGCAAAAATCCAGAGAAATTTTTAGGGAACCATAAACGGGCCGGGGAAAAAGCCTCTGCCCCGAAGGAACGTTTTCAGCAACA  -481
                                      ----------------------ae2a

-480  GTTTACAGTTTTTATGTCTTTATGATTATTGCAATTAGAGGAGATCGGCTGAGAGTCGCGCCCTCTCGCTCTGCGCACCTCATAGGTAGG  -391
           --------------------ae2b                ----------------------------------------ae3

-390  CACCTCATGGCCGTAATTACTGCAGCACCGTCTCAAGGTCGCCGAGTAGGAGAAGCGCGCGGGCGGATAAATCGCGATGATAATGGGCGC  -301
              ---------------rel  --------**----------del=ftz-f1 -----------------------------re2

-300  GATGGGTAGGTAATAAGCCGCGCAGCAGGTAGGCACCGTACGGATAAAGTTGCCAGGACCTCGGATAACTTCCCCTCTCCGTGCCTGCAA  -211
                                                ------------------**----------------re3=f2 I --------
                                                                                             -------*--- ftz-f3

-210  GGACATTTCGCCGGAGGGGTGGCTGCGAACAGCAGCCGGCAAAGTGTCATGCGCAGGGATATTTATGCGCTATAACGGCGAGCGTGTGCC  -121
      **------de2=ftz-f2 II

                                                                .-->-69
-120  GAGGGCTCTCTGATTTTGCTATATATGCAGGATCTGCCGCAGGACCAGCTCATTCGCAAACTCACCAGCGTTGCGTGCACATCGCAGAGT  -31
                         -----

-30   TAGAGAAGAAATCTAGCAATACACATCCGATATGGCTACCACAAACAGCCAGAGCCACTACAGCTACGCCGACAACATGAACATGTACAA  59
                                       MetAlaThrThrAsnSerGlnSerHisTyrSerTyrAlaAspAsnMetAsnMetTyrAs  (20)

60    CATGTATCACCCCCACAGCCTGCCGCCCACCTACTACGATAATTCAGGCAGCAATGCCTACTATCAGAACACCTCCAATTACCACAGCTA  149
      nMetTyrHisProHisSerLeuProProThrTyrTyrAspAsnSerGlySerAsnAlaTyrTyrGlnAsnThrSerAsnTyrHisSerTy  (50)

150   TCAGGGCTACTATCCCCAGGAGAGTTACTCGGAGAGCTGCTACTACTACAACAATCAGGAGCAGGTGACCACCCAGACTGTACCGCCCGT  239
      rGlnGlyTyrTyrProGlnGluSerTyrSerGluSerCysTyrTyrTyrAsnAsnGlnGluGlnValThrThrGlnThrValProProVa  (80)

240   GCAACCCACCACCCCGCCGCCCAAGGCCACCAAGCGCAAGGCCGAAGATGATGCTGCTTCCATCATCGCCGCCGTGGAGGAGCGACCCAG  329
      lGlnProThrThrProProProLysAlaThrLysArgLysAlaGluAspAspAlaAlaSerIleIleAlaAlaValGluGluArgProSe  (110)

330   CACACTGAGGGCTCTGCTCACCAATCCCGTGAAGAAGCTGAAGTACACCCCCGACTATTTCTACACAACCGTCGAGCAGGTGAAGAAGGC  419
      rThrLeuArgAlaLeuLeuThrAsnProValLysLysLeuLysTyrThrProAspTyrPheTyrThrThrValGluGlnValLysLysAl  (140)

420   TCCCGCCGTAACCACCAAGGTCACCGCCAGCCCCGCTCCCAGCTACGACCAAGAGTACGTGACTGTGCCCACGCCCAGCGCCTCCGAGGA  509
      aProAlaValThrThrLysValThrAlaSerProAlaProSerTyrAspGlnGluTyrValThrValProThrProSerAlaSerGluAs  (170)

510   TGTCGACTACTTGGACGTCTACTCGCCCCAGTCGCAGACGCAGAAGCTGAAGAATGGCGACTTTGCCACCCCTCCGCCAACCACGCCCAC  599
      pValAspTyrLeuAspValTyrSerProGlnSerGlnThrGlnLysLeuLysAsnGlyAspPheAlaThrProProProThrThrProTh  (200)
```

(continued)

```
             .        .        .        . T=Ual2  .    T=Ual1.        .        .        .
 600  CTCTCTGCCGCCCCTCGAAGGCATCAGCACGCCACCCCAATCGCCGGGGGAGAAATCGTCGTCAGCTGTCAGCCAGGAGATCAATCATCG   689
      rSerLeuProProLeuGluGlyIleSerThrProProGlnSerProGlyGluLysSerSerSerAlaValSerGlnGluIleAsnHisAr   (230
                                           Leu         Leu

             .        .        .        .        .        .        .        .        .        .
 690  AATTGTGACAGCCCCGAATGGAGCCGGCGATTTCAATTGGTCGCACATCGAGGAGACTTTGGCATCAGGTAGGCATCACACACGATTAAC   779
      gIleValThrAlaProAsnGlyAlaGlyAspPheAsnTrpSerHisIleGluGluThrLeuAlaSerA                          (253
             .        .        .        .        .        .        .        .        .        .
 780  AACCCCTAAAAATACACTTTGAAAATATTGAAAATATGTTTTTGTATACATTTTTGATATTTTCAAACAATACGCAGTTATAAAAGCTCA   869
             .        .        .        .        .        .        .        .        .        .
 870  TTGAGCTAACCCATTTTTTCTTTTGCTTATGCTTACAGATTGCAAAGACTCGAAACGCACCCGTCAGACGTACACCCGCTACCAGACCCT   959
                                                    spCysLysAspSerLysArgThrArgGlnThrTyrThrArgTyrGlnThrLe   (270
                                                                 |    *    *         *  ------*-------
             .        .        .        .        .        .        .        .  T=f47ts .        .
 960  GGAGCTCGAGAAGGAGTTCCACTTCAATAGATACATCACCCGGCGTCGTCGCATCGATATCGCCAATGCCCTGAGCCTGAGCGAAAGGCA   1049
      uGluLeuGluLysGluPheHisPheAsnArgTyrIleThrArgArgArgArgIleAspIleAlaAsnAlaLeuSerLeuSerGluArgGl   (300
      ----*-----------------H1        *         -------------------*--Val---------H2 *     --------

             .        .        .        .        . ||=Rpl  .        .        .        .
1050  GATCAAGATCTGGTTCCAAAACCGACGCATGAAGTCGAAGAAGGATCGCACGCTGGACAGCTCCCCGGAGCACTGTGGTGCCGGCTACAC   1139
      nIleLysIleTrpPheGlnAsnArgArgMetLysSerLysLysAspArgThrLeuAspSerSerProGluHisCysGlyAlaGlyTyrTh   (330
      -*------*--*--*-----*--*H3*     *     *         | HOMEODOMAIN
             .        .        .        .        .        .        .        .        .        .
1140  CGCGATGCTGCCGCCACTGGAGGCCACAAGCACCGCCACCACCGGGGCACCATCGGTGCCAGTGCCCATGTACCACCACCACCAAACCAC   1229
      rAlaMetLeuProProLeuGluAlaThrSerThrAlaThrThrGlyAlaProSerValProValProMetTyrHisHisHisGlnThrTh   (360
             .        .        .        .        .        .        .        .        .        .
1230  CGCCGCCTACCCCGCTTACAGCCACAGTCACAGTCATGGTTATGGCCTGCTCAATGATTACCCTCAGCAGCAGACCCACCAGCAGTACGA   1319
      rAlaAlaTyrProAlaTyrSerHisSerHisSerHisGlyTyrGlyLeuLeuAsnAspTyrProGlnGlnGlnThrHisGlnGlnTyrAs   (390
             .        .        .        .        .        .        .        .        .        .
1320  TGCCTACCCGCAGCAGTACCAACAGCAGTGCAGCTACCAGCAACATCCACAGGACCTCTACCATCTGTCTTGAGGTCCGGCGATGCTCAG   1409
      pAlaTyrProGlnGlnTyrGlnGlnGlnCysSerTyrGlnGlnHisProGlnAspLeuTyrHisLeuSerEnd                     (413
             .        .        .        .        .        .        .        .        .        .
1410  TTACTCTCTTCCCCAGAGCGGAACCGAAAGCCGTACCGCCACGAAACCGAAGCGCACTTCTCTCGACCATTTGTAGGTGACACGCAAATG   1499
             .        .        .        .        .        .        .        .        .        .
1500  ACACAGCCGAGAACGAAGCTGCGACGCGATGAGTTGCACAGTAGAGGGCGCACTCCCTACGGTGCCCAGGACATTTTGGGCACAAGGACG   1589
             .        .        .        .        .        .        .        .        .        .
1590  AGTGCGCAAGTGCAGAAGGCAGAGGCAAAAGAGGCAGCGCAAACAGAAAAGGAGCCTTGCTGCGCGCGGAACCCAGTGGCTGGCCATGAT   1679
             .        .        .        .        .        .        .        .        .        .
1680  GGGTTCTCAGCGATCGATTAGCTGCGGCCAAACACAAGCCCAAAACACTCAGCTGGGAGTGATAATGGCCAAGAGACTTGGAGACTGACA   1769
             .        .        .        .        .        .        .        .        .        .
1770  CACATGTTTTTGTACATATAGTAGTTAAGATATTCCTATCATAGAATTCTATTTATTAAAATATACGAGTAAAGTAAATCGATCGAATTT   1859
                                                                                       ------
             .        .        .        .        .        .        .        .        .        .
1860  AAAACAAATCAAGTTGAACATTCATTTGGCAATTTGTGAAGAAGAGTCTTGGGCATGCTGCAATTTGACTGCTTTAAAATTTTAAACTTA   1949
             .        .        .        .        .        .        .        .        .        .
1950  TAGGCCGTGGCGCGTATGTGGAATACATTTCATATGTATATGTGTTGAAATACAATTAAATGCCTTTCAATGATAACTACTCAATAAACT   2039
             .        .        .        .        .        .        .        .        .        .
2040  TCCGAACTTATACGAAACGCAAACGATTTAATGTTGAGCACGAATCGTACAAATTCGAGCAGCTGCATTTTGTCGCTTCAGTCCCCCTCA   2129
             .        .        .        .        .        .        .        .        .        .
2130  TCCCTGACCCATTGCTGTCTCCCGGATTTTCTATTAAATGCACTCTTTTCGCCAGAGAAAATGTCACATTTTGGTCTGGCTTCGGGGCAT   2219
             .        .        .        .        .        .        .        .        .        .
2220  ATCTACCACCGCATCCCTGCTCCCTTCCTCCCTCCGACGCTGCACGTTCCTCTATTGAAGTGAGACATTGATTGGTAATTTTTCATTGCA   2309
             .        .        .        .        .        .        .        .        .        .
2310  CATCCGTGACAGTTATGGGTAACGCAACGCAAAAGGAAAAGCCCGGTGCGGAATCGGATTCGGAATCAGAATCAATATCAAAGGCAAAGG   2399
```

(P2) and *Ultrabithorax* (*Ubx*) (Ingham and Martínez-Arias 1986). In experiments carried out using cultured cells, FTZ stimulates *Ubx* transcription if a segment of *Ubx* that extends from 225 to 292 bp downstream of the transcription initiation site (*Ubx* downstream element U-B) is present (Winslow et al. 1989). Two FTZ-HD binding sites were detected in *Antp* by DNase I protection assays; these are approximately 500 bp upstream of the P2 transcription initiation site. Other homeodomain binding sites were detected near the distal transcription initiation site, but it is less certain that they are functional. The consensus sequence of these binding sites is CAATTA (Nelson and Laughon 1990).

FTZ is also required for expression of the *ftz* gene itself (see *Promoter*) (Hiromi and Gehring 1987; Ish-Horowicz et al. 1989), and it is also involved in the regulation of the segment polarity genes *en* and *wingless* (*wg*) (Howard and Ingham 1986; Lawrence et al. 1987; Ingham et al. 1988).

ftz is one of the pair-rule segmentation genes. Its overall function is thought to be to define the anterior border of even-numbered parasegments (Lawrence et al. 1987).

Tissue Distribution

FTZ is first detectable by antibody-staining after the 13th nuclear division; it is localized in nuclei in seven stripes each approximately four nuclei wide (the spacing between stripes is also four nuclei). During gastrulation, the stripes narrow to three nuclei wide; FTZ stripes disappear just before the germ band is fully extended (Carroll and Scott 1985). FTZ and EVE (product of *even-skipped*) accumulate during approximatey the same time in development. At first, the areas of EVE and FTZ accumulation overlap somewhat; but, as the stripes become narrower and better defined, the two products end up in an alternating pattern, FTZ in even- and EVE in odd-numbered parasegments (Appendix, Fig. A.3) (Frasch and Levine 1987). In embryos with fully extended germ bands, antibody staining is visible in 15 metameric clusters of nuclei within the developing ventral nervous system; this staining disappears soon after germ band shortening is completed (10–12 h of development). In 12–15 h embryos, FTZ reappears in the developing hindgut (Carroll and Scott 1985; Krause et al. 1988).

---

ftz SEQUENCE (*opposite*). Strain carrying marker p^{P}. Accession, X00854 (DROANTCF) (modified by adding a G at −259 as per Brown et al. 1991). The following mutations are indicated: ftz^{Ual3}, Pro-215 to Ser-215; ftz^{f47ts}, temperature-sensitive, Ala-291 to Val-291; and the chromosome 2 translocation ftz^{Rp1}, with a breakpoint after position 1,091 (Laughon and Scott 1984). The "zebra" element regulatory sites ftz-f1, ftz-f2 and ftz-f3 are from Ueda et al. (1990) and Brown et al. (1991). Sites ae2–ae3 (to which activators bind), re1–re3 (to which repressors bind) and de1–de2 (to which both activators and repressors bind) are from Topol et al. (1991). ae2a and ae2b correspond to the CAD-binding sites, cdre of Dearolf et al. (1989b).

Mutant Phenotypes

Homozygotes for null alleles of *ftz* show severe developmental abnormalities that become evident at about the time *ftz* should be expressed. These mutants have half the correct number of segments due to the absence of regions corresponding to even-numbered parasegments (Wakimoto et al. 1984).

The dominant gain-of-function mutations Ual1, Ual2 and Ual3 cause substitutions in Pro-211 and Pro-215 (*ftz* Sequence) which increase the half-life of FTZ from <10 min to 40 min. The increase in level and persistence of the protein and the concomitant expansion of its domain result in the corresponding suppression of *eve* expression in odd-numbered parasegments; this leads to abnormalities in the parasegments where *ftz* is normally not expressed, the anti-*ftz* phenotype. The Ual mutations affect a segment of the polypeptide (Thr-210 to Ser-221) that seems to be conserved in other early development genes (*hb*, *eve* and *prd*) as well as in *myc*. It has been suggested that those 12 residues serve as a signal for protein degradation (Kellerman et al. 1990). PEST-like sequences, also thought to be involved in protein degradation, are present in that region (Rogers et al. 1986).

Gene Organization and Expression

Open reading frame, 413 amino acids; expected mRNA length, approximately 1,770 bases (assuming it extends for 20–30 bases beyond the putative poly(A) signal highlighted in the *ftz* Sequence), in agreement with an observed RNA of 1.8 kb (Laughon and Scott 1984). Primer extension analysis was used to identify the 5′ end (Dearolf et al. 1989a; Ueda et al. 1990). The 3′ end was not determined. There is an intron in the Asp-253 codon (Laughon and Scott 1984).

ftz is centromere-proximal to *Antp*, separated from it by about 30 kb, and transcribed in the opposite orientation (Weiner et al. 1984; Wakimoto et al. 1984).

Developmental Pattern

ftz transcripts appear in embryos after the 11th nuclear division; they accumulate along the periphery of the embryo between 65% and 15% egg length. Between this stage and the end of nuclear cycle 13, the signal intensifies and becomes less uniform along the antero-posterior axis. Eventually, in nuclear-elongation-stage embryos (cycle 14), *ftz* RNA becomes localized in seven stripes positioned between 65% and 15% egg length, and it remains so through the completion of blastoderm cellularization. The anterior-most stripe is positioned posterior to the cephalic furrow. Stripes are 3–5 cells wide, and they are separated from one another by 3–5 cells. This segmented pattern persists through the early stages of gastrulation, but by the time the germ band is fully extended, *ftz* transcripts are no longer detectable.

The strongest embryonic expression of *ftz* is restricted to the period between 2 h and 4 h of development approximately (Hafen et al. 1984). The turnover rate of *ftz* mRNA is extremely high (half-life, 7–14 min) (Edgar et al. 1986); and the phenotype of a gain-of-function mutation *T(2; 3)ftz*^{Rp1} seems to be the result of increased mRNA stability, possibly because of the loss of degradation signals in the 3′ untranslated region (Kellerman et al. 1990).

The developmental pattern of *ftz* expression was also studied using the promoter region of *ftz* and *β*-galactosidase as a reporter enzyme. This method demonstrates that the seven stripes are sharper and more intense at the anterior border and that they fade posteriorly. The sharp anterior edge of each stripe coincides with the anterior edge of *en* expression in even-numbered parasegments, and it thus defines the anterior edge of these parasegments. (The same kind of pattern is observed for *eve* expression, except that it is the odd-numbered parasegments that are involved.) The *β*-galactosidase method also demonstrates segmental staining of prospective ventral ganglia neuroblasts in fully extended germ-band embryos (Hiromi et al. 1985; Lawrence et al. 1987).

Promoter

Approximately 6 kb of 5′ sequences are required for normal *ftz* expression (as measured by the ability of fragments of various sizes to rescue *ftz* mutant embryos in transgenic experiments). However, fusions of the promoter to the reporter gene *lacZ* showed that the most proximal 0.62 kb of 5′ sequences ("zebra" element) are sufficient to produce the "zebra" pattern of expression. A segment between 2.45 and 0.62 kb upstream of the transcription initiation site is required for expression in the ventral nervous system. A segment further upstream, between 6.1 and 2.45 kb of the transcription initiation site, functions as an enhancer of expression of the "zebra" pattern. In the absence of this distal enhancer element, the striped pattern of expression is weaker, mostly restricted to the mesoderm and extended anteriorly, so that one or two extra stripes appear anterior to the cephalic furrow (Hiromi et al. 1985).

The "Zebra" Element The striped pattern of *ftz* expression seems to be established through a combination of generalized activation of the gene throughout the embryo and a specific pattern of repression. Two systems of repression contribute to the *ftz* expression pattern: one system represses expression in the anterior and posterior poles of the embryo, and the other represses in the inter-stripe regions of the "zebra" pattern (Edgar et al. 1986). Several activator and repressor sub-regions were identified within the "zebra" element by promoter deletion analysis, and they were found to correspond to protected regions in footprinting analysis (*ftz* Sequence) (Dearolf et al. 1989a; Topol et al. 1991).

A search for *ftz*-promoter-binding proteins yielded three fractions: FTZ-F1, FTZ-F2 and FTZ-F3.

FTZ-F1 first appears in 1.5–4.0 h embryos (at the time the *ftz* stripes occur);

it then diminishes, to reappear after 13 h of development and in larval and adult stages. FTZ-F1 binds to four sites in the *ftz* gene: site I is a 21-bp segment from −362 to −343 (*ftz* Sequence), sites II and III are in the coding region, and site IV (to which binding is 10 times weaker) partly overlaps the binding site of FTZ-F2 (see below). Sites I, II and III have the consensus sequence YCAAGGYCRCCR. Close contact with FTZ-F1 seems to be made by the two consecutive Gs of the top strand (marked by an asterisk in the *ftz* Sequence) and the two Gs on the bottom strand that are opposite the Cs at positions 8 and 10. Expression of a construction containing the "zebra" element attached to *lacZ* in transgenic embryos showed that mutations of site I that abolish FTZ-F1 binding lead to overall reduced expression of *ftz*, in particular in stripes 1, 2, 3, and 6 (Ueda et al. 1990). The sequence of FTZ-F1 has similarities with proteins of the steroid receptor superfamily both in the putative DNA-binding region and in the putative ligand-binding domain (Lavorgna et al. 1991).

FTZ-F2 is present at low levels in 1.5–4.0 h embryos, and its concentration rises after 4.0 h as expression of *ftz* diminishes. FTZ-F2 affords protection against nuclease digestion to two sites within the "zebra" element that share the sequence TGCNAGGACNT (*ftz* Sequence): ftz-f2 I (abbreviated f2 I) and ftz-f2 II, located between −260 and −200. The two adjacent Gs marked with asterisks seem to interact directly with an FTZ-F2 residue as indicated by methylation interference. Mutant ftz-f2-binding sites are unable to bind FTZ-F2. When such mutations are part of a "zebra"-element-*lacZ* construction, there is continuous *lacZ* expression along the antero-posterior axis; i.e., the repression of the *ftz* promoter in the inter-stripe regions fails. These mutations also lead to precocious expression of *ftz*, as early as the third nuclear division (Brown et al. 1991). FTZ-F2 is probably the product of *tramtrack* (*ttk*), a Zn-finger protein (Harrison and Travers 1990; Brown et al. 1991; Read and Manley 1992).

FTZ-F3 also bind to the "zebra" element, partly overlapping the FTZ-F2 binding sites (Brown et al. 1991).

CAD, the product of the segmentation gene *caudal* (*cad*), a homeodomain protein that forms a gradient of increasing concentration from the anterior to the posterior pole, participates in the regulation of *ftz* expression. CAD activates expression of *ftz* in the posterior regions of the embryo through its binding to the hexanucleotide TTTATG that is present in the protein binding sites ae2a and ae2b of the "zebra" element (*ftz* Sequence) (Dearolf et al. 1989a, 1989b).

Distal Upstream Enhancers A DNA segment that extends from approximately 6.1 to 3.5 kb upstream of the transcription initiation site can direct transcription of the basal *Hsp70* promoter and an associated reporter gene in a seven-stripe pattern in both ectoderm and mesoderm. The 2,574-bp segment contains multiple regulatory regions; from distal to proximal they are: (1) the most upstream 330 bp portion of this segment, which seems to be an activator of parasegment 4 expression; (2) the Distal Enhancer, extending from 331 to 1,502, which is capable of directing expression in seven mesodermal stripes; (3) the 583-bp Element A of the Proximal Enhancer, between 1,780 and 2,363, which can direct expression in seven stripes in the ectoderm and mesoderm; (4) the 211-bp Element B of the Proximal Enhancer, which is required, in conjunction with element A, for ectodermal expression (Pick et al. 1990). There is also a scaffolding attachment region that occurs in an AT-rich segment between positions 575 and 763 of this distal upstream regulatory region (Amati et al. 1990).

FTZ itself seems to interact with the Distal Enhancer region to activate transcription (Hiromi and Gehring 1987; Ish-Horowicz et al. 1989). Numerous FTZ-binding sites are found within the Distal and Proximal Enhancers, and two independent autoregulatory loops seem to control expression (Harrison and Travers 1988; Pick et al. 1990). The product of *ttk* binds to DNA in the distal upstream region (Harrison and Travers 1988, 1990).

The pattern of *ftz* expression also depends on the products of gap genes and other pair-rule genes, *eve* and *h* in particular (Carroll and Scott 1986; Howard and Ingham 1986; Harding et al. 1986; Frasch and Levine 1987; Ish-Horowicz and Pinchin 1987).

References

Amati, B., Pick, L., Laroche, T. and Gasser, S. M. (1990). Nuclear scaffold attachment stimulates but is not essential for ARS activity in *Saccharomyces cerevisiae.* Analysis of the *Drosophila ftz* SAR. *EMBO J.* **9**:4007–4016.

Brown, J. L., Sonoda, S., Ueda, H., Scott, M. P. and Wu, C. (1991). Repression of the *Drosophila fushi tarazu* segmentation gene. *EMBO J.* **10**:665–674.

Carroll, S. B. and Scott, M. P. (1985). Localization of the *fushi tarazu* protein during Drosophila embryogenesis. *Cell* **43**:47–57.

Caroll, S. B. and Scott, M. P. (1986). Zygotically active genes that affect the spatial expression of the *fushi tarazu* segmentation gene during early *Drosophila* embryogenesis. *Cell* **45**:113–126.

Dearolf, C. R., Topol, J. and Parker, C. S. (1989a). Transcriptional control of *Drosophila fushi tarazu* zebra stripe expression. *Genes Dev.* **3**:384–398.

Dearolf, C. R., Topol, J. and Parker, C. S. (1989b). The *caudal* gene product is a direct activator of fushi tarazu transcription during embryogenesis. *Nature* **341**:340–343.

Edgar, B. A., Weir, M. P., Schubiger, G. and Kornberg, T. (1986). Repression and turnover pattern of *fushi tarazu* RNA in the early Drosophila embryo. *Cell* **47**:747–754.

Florence, B., Handrow, R. and Laughon, A. (1991). DNA binding specificity of the *fushi tarazu* homeodomain. *Mol. Cell. Biol.* **11**:3613–3623.

Frasch, M. and Levine, M. (1987). Complementary patterns of *even-skipped* and *fushi tarazu* expression involve their differential regulation by a common set of segmentation genes in *Drosophila. Genes Dev.* **1**:981–995.

Hafen, E., Kuroiwa, A. and Gehring, W. J. (1984). Spatial distribution of transcripts from the segmentation gene *fushi tarazu* during Drosophila embryonic development. *Cell* **37**:833–841.

Harding, K., Rushlow, C., Hoyle, H. and Levine, M. (1986). Cross-regulatory interactions among pair-rule genes in *Drosophila. Science* **233**:953–959.

Harrison, S. D. and Travers, A. A. (1988). Identification of the binding sites for potential regulatory proteins in the upstream enhancer element of the *Drosophila fushi tarazu* gene. *Nucl. Acids Res.* **24**:11403–11416.

Harrison, S. D. and Travers, A. A. (1990). The *tramtrack* gene encodes a *Drosophila* finger protein that interacts with the *ftz* transcriptional regulatory region and shows a novel embryonic expression pattern. *EMBO J.* **9**:207–216.

Hiromi, Y. and Gehring, W. J. (1987). Regulation and function of the Drosophila segmentation gene *fushi tarazu. Cell* **50**:963–974.

Hiromi, Y., Kuroiwa, A. and Gehring, W. J. (1985). Control elements of the Drosophila segmentation gene *fushi tarazu. Cell* **43**:603–613.

Howard, K. and Ingham, P. (1986). Regulatory interactions between the segmentation genes *fushi tarazu, hairy,* and *engrailed* in the Drosophila blastoderm. *Cell* **44**:949–957.

Ingham, P. W. and Martínez-Arias, A. (1986). The correct activation of Antennapedia and bithorax complex genes requires the *fushi tarazu* gene. *Nature* **324**:592–597.

Ingham, P. W., Baker, N. E. and Martínez-Arias, A. (1988). Regulation of segment polarity genes in the *Drosophila* blastoderm by *fushi tarazu* and *even skipped. Nature* **331**:73–75.

Ish-Horowicz, D. and Pinchin, S. M. (1987). Pattern abnormalities induced by ectopic expression of the *Drosophila* gene *hairy* are associated with repression of *fushi tarazu* transcription. *Cell* **51**:405–415.

Ish-Horowicz, D., Pinchin, S. M., Ingham, P. W. and Gyurcovics, H. G. (1989). Autocatalytic *ftz* activation and metameric instability induced by ectopic *ftz* expression. *Cell* **57**:223–232.

Kellerman, K. A., Mattson, D. M. and Duncan, I. (1990). Mutations affecting the stability of the *fushi tarazu* protein of *Drosophila. Genes Dev.* **4**:1936–1950.

Krause, H. M. and Gehring, W. J. (1989). Stage-specific phosphorylation of the *fushi tarazu* protein during *Drosophila* development. *EMBO J.* **8**:1197–1204.

Krause, H. M., Klemenz, R. and Gehring, W. J. (1988). Expression, modification, and localization of the *fushi-tarazu* protein in *Drosophila* embryos. *Genes Dev.* **2**:1021–1036.

Laughon, A. and Scott, M. P. (1984). Sequence of a *Drosophila* segmentation gene: protein structure homology with DNA-binding proteins. *Nature* **310**:25–31.

Lavorgna, G., Ueda, H., Clos, J. and Wu, C. (1991). FTZ-F1, a steroid hormone receptor-like protein implicated in the activation of *fushi tarazu. Science* **252**:848–851.

Lawrence, P. A., Johnston, P., Macdonald, P. and Struhl, G. (1987). The *fushi tarazu* and *even-skipped* genes delimit the borders of parasegments in *Drosophila* embryos. *Nature* **328**:440–442.

Nelson, H. B. and Laughton, A. (1990). The DNA binding specificity of the *Drosophila fushi tarazu* protein: a possible role for DNA bending in homeodomain recognition. *New Biol.* **2**:171–178.

Percival-Smith, A., Müller, M., Affolter, M. and Gehring, W. J. (1990). The interaction with DNA of wild-type and mutant *fushi tarazu* homeodomains. *EMBO J.* **9**:3967–3974.

Pick, L., Schier, A., Affolter, M., Schmidt-Glenewinkel, T. and Gehring, W. J. (1990). Analysis of the *ftz* upstream element: germ layer-specific enhancers are independently autoregulated. *Genes Dev.* **4**:1224–1239.

Read, D. and Manley, J. L. (1992). Alternatively spliced transcripts of the *Drosophila tramtrack* gene encode zinc finger proteins with distinct DNA binding specificities. *EMBO J.* **11**:1035–1044.

Rogers, S., Wells, R. and Rechsteiner, M. (1986). Amino acid sequences common to rapidly degraded proteins: The PEST hypothesis. *Science* **234**:364–368.

Topol, J., Dearolf, C. R., Prakash, K. and Parker, C. S. (1991). Synthetic oligonucleotides recreate *Drosophila fushi tarazu* zebra-stripe expression. *Genes Dev.* **5**:855–867.

Treisman, J., Gonczy, P., Vashishtha, M., Harris, E. and Desplan, C. (1989). A single amino acid can determine the DNA binding specificity of homeodomain proteins. *Cell* **59**:553–562.

Ueda, H., Sonoda, S., Brown, J. L., Scott, M. P. and Wu, C. (1990). A sequence-specific DNA-binding protein that activates *fushi tarazu* segmentation gene expression. *Genes Dev.* **4**:624–635.

Wakimoto, B. T., Turner, F. R. and Kaufman, T. C. (1984). Defects in embryogenesis in mutants associated with the *Antennapedia* gene complex of *Drosophila melanogaster*. *Dev. Biol.* **102**:147–172.

Weiner, A. J., Scott, M. P. and Kaufman, T. C. (1984). A molecular analysis of *fushi tarazu*, a gene in *Drosophila melanogaster* that encodes a product affecting embryonic segment number and cell fate. *Cell* **37**:843–851.

Winslow, G. M., Hayashi, S., Krasnow, M., Hogness, D. and Scott, M. P. (1989). Transcriptional activation by the *Antennapedia* and *fushi tarazu* proteins in cultured Drosophila cells. *Cell* **57**:1017–1030.

15

hairy: h

Chromosomal Location: 3L, 66D8-15

Map Position: 3-26.5

Product

DNA-binding regulatory protein of the basic helix-loop-helix (bHLH) type involved in embryonic segmentation and neurogenesis.

Structure

Sequence comparisons indicate that the HLH motif extends from Ala-45 to Arg-90 (*h* Sequence), with 26% identity to a region of the mammalian oncogene *myc*. The helices have the amphipathic nature characteristic of dimer-forming regulatory proteins such as the products of the genes *daughterless* (*da*), *Enhancer of split* [*E*(*spl*)], *extramacrochaetae* (*emc*), *twist* and of the *achaete-scute* complex genes *ac*, *sc*, *lsc* and *ase*. In the *h* and *E*(*spl*) proteins, the sequences of the basic regions, adjacent to and upstream of the HLH domain are more closely related to each other than to those in the *da* and AS-C proteins (Rushlow et al. 1989; Harrison 1991; Van Doren et al. 1991).

Within the *h* sequence, the OPA (CAG) repeat occurs several times and results in stretches of Gln, Ala or Ser in the C-terminal half of the protein. Near the C-terminus there are also regions of similarity to PEST sequences (segments rich in Pro, Glu, Ser and Thr that may be degradation signals). There are three potential glycosylation sites, at Asn-9, Asn-209 and Asn-296 (Rushlow et al. 1989).

Function

During embryonic development, the HAIRY product seems to act as a repressor of *fushi tarazu* (*ftz*) helping to define the posterior border of *ftz* stripes. In *h* mutants, *ftz* expression occurs, but the striped pattern fails to develop (Howard and Ingham 1986; Carroll and Scott 1986; Ish-Horowicz and Pinchin 1987).

h

```
-3210  CGGCGCGTGGGGGTTTCTGTCGCTGTTAAACTCGCAACGTTGCTGTTAAAAGAGCCCTACGATCAACAGTATACATAGTATAGTATATAT  -3121
-3120  AGTATAGTTTATGGATACTATATATAAATAATATCAACATAGTTAGTATCTAAGATAAGATTTTCTGTATATTTTAAACTTAATAGTCAA  -3031
-3030  AACTATATGGTATTTTGAGTCTAGTGAATAACCATTTTGAATGATAATGGCACACAAATTGAATTCATTGATCTTATAAAATACAAGCAA  -2941
-2940  ATAATAATACCTATAATATTATACATATGCTATAATGTTATTATCAACGCCTTTACGATTATTAAATAGTTAACCAACATGGTCCAAAAT  -2851
-2850  GATTCGAAATACCTTCAAGGGGTTCTTTATCGTACCACCAACGTGTTTGTTTTGTTTTCAGTGATCAGGGCTCCCGGGGCTTATGGGGAA  -2761
-2760  TCTGGGGGATCTGGCGCTGTCTAATTTTAGACGCAATTAGCAATGCGCACATTTTTGTTGTTGCTTGCGCCTTTTCGACTATAAATTTTT  -2671
-2670  GCCACAGTTTATTTTAGAAGCTGCATGTGATCGGGTCCGCCAACAACAACAATGGGGCGTCAAATTGGGCGTTCAACGCACAAACAACTG  -2581
-2580  CGAGTGTATCTGTATCTGTGACTGTATCTTTAGCGTTGTATCCGTGAGATACATCCACACCTTTGGCTGTTTTTTGGCCAGCTAGCATGA  -2491
-2490  TGTAGCTAGCATGATGTAAAACGCCGCCAACGTTTTCCGACCTCTCGTTTTTTTTTCTTTTTTGTTTATTTTCTTTTTTGCTTTTCGTTG  -2401
-2400  CAATTAAATTGGCATGCACAAGTGCCGCCTCTGCCGCCGACACCGCCTCTGCCGACGCTGACCGCGGCGGGCCGCCTTTTGATCGGCTGC  -2311
-2310  CAAATTGTAATTGGAACGCGAAGGTGTTGTCGACGTCCGCCACTACCGTCTATATATATATGTATCCATATTGTTGGGGCATGTGTTCTC  -2221
-2220  GGCATAACAACCTTCTCTGGCGACCACAAAACGCACAACACTTTAGACAACCCTCAAAAATTTCAGAAATTCCCCTAACTTTTTGAGTAT  -2131
-2130  TTTCACGAATCGATAGATATGCATATTTGTAAGACGTGATTGTTGATTAAGTTTAATTTCATTTAGTTATTAAGCGGAAATTAAGTGTAG  -2041
-2040  TAAAATCAAATTAACTTCTAAACGTTTTTTACTCATCTTCATTAGAGTCAACTTTATTAGTTTCTATAAAAACACTGCCAGGTGGTTTC  -1951
-1950  GTTATAAAAAAAATATTGTAAACACCCGTTTTTAGCCAACTTTAATGTTTAAAGCCTGACTGACTCATTCCAATGTAACTTTGTTGACGA  -1861
-1860  TTCGTGGTTTTGGTATAACTTCACTAATCAGTGGTCAGAGTCCAAGTCAGGCTTTAAAAATATTTCCCAAGAACAAACGTCAAAGATAAC  -1771
-1770  GTAATTTCTCTTTATAGATCGTGTAACCTAAATATGTGTCATCTACCTTTACTGAGCTCAGCCTGGTTAAACTAATTACATGGTTATTAC  -1681
-1680  CATTTCTTAGAACTTAACCCATATTTTGTAGATAATAGAAGGCTTAAGCAGTTATTTAAAATATCACTTTCGGTTGTAACCAAATGTGTG  -1591
-1590  TGACGCACTTTGGCTTTTTACTACCAAATAAACAATATAATTTAAGCTTCATTTTCACCGTAATATTCCCAGTTTTCACAGCAATGCCCC  -1501
-1500  TCTTCTCATTCTGCTAATGAATGGTTAGTTTTCTGATGCCCGACTATTCCGCGTGTCGCGTAATTATAGTCAACCTTCGATTAATCATTA  -1411
-1410  CTCCAAAACAAAACAACAAATAATATATGAAAAACGTGAAAATCCAACGCTGCACGTAGAAGCCATCAAGCTGAATCTAAGCGTCCGGCG  -1321
-1320  GAGCACGTGTGATCCACGCAGCCTTGTCCACAGCGATTTCCATTTCATTTAGCCCGTTGGCGGCTATCGATCAAAAGCCAAAAGGGCGAC  -1231
-1230  CTTCACTTAATTGAGGCGTACGGCATGCTGAATGAGTCGGTTGTACAGACTGGTCTGGAAAATGCTAGGGGGATAACTATAGCCACCACC  -1141
-1140  CACTGCCCGATCGCCCAACCACCCAACCACCCACTTCCGCCTAGCGTGCGCACAACCTTGTGATCTTGTTTACTGTTTAGCGACCCCCGA  -1051
-1050  GCCGCAGATACACAGTACACAGCACAAAAAACCGAACCTGTCGCACTGGGGTGGCGTCATATAGCCAGCTATTTTCACCTTCTATGGGAC  -961
 -960  GTCGTCGCGTTGGCCGCATGAATCAGCAAACCACGAACGGCGAGCCACCAGAAACCACCGCAGAAGCAGCAACAACACCAACACCACCGC  -871
 -870  GACCATCACCAACAGCACAGCCAGAAACACAGCCTCTTGTGAATCCCTCAGTTAGCAGAGCCCAGCAGAGTCAAGCCAAACCGATCGCTG  -781
```

(*continued*)

```
-780  ATCGACCGACCGACCGACGATCACCAATGGGGGTTTCGCAGTGTGATTTCCAAAAAGGAAGAAATGCCCATTCCCGCGAGCCACGGGGGC  -691

-690  GTATGAGTAACGCGGTGTGTACTTTAGGTACCGCTCACAAGCGCGAGCACGCACACACACACGCACACACTGAAGCGAGCAGGTAGCGCA  -601

-600  TAATGTATACCCCTAGGTAGCCGCAATGCCAGTGCAATTGTATTGGTGCGGTCGTGTGGCTCGCGCTCGCGTCTCGCAGCGGCGATTTAG  -511
                             ----          ---- ----
                           -->-490
-510  CCTACGAACCTGTCGATCAATCGTCAGTCTTCCGCCGAGAGCCCAGCGATAAGGTAGTCCCGCTACGCTCCGCAACATCCAGACCGAGTA  -421

-420  AAGCAAATACTTATATATATATATACATACATATATATAGCGCAACCATCCGAAAGCCGAACGATCGATCTTCTGCTTCGTGCCGAGTCG  -331

                                              -->-294
-330  TTTCGGTCTTTTAATCGCACTGCAGCCAGAACCTGCTGCTCATTCGCCTGCCGTATTTCGTAGCGTGCGGTTCTATCGCTCCGCTTTGAT  -241

-240  AAACCGAATCGAAATCTAGAGAACCCCCCCAGACACAATACCATTTTACGTGCTTCTCTGCGACGCTGCGCGAAAGTGAAACCACCAAGT  -151

-150  GAACTTGAAAAAAAAAAAAACTGACAACTTGAGTTATTCTAAAAAAAGCAAAAAAGCAGTGAACTTATATTGCAAAGAGCAGCAAATTCA  -61

 -60  GATTTGCTGCCAAGTGAAAACCAAACTGTGCCTCAAACTCAACAAACAATATCTGACCGAAATGGTTACCGGCGTAACAGCAGCCAACAT  29
                                                                      MetValThrGlyValThrAlaAlaAsnMe  (10)

  30  GACCAACGTTCTGGGCACCGCCGTTGTGCCGGCCCAGCTCAAGGAGACGCCGCTCAAAAGTGACCGTCGGGTAAGTTTCTTTCGAGAAAA  119
      tThrAsnValLeuGlyThrAlaValValProAlaGlnLeuLysGluThrProLeuLysSerAspArgArg                  (33)

 120  ATAAGACTCGAAAAAAAAAAACCAAAGCCAAAAACAAAAAACCCCGTCAATGGATTAAATCAGAACCTCTAGTTCCCCGAAATCTGTGGA  209

 210  TTAACTGAAGCGAACCAAACCGAGAATTCCCCTAATTGAGAGCCACCCACTTCGAGCTCAAGTTGATTTCCATTCGCGACTGTGGGCGAC  299

 300  TGGCCCCATCAATCCCGCTGCCCAAATGCATTTCCTTTTTAGCCATCTCCCACATGGCTGGCTGGGAATAAAAATACGAAATAAAGAAAA  389

 390  ACACTCTGAGCCAGACCAAAAAAAGGCCGCAACTGCCGTCGCGCGCCAACACAAAGCGAATTTATCTCGCGTCGCGTTGGTGGCATTTAC  479

 480  TATATGGCATATGGCATACTACTCCGACTACACACACGCTCCATCCATTCATGAGTGCCGCCCAAAATTGGCTGGCGTAGCTGCCCACCA  569

 570  GCTCCCCAGATTCGGACTCGGATTCGGATTTGGCTGCCACTTGGCGCGTGCGTCGCGTGGTGGCTGCAATTGTCGGTCATCAGCCGCCGT  659

 660  TTTTGGCCATTCGAACGGCACCGGTTCAATGAGTTGGCCAGAAAAAAAAAAACTGACTTCATGCGAGTCCAAATTTGGCAACTTGTGCTCC  749

 750  CTAGATAATAATGTACCTAGTTATGGCATCTTCGAATTTCCCGCTCTGACCTTGTCACATTCTCTTTATTTTTGCTCATACCTGGTTGGT  839

 840  TTATTTATAGCCACTGGCTGTAATTTATAACCGCCAAACTATTTTTAAATAAATGCCTCGGCCGAGTGGCGCTATAAATAGAGCACGCGC  929

 930  CGTGCGACTAAATTTGGCCGCCAGCCAGTCAATCCGCTCCCCAACCTACGCCGCCTCCTCCTTGATCTCCTCCAATCCAATTGAAGACCC  1019

1020  ATGCAGCTTCCTCTATTTTTGGGTCGTTGCATGAGGTCAAATTAGCCGTGCAAAAGCCGTGACTAATCGATGTTTTTTTTTCCATTTCTT  1109

1110  GCCTCTTGCAGTCGAACAAGCCCATCATGGAGAAACGCCGACGTGCCCGTATTAACAACTGTCTCAATGAACTCAAGACTCTGATTCTGG  1199
                SerAsnLysProIleMetGluLysArgArgArgAlaArgIleAsnAsnCysLeuAsnGluLeuLysThrLeuIleLeuA  (60)
                                                        ------*-----------*--------*--------*--*-- Helix 1

1200  ATGCCACCAAAAAAGACGTAAGTATCAGACAAAAATATAAGTTAAAAGCCACAAAAAAATAAAAAAAACTACATATTTGAAATTATCTCA  1289
      spAlaThrLysLysAsp                                                                           (65)
```

```
1290  AATATTTTTTAAAAGTTTAAATCATTGACTAATTTCCAAATTATTTTCTCTCTCTACTTTTAGCCGGCTCGCCACTCCAAATTGGAAAAG  1379
                                                                    ProAlaArgHisSerLysLeuGluLys  (74)

1380  GCCGACATTCTGGAGAAGACAGTAAAGCATCTGCAGGAGCTGCAGCGCCAGCAGGCAGCCATGCAGCAGGCCGCCGATCCCAAGATTGTG  1469
      AlaAspIleLeuGluLysThrValLysHisLeuGlnGluLeuGlnArgGlnGlnAlaAlaMetGlnGlnAlaAlaAspProLysIleVal  (104)
         *--*--------*--*--------*--------*-------- Helix 2

1470  AACAAATTCAAGGCCGGATTCGCCGACTGTGTGAACGAGGTTAGCCGCTTTCCCGGCATCGAGCCCGCCCAGCGTCGTCGCCTGCTACAG  1559
      AsnLysPheLysAlaGlyPheAlaAspCysValAsnGluValSerArgPheProGlyIleGluProAlaGlnArgArgArgLeuLeuGln  (134)

1560  CACCTGAGCAACTGCATCAATGGCGTTAAGACAGAGCTGCACCAGCAGCAGCGCCAGCAGCAACAGCAGTCCATCCACGCCCAGATGCTG  1649
      HisLeuSerAsnCysIleAsnGlyValLysThrGluLeuHisGlnGlnGlnArgGlnGlnGlnGlnGlnSerIleHisAlaGlnMetLeu  (164)

1650  CCCTCGCCGCCCAGCTCGCCGGAGCAGGATAGCCAGCAGGGAGCAGCGGCACCCTACCTCTTTGGTATCCAGCAGACGGCCAGCGGTTAC  1739
      ProSerProProSerSerProGluGlnAspSerGlnGlnGlyAlaAlaAlaProTyrLeuPheGlyIleGlnGlnThrAlaSerGlyTyr  (194)

1740  TTTCTGCCCAATGGCATGCAGGTGATCCCCACCAAGCTGCCCAACGGTAGCATTGCCCTCGTGTTGCCCCAGAGCCTGCCCCAGCAGCAG  1829
      PheLeuProAsnGlyMetGlnValIleProThrLysLeuProAsnGlySerIleAlaLeuValLeuProGlnSerLeuProGlnGlnGln  (224)

1830  CAGCAACAGTTGCTGCAGCACCAACAGCAGCAGCAGCAACTCGCCGTCGCAGCAGCAGCAGCGGCCGCAGCAGCAGCACAACAGCAACCC  1919
      GlnGlnGlnLeuLeuGlnHisGlnGlnGlnGlnGlnLeuAlaValAlaAlaAlaAlaAlaAlaAlaAlaAlaAlaGlnGlnGlnPro  (254)

1920  ATGTTGGTTAGCATGCCCCAGCGTACAGCCAGCACCGGATCCGCCAGCTCGCACTCCTCCGCCGGATACGAGTCGGCGCCCGGAAGCAGC  2009
      MetLeuValSerMetProGlnArgThrAlaSerThrGlySerAlaSerSerHisSerSerAlaGlyTyrGluSerAlaProGlySerSer  (284)

2010  AGCAGCTGCAGCTACGCCCCGCCCAGTCCGGCCAACTCTAGCTACGAGCCCATGGACATCAAGCCATCGGTCATCCAGCGCGTGCCCATG  2099
      SerSerCysSerTyrAlaProProSerProAlaAsnSerSerTyrGluProMetAspIleLysProSerValIleGlnArgValProMet  (314)

2100  GAACAGCAGCCCCTGTCGCTGGTGATCAAGAAGCAGATCAAGGAGGAGGAGCAGCCCTGGCGGCCCTGGTAGAGGGTGTCTGCATATGCA  2189
      GluGlnGlnProLeuSerLeuValIleLysLysGlnIleLysGluGluGluGlnProTrpArgProTrpEnd                    (337)

2190  TATCATATAGCATAGCCACCCCTATCGAATCTCCCGCTTTTAAGACTGACCCCCCACAACTCATCCAACTCACACACATGCGCAGGCGTT  2279

2280  GTGCGCATGCGCGTAGACATTTCACATCATTCGCCGGGATTGCGCAAATGTTGCTTTGAAGTGTTGCAAACATGCGAATCCTAAACTCGG  2369

2370  TTCACAACTTCGTTGGCTTAGTTTCCTGGCTTATATCCTGGAAACCCGTCGACGAGGCTAAGGACCTTCATCAGACGCACCCACACACAA  2459

2460  ACACACACACGCAAACGTTGTTATAATTTATTTATTATTATATTATGTAATCGATTTGAAAGAACGGTATTCTACCAGGACATCGCCAAA  2549

2550  CTACCTCAGTCCAAGTACTTGGTGTTGAATTGCCTCATGTATTATGTATTACTCTTTGAATAACAGCAAATCAGCAAAAGTCTTCCAAAC  2639

2640  ACAGAAAATGAAAATGCGAAAATAAGCACCTGAAAAGCTGAAATACTTTTTATGAAAAAGATAAACGCAAAAGCATAACTCTTACACGTA  2729

2730  GTCGTACATCTCCATTTAAGTATAGGTTTTGTACCATAGCCAGCTAAGCCGCTTAGGGTTTCTCTCGCTCTTAAGTCTAATCAAAGAATA  2819

2820  ATTATATTTATAAAACACACAAATCTATTCGTAAGGCCACGTGATATAGTGAACATAATGAGCTTCTAAGAAAACAAAACAAGAATTTGA  2909

2910  TGCAAGCAAAAGCAAAAAAATCAACAAGAAAGAAAAAACAAACAAACAACACACTAAAAAGAAATAAATTTAAAAGATTCTACTAAAAAA  2999
                                                                        ------

3000  TCAAATACGCAAAACGGATTTGTTATTGTGGTTGGGGTATCTTTTCCTGGGTTTTTTTTTCATTCGGGTGAAAGTCCGATTATGGTTATT  3089
      |(A)n

3090  TTTTTTGTTTTAGAGGTCAAACGCTTTGAGTGACAGGAAACTTATCGAGCCCCCCGACTTATCGTAGCAAATTTCGACGCTAATTATTAT  3179
```

During adult development, *h* seems to counteract the function of *ac-sc* complex genes in the development of sensory organs (Botas et al. 1982; Ingham et al. 1985b). It has not been possible, however, to demonstrate direct interaction of HAIRY with any of the *ac-sc* complex products involved (Van Doren et al. 1991).

Tissue Distribution

HAIRY is intranuclear, as revealed by antibody staining. In cellular-blastoderm-stage embryos HAIRY-containing nuclei are distributed in eight stripes. After the onset of germ band extension, HAIRY rapidly disappears from the seven posterior stripes. (This pattern of occurrence is quite similar to that observed at the RNA level, see below.) A little later, in embryos having fully extended germ bands, HAIRY is transiently detectable in cells associated with pairs of tracheal pits (parasegments 4–13); still later, during germ band retraction and the following stages, HAIRY appears in the mesoderm, proctodeum and anal plates (Carroll et al. 1988; Hooper et al. 1989; a detailed comparison of the metameric distribution of HAIRY and other pair-rule-gene products is presented in these references). HAIRY also occurs in the imaginal discs of third-instar larvae and early pupae. In the eye-antennal disc, HAIRY is transiently present in a band of cells just anterior to the morphogenetic furrow. In leg discs, HAIRY is localized in groups of cells that evolve into longitudinal rows during disc eversion. In wing discs, expression occurs along presumptive wing veins. In all these imaginal structures, HAIRY is excluded from peripheral neurons and sensory organs (Carroll and Whyte, 1989).

Mutant Phenotypes

h belongs to the pair-rule class of segmentation genes. In amorphic *h* embryos, certain metameric elements fail to develop in alternating segments. The missing structures correspond to the regions where gene expression is detectable (see below). Hypomorphic mutations are viable; they result in extra microchaetae and other sensory organs in the adult epidermis. In these hypomorphic mutants, the adult phenotype can be rescued by expression of *h* coding sequences under the control of a heat-shock promoter 6–11 h after pupariation (Ingham et al. 1985b; Carroll et al. 1988; Hooper et al. 1989; Rushlow et al. 1989).

Gene Organization and Expression

The open reading frame that is thought to produce active protein is 337 amino acids long. However, the Met at position 10 occurs within a very good

---

h SEQUENCE (*previous pages*). Accession, X15904 (DROHAIRG). The amino acids underlined constitute the HLH domain. Asterisks mark the hydrophobic residues thought to participate in the formation of dimers.

translation initiation context and so may serve as an alternative initiation site. There are two mRNAs, α1, with an expected size of 2,335 bases and α2, with an expected size of 2,139 bases; this is in agreement with the results of northern analysis. The two different sites of transcription initiation involved in production of the two mRNAs are 196 bp apart, and neither one has a canonical TATA box. Primer extension and S1 mapping were used to define the 5′ ends. The 3′ end was obtained from a cDNA sequence that included a poly(A) tail. There are introns after the Arg-33 and Asp-65 codons (*h* Sequence) (Rushlow et al. 1989).

Developmental Pattern

h transcripts are first detectable in 2–4 h embryos, and they remain present throughout larval development. α1 mRNA is prevalent up to 4 h, then both mRNAs are equally represented until the end of larval development, except for late third-instar larvae when α2 becomes more abundant. In pupae, *h* mRNAs nearly disappear, but they become abundant in adults, with the two RNAs occurring in nearly equal amounts (Rushlow et al. 1989).

In situ hybridization shows that *h* mRNA in cell-cycle 12 embryos (syncytial blastoderm) is nearly uniformly distributed around the periphery of the embryo. Labeling then differentiates an anterior dorsal region (region 0 or AD) that extends for 12–15 nuclei at 95–85% egg length (Appendix, Fig. A.3) and a region of continuous labeling from 75% to 20% egg length. In the latter region labeling becomes discontinuous; before the completion of cellularization (mid-cycle 14), the labeling is distributed in seven evenly spaced stripes, each approximately 3–4 nuclei wide. In the abdominal region, the stripes of expression correspond to the posterior portion of the odd-numbered segments and the anterior portion of the even-numbered ones. This pattern is carried forward into the thoracic and cephalic regions, with the AD patch corresponding to the labrum (Appendix, Fig. A.3). When gastrulation starts, the cephalic fold invaginates between *h* stripes 1 and 2. Soon thereafter, the striped pattern disappears; and, by the time of germ band elongation, *h* transcripts are most evident in the hindgut and the foregut (Ingham et al. 1985a).

Promoter

As in the case of *even-skipped* (*eve*), the *cis*-acting regulatory region is very extensive, >14 kb; and the striped pattern is the result of independent regulation of the individual stripes by various segments of the regulatory region. Thus each section of the regulatory region responds to unique positional cues along the antero-posterior axis of the embryo to activate transcription and produce a particular stripe. The positional cues are given by maternal products, gap genes and other pair-rule genes such as *eve*.

A construction that carries 14 kb of upstream sequences and the coding region of *h* is sufficient, in germline transformants, to rescue the embryonic mutant phenotype, but the adults that result exhibit a severe hairy phenotype.

This suggests that the region controlling *h* expression in adults is located more than 14 kb upstream of the transcription initiation site (TIS) (Rushlow et al. 1989). The 14 kb of upstream sequence was further subdivided into stripe-specific segments using *lacZ* as a reporter gene in germline transformation experiments. The whole 14 kb segment resulted in expression in the seven posterior stripes but not in the AD zone. Expression in individual stripes requires the following segments: stripe 1, 4.9–4.0 kb upstream of the TIS; stripe 2, several elements dispersed between 9.4 and 4.0 kb upstream of the TIS; stripes 3 and 4, several elements dispersed between 14.0 and 6.4 kb upstream of the TIS and elements further upstream; stripe 5, a segment between 6.8 and 4.0 kb upstream of the TIS; stripe 6, a segment between 9.1 and 5.2 kb upstream of the TIS; stripe 7, a segment between 11.0 and 9.4 kb upstream of the TIS. The positions of stripes produced by many of these artificial promoters are shifted slightly relative to positions of normal *h* stripes. Thus, with the possible exception of stripe 1, sequences other than those listed here for each stripe are required for normal expression (Howard and Struhl 1990; Pankratz et al. 1990; Riddihough and Ish-Horowicz, 1991).

The products of the gap genes *knirps* (KNI) and *Krüppel* (KR) bind, with varying affinities, to several regions of the *h* promoter. For example, KR, which is thought to act as a repressor, binds with high affinity to the region responsible for stripe 6, and KNI, which is thought to act as an activator, binds with low affinity to the same region. The formation of stripe 6 then, probably results because there is only one zone along the axis of the embryo where KNI is in high enough concentration to stimulate *h* transcription while KR concentration is so low that it does not repress *h*; this zone is in the posterior region of the embryo that corresponds to stripe 6 (Appendix, Figs A.2 and A.3). By this argument, the anterior border of stripe 6 is defined by the posterior slope of KR's bell-shaped concentration distribution. More posteriorly, stripe 7 may arise by similar interactions involving KNI and the product of the gap gene *tailless* (TLL). In this case TLL would act as a positive regulator at high concentration; and KNI would act as a repressor, defining the anterior border of stripe 7 through its posterior concentration gradient (Pankratz et al. 1990).

References

Botas, J., Moscoso del Prado, J. and García-Bellido, A. (1982). Gene dose titration analysis in the search for trans-regulatory genes in *Drosophila*. *EMBO J.* **1**:307–310.

Carroll, S. B. and Scott, M. P. (1986). Zygotically active genes that affect the spatial expression of the *fushi tarazu* segmentation gene during early *Drosophila* embryogenesis. *Cell* **45**:113–126.

Carroll, S. B. and Whyte, J. S. (1989). The role of the *hairy* gene during *Drosophila* morphogenesis: stripes in imaginal discs. *Genes Dev.* **3**:905–916.

Carroll, S. B., Laughon, A. and Thalley, B. S. (1988). Expression, function, and regulation of the *hairy* segmentation protein in the *Drosophila* embryo. *Genes Dev.* **2**:883–890.

Harrison, S. C. (1991). A structural taxonomy of DNA-binding domains. *Nature* **353**:715–719.

Hooper, K. L., Parkhurst, S. M. and Ish-Horowicz, D. (1989). Spatial control of *hairy* protein expression during embryogenesis. *Development* **107**:489–504.

Howard, K. and Ingham, P. (1986). Regulatory interactions between the segmentation genes *fushi tarazu, hairy*, and *engrailed* in the Drosophila blastoderm. *Cell* **44**:949–957.

Howard, K. R. and Struhl, G. (1990). Decoding positional information regulation of the pair-rule gene *hairy*. *Development* **110**:1223–1232.

Ingham, P. W., Howard, K. R. and Ish-Horowiza, D. (1985a). Transcription pattern of the *Drosophila* segmentation gene *hairy*. *Nature* **318**:439–445.

Ingham, P. W., Pinchin, S. M., Howard, K. R. and Ish-Horowicz, D. (1985b). Genetic analysis of the *hairy* locus in *Drosophila melanogaster*. *Genetics* **111**:463–486.

Ish-Horowicz, D. and Pinchin, S. M. (1987). Pattern abnormalities induced by ectopic expression of the *Drosophila* gene *hairy* are associated with repression of *fushi tarazu* transcription. *Cell* **51**:405–415.

Pankratz, M. J., Seifert, E., Gerwin, N., Billi, B., Nauber, U. and Jäckle, H. (1990). Gradients of *Krüppel* and *knirps* gene products direct pair-rule gene stripe patterning in the posterior region of the *Drosophila* embryo. *Cell* **61**:309–317.

Riddihough, G. and Ish-Horowicz, D. (1991). Individual stripe regulatory elements in the *Drosophila hairy* promoter respond to maternal, gap, and pair-rule genes. *Genes Dev.* **5**:840–854.

Rushlow, C. A., Hogan, A., Pinchin, S. M., Howe, K. M., Lardelli, M. and Ish-Horowicz, D. (1989). The *Drosophila hairy* protein acts in both segmentation and bristle patterning and shows homology to N-*myc*. *EMBO J.* **8**:3095–3103.

Van Doren, M., Ellis, H. M. and Posakony, J. W. (1991). The *Drosophila extramacrochaeta* protein antagonizes sequence-specific DNA binding by *daughterless/achaete-scute* protein complexes. *Development* **113**:245–255.

16

hunchback: hb

Chromosomal Location:
3R, 85A3-B1

Map Position:
3-48

Product

DNA-binding regulatory protein of the Zn-finger type involved in the earliest stages of embryonic pattern determination.

Structure

The amino acid sequence of HB suggests that there are two Zn-finger domains, one with four fingers, the other with two. Two short segments near the N-terminus (boxes A and B, *hb* Sequence) have some similarity to the *Krüppel* (*Kr*) protein (KR, another finger protein) and to the retrovirus HIV-1 *pol* product (Tautz et al. 1987; Evans and Hollenberg 1988; Harrison 1991).

Functions

HB participates in the transcriptional regulation of several developmentally important genes; it recognizes a sequence distinguished mainly by a run of 6 As:

1. HB binding sites have been found in the *hb* promoter itself where it is thought to stimulate transcription (Treisman and Desplan 1989).

2. Binding sites for HB have also been demonstrated in the *even-skipped* (*eve*) promoter elements responsible for two of the embryonic stripes (#2 and #3) in which *eve* is expressed; here again HB probably acts as a positive regulator of transcription. (Stanojevic et al. 1989, 1991; Small et al. 1991).

3. HB has been demonstrated to repress *Kr* expression (Hoch et al. 1991). *Kr* expression normally occurs in an embryonic band immediately posterior to the area of HB accumulation (Appendix, Fig. A.2). HB is thought to be a repressor of *Kr* at high concentration (thus defining the anterior edge of the *Kr* zone of expression), and an inducer at low concentrations (Hülskamp et al. 1990).

hb

```
-4682  CATACAAATAATAAGTTATCCTTTTGTATTGTATAGAGAAAAAAAGGTTTTTACCAATGAACTATGAATAATGAATAATAATAATAGTTT  -4593
                                                                                               <-

-4592  TTTTTTTTAGTCCAAAATTTGTCATTAAACCTAGTTAGAACAATCGCTCCTAATTTATCATTCTAAAAGCGAACATTCCGCTTGGGAAAA  -4503
       ----------- hb8                                                                  -------

-4502  AAATTGGTCTAAACCGAATGATACTATTAATGATATGCATTTATTGCTAACCATAATCCTTGTCAAGCTAACAATGATACATTTTCCGAA  -4413
       -------> hb7

-4412  ATTAGCTTAAAAAGGTGGAATACACCCAATATGCACAAACTACCTTAAGGAGATTTGGAATTTCGAATGCTAATTGTGGCAAAGCTTTGC  -4323
           -------------> hb6

-4322  CCAAATTAAGTTAACACGCACAGCAACAGGAAAATGTGTTAAAGCAACAAGGAATCTCCTCGGCCCAAACTTCCATCGTCCCAATTGCAG  -4233

-4232  TTGGCTAAGTTGTTAATGTGTCTGGGCTTAAAGTTGCCCAAAAAACAATTGGCGAAGGCCCCCATCTTCCTCCATTTCCGCTCTCTCAC  -4143
                                          -------------> hb5

-4142  TTTTGGGCCAGAAATCAATAATCAATAGTGAAGCGGAGATGCCAAAAAACGGCAAAGAGCCAAAAAGGCAGCTGCATTCGGCCAAAATGC  -4053
                                              -------------> hb4

-4052  AGCGCCAGAAAATGCAAAAGGATAAAATGAGCGAGTCAGAGCGAGAGAGTGGGTGAGTGAGTGAAAGAGCGAACGCCACGAAGGGGATGC  -3963

-3962  TGTTTAGTTATTGCTTTTGGGGATGGGGAAAGTCACTCAGATTTACAGCTAGCATCCGTATCCGTTTTGAGTGAGTTTCGTCGTTACGTT  -3873

-3872  GTTGATGCTCTCCGGCTGCTCTCTCATTTCGATTTCTGCTTCTCCGTGTAACGGCTCTCGTGCGCCTTTGTGTTGTTGCACTTCTGGCAT  -3783

-3782  TTGTTTTTGCTTTGCAATTGAGATTTTAGATTTGAGTTCTTTTTTGGCGAGTGCACTGTTGTGAGGAGCAGTGAGGAGATTTTCAGCTAT  -3693
                                  =====                                                     ===
                                   -->-3662

-3692  TAGAAGAGCCCGCTGAGCGTGAGTTTGGTCAGTTGTGCTCCGAGTCCCGAAAACGAAAGTCGCCAGCATTGACAGGCAGCCACGGAAATA  -3603
       ==

-3602  CAAAAATAACCAAACATCCAAAGGACGAAACGTAACTGCTATCAAAACAAATATTGCCATTAAATACAATTAAACTTCGTGCTTGTGCTA  -3513

-3512  AAAGATAACCAATTGCAAAAAGACTTTTGTCCCGAAAACTTATTTTTTGGCAAAGACCACATCCCGCACATGCGCGAATTCCGCGAAAAA  -3423

-3422  GAAAGCACAAAAGCAAGCTAAAAAGCGAGGCCCAAAAAATAGACAAAAACGAAGAGCAAGGAGCCCCCACATCGCCGCTCCCCCTCGCAC  -3333

-3332  GCACTGTGCGTGTTGGTCTAACGGTAACCGTGCCGCGGTCAAGCGAGAGAGTGGGAAAGAGAGAGTCCGCGAGCGAAAAGCGAAAACTTA  -3243

                                                                              _
                                                                               |
-3242  CGTGCTCCTGCTCCTGCTCCCGGTTCCCGTTTCCAAATTCTCCCGCAGAATTGCACTTGGACTGTCCGCAAGGTAAGTAACTGGGCGTAC  -3153
                                                                              _|

-3152  CATCCAATATCCTAGTTATACACTGCATTCGACCTCCAATAAATCGTAAAAAACACATGGAGGTAGAAATTCGCAAAAGCTTTCCGCGGA  -3063

-3062  TAAACAAATAAACAAGAATTACAAATCGCTTTTGCGGGAGCAAATGCCAAAATGTTGGGGTATCCAGAATATACACAGTTTTTGTGAGGA  -2973

-2972  TTTACATACGCCCTGTAAATTTTAATTTAGTTCTCAATTGATACGAAATCTGTTTTTTTTTATAGAATATATTCGATATATAGTTGTTTA  -2883
                                                                                 <----------

-2882  ATTATAATTCATCATGTGATATACTTTCAAAAGAAACAGATTTAAATAGTTCGTTTATATGCTATTATGCACTATGCTTAATGTATTTTA  -2793
       --- hb3
```

(continued)

```
-2792  CTTTATTAATTCATGCTAATCTGATGACTGATAACCAATTTGCTTATTCTATGTCATAATCACCTTTAATCCCAAGTACTCAACTTTCTT  -270
                   --------> bcd-B1                                           --------> bcd-B2

-2702  CTAGTTCTGCACATTTTCTTGTTCTCTTCTTGTTGTTTGTAATCGAGTGCTTCTTTTTCTGTCTAACCTTAGGAACACAAACAAGAGATC  -261

-2612  TCACACACATGCACACACACAACCACACAGAGTGGTAAATCATTGCTAAAAATGCAAATGGCAAAAATTCTAAAACAATATTTTAAAATC  -252

-2522  TGTCTGGCTTTTGGCCGATCTTCGGGTGAACTTGTTTTTTGCCCGCTCTGTCTGTTTTTGCCTAGAACTGCAGCGTATATCCAACCCCCA  -243

-2432  ACCGACCCCCCCCCCTCCTCATTTGCTCCTCCTAAGCAACGCCCCTGGCCACGCCCCCCTCTACGCCTGACAGCTAATTTTATTTGTTTA  -234

-2342  CATGTCGACTTTCATGTTGTTTTTTTTTCGCCTTCCTGCCCCCCAAATAACCCCTTCAATTTTTAGCATTTCTTTTCGCTCTTCTTTCGG  -225

-2252  CAAATGCATTTTCGACTTTTCTTTTTTAATAATTTAAATTTAGACATCGAACTATAATTCAAAGTCAACGTGGCAATTTTTTAATGGCAA  -216

-2162  ATAGTTCGGCACTTGCCAAAATGCATTTCAAATATAATGAATACACCTATGTGACGCTCGCAGGCTTTGTTTTTTTTTTTTGTTTAGGAT  -207

-2072  GATGATGGGAATCATTATTGCACGATCTTTTGCAGAATTGATTGTCATCCGACAGTTTTAATGTTTGCCGACAAAAAGAACTGAGCGCAT  -198

-1982  TTCAAACAAATGGGGCTTTGTTAGGCGCATAAAAATAAATATGCACATATTATACGGGATATATTTATAAACCGCAAATGACATTGAATC  -189

-1892  ACTTCAGTCAGCAGCAGCAGTGCAGTTTTCCCTAAACGAATGCAGCTCCAAAAAACAACAAGTTGTTCTAAGCCAACAACATAACGGTTG  -180

-1802  CAACTACACATATGTATACGCATACATACATATATTTTATATACGCCTGCAAACAAGCAGGCATATCCTGCTCTTGGCTTGCTTTTTGTA  -171

-1712  CGGATTTTTCAACAAAAACATTTTTTGTGTGGCGCATTTTCTGCGTTTTCGAATTTTTCCATTTTTGCCATTTTCATGATATTTTTAGAG  -162

-1622  GGTTAAAAAGGGGGCGGTGTTGGGGTGGGGGGGGAGGCACCAAAAACGACTAGCACATGTTTAGTTTACTTTTGCAGTTTCGGTTGCCTG  -153

-1532  GCTGCACTCGTTGTGGCCACACCACGCCACTAACCCCCACTGCATGGTCCCCATGCCCCCACCCCTGCCCCCTTGGCCCATGCCCCTGCC  -144

-1442  CGGCGTAATATTAATTTTACACTTGGCCTTTAGTTTGGCTTTGTTGCTGTTGTTGGGCATGGCACAAAAAAGCCCAGACGAAAGGCGAAA  -135

-1352  ATTCTCTTTGTTTCATAGTGCGCACACACACTAACACACTTGCCGAGCAATTAATTTTGGCTTTCTTTAAAAATATCCACAAAAGCGAAT  -126

-1262  TTAAATAAATATGCTAAGCTTCATTTTGTGTGGTGCACTTTCTGTTTCCTGAACCATCAATTATGCCTAAGTATTGTAGATATTTTAGCT  -1173

-1172  GCCAGATAGCACAGCACCATCCTCCCATAATAATATTCCGTAAATGCCCCTTTTTCCCGTTTTGCGTTTTTAATAATATTTACTTGAAAG  -1083
                                              kr2 <-----------<---------------- hb2

-1082  CACAAACAATTAGCCAAAAATGCAGCAACTGCACAATTTTTCAGTGTGAAATTGGAAATGGAAAAAATATAGGCAACAAGCAATTTTAAT  -993

-992   GCGAGAATTATTAGAAAAACTACGCAAATCAAAGTGAAATGTCTGGCGGAAAATTGTTGTCAGGAAAATGTTTTTCAAATGGGTGTGTAA  -903

-902   TAATTATACTGCACATATTATGCATATAGTTTAGTTGGTCCTTAGAGTTTTCCCGCAGGTGTAAGCAGTTCTGATCCGTTAATTTAGTTA  -813
                                                                         ------ da/lsc         <-

-812   AGTCCCGCAATCCTTTTTACTTTTTATTATTAACTACGAAACTGCCCACGCTAGCTGCCTACTCCTGCTGTCGACTCCTGACCAACGTAA  -723
       --------- kr1                                                                         -----

-722   TCCCCATAGAAAACCGGTGGAAAATTCGCAGCTCGCTGCTAAGCTGGCCATCCGCTAAGCTCCCGGATCATCCAAATCCAAGTGCGCATA  -633
       ---> bcd-A1                                                                              <-
```

```
-632  ATTTTTTGTTTCTGCTCTAATCCAGAATGGATCAAGAGCGCAATCCTCAATCCGCGATCCGTGATCCTCGATTCCCGACCGATCCGCGAC   -543
      -----------hb1 --------> bcd-A2

-542  CTGTACCTGACTTCCCGTCACCTCTGCCCATCTAATCCCTTGACGCGTGCATCCGTCTACCTGAGCGATATATAAACTAATGCCTGTTGC   -453
                                         --------> bcd-A3                          ------
               --> -->-444/-440
-452  AATTGTTCAGTCAGTCACGAGTTTGTTACCACTGCGACAACACAACAGAAGCAGCACCAATAATATACTTGCAAATCCTTACGAAAATCC   -363

-362  CGACAAATTTGGAATATACTTCGATACAATCGCAATCATACGCACTGAGCGGCCACGAAACGGTAGGATATTGTTAGCCATTACCAAGTG   -273

-272  TCTCCATTTTGAACACAAAATCACTCAAATCGCCTTCAGGGGGTGGGTGCCGCCCAGCCACCCCTGACGTATTTTTTGTTAGGGGTGGTG   -183

-182  CCGCAAGCACACCAAAAAAAGAGAAAAAAAAAATAAAAGCGAGGAAAAATAAAATGAAAAACAAGCGGAAAAAAAGAGGAAAAAACTCGA   -93

 -92  CGCAGGCGCAGTGCATGAATGAATAAATGAATATGCCCACTAACCCCACTCTCTCTGTTTTCTTATCCATTACAGCCGTCTAGAGCCGCC   -3

  -2  AAGATGCAGAACTGGGAGACGACAGCCACGACCAACTACGAGCAGCACAACGCCTGGTACAACAGCATGTTCGCGGCAAATATCAAACAG   87
         MetGlnAsnTrpGluThrThrAlaThrThrAsnTyrGluGlnHisAsnAlaTrpTyrAsnSerMetPheAlaAlaAsnIleLysGln   (29)
                                                                                |-        box A

  88  GAGCCAGGTCATCATCTCGACGGGAATAGCGTGGCCAGCAGTCCGCGCCAATCGCCCATTCCCTCGACCAATCACCTGGAACAGTTCCTC   177
      GluProGlyHisHisLeuAspGlyAsnSerValAlaSerSerProArgGlnSerProIleProSerThrAsnHisLeuGluGlnPheLeu   (59)
                          -|

 178  AAGCAGCAGCAGCAGCAGCTTCAGCAGCAACCCATGGATACCCTGTGCGCCATGACCCCATCACCCAGCCAAAACGATCAAAACAGCCTG   267
      LysGlnGlnGlnGlnGlnLeuGlnGlnGlnProMetAspThrLeuCysAlaMetThrProSerProSerGlnAsnAspGlnAsnSerLeu   (89)
                                |-         box B          -|

 268  CAGCATTACGATGCTAACTTGCAGCAACAGTTGCTGCAGCAACAGCAGTACCAGCAGCATTTCCAGGCAGCCCAGCAGCAACATCATCAC   357
      GlnHisTyrAspAlaAsnLeuGlnGlnGlnLeuLeuGlnGlnGlnGlnTyrGlnGlnHisPheGlnAlaAlaGlnGlnGlnHisHisHis   (119)

 358  CATCACCATCTGATGGGTGGATTCAATCCGCTGACGCCACCTGGTCTGCCCAATCCCATGCAGCACTTCTATGGCGGCAATCTGCGACCC   447
      HisHisHisLeuMetGlyGlyPheAsnProLeuThrProProGlyLeuProAsnProMetGlnHisPheTyrGlyGlyAsnLeuArgPro   (149)

 448  AGTCCGCAGCCCACGCCCACATCTGCCTCCACAATTGCGCCCGTTGCAGTTGCCACTGGCAGCAGCGAGAAGTTGCAGGCACTAACACCA   537
      SerProGlnProThrProThrSerAlaSerThrIleAlaProValAlaValAlaThrGlySerSerGluLysLeuGlnAlaLeuThrPro   (179)

 538  CCCATGGATGTCACACCGCCTAAGTCGCCGGCCAAGTCGAGTCAGTCGAATATTGAGCCGGAGAAGGAGCACGATCAGATGTCGAACTCC   627
      ProMetAspValThrProProLysSerProAlaLysSerSerGlnSerAsnIleGluProGluLysGluHisAspGlnMetSerAsnSer   (209)

 628  AGCGAGGACATGAAGTACATGGCCGAGTCCGAGGACGATGATACCAACATCCGGATGCCCATCTACAATTCGCACGGCAAGATGAAGAAC   717
      SerGluAspMetLysTyrMetAlaGluSerGluAspAspAspThrAsnIleArgMetProIleTyrAsnSerHisGlyLysMetLysAsn   (239)

 718  TACAAGTGCAAGACCTGCGGCGTGGTGGCCATCACCAAGGTGGACTTCTGGGCGCACACCCGCACCCACATGAAACCAGACAAGATCCTG   807
      TyrLysCysLysThrCysGlyValValAlaIleThrLysValAspPheTrpAlaHisThrArgThrHisMetLysProAspLysIleLeu   (269)
            ---     ---                                             ---      ---
```

(continued)

```
808  CAGTGCCCGAAGTGCCCGTTCGTCACCGAGTTCAAGCACCACTTGGAGTACCATATCCGGAAGCACAAGAACCAAAAGCCCTTCCAGTGC  897
     GlnCysProLysCysProPheValThrGluPheLysHisHisLeuGluTyrHisIleArgLysHisLysAsnGlnLysProPheGlnCys  (29
        ---      ---                                    ---         ---                     ---

898  GACAAATGCAGCTACACGTGTGTCAACAAATCCATGCTAAACTCGCACCGCAAGTCGCACAGTTCTGTGTATCAGTACCGTTGTGCGGAT  987
     AspLysCysSerTyrThrCysValAsnLysSerMetLeuAsnSerHisArgLysSerHisSerSerValTyrGlnTyrArgCysAlaAsp  (32
           ---                                    ---         ---                     ---

988  TGTGATTACGCCACCAAGTATTGCCACAGCTTCAAGCTGCATCTGCGCAAGTATGGTCACAAGCCCGGCATGGTTTTGGACGAGGATGGC  107
     CysAspTyrAlaThrLysTyrCysHisSerPheLysLeuHisLeuArgLysTyrGlyHisLysProGlyMetValLeuAspGluAspGly  (35
     ---                                    ---               ---

1078 ACCCCGAATCCCTCGTTGGTCATCGATGTTTACGGCACGCGTCGTGGTCCGAAGAGCAAGAATGGTGGACCGATTGCCAGTGGAGGAAGT  116
     ThrProAsnProSerLeuValIleAspValTyrGlyThrArgArgGlyProLysSerLysAsnGlyGlyProIleAlaSerGlyGlySer  (38

1168 GGCAGCGGCAGCCGGAAGTCAAATGTTGCAGCTGTCGCTCCGCAGCAACAGCAATCTCAGCCAGCTCAGCCAGTCGCCACATCTCAGCTG  125
     GlySerGlySerArgLysSerAsnValAlaAlaValAlaProGlnGlnGlnGlnSerGlnProAlaGlnProValAlaThrSerGlnLeu  (41

1258 AGTGCCGCCCTGCAAGGATTCCCTCTGGTTCAAGGCAACTCCGCTCCTCCGGCGGCATCTCCAGTGCTCCCGCTGCCCGCCTCTCCTGCC  134
     SerAlaAlaLeuGlnGlyPheProLeuValGlnGlyAsnSerAlaProProAlaAlaSerProValLeuProLeuProAlaSerProAla  (44

1348 AAGAGTGTGGCCAGTGTGGAACAGACGCCCAGCTTGCCCAGTCCAGCCAATCTTCTGCCTCCTCTGGCCAGCCTTCTGCAGCAGAACCGC  143
     LysSerValAlaSerValGluGlnThrProSerLeuProSerProAlaAsnLeuLeuProProLeuAlaSerLeuLeuGlnGlnAsnArg  (47

1438 AACATGGCCTTCTTCCCCTACTGGAACCTCAATCTCCAGATGCTGGCCGCCCAACAACAGGCCGCTGTCTTGGCCCAATTGTCGCCAAGA  152
     AsnMetAlaPhePheProTyrTrpAsnLeuAsnLeuGlnMetLeuAlaAlaGlnGlnGlnAlaAlaValLeuAlaGlnLeuSerProArg  (50

1528 ATGCGAGAGCAACTGCAGCAACAGAACCAGCAGCAGAGCGACAATCAGGAGGAGGAGCAGGACGATGAGTACGAGCGTAAGTCAGTGGAC  161
     MetArgGluGlnLeuGlnGlnGlnAsnGlnGlnGlnSerAspAsnGlnGluGluGluGlnAspAspGluTyrGluArgLysSerValAsp  (53

1618 TCTGCCATGGATCTGTCCCAAGGAACGCCAGTGAAGGAGGATGAGCAGCAGCAACAACCGCAGCAGCCGCTGGCCATGAATCTCAAGGTG  170
     SerAlaMetAspLeuSerGlnGlyThrProValLysGluAspGluGlnGlnGlnGlnProGlnGlnProLeuAlaMetAsnLeuLysVal  (56

1708 GAGGAGGAGGCCACGCCTCTGATGAGCAGCTCGAATGCCTCGAGACGCAAGGGACGCGTCCTCAAGCTGGACACCCTGTTACAACTGCGA  179
     GluGluGluAlaThrProLeuMetSerSerSerAsnAlaSerArgArgLysGlyArgValLeuLysLeuAspThrLeuLeuGlnLeuArg  (59

1798 TCGGAGGCCATGACATCTCCCGAGCAACTGAAAGTACCCAGCACACCCATGCCAACTGCATCCTCGCCCATTGCCGGACGCAAACCCATG  188
     SerGluAlaMetThrSerProGluGlnLeuLysValProSerThrProMetProThrAlaSerSerProIleAlaGlyArgLysProMet  (62

1888 CCCGAGGAGCACTGCTCGGGCACCAGTTCGGCAGATGAGTCGATGGAGACGGCCCATGTGCCGCAGGCCAATACCAGTGCCAGTTCGACG  197
     ProGluGluHisCysSerGlyThrSerSerAlaAspGluSerMetGluThrAlaHisValProGlnAlaAsnThrSerAlaSerSerThr  (65

1978 GCGTCCAGCTCGGGGAACAGCTCCAATGCCAGCAGCAATAGCAACGGCAACAGCAGCAGCAATTCCAGCAGCAATGGAACCACCTCAGCG  206
     AlaSerSerSerGlyAsnSerSerAsnAlaSerSerAsnSerAsnGlyAsnSerSerSerAsnSerSerSerAsnGlyThrThrSerAla  (68

2068 GTTGCAGCTCCTCCATCCGGAACTCCGGCGGCGGCGGGTGCCATCTACGAGTGCAAGTACTGTGATATCTTCTTCAAGGACGCCGTGCTC  215
     ValAlaAlaProProSerGlyThrProAlaAlaAlaGlyAlaIleTyrGluCysLysTyrCysAspIlePhePheLysAspAlaValLeu  (71
                                                        ---      ---

2158 TACACCATTCACATGGGCTACCACAGCTGCGACGATGTGTTCAAGTGCAACATGTGCGGCGAGAAGTGCGACGGACCCGTCGGCCTCTTC  224
     TyrThrIleHisMetGlyTyrHisSerCysAspAspValPheLysCysAsnMetCysGlyGluLysCysAspGlyProValGlyLeuPhe  (74
              ---         ---                     ---      ---
```

```
2248 GTTCACATGGCCAGGAATGCTCACTCCTAAGTTCCCCATCACCATCACCTTGTTATTATTATTTATCACTATTATCATATAATCGTTGTC 2337
     ValHisMetAlaArgAsnAlaHisSerEnd                                                            (758)
       ---         ---

2338 CAGAATTGTATATATTCGTAGCATAAGTTTTCCAAACATTATTTTGTTGTCGAAAATTGTACATAAGCCAATTAAGCCGCTAATTCTAGA 2427

2428 CCTAAGTTTATCTAACTATCCTAACTGTATTGAACTGTAGCCACCTTTCAATCTGTCTCCTATACACTCTTGTATTTTCGAAATCGACTA 2517

2518 AAAACCCTGAAAACGGTTTAAAAACTATCATAAATGCATGGAGAAACATAAGCCTAAGTTAAATCTAATTTGTAAGTTGAGTCAAGCGAA 2607

2608 ACAACCAAACAATACCAACAGTCCAAAGTCAAATTAATAAAATATAGTTTATAACATATACATAATGAGTATGTTTTCTAAAATAATTAA 2697
                                         ------

2698 TTAGTCTTATTTAACCTAACATATTCGTATATGCGCATAACACTCAGTTCTTTCTCTGATATTATTCTCTCAGTATTTTGTTAGTTGAAA 2787

2788 GCGAATTCGAATCGAACGAAATCAAATCAAATAAATCCAATTATTCAATATAATTTCACAAGTTTTTCGCTTTTTTTTTTGTTGTTAACC 2877
                               ------                 |(A)n

2878 TTTTGGCCAATAATGACAATATTTTCGATGCAACTGAAACTGACGAAAGAAGAAGTACAAATTTAGAGATTTTTAAAGAGTAGCTAAGAT 2967

2968 GCGCGAAATCTGAGCAACGGATCAAATTAG  2997
```

hb SEQUENCE. Strain, *Canton S*. Accession, Y00274 (DROHBG). Binding sites for DA/LSC (*da*, *lsc* products heterodimer), BCD, HB and KR are indicated by underlining the short sequences that match the consensus (or its complement) in each binding site. The presumptive Zn-binding Cys and His residues are underlined.

4. HB-binding sites exist in the *knirps* (*kni*) regulatory region (Pankratz et al. 1992) where HB may act as a repressor at intermediate concentrations, thus positioning the anterior border of *kni* expression more posteriorly than the anterior border of *Kr* expression (Hülskamp et al. 1990).

5. HB binds to a *bithorax* region enhancer (BRE) and thereby represses the expression of *Ultrabithorax* in the anterior half of the embryo (Qian et al. 1991).

Tissue Distribution

HB is a nuclear protein localized initially in the anterior half of the embryo. It does not appear until the *hb* RNA antero-posterior gradient is apparent (see below), and thereafter it follows the general distribution of this RNA (Tautz 1988). After gastrulation, HB is detectable in four longitudinal rows of cells (6–8 cells per row per segment) that correspond to the first wave of differentiating neuroblasts (Cabrera and Alonso 1991).

Mutant Phenotype

This gene belongs to the gap class of segmentation genes. In amorphic *hb* embryos, gnathal and thoracic segments are absent, and there are abnormalities in abdominal segments 7 and 8; it is an embryonic lethal (Nüsslein-Volhard and Wieschaus 1980; Ingham 1988).

Gene Organization and Expression

Open reading frame, 758 amino acids. There are three mRNAs: the two transcribed from a proximal promoter have an expected size of 2,996 and 3,000 bases, and the third, transcribed from a distal promoter, has an expected size of 3,348 bases. These expected sizes are consistent with the two RNAs of approximately 2.9 kb and 3.2 kb detectable by northern analysis. Of the three transcription initiation sites, the most upstream was deduced from Southern analysis and sequence features while the other two were defined by S1 mapping and sequence features (*hb* Sequence and Fig. 16.1) (Tautz et al. 1987).

The two proximal initiation sites are under the control of a single promoter included within the leader intron of the distal transcription unit that extends between −3,170 and −18. The proximal transcripts have leader introns that extend between −300 and −18. There are no introns in the coding region (Tautz et al. 1987).

The 3′ end was deduced from Southern analysis and sequence features. All transcripts have the same protein-coding capacity (Tautz et al. 1987).

The proximal breakpoint of the deficiency *Df(3R)p-XT104* is within the transcribed region and transcription is toward the centromere (Tautz et al. 1987).

Developmental Pattern

Overall, expression of *hb* is restricted to oogenesis and the first 8 h of embryonic development.

The distal promoter is first expressed during oogenesis, and the mRNA persists after fertilization. The 3.2 kb maternal RNA is uniformly distributed in newly laid eggs. Between the 8th and 11th rounds of embryonic nuclear divisions (Appendix, Fig. A.1), an anterior–posterior gradient develops, probably by differential degradation, and under the control of *oskar* (Tautz et al. 1987; Tautz 1988).

The first embryonic expression of *hb* is from the proximal promoter, and it starts at the 11th or 12th nuclear divisions under the control of the *bicoid* gene (*bcd*) product (BCD). A combination of threshold effect and BCD gradient leads to uniform transcription of the 2.9 kb RNA in the anterior 45% of the embryo, with a sharp posterior boundary. Initiation of transcription by the proximal *hb* promoter is one of the earliest transcriptional events in embryogenesis. After cycle 14, with the beginning of gastrulation, the 2.9 kb RNA

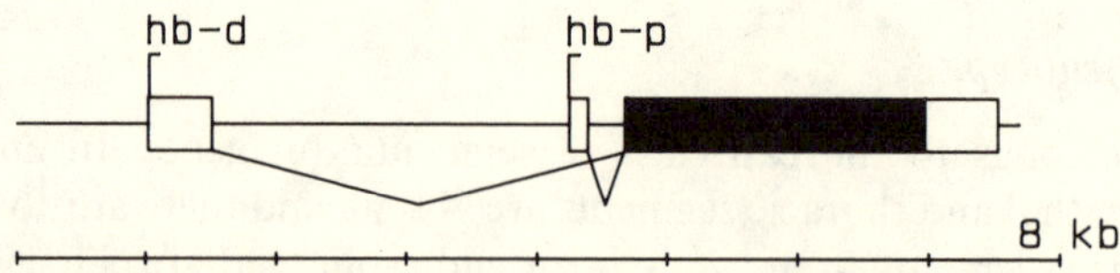

FIG. 16.1. Gene organization

disappears (Driever and Nüsslein-Volhard 1988, 1989; Schröder et al. 1988; Struhl et al. 1989).

Beginning at cycles 13–14 the 3.2 kb RNA is transcribed in a band at approximately 53% egg length (Appendix, Figs A.2 and A.3) and in a region of the embryo that corresponds to abdominal segments 7 and 8. During gastrulation, the spatial distribution of the 3.2 kb RNA increases in complexity; and after germ band extension, it becomes undetectable (Tautz et al. 1987; Schröder et al. 1988).

Promoter

A 1.5-kb segment of DNA upstream of the distal transcription initiation site is insufficient for correct expression of the 3.2 kb transcript. On the other hand, a considerably smaller segment, one that extends between 50 and 300 bp upstream of the proximal site of transcription initiation is sufficient for correct developmental expression of the 2.9 kb transcript (Schröder et al. 1988; Driever and Nüsslein-Volhard 1989). The active core of the proximal promoter is a 100 bp segment that extends between −540 and −640 bp (Struhl et al. 1989), although some binding sites for the homeodomain protein BCD, as well as for the finger proteins HB and KR, are found further upstream in the proximal promoter region (*hb* Sequence).

The consensus sequence of the BCD-binding sites is TCTAATCCC. While the central TAAT seems to be the most conserved element (Driever and Nüsslein-Volhard 1989), the terminal CCC is important for discrimination between the BCD and Antennapedia homeodomains (Hanes and Brent 1991). Transcription from the proximal promoter in the posterior half of the embryo is repressed by KR, for which there are two binding sites in this promoter (Treisman and Desplan 1989; Licht et al. 1990). The existence of numerous binding sites for *hb* product (hb1–hb8) seems to indicate that this gene is also autoregulated (Stanojevic et al. 1989; Treisman and Desplan 1989).

At −847 there is a binding site for heterodimers of the helix-loop-helix proteins DA (*daughterless*) and products of the *achaete-scute* complex. This site may be responsible for activation of *hb* in neuroblasts, an activation that requires the *lethal of scute* product (Cabrera and Alonso 1991).

References

Cabrera, C. V. and Alonso, M. C. (1991). Transcriptional activation by heterodimers of the *achaete-scute* and *daughterless* gene products of *Drosophila*. *EMBO J.* **10**:2965–2973.

Driever, W. and Nüsslein-Volhard, C. (1988). The *bicoid* protein determines position in the *Drosophila* embryo in a concentration-dependent manner. *Cell* **54**:95–104.

Driever, W. and Nüsslein-Volhard, C. (1989). The *bicoid* protein is a positive regulator of *hunchback* transcription in the early *Drosophila* embryo. *Nature* **337**:138–143.

Evans, R. M. and Hollenberg, S. M. (1988). Zinc Fingers: Gilt by association. *Cell* **52**:1–3.

Hanes, S. D. and Brent, R. (1991). A genetic model for interaction of the homeodomain recognition helix with DNA. *Science* **251**:426–430.

Harrison, S. C. (1991). A structural taxonomy of DNA-binding domains. *Nature* **353**:715–719.

Hoch, M., Seifert, E. and Jäckle, H. (1991). Gene expression mediated by *cis*-acting sequences of the *Krüppel* gene in response to the *Drosophila* morphogens *bicoid* and *hunchback*. *EMBO J.* **10**:2267–2278.

Hülskamp, M., Pfeifle, C. and Tautz, D. (1990). A morphogenetic gradient of *hunchback* protein organizes the expression of the gap genes *Krüppel* and *knirps* in the early *Drosophila* embryo. *Nature* **346**:577–580.

Ingham, P. W. (1988). The molecular genetics of embryo pattern formation in *Drosophila. Nature* **335**:25–34.

Licht, J. D., Grossel, M. J., Figge, J. and Hansen, U. M. (1990). *Drosophila Krüppel* protein is a transcriptional repressor. *Nature* **346**:76–79.

Nüsslein-Volhard, C. and Wieschaus, E. (1980). Mutations affecting segment number and polarity in *Drosophila. Nature* **287**:795–801.

Pankratz, M. J., Busch, M., Hoch, M., Seifert, E. and Jäckle, H. (1992). Spatial control of the gap gene *knirps* in the *Drosophila* embryo by posterior morphogen system. *Science* **255**:986–989.

Qian, S., Capovilla, M. and Pirrotta, V. (1991). The *bx* region enhancer, a distant *cis*-control element of the *Drosophila Ubx* gene and its regulation by *hunchback* and other segmentation genes. *EMBO J.* **10**:1415–1425.

Schröder, C., Tautz, D., Seifert, E. and Jäckle, H. (1988). Differential regulation of the two transcripts from the *Drosophila* gap segmentation gene hunchback. *EMBO J.* **7**:2882–2887.

Small, S., Kraut, R., Warrior, R. and Levine, M. (1991). Transcriptional regulation of a pair-rule stripe in *Drosophila. Genes Dev.* **5**:827–839.

Stanojevic, D., Hoey, T. and Levine, M. (1989). Sequence-specific DNA-binding activities of the gap proteins encoded by *hunchback* and *Krüppel* in *Drosophila. Nature* **341**:331–335.

Stanojevic, D., Small, S. and Levine, M. (1991). Regulation of a segmentation stripe by overlapping activators and repressors in the *Drosophila* embryo. *Science* **254**:1385–1387.

Struhl, G., Struhl, K. and Macdonald, P. M. (1989). The gradient morphogen bicoid is a concentration-dependent transcriptional activator. *Cell* **57**:1259–1273.

Tautz, D. (1988). Regulation of the *Drosophila* segmentation gene hunchback by two maternal morphogenetic centers. *Nature* **332**:281–284.

Tautz, D., Lehmann, R., Schnürch, H., Schuh, R., Seifert, E., Kienlin, A., Jones, K. and Jäckle, H. (1987). Finger protein of novel structure encoded by hunchback, a second member of the gap class of *Drosophila* segmentation genes. *Nature* **327**:383–389.

Treisman, J. and Desplan, C. (1989). The products of the *Drosophila* gap genes *hunchback* and *Krüppel* bind to the *hunchback* promoters. *Nature* **341**:335–337.

17

The Heat-shock Gene Cluster at 67B: *Hsp22, Hsp23, Hsp26, Hsp27, HspG1, HspG2, HspG3*

Chromosomal Location: **Map Position:**
2L, 67B 2-[28]
Synonyms for *HspG1*, *HspG2* and *HspG3*: *Gene1*, *Gene2* and *Gene3*

Products

Small heat-shock proteins (HSPs): proteins of 22, 23, 26, and 27 kD, and three other small heat-inducible proteins.

Structure

The small HSPs of *Drosophila* are thought to be homologous to those of many other species, from bacteria to mammals and higher plants. Although diverse in sequence, they all share the following features: (1) heat-inducibility; (2) some structural characteristics; and (3) the ability to form polymeric aggregates. In some species, *Drosophila* included, these proteins are phosphorylated and associated with RNA (see Lindquist and Craig (1988) for a review).

The polypeptides encoded by six of the genes in this cluster (*HspG2* is the exception) have two regions of similarities: (1) the 15 N-terminal amino acids, a hydrophobic segment with some resemblance to signal peptides; and (2) a segment of approximately 108 amino acids near the C-terminus with sequence similarities that range between 45% and 75%. In the latter segment, the first 83-amino-acid stretch matches approximately 50% of the mammalian α-crystallin B2 chain; in *HspG3*, the crystallin-like region is only 50 amino acids long (Fig. 17.1) (Ayme and Tissières 1985; Ingolia and Craig 1982; Southgate et al. 1983; Pauli and Tonka 1987). The sequence similarities exhibited by these six genes, their uniform lack of introns, and their clustering, suggest that they are evolutionarily related to one another.

```
       1                                                  50                                                 100
Hsp22  MRSLPMFWRM AEEMARMPRL SSPFHAFFHE PPVWSVALPR NWQHIARWQE QELAPPATVN .......... .......... ......KDGY KLTLDVKDY.
Hsp23  MANIPLLLSL ADDLGRMSMV PFYEPYYCQR QRNPYLALVG PMEQQLRQLE KQVGASSGSS GAVSKIG... .......... ......KDGF QVCMDVSHFK
Hsp26  MSLSTLLSLV DELQEPRSPI YELGLGLHPH SRYVLPLGTQ QRRSINGCPC ASPICPSSPA GQVLALRREM ANRNDIHWPA TAHVG.KDGF QVCMDVAQFK
Hsp27  MSIIPLLHLA RELDHDYRTD WGHLLEDDFG FGVHAHDLFH PRRLLLPNTL GLGRRRYSPY ERSHGHHNQM SRRASGGPNA LLPAVGKDGF QVCMDVSQFK
HspG1  MSLIPFILDL AEELHDFNRS LAMDIDDSAG FGLYPLEATS QLPQLSRGVG AWECNDVGAH QGSVGGHRSI AIIRTIVWPE PRLLAAISRW WSWKRNWAIR
HspG3  MPDIPFVLNL DSPDSMYYGH DMFPNRMYRR LHSRQHHDLD LHTLGLIARM GAHAHHLVAN KRNGELAALS RGGASNKQGN FEVHLDVGLF QPGELTVKLV
CON    MS-IPLLL-L AE-------- ---------- -------L-- -----LR--- ---------- ---------- ---------- ------KDGF QVCMDV--FK

       101                                                150                                                200
Hsp22  .SELKVKVLD ESVVLVEAKS EQQEAEQGGY SSRHFLGRYV LPDGYEADKV SSSLSDDGVL TISVPNPPGV QET....... .......... ..........
Hsp23  PSELVVKVQD NSV.LVEG.N HEEREDDHGF ITRHFVRRYA LPPGYEADKV ASTLSSDGVL TIKVPKPPAI EDK....... .......... ..........
Hsp26  PSELNVKVVD DSI.LVEGK. HEERQDDHGH IMRHFVRRYK VPDGYKAEQV VSQLSSDGVL TVSIPKPQAV EDK....... .......... ..........
Hsp27  PNELTVKVVD NTVV.VEGK. HEEREDGHGM IQRHFVRKYT LPKGFDPNEV VSTVSSDGVL TLKAPPPPSK EQA....... .......... ..........
HspG1  ARPGQAARPV ANGASKSAYS VVNRNGFQVS MNVKQFAANE LTVKTIDNCI VVEGQHDEKE DGHGVISRHF IRKYILPKGY DPNEVHSTLS SDGILTVKAP
HspG3  NECIVVEG.. .........K HEEREDDHGH VSRHFVPAVS AAQGVRFGCH CFHFVGGWSS QYHGSTISFQ GGAQGAHHTH *......... ..........
CON    PSEL-VKV-D -SV-LVEGK- HEER-DDHG- I-RHFVRRY- LP-GY-A--V VS-LSSDGVL T---P-PP-- E-K------- ---------- ----------

       201                                                250
Hsp22  .......... .LKEREVTIE QTGEPAKKSA EEPKDKTASQ *......... ....
Hsp23  .......... .GNERIVQIQ QVGPAHLNVK ENPKEAVEQD NGNDK*.... ....
Hsp26  .......... .SKERIIQIQ QVGPAHLNVK ANESEVKGKE NGAPNGKDK* ....
Hsp27  .......... .KSERIVQIQ QTGPAHLSVK APAPEAGDGK AENGSGEKME TSK*
HspG1  QPLPVVKGSL ERQERIVDIQ QISQQQKDKD AHRQSRQR*. .......... ....
HspG3  .......... .......... .......... .......... .......... ....
CON    ---------- ---ERIVQIQ Q-GPAHL-VK A---E----- ---------- ----
```

FIG. 17.1. Comparison of six of the sequences in the 67B cluster. A residue is indicated in the CON(sensus) if three or more polypeptides agree in that position.

Function

The specific function of small HSPs is unknown, but they seem to protect cells from heat damage. An extensive mutagenesis screen focused on the 67A–D region failed to uncover mutations in any of the small HSP genes. This failure and the sequence similarities among the genes in the cluster suggest functional equivalency and redundancy (Leicht and Bonner 1988).

HSP27 is localized in nuclei (Beaulieu et al. 1989). During development, the level of this protein parallels the transcription profile of *Hsp27* (Arrigo and Pauli 1988).

Organization and Expression of the Cluster

The seven heat-inducible genes are clustered within 13–14 kb (Fig. 17.2).

In the absence of heat shock, all seven genes are expressed late in the third larval instar and during early pupation under the control of β-ecdysone (Thomas and Lengyel 1986). The level of expression is not uniform for the various genes: *Hsp23* is the most active gene; *Hsp26*, *Hsp27*, *HspG1* and *HspG3* are intermediate in activity and *Hsp22* and *HspG2* are the least active (Sirotkin and Davidson 1982; Mason et al. 1984; Ayme and Tissieres 1985). The seven genes are also expressed individually at other times in development.

All seven genes respond to heat shock (optimal temperature 35–36°C) at every stage of development except for early embryogenesis (Zimmerman et al. 1983), with *HspG2* response being lower than that of the others.

Transcriptional response to heat shock depends on the presence of at least two copies of a short, nearly palindromic sequence known as the heat-shock element, hse: CTNGAANNTTCNAG (Pelham 1985). In different genes, the position of the hse's varies considerably; their effect is independent of position so long as they lie within several hundred bp of the TATA box (see *Hsp70*).

Hsp22

Gene Organization and Expression

Open reading frame, 174 amino acids; expected mRNA length, 957 bases. Primer extension and S1 mapping were used to define the 5′ end. S1 mapping was used to define the 3′ end. There are no introns (*Hsp22* Sequence)

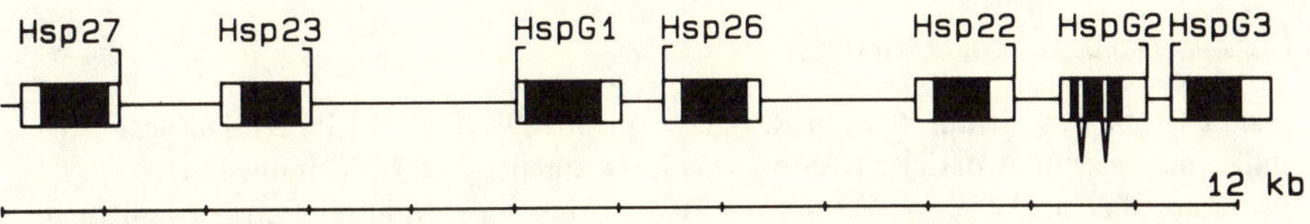

FIG. 17.2. Cluster organization

Hsp22

```
         .         .         .         .         .         .         .         .         .
-764  GAATAAATGAAGATTTTAATATTAATAGCTAAAAAAAAACAGAAAACTTAAATTATTTGTTAATATTAAGCTGTATTTTTCATATATCTC   -6
      ------                                                                |(A)n Gene2
         .         .     ______HindIII .         .         .         .         .         .
-674  AAGTTCTAGACTGCCCATGCAAGCTTATCAATACACACACGTATACACTCGCACTCAGAAAGCTGTGCACTCCCACAAAACTCTCTCTCC   -5
         .         .         .         .         .         .         .         .         .
-584  CACTCTCTAATCGAGCTCTCTCAATGTGTCTCTCTGCGTATGGAAACTGACCTTCCCCAAGGCGCAACAGCGAGAGAGAACTTCGCTAAA   -49
         .         .         .         .         .         .         .         .         .
-494  TGCTAAAATAAAAGGTAAATAAAGTAATATTTGGACACCCAGAGAGCCCCAGAAACTTCCACGGAGTTCGCTAAAGAACAGTGAACAACC   -4
                                                                    -  ---   --- - hse3

         .         .         .         .         .         .         .         .         .
-404  CCTAACTAAATGCCATTGCCCGATTTCAGGCAAAGCGGAAAATTGCATCAGCAAAGGGCGAAGAAAATTCGAGAGAGTGCCGGTATTTTC   -31
                                                                              ---  --- -- hse2 -  - -  ---
         .         .         .         .         .         .       -->-250   .         .
-314  TAGATTATATGGATTTCCTCTCTGTCAAGAGTATAAATAGCCACCGGTTGGACACTACGCTCTCAGTTCAAAAAAACCAAACCAACTGCT   -22
      --- hse1                      ------

         .         .         .         .         .         .         .         .         .
-224  AACAACTCGAAGAAAGTCAACTAAATTAAAATTTCGCCAGCTAAATAGAAATTTCATACGATTGAAACCTCAGACAACAAGATTATCTTC   -13
         .         .         .         .         .         .         .         .         .
-134  GAAACATAGAGGAAAAATTTAAAAAAAAAGCCAAGAAGTATTTCAAAGATAACAATTGGACGGAATTTCATCAAATTATTCGAATTTGCA   -45
         .         .         .         .         .         .         .         .         .
 -44  TAAGAAGCTTTATTTGGAAAAACCCAAGTTACCTTATCAACTACAATGCGTTCCTTACCGATGTTTTGGCGGATGGCCGAGGAGATGGCA   45
                                                   MetArgSerLeuProMetPheTrpArgMetAlaGluGluMetAla   (15)

         .         .         .         .         .         .         .         .         .
  46  CGGATGCCACGCCTCTCCTCGCCCTTTCACGCCTTCTTCCACGAGCCGCCCGTTTGGAGTGTGGCGCTACCGAGGAACTGGCAGCATATT   135
      ArgMetProArgLeuSerSerProPheHisAlaPhePheHisGluProProValTrpSerValAlaLeuProArgAsnTrpGlnHisIle   (45)
         .         .         .         .         .         .         .         .         .
 136  GCCCGCTGGCAGGAGCAGGAGTTGGCTCCGCCGGCCACCGTCAACAAGGATGGCTACAAACTCACCCTGGACGTCAAGGACTACAGCGAG   225
      AlaArgTrpGlnGluGlnGluLeuAlaProProAlaThrValAsnLysAspGlyTyrLysLeuThrLeuAspValLysAspTyrSerGlu   (75)
         .         .         .         .         .         .         .         .         .
 226  CTGAAGGTCAAGGTGCTGGACGAGAGCGTGGTCCTGGTGGAGGCAAAATCGGAGCAGCAGGAGGCCGAACAAGGTGGCTATAGTTCCAGG   315
      LeuLysValLysValLeuAspGluSerValValLeuValGluAlaLysSerGluGlnGlnGluAlaGluGlnGlyGlyTyrSerSerArg   (105
         .         .         .         .         .         .         .         .         .
 316  CACTTCCTCGGCCGATACGTTCTGCCGGATGGATACGAGGCGGACAAGGTGTCCTCGTCGCTGAGCGACGACGGCGTTCTGACCATCAGT   405
      HisPheLeuGlyArgTyrValLeuProAspGlyTyrGluAlaAspLysValSerSerSerLeuSerAspAspGlyValLeuThrIleSer   (135
         .         .         .         .         .         .         .         .         .
 406  GTGCCCAATCCTCCAGGCGTGCAGGAGACACTCAAGGAGCGTGAGGTGACCATCGAGCAGACTGGCGAGCCGGCAAAGAAGTCCGCCGAG   495
      ValProAsnProProGlyValGlnGluThrLeuLysGluArgGluValThrIleGluGlnThrGlyGluProAlaLysLysSerAlaGlu   (165
         .         .         .         .         .         .         .         .         .
 496  GAGCCAAAAGACAAAACCGCCAGTCAGTAGAAATAAGTTGAGATTATACTAAAACCGATAAAATGCTAGTGAACTCCTATGTTTAGATAT   585
      GluProLysAspLysThrAlaSerGlnEnd                                                              (174
         .         .         .         .         .         .         .         .         .
 586  TCCAAAACCTATCAAATTTAAGTTCTTGTTAAATTAACAAGTTAATTTTAAAACAATTGTGATTCGGTAGCCCGCAAGCCCAATAATTTT   675
         .         .         .         .         .         .         .         .         .
 676  ATTTAGAAGAAAATAAATATTTGAAAAGACTATGATCAAAATATTTACTTTNATTGGTTGGGTTGGGAACACATTTGATATGGATAGTAT   765
                ------                  |(A)n
         .         .         .
 766  TATAGATTATTATATATATCTGTCAAGTCT  795
```

Hsp22 SEQUENCE. Strain, *Oregon R*. Accession, J01098 (DROHSP671). Dashes underline bases that match the consensus hse sequence. *HspG2* is immediately upstream of *Hsp22*: its poly(A) signal (−763) and last poly(A) site (−702) are indicated.

(Holmgren et al. 1981; Ingolia and Craig 1981, 1982; Southgate et al. 1983).

Developmental Pattern

Hsp22 is expressed in the third larval instar and, at barely detectable levels, in early pupae (Mason et al. 1984).

Promoter

A 209-bp segment upstream of the transcription initiation site (to position −458) includes three hse's, and is necessary for full developmental and heat-inducible expression, as was demonstrated by study of 5′ deletions (Klemenz and Gehring 1986). These studies also suggest that the segment between −443 and −383 is involved with hormonal induction. The first 26 bp of the leader seem to be important for transcription and for the preferential translation of *Hsp22* mRNA at high temperature (Hultmark et al. 1986).

Hsp23

Gene Organization and Expression

Open reading frame, 186 amino acids; expected mRNA length, 874 bases. Primer extension and S1 mapping were used to define the 5′ end. S1 mapping was used to define the 3′ end. There are no introns (*Hsp23* Sequence) (Holmgren et al. 1981; Ingolia and Craig 1981, 1982; Southgate et al. 1983).

Developmental Pattern

Hsp23 is expressed in late third instar larvae as well as in early pupae, when it is the *Hsp* gene that is most abundantly transcribed. *Hsp23* transcript reappears transiently in newly eclosed adults (Mason et al. 1984; Ayme and Tissières 1985).

Promoter

Deletion analysis of the promoter region suggests that heat inducibility is controlled by a segment of the promoter region between −260 and −729. This segment includes five of the six hse's that occur within the promoter (Pauli et al. 1986). A segment between −250 and −490 is responsible for ecdysterone induction (Mestril et al. 1986).

Hsp23

```
-613  TTTCCCCACTACAGAGCCCCATTCTTGGATATTAATTAAAGTTAATAGCTTAAATGCCAGGCCATAAAAAGAAGAACTGTTCTGCTGTCT  -52
                                                                                              --

-523  CGAAGTTTCGCGAATTTACTCCATCCTTCGTGGAATATACTCCAACCTTCCTATCTGCTATGTATGTACATACATACGTGCTTACATACG  -4
       ---  ---  - hse6 --  -    ---  - hse5  --  --  --- hse4

-433  TACATCTATACATACACATAATATTTGCCGGTGCTGATGCGACTTATCACTCCACCAGGCCTTTTCATTCCCACTCCCCTAGGAGATTGC  -34

-343  TCATTTTCCATAGCGATACTCTCACTTTCAATGGCAGATAATGCGTAATTGCGGCAAATTCGAGAACTCTGCGATATTTTCAGCCCGAGA  -25
                        --   -  --- - hse3                      - -  --- -- hse2                - --

-253  AGTTTCGTGTCCCTTCTCGATGTCGATGTTTGTGCCCCCTAGCACACAGACACGACGCGCACACACACAGCGCCGACGGGCGCCGCACAC  -16
      -  ---  - hse1
                                                          -->-111
-163  TTCGACAGCAAGCGGTTGTATAAATATCCGGCACTTTCGTGCAACCGGCGTCAGTTGAATTCAAAAAGCCAAAGCGATAACAGCTAAAGC  -74
                        ------

 -73  GAAAGTAACCTATTAACAAAAGAAGTTTATTCTTTGAAGGAGGAGAATCATCTTGAAGCAATTAAAAAACAAAAATGGCAAATATTCCAT  16
                                                                                MetAlaAsnIleProL  (6)

  17  TGTTGTTGAGCCTTGCCGACGATTTGGGCCGAATGTCGATGGTGCCCTTCTATGAGCCCTACTACTGCCAGCGCCAGAGGAATCCCTACT  106
      euLeuLeuSerLeuAlaAspAspLeuGlyArgMetSerMetValProPheTyrGluProTyrTyrCysGlnArgGlnArgAsnProTyrL  (36)

 107  TGGCCCTGGTTGGACCGATGGAGCAGCAGCTGCGCCAGCTGGAGAAACAGGTGGGCGCCTCGTCGGGATCGTCGGGAGCCGTGTCGAAAA  196
      euAlaLeuValGlyProMetGluGlnGlnLeuArgGlnLeuGluLysGlnValGlyAlaSerSerGlySerSerGlyAlaValSerLysI  (66)

 197  TCGGAAAGGATGGCTTCCAGGTCTGCATGGATGTGTCGCACTTCAAGCCCAGCGAACTGGTGGTCAAAGTGCAGGACAACTCCGTCCTGG  286
      leGlyLysAspGlyPheGlnValCysMetAspValSerHisPheLysProSerGluLeuValValLysValGlnAspAsnSerValLeuV  (96)

 287  TGGAGGGCAACCATGAGGAGCGCGAAGATGACCATGGCTTCATCACTCGTCACTTTGTCCGCCGCTATGCTCTGCCACCCGGTTATGAGG  376
      alGluGlyAsnHisGluGluArgGluAspAspHisGlyPheIleThrArgHisPheValArgArgTyrAlaLeuProProGlyTyrGluA  (126

 377  CTGATAAGGTGGCCTCCACCTTGTCCTCCGATGGTGTCCTGACCATCAAGGTGCCCAAGCCACCGGCAATCGAGGATAAGGGCAACGAGC  466
      laAspLysValAlaSerThrLeuSerSerAspGlyValLeuThrIleLysValProLysProProAlaIleGluAspLysGlyAsnGluA  (156

 467  GCATCGTTCAGATCCAGCAGGTGGGACCCGCCCATCTCAATGTGAAGGAGAATCCCAAGGAGGCGGTGGAGCAGGACAATGGCAACGATA  556
      rgIleValGlnIleGlnGlnValGlyProAlaHisLeuAsnValLysGluAsnProLysGluAlaValGluGlnAspAsnGlyAsnAspL  (186

 557  AGTAGAGGACTCGTTCCGGGAGATGCCCTGCATTATTTAACCATTATCAAAGTCATACATCTGTTTTATAAGCTGTAGTTATCCAAGGAC  646
      ysEnd

 647  ACTTCACTCATACACAATAGCCATTAAGGGTGTCCTGCTTTAATCTTAGTTTGGAATATGTATTACTAAATTGGCGAAATTAATATTACC  736

 737  CATAAAAATAAATAACAAGTACACTTACTTATAATTGTGTTTGGTCTGTTTTTCTGGTTGGTTATGGGTTACTATTACTATTACTATTAC  826
            ------                  |(A)n

 827  TTCGGGAATTGTTTGGGTAGCTCGGCCCTTTTTCCTGTGATCCCGGTTCTAGATTTACTTTCTGCATTGTATATTGCATTGTTGTGTCAC  916

 917  GTAAAATGGCATTTTTATTTAATTGTTGTTTGTGTACATAACTGACTTTTTACATTACTTCGGTAAAGAGTCTTGAAGCTATGAATGTAA  10

1007  GGAACTCCAGTCAAGGTTAAATCCTTATGTAAAGCATGCATAAGTACATCATGTACATACATACGTACATAAAAATATACATCCCTTTTC  109
```

Hsp26

Gene Organization and Expression

Open reading frame, 208 amino acids; expected mRNA length, 949 bases. Primer extension and S1 mapping were used to define the 5′ end. S1 mapping was used to define the 3′ end. There are no introns (*Hsp26* Sequence) (Holmgren et al. 1981; Ingolia and Craig 1981, 1982; Southgate et al. 1983).

Developmental Pattern

In addition to being expressed in late third instar and early pupae, this gene is active in ovarian nurse cells in egg chambers at stages 7–10; the transcripts are transferred to the oocyte where they persist until the blastoderm stage of embryogenesis (Zimmerman et al. 1983; Mason et al. 1984). *Hsp26* promoter expression in several other tissues, including spermatocytes was detected using *lacZ* as a reporter gene (Glaser et al. 1986).

Promoter

The effects of partial promoter deletions on *Hsp26* gene expression, as well as the localization of DNA-binding proteins suggest that hse1–2 and hse6 (*Hsp26* Sequence) are the *cis*-acting sequences responsible for heat-inducible expression of *Hsp26* (Cohen and Meselson 1985; Pauli et al. 1986; Simon and Lis 1987; Thomas and Elgin 1988). Nuclease protection studies identified (1) a constitutive footprint overlapping the TATA box and a fixed-position nucleosome between hse1–2 and hse6 (Thomas and Elgin 1988) and (2) a footprint produced by the GAGA-binding factor that extends from −312 to −264 (Gilmour et al. 1989).

Further upstream, from −704 to −534 there occurs a *cis*-acting region necessary for ovarian expression. All of the necessary information for ovarian, larval, pupal and heat shock expression is contained within the segment −910 to −169 (Cohen and Meselson 1985). Within that segment, two copies of the ovary-specific regulatory sequence (−704 to −534) are required to stimulate transcription of a basal-promoter/reporter-gene. Stimulation is very specific to the nurse cells and oocytes in egg chambers from stage 6 onwards. Footprinting experiments with ovarian nuclear proteins identified two binding sites in this 171-bp fragment (onf1a and onf2a in the *Hsp26* Sequence). Integrity of these sites is required for maintenance of the regulatory activity of the 171-bp

---

Hsp23 Sequence (*opposite*). From −613 to 995: strain, *Oregon R*. Accession, J01100 (DROHSP673) with additions from Pauli et al. (1986) (see also V00210, DROHS09). From 996 to 1461, strain, *Canton S*, Hoffman and Corces (1986). Dashes underline bases that match the consensus hse sequence.

Hsp26

```
       ______PstI  .         .         .         .         .         .         .         .
-942   TGCAGCAAAACCGAGGAACTGGCCAAGTGAAGTCGAACTAAAAGAAAGAACATAAATAGTAATTAAGACAAAATAATAATCTGCACGGGT   -853
                                        |(A)n(minor) Gene1

         .         .         .         .         .         .         .         .         .
-852   AGGCGTGCGTTTTATTCCATACGTGTTTCTTGTGGTTTTCTTTTCTTGCATTTCACACAAAAAAAAAGAAGCGAGAAAGCTGACGGGAAA   -763
                                                             -------------onf1b

         .         .         .         .         .         .         .         .         .
-762   AGCACTCAATTACTAATAGTGGGAGATTGCGGGCGTTATATGTATGTATGATTTCCTAAAAACATATGTGACAACAACTACAAGTATTCC   -673
                                                                            --------------onf2a

         .         .         .         .         .         .         .         .         .
-672   CAAAGTAAAACTTAAAGACAGAAACACGAAATAATGTACTTAATAAAGAGGAAAACCAGAATAAAAAAAACTGACGTTTTGTTTTGTTGC   -583
                                                              ======                 |(A)n(maj) Gene
                                                             --------------onf1a

         .         .         .         .         .         .         .         .         .
-582   CGTTAGCCGGCTGTTTCTTTTGCGCTCTTTCTAGAAAATTGCAACAACTCTCTAGAAACTTCGGCTCTCTCACTCATACAGGCGCACTAG   -493
                                     -- ---  --  - hse7   -- ---  --- hse6
                                                     |-         f6           -|

         .         .         .         .         .         .         .         .         .
-492   CTCTGCTTTTGCGCGTACGACAACAACTACTTTAAAATTTCTCGAAACTCATGGCATTTATTGGGAAAGGTTAGTTAGTTTTATTTTTTG   -403
                                      -  --  ---- ---  -   - hse4-5
                         --------------onf2b

         .         .         .         .         .         .         .         .         .
-402   TTTTTAGAGCAGCATTCAATTTAGACTTTTATAAAAGAAATTTCTAATTTGATCCCTCGTTTATCAAACGATACAAAGCTATATTCATAA   -313
                                           ---  --- hse3

         .         .         .         .         .         .         .         .         .
-312   TTTTTTCTCTCTGTGCACGTTCTCTCTCTTCTCTTCTCTCTCTCTACTCTTTCCTTTTTCTGTCACTTTCCGGACTCTTCTAGAAAAGCT   -223
                                                                  --   -  ---  --   ---  --- hse1-2
       |-               GAGA   protein                 -|    |-              f1-2              -|

         .         .         .         .      -->-183      .         .         .         .
-222   CCAGCGGGTATAAAAGCAGCGTCGCTTGACGAACAGAGCACAGATCGAATTCAAAAATCGAGCAGTGAACAACTCAAAGCAACTTTGCGC   -133
               ======
                 |- fT -|

         .         .         .         .         .         .         .         .         .
-132   AAAAGCAAAACTTCAAACGAGAAAAAAAAAGGATTAAAAACCTTTGCTTACAAGTCAAACAAGTTCATTCAACTTAACCAAAGAAAAAATA   -43

         .         .         .         .         .         .         .         .         .
 -42   TTTCAATCTCGCAAAAGGAACATAACCTAAAGGAAACGTAAAAATGTCGCTATCTACTCTGCTTTCGCTTGTGGATGAACTCCAGGAGCC   47
                                                  MetSerLeuSerThrLeuLeuSerLeuValAspGluLeuGlnGluPr   (16)

         .         .         .         .         .         .         .         .         .
  48   CCGCAGCCCCATCTACGAGCTTGGACTGGGATTGCATCCGCATTCCCGCTACGTGCTGCCCCTTGGCACTCAGCAGCGCCGTTCCATCAA   137
       oArgSerProIleTyrGluLeuGlyLeuGlyLeuHisProHisSerArgTyrValLeuProLeuGlyThrGlnGlnArgArgSerIleAs   (46)

         .         .         .         .         .         .         .         .         .
 138   CGGATGCCCTTGCGCATCGCCGATATGCCCATCGTCGCCCGCCGGCCAGGTTTTGGCTTTGCGGCGCGAGATGGCCAACCGCAACGACAT   227
       nGlyCysProCysAlaSerProIleCysProSerSerProAlaGlyGlnValLeuAlaLeuArgArgGluMetAlaAsnArgAsnAspIl   (76)

         .         .         .         .         .         .         .         .         .
 228   TCACTGGCCGGCAACCGCCCATGTGGGCAAGGATGGATTCCAGGTGTGCATGGACGTCGCCCAGTTCAAGCCCAGTGAGCTCAACGTGAA   317
       eHisTrpProAlaThrAlaHisValGlyLysAspGlyPheGlnValCysMetAspValAlaGlnPheLysProSerGluLeuAsnValLy   (106)

         .         .         .         .         .         .         .         .         .
 318   GGTGGTGGACGACTCCATCTTGGTCGAGGGCAAGCATGAGGAACGCCAGGACGACCATGGTCACATCATGCGCCACTTTGTGCGCCGCTA   407
       sValValAspAspSerIleLeuValGluGlyLysHisGluGluArgGlnAspAspHisGlyHisIleMetArgHisPheValArgArgTy   (136)

         .         .         .         .         .         .         .         .         .
 408   CAAGGTTCCCGATGGCTACAAGGCGGAGCAAGTGGTCTCGCAGCTGTCGTCGGATGGCGTGCTCACCGTCAGTATTCCCAAGCCGCAGGC   497
       rLysValProAspGlyTyrLysAlaGluGlnValValSerGlnLeuSerSerAspGlyValLeuThrValSerIleProLysProGlnAl   (166)
```

```
 498  CGTCGAGGACAAGTCCAAGGAGCGCATCATTCAAATTCAGCAAGTGGGACCCGCTCACCTCAACGTTAAGGCAAATGAAAGCGAGGTGAA   587
      aValGluAspLysSerLysGluArgIleIleGlnIleGlnGlnValGlyProAlaHisLeuAsnValLysAlaAsnGluSerGluValLy  (196)

 588  GGGCAAGGAGAACGGAGCACCCAACGGCAAGGACAAGTAAAGGAGCCATCATCATCCAACATCATCCATCATCATTCCCCTACTTAATTG   677
      sGlyLysGluAsnGlyAlaProAsnGlyLysAspLysEnd                                                    (208)

 678  TTCCTAATTTATTGCATTGTATTTGTAATGAGCTAAAGACTAGAATACTCATATTAATTTAATAAATCCTTTTGTTCACCTGGTGTGGAA   767
                                                                ------                      |(A)n

 768  AATTAAAATTGTTGCGACTTTTGTATATGAAAGTTGGTTTTTGAAAGAGGCAAATATTTGGAAATCGATCCGAAGATTTGAATTGGGCGC   857

 858  GACGAGGTGAAGACCCATTCGTAAACACCAGTGTTTCTACCAAATATTTATTGCGCATTTATTATATCAACTACGGGTACAATTTGTATT   947

 948  TTATTTATGTTTGAATCCAATTTAAATGTTCGGCTGCAATTGCTTGGTGTCCGAAAATAGTTCACCTTGAGTTAGGCGCATTCGATGGTT  1037

1038  GGGATTTGGGTTTGGTAAACACACATTCACTGCTTGCCTTCCTGATTTCTGACACATGGTCCACTATTTCCAGGGCAGGGCCAGCTTTCC  1127

1128  GGTTTCATGAACGCGGACCAATCTCTCTCCGGGCGTGTAGTACTTGGCTGGCGGCGGTGGTGGAGCCTTGATGGTGAGGATGCCATCGCT  1217

1218  GGATATGTCCGAGATTACCTCATTGGCATTGTATCCGCGGGGCAGAAGGTACTTCCTCACAAAGTGCCGCTCCACTAGGCCATTGGAACC  1307

1308  CTCGTCGCGACGATTGTGATTTCCCTGGACGATGACATAGTCGTCATTGGTTTTGACCACAATGTCGTGGGGATGAAATTGTCGTATCGA  1397

1398  T  1398
```

Hsp26 SEQUENCE. Strain, *Oregon R*. The segment −672/1,398 is from GenBank: Accession, J01099 (DROHSP672), as modifed by Thomas and Elgin (1988) (see also X03890, DROHSP26G). The segment −942/−839 is from *HspG1* Sequence. The segment −838/−673 was kindly supplied by R. S. Cohen. Dashes underline bases that match the consensus hse sequence. fT is the footprint associated with the TATA box; f1–2 and f6 are footprints associated with hse's. The onf's are ovarian-nuclear-factor binding sites. The polyadenylation sites of *HspG1* are indicated.

fragment. Second copies of these binding sites occur at −798 (onf1b) and −474 (onf2b). The nuclear factors that bind to onf1 and onf2 are ovary-specific (Frank et al. 1992).

Hsp27

Gene Organization and Expression

Open reading frame, 213 amino acids; expected mRNA length, approximately 1 kb. Primer extension and S1 mapping were used to define the 5′ end. The 3′ end has not been defined. There are no introns (*Hsp27* Sequence) (Holmgren et al. 1981; Ingolia and Craig 1981, 1982; Southgate et al. 1983).

Hsp27

```
-698  CGGCAAACATGAGGAGCAGCACGAAGCGAGACAAGGGTTCAATGCACTTGTCCAATGAAAATACAAGCTCTGTTGCACTCTGAAAAGACT   -609

-608  GCTTTTAAAAGCGCGATAAGAGAAGAAAATGTTTTAAATAAATACATATATCTGCATATATACGTACATGTACATATGTATGTACTGCAT   -519

-518  TTTAACTGTTCGTTTTGCTTTTTATTCGCAAAGAGAAACTCCCAGAAAAGAAATGTCAAGAAGTTTCTGGTTCTTTCTCCCTCTCTCTAT   -429
                            hse5 ---  - - -    ---   --  ---  ---  - hse3-4

-428  GAAAAGCCGCTGTGCCAGAAAGAGCCAGAAGATGCGAGAGAAAACTGTTTGTTGAATTACGGGGCGTATTCAAAGGGGCTTTTAAATGTC   -339
           -  ---    -  ---  - -  - hse1-2

-338  GCTTAAATTTTAAGTTTGACAGGCTAATAATTGCTTGCCTATATCTAAATATTATTATATTTGCATTAGGGGATCATAGGGAAAACCTTC   -249

-248  TCTGCAGGCAAAATCTAACGAAGATGGCAACCCCCCATCATTTTATTAAAGTTCCGTCCCTGGTTGCCATGCACTAGTGTGTGTGAGCCC   -159

                                     -->-118
-158  AGCGTCAGTATAAAAGCCGGCGTCAACGTCGCCCGAGCACAGTCTAAACTGAAAAATTGAAGGCAAACGTTGAAGCAAACTTCGCTAAAA   -69
          -----

-68   AAATTCGAAAAAGCAAAAAAAATTCCTTTGTCTAGACAGGGTTGTGAATAAAGAGAAAAAAAATCAAAAATGTCAATTATACCACTGCTG    21
                                                                            MetSerIleIleProLeuLeu   (7)

22    CACTTGGCCCGGGAGTTGGATCATGACTACCGCACCGACTGGGGGCATTTGCTGGAGGATGACTTCGGTTTTGGCGTCCATGCCCACGAT   111
      HisLeuAlaArgGluLeuAspHisAspTyrArgThrAspTrpGlyHisLeuLeuGluAspAspPheGlyPheGlyValHisAlaHisAsp  (37)

112   CTGTTCCATCCGCGTCGCCTGCTACTGCCCAACACCCTGGGACTGGGTCGTCGTCGTTATTCGCCGTACGAGAGGAGCCATGGCCACCAC   201
      LeuPheHisProArgArgLeuLeuLeuProAsnThrLeuGlyLeuGlyArgArgArgTyrSerProTyrGluArgSerHisGlyHisHis  (67)

202   AATCAAATGTCACGTCGCGCGTCGGGGGGTCCAAACGCTCTGCTGCCCGCCGTGGGCAAAGATGGCTTCCAGGTGTGCATGGATGTGTCG   291
      AsnGlnMetSerArgArgAlaSerGlyGlyProAsnAlaLeuLeuProAlaValGlyLysAspGlyPheGlnValCysMetAspValSer  (97)

292   CAGTTCAAGCCCAACGAGCTGACCGTCAAGGTGGTGGACAACACCGTGGTGGTAGAGGGGAAGCACGAGGAGCGCGAGGACGGCCATGGA   381
      GlnPheLysProAsnGluLeuThrValLysValValAspAsnThrValValValGluGlyLysHisGluGluArgGluAspGlyHisGly  (127)

382   ATGATCCAGCGTCACTTTGTGCGCAAGTATACCCTGCCCAAGGGCTTTGACCCCAACGAGGTAGTGTCCACTGTCTCATCCGACGGTGTG   471
      MetIleGlnArgHisPheValArgLysTyrThrLeuProLysGlyPheAspProAsnGluValValSerThrValSerSerAspGlyVal  (157)

472   CTGACCCTCAAGGCCCCGCCGCCGCCCAGCAAGGAACAGGCCAAGTCGGAGCGCATTGTCCAGATCCAGCAAACGGGGCCTGCCCATTTG   561
      LeuThrLeuLysAlaProProProProSerLysGluGlnAlaLysSerGluArgIleValGlnIleGlnGlnThrGlyProAlaHisLeu  (187)

562   AGCGTCAAGGCACCGGCACCCGAGGCTGGCGATGGAAAAGCCGAAAATGGCAGCGGCGAGAAAATGGAGACTAGCAAGTAAAAGACGAAA   651
      SerValLysAlaProAlaProGluAlaGlyAspGlyLysAlaGluAsnGlySerGlyGluLysMetGluThrSerLysEnd         (213)

652   AGAGGAAGAAGACTAGGAGATGAAGAAGACGAGAAGAGGAAGAAGACTAGAAGAGGAAGAAGTCGTGAAGGAGGAAGAAGACGACGAGAT   741

742   TCGCTGGCGAAGCACGAGAGAAAGAAGAATTTAAAAGAAGAAACGGGAGTGTTGCCGCGCTGCTCGGAGAGAGCAAGACTAAAAAGGACA   831

832   CACCACAACACACCCAATGTATTACATTCACACACATCACATCATTACATCATCATCATAACATCCTAGACTAAGTGATTTTAAACTCCA   921

922   TTTATCATAATGCATAAAAAAAACAAAATTTT  953
```

Developmental Pattern

The pattern of expression during late third instar, early pupal stages and oogenesis is similar to that of *Hsp26* (Zimmerman et al. 1983; Mason et al. 1984).

Promoter

Studies of 5′ deletions established that the 579 bp upstream of the transcription initiation site (to position −696) are sufficient for full response to induction by ecdysterone (late third instar expression) or heat. The effect of the two treatments is mediated by two independent regions of the promoter: the hormonal-response segment extends from −696 to −572, and the heat-induction segment from positions −486 to −345. This latter segment includes five hse's in two clusters (Riddihough and Pelham 1986; slightly different results were reported by Hoffman and Corces 1986). The positions of the hormonal and heat-inducible regulatory regions are well correlated with DNase hypersensitive sites (Costlow and Lis 1984).

HspG1

Gene Organization and Expression

Open reading frame, 238 amino acids; from the major 5′ end (at −92), the expected mRNA lengths are 1,423 and 1,904 bp, in agreement with bands of 1.6 and 1.9 kb seen in gels (the 1.9 kb band is the stronger). Primer extension and S1 mapping were used to define the 5′ ends. S1 mapping and cDNA sequences were used to define the 3′ ends. There are no introns (*HspG1* Sequence) (Ayme and Tissières 1985; Vázquez 1991).

Developmental Pattern

HspG1 is expressed in late third instar larvae, in white pupae and in freshly eclosed adults (Ayme and Tissières 1985). Heat shock causes a weak response in embryos and adults but a 10–100 times stronger response in pupae. This developmental response seems to be hormonally controlled, because cells in culture respond much more strongly to heat shock if ecdysterone is present (Vázquez 1991).

Hsp27 SEQUENCE (*opposite*). Strain, *Oregon R.* Accession, J01101 (DROHSP674) as modified by Riddihough and Pelham (1986). This sequence follows immediately after position 1,461 in the *Hsp23* Sequence (Hoffman and Corces 1986). Dashes underline bases that match the consensus hse sequence.

HspG1

```
-1196  TTTTATTACTATGTACAAGGGGGCATCTCGTACGCAGCATGCTCTGAAGTTTTGCTCTTTCCGACTGCAGCTGGCATATACACCATATCA  -1107

-1106  ATACAAACATACTATATAATATAATATATGGCCATACAGAATTGTATCCCGCAGCTGAGTTCGGGGCCCCAGTAAATTTTTAGCAAAGTC  -1017

-1016  TCCACTGTCTGGCCTCCGTCTGGATGTTGTTGGTGTTGTTGTTGTTTTTTGCATTTTGGAGCTTTTCAACCGGTTGCCATCGCTTGCACT  -927

 -926  TGGCTATGTAACCACATACGAATCCAGCAATATCATCATCATCTGTGTGGCAGGGTACATACATATGTATGTAGATACAAATGTATATGC  -837

 -836  CGACACCATATGTATGGTTGCCCCAGACGCTGTCACTGCGCATGTTTACGCGACGCCGGTTGCCAATCCTCCAGCTCTGACAACAGCGGA  -747

 -746  TTTGTAGCTTCCAGGCGGCCTGCCAGCCAGCCAGCCAGCCAGCTGTTGTTGTAGTTGTTTATCGCCGGCGGACTCGAATTCGCATCGGCA  -657

 -656  AGCCGGCACGAGACTCAGACCTCTCAGCTGTTCGCTCAATGCCGGCAGTGGAAATTCAGCTGCAACACGGACCACTTTACATATACCCCG  -567

 -566  TCTATATGGATATTTGTATATATGAGTACATATATGTATATCGCCGGTACAAGGAAGATGGCATCTCTTGGGGGGGATATTCGTGCATAT  -477

                                                                              -->(minor).
 -476  ATGCTTCGATTTCAAGCCGGTTTGCCTCTTTTACTTATCTTTTTTTATTTGGTTTTGCAACGTTGCAGTTTGGTTTGTCTGGTTTCTCGC  -387
                                                                      -  --  --  --

 -386  CGACTACGAGTACGAGTACTTTCTTTGTTCTCTGGCTATCTGCGGTAGAGGAAAAGTATCTCTTATTTCGTGTATATAGCAGAAAATGGC  -297

 -296  ATAGTACATGGCTTGACTGACTGTTTTAATGGGTAGCCCTTCCCCTTGGCTGAGGCTTCTCTGGAGGAGTTGCATTAGTTTTTCGCCTGG  -207

                                                         -->(minor) .
 -206  GAGCTGGCCTGGAAGCCGACTGGAAGTGACCAGGTTTTCCATTCAGCGCTGCACAGCCGCTTAAAAGCGTCGACATTCAGCCATAAGGGC  -117
                  -- ---    - --                                                       =====
                      -->-92              .-->(minor)
 -116  TCAAACGCAGTCCAGTTGGAGGCCAGAACGGATCGCCGCCGGTTCCCAGACGACACCAATCCCCGCAAGACCTAAAAAATAAAAGATATA  -27

  -26  TCTTAGCCAGATAGGAAGAAAGTGAAAATGTCGCTGATACCGTTCATACTAGATTTGGCCGAGGAGCTGCACGATTTCAATCGCAGCCTG  63
                                    MetSerLeuIleProPheIleLeuAspLeuAlaGluGluLeuHisAspPheAsnArgSerLeu  (21)

   64  GCAATGGATATAGATGATTCGGCCGGATTCGGGTTGTATCCACTGGAGGCCACCTCACAGTTGCCACAGCTGAGTCGTGGCGTTGGGGCG  153
       AlaMetAspIleAspAspSerAlaGlyPheGlyLeuTyrProLeuGluAlaThrSerGlnLeuProGlnLeuSerArgGlyValGlyAla  (51)

  154  TGGGAATGCAATGATGTGGGTGCCCATCAAGGGTCAGTCGGCGGCCATCGCAGCATCGCCATCATCCGTACAATCGTGTGGCCGGAGCCA  243
       TrpGluCysAsnAspValGlyAlaHisGlnGlySerValGlyGlyHisArgSerIleAlaIleIleArgThrIleValTrpProGluPro  (81)

  244  AGACTGCTTGCTGCAATAAGTCGCTGGTGGAGCTGGAAAAGGAATTGGGCGATAAGGGCACGTCCGGGGCAAGCGGCACGACCAGTGGCC  333
       ArgLeuLeuAlaAlaIleSerArgTrpTrpSerTrpLysArgAsnTrpAlaIleArgAlaArgProGlyGlnAlaAlaArgProValAla  (111

  334  AACGGGGCCAGCAAATCCGCCTACTCCGTGGTGAATAGGAACGGCTTCCAGGTGAGCATGAATGTGAAGCAGTTCGCCGCCAACGAACTG  423
       AsnGlyAlaSerLysSerAlaTyrSerValValAsnArgAsnGlyPheGlnValSerMetAsnValLysGlnPheAlaAlaAsnGluLeu  (141

  424  ACCGTCAAGACCATCGATAACTGCATCGTGGTCGAGGGTCAGCACGACGAGAAGGAGGATGGCCACGGGGTGATCTCGCGCCACTTCATC  513
       ThrValLysThrIleAspAsnCysIleValValGluGlyGlnHisAspGluLysGluAspGlyHisGlyValIleSerArgHisPheIle  (171

  514  CGCAAGTACATCCTGCCCAAGGGCTATGATCCCAACGAGGTGCACTCGACCCTCTCCTCGGACGGCATTCTGACGGTGAAGGCGCCGCAG  603
       ArgLysTyrIleLeuProLysGlyTyrAspProAsnGluValHisSerThrLeuSerSerAspGlyIleLeuThrValLysAlaProGln  (201

  604  CCACTTCCAGTCGTCAAAGGCAGCCTGGAACGACAGGAGCGCATCGTAGACATCCAGCAGATATCGCAGCAGCAGAAGGATAAGGATGCG  693
       ProLeuProValValLysGlySerLeuGluArgGlnGluArgIleValAspIleGlnGlnIleSerGlnGlnGlnLysAspLysAspAla  (231
```

```
                .         .         .         .         .         .         .         .         .
 694  CACCGCCAAAGCCGTCAGAGGTAGAGCAGCAGGCGCACGTAGTGCCACCACTTCCACTTTAAATCCGACTGCACCCACACCACTCCTTCG   783
      HisArgGlnSerArgGlnArgEnd                                                                 (238)

                .         .         .         .         .         .         .         .         .
 784  CTCTCGCTCACTCTCGCCGAGAGCAACGGCAAGGTCAGGAAGAGACAGAGATGGAAATGCCGGCTGTTTCGCCCATTTCAATGAGGCTGC   873

                .         .         .         .         .         .         .         .         .
 874  TGCTGCCGCTGCTGTTGCGATGGAAGCCCTTCCACCGCAGGAACCACTTCCAGTGCCAACAATGGCGTTGCAGAACCAGAATCAGAGTCC   963

                .         .         .         .         .         .         .         .         .
 964  ATGGAAGTGGCGTTGGCCAAAAACGAAGAGACTGCCAATGTGGATGAACCCACACCCAATCCCGTTATAAGCTACGAAGAGGAGCAAAAG  1053

                .         .         .         .         .         .         .         .         .
1054  GCAGAGGATGCAAATGCCAACGAAGTGCCCGTTGCCTCGAATAACGGCAATGGAGCAGTCGCAGCAGCCGAGGATGTGAAATGCCGCTGG  1143

                .         .         .         .         .         .         .         .         .
1144  CCAAGAAACCGAAATCTCCACGGAAGACAGCAAAGAGGAGCAGGCGGAGAAGTTGATAAAGTAGAGAAATGGAGGAGAAGGGCGGCGAGG  1233

                                                                         _____PstI   .         .
                .         .         .         .         .         .
1234  CAACTGGCAGCCGTAGAATGCGGCCATTCTACTGGCCAAAAACCAAGGCGAAAATGGAGCCACTGCAGCAAAACCGAGGAACTGGCCAAG  1323

                .         .         .         .         .         .         .         .         .
1324  TGAAGTCGAACTAAAAGAAAGAACATAAATAGTAATTAAGACAAAATAATAATCTGCACGGGTAGGCGTGCGTTTTATTCCATACGTGTT  1413
             |(A)n(minor)

                .         .         .         .         .         .         .         .         .
1414  TCTTGTGGTTTTCTTTTCTTGCATTTCACACAAAAAAAAAGAAGCGAGAAAGCTGACGGGAAAAGCACTCAATTACTAATAGTGGGAGAT  1503

                .         .         .         .         .         .
1504  TGCGGGCGTTATATGTATGTATGATTTCCTAAAAACATATGTGACAACAACTACAAGTATTCCATACGTGTTTCTTGTGGTTTTCTTTTC  1593

                .         .         .         .         .         .         .         .         .
1594  TTGCATTTCACACAAAAAAAAAGAAGCGAGAAAGCTGACGGGAAAAGCACTCAATTACTAATAGTGGGAGATTGCGGGCGTTATATGTAT  1683
                -------------onf1b/Hsp26

                .         .         .         .         .         .         .         .         .
1684  GTATGATTTCCTAAAAACATATGTGACAACAACTACAAGTATTCCCAAAGTAAAACTTAAAGACAGAAACACGAAATAATGTACTTAATA  1773
                                ---------------onf2a/Hsp26

                .         .         .         .
1774  AAGAGGAAAACCAGAATAAAAAAAACTGACGTTTTGTTTTGTTGC   1818
                 ======                   |(A)n (major)
                -------------onf1a/Hsp26
```

HspG1 SEQUENCE. Strain, *Oregon R*. The segment $-1{,}196/1{,}400$ is from GenBank: Accession, M26267 (DROHSP1). Downstream of the *Pst*I site, the sequence continues in the *Hsp26* Sequence. Several changes were introduced between 1,311 and 1,400 following R. S. Cohen (personal communication). The binding sites of *Hsp26* ovarian nuclear factors are indicated. Dashes underline bases that match the consensus hse sequence.

HspG2

Gene Organization and Expression

Open reading frame, 111 amino acids; expected mRNA length, 465–622 bases depending on which of the multiple polyadenylation sites and promoters are used, or approximately 2 kb when a polycistronic mRNA is made in response to heat shock. The corresponding RNA bands are observed in northern blots. Two transcription initiation sites were defined by primer extension analysis and

HspG2

```
               .         .         .         .         .         .         .         .         .
-489  GTGGAGTTTAACGGTTTGTCTGCGCCCTTTTATAGAGACGGAAGAGCTTTGCCCATTGCCACAGAGCTTTTCTGGAGCAGCAACTCGTTG   -40
                              <<  ------Gene3

               .         .         .         .         .         .         .         .         .
-399  TTTCGTTGATTCTAGGGAGACAACTGGGAACCTTCTGGGGGCCAAGCTTTCGTAGACCGTAAACTGTTATATGTGATCTGCTTTAAGGTA   -31
        - -    --- -- hse4      ---  ---  - hse3


                                                               -->-252
               .         .         .         .         .         .         .         .         .
-309  TGTACATACATTGTATGTATAAAGTGGGTACAGATAGCAAGCTCTGTATTGGAGATCATACCATAAGATTTTAATTTTAAATTCAAAGTG   -22
                       -----
                                                                                          _
               .         .         .         .         .         .         .         .  |      .
-219  AAATCGGATACGGATGAGAGACGCAATCTCCAGTGTTAGCTGGACAAACAAGACGACACCACGATAATCAAAATGCGATAAGCAAGTACG   -13
                                                                          _                 |
                                                                                           _|
               .         .         .         .         .         .       |  .-->-59 .         .
-129  GACATACAAATGTACATACCCGAATCTTTAATCTTGAACCTCATAAATGGATCATCTGCGCCCAGCTGGCAAGTCAGTTGTTATTCAGCT   -40
                   hse2 -  ---  --  - -- ---  ----x-- hse1               |_

               .         .         .         .         .         .         .         .         .
 -39  GGCGAACCGGTTGAAATTCGTGCTCCGCCCCATTACTACAATGGCCACGTACGAACAGGTTAAGGATGTTCCCAACCATCCGGATGTGTA    50
                                                    MetAlaThrTyrGluGlnValLysAspValProAsnHisProAspValTy  (17)

               .         .         .         .         .         .         .         .         .
  51  TCTTATCGACGTTCGACGGAAGGAAGAGCTCCAGCAGACGGGCTTCATTCCAGCCAGCATCAATATACCCTGTAATAAACTACTTTCCCT   140
      rLeuIleAspValArgArgLysGluGluLeuGlnGlnThrGlyPheIleProAlaSerIleAsnIleProL                       (41)

               .         .         .         .         .         .         .         .         .
 141  AGTATTTGCTTTATTACCATTTGTTTTATTACTATTTTTTTTACTTAGTGGATGAACTGGACAAGGCTCTAAATCTGGATGGATCTGCTT   230
                                                                  euAspGluLeuAspLysAlaLeuAsnLeuAspGlySerAlaP  (55)

               .         .         .         .         .         .         .         .         .
 231  TTAAAAACAAGTACGGAAGATCGAAACCGGAGAAGCAGTCGCCAATCATATTCACCTGCCGGTCGGGAAATCGAGTCTTGGAAGCAGAGA   320
      heLysAsnLysTyrGlyArgSerLysProGluLysGlnSerProIleIlePheThrCysArgSerGlyAsnArgValLeuGluAlaGluL  (85)

               .         .         .         .         .         .         .         .         .
 321  AAATTGCCAAAAGTCAGGGATACAGCAAGTGAGCTTTAAAAGTTTATTATAGTTGCAATACTTTATATCGGATACATATACATATGTATG   410
      ysIleAlaLysSerGlnGlyTyrSerAs                                                                (94)

               .         .         .         .         .         .         .         .         .
 411  CTCATTTTAGTGTGGTGATCTACAAAGGCTCCTGGAATGAATGGGCTCAAAAGGAGGGACTTTAACGATAAACGTCGCTATATTTCTGAA   500
                nValValIleTyrLysGlySerTrpAsnGluTrpAlaGlnLysGluGlyLeuEnd                         --  (111

               .         .         .         .         .         .         .         .         .
 501  TAAATGAAGATTAATATTAATTAATTAATTAATTATTAATAGCTAAAAAAAAACAGAAAACTTAAATTATTTGTTAATATTAAGCTGTAT   590
      ---                                         |(A)n              |(A)n              |(A)n

               .         .         .         .
 591  TTTTCATATATCTCAAGTTCTTGACTACGCCCATGGCAAGCTT  633
                                           ------HindIII
```

HspG2 SEQUENCE. Strain, *Oregon R*. Accession, X07311 (DROHGSG2). In the first line, double underlining marks the inverse complement of the TATA box of *HspG3*. The sequence ends in the *HindIII* site located at −650 in the *Hsp22* Sequence. Dashes underline bases that match the consensus hse sequence. hse3 and hse4 are the same segments labeled hse2 and hse1, respectively, in *HspG3*. The x-mark under the TATA box marks a nucleotide that also belongs to hse1–2 (two partly overlapping hse's).

HspG3 SEQUENCE (*opposite*). Strain, Schneider cell line 3. Accession, X06542 (DROHSPG3). The inverse complement of the *HspG2* distal TATA box is at −370/−365. Dashes underline bases that match the consensus hse sequence; hse1 and hse2 correspond to hse4 and hse3, respectively, of *HspG2*.

HspG3

```
-374  CCACTTTATACATACAATGTATGTACATACCTTAAAGCAGATCACATATAACAGTTTACGGTCTACGAAAGCTTGGCCCCCAGAAGGTTC   -285
      <<  ------Gene2                                                              -  ---   --- hse2

-284  CCAGTTGTCTCCCTAGAATCAACGAAACAACGAGTTGCTGCTCCAGAAAAGCTCTGTGGCAATGGGCAAAGCTCTTCCGTCTCTATAAAA   -195
                 -- ---     - - hse1                                                        ------
                            -->-167.
-194  GGGCGCAGACAAACCGTTAAACTCCACATTCGAGTCGGAAAAGTCAAGGTGAATTGTGCGCCAAACTGCAGCTGAGATTGTGGATTCACA   -105

-104  CCGCTGGCAAATCAACCCTTGGATACTTTTGAAAGGAAAACAGGTCGTCGGTTCAACGAACCTCCTTCCGCCAGATACACAAGAAGTTAA   -15

 -14  GCAAAGAAAAGTAAAATGCCAGATATTCCCTTTGTCTTGAATTTGGACTCCCCGGACTCCATGTACTACGGCCACGATATGTTCCCGAAT   75
                      MetProAspIleProPheValLeuAsnLeuAspSerProAspSerMetTyrTyrGlyHisAspMetPheProAsn  (25)

  76  CGCATGTACAGGCGATTGCATTCGCGGCAGCATCATGATCTTGATTTGCACACCCTGGGTCTGATTGCCCGGATGGGTGCACATGCCCAT   165
      ArgMetTyrArgArgLeuHisSerArgGlnHisHisAspLeuAspLeuHisThrLeuGlyLeuIleAlaArgMetGlyAlaHisAlaHis  (55)

 166  CACCTGGTGGCCAATAAAAGGAACGGAGAGCTGGCTGCATTGAGCCGCGGTGGAGCCTCAAATAAGCAGGGCAATTTCGAGGTCCATCTG   255
      HisLeuValAlaAsnLysArgAsnGlyGluLeuAlaAlaLeuSerArgGlyGlyAlaSerAsnLysGlnGlyAsnPheGluValHisLeu  (85)

 256  GATGTGGGACTCTTTCAGCCAGGTGAACTGACCGTCAAACTGGTCAACGAGTGCATTGTGGTCGAGGGAAAACACGAGGAGCGCGAGGAC   345
      AspValGlyLeuPheGlnProGlyGluLeuThrValLysLeuValAsnGluCysIleValValGluGlyLysHisGluGluArgGluAsp  (115)

 346  GATCATGGACATGTATCCCGGCATTTTGTTCCGGCCGTATCCGCTGCCCAAGGAGTTCGATTCGGATGCCATTGTTTCCACTTTGTCGGA   435
      AspHisGlyHisValSerArgHisPheValProAlaValSerAlaAlaGlnGlyValArgPheGlyCysHisCysPheHisPheValGly  (145)

 436  GGATGGAGTTCTCAATATCACGGTTCCACCATTAGTTTCCAAGGAGGAGCTCAAGGAGCGCATCATACCCATTAAGCATGTGGGTCCATC   525
      GlyTrpSerSerGlnTyrHisGlySerThrIleSerPheGlnGlyGlyAlaGlnGlyAlaHisHisThrHisEnd                 (169)

 526  GGATCTCTTCCAGGAATGGAAACGGTCATAAGGAGGCCGGTCCGGCAGCTTCTGCTTCAGAGCCAGAAGCCAAGTGAAGAGCCCCCTCCT   615

 616  AAAGATTGCAGCCTAAGCAGCCAAGTGATTTCCCAAGACTCTCGTTTATCGTTGCACCAAAAAAAAAGTCCAAGAAAGTATCGCACAAAT   705

 706  CGTATTATATTTATTAATTTATTATTAGCTACATTTTAAACAGTCCAATCAAATTTTTAAGACTAATCGAAATCCAGTATTAATAAAGGA   795
                                                                                        ------

 796  ATATGAATGTCTCAGTAATCAAAAGACTTTTACTAATATTTAAGAGCTTAATTCATATCAAAAAGCACGAAATCCAATTTTGGGTACAAT   885
                     |(A)n

 886  ATTAACTTTCCTTTGTTCGATTAGACAGGTATTAAAAGCTGTGCATATTAAAAATAGGTCCCCGGATGTCAATCCTACTTAAAAAAGCTT   975

 976  TGGTTAGCCTTTTCCCAGGTGCGATTGAGTGAACTTTTGAACTTTGAAATGAAAGCCCGCCATAAGTGTAATTATCGATAGCTTTTAGTC   1065

1066  ATCTTTCCAAAACTATCTATCGAAGTAACAGTTTTTAACAAGTGGTAAGTCAACGATAAATTTAATAAAAGAAACTAACATTTAATTACA   1155

1156  CAAAGTATATATATTATTTTTAAAGTTATTTAGCAGATGGAGTACATTATAAACTAATTTATTTTGTTTGGGATTATAATCTGAATAATA   1245

1246  AAAACCCGTGACATATTGCATGTTGCCCATCTCCAGCTGGCACTGTAACCTCAAAAAAATGTTTTGTTTACTTTTGCGCCGCTCTGCAGT   1335

1336  TCATAATTCCTGCAAATTAATCAGTAAAACAGATTGCCAAGCCCGCGTTCTAACAACACCCCAACAATGCTCTGCACAACCACAATACGT   1425

1426  AAGTGGGAGCCTTTAAACCTACAGAATCATCACTATATTATGCCGAAAACCCCCACTGATTTATGAAATTCGGTTGATTTTACAGCGCGG   1515

1516  CGGCATGGCGAGTTCGAATGGCAGGATCCCAAGTCCACGGATGAAATGTAAGTCCCTAGAGAGAAGCTAATTGTACACAATATAACCAAG   1605
```

cDNA sequences. Sequencing of multiple cDNA clones suggested three different 3′ termini; at least one of which may be an artifact of cDNA cloning resulting from the presence of a stretch of As. Transcripts from the distal promoter have an intron between −135 and −64, and transcripts from both promoters have introns within the Leu-41 and Asn-94 codons (*HspG2* Sequence) (Pauli et al. 1988).

Developmental Pattern

The distal transcript is testes specific; it appears first in early pupae and persists in adult males. The proximal transcript appears first in 7-h embryos, reaches a maximum at 10–12 h and persists through the second larval stage. It drops to very low levels in third instar larvae and adults, but rises to a second pronounced peak in early pupae.

Heat shock induces transcription from the proximal promoter but normal termination fails so that the *HspG2* heat-shock transcripts extend to the next polyadenylation site, that of *Hsp22* (*Hsp22* Sequence and Fig. 17.2). However, the amount of *Hsp22* transcript in the 2 kb RNA is a small fraction of that derived from the *Hsp22* promoter. Whether or not the polycistronic nature of this mRNA is functionally significant is not known. The two introns are properly excised. Of the genes in the cluster, *HspG2* is the least responsive to heat shock (Pauli et al. 1988).

Promoter

There are 430 bp between the divergent, heat-inducible, transcription initiation sites of *HspG2* and *HspG3*, and there are four putative hse's in that region. Whether some hse's are allocated to one gene and some to the other or whether they are shared is not known.

HspG3

Gene Organization and Expression

Open reading frame, 169 amino acids; expected mRNA length, 979 bases, in agreement with the observed 1.0 kb major RNA. Upon induction, two minor RNA bands (1–2% of the major band) are also detectable; they are 1.6 kb and 2.3 kb long and appear to result from downstream extension of the major RNA. The site of transcription initiation was determined by primer extension, S1 mapping and the sequence of a cDNA clone. The polyadenylation site was obtained from the sequence of a cDNA clone. There are no introns (*HspG3* Sequence) (Pauli and Tonka 1987).

Developmental Pattern

During embryogenesis, the expression of *HspG3* is first detectable at 7–8 h and reaches a peak at 10–12 h. No mesage is detectable through most of the larval period, but it reappears in the late third instar and peaks in early pupae. *HspG3* responds strongly to heat shock (Pauli and Tonka 1987).

References

Arrigo, A-P. and Pauli, D. (1988). Characterization of the HSP27 and three immunologically related polypeptides during *Drosophila* development. *Exp. Cell Res.* **175**:169–183.

Ayme, A. and Tissières, A. (1985). Locus 67B of *Drosophila melanogaster* contains seven, not four, closely related heat shock genes. *EMBO J.* **4**:2949–2954.

Beaulieu, J-F., Arrigo, A-P. and Tanguay, R. M. (1989). Interaction of *Drosophila* 27,000 M-r heat-shock protein with the nucleus of heat-shocked and ecdysone-stimulated cultured cells. *J. Cell Sci.* **92**:29–36.

Cohen, R. S. and Meselson, M. (1985). Separate regulatory elements for the heat-inducible and ovarian expression of the *Drosophila hsp26* gene. *Cell* **43**:737–746.

Costlow, N. and Lis, J. T. (1984). High resolution mapping of DNase I-hypersensitive sites of *Drosophila* heat-shock genes in *Drosophila melanogaster* and *Saccharomyces cerevisiae. Mol. Cell. Biol.* **4**:1853–1863.

Frank, L. H., Cheung, H.-K. and Cohen, R. S. (1992). Identification and characterization of *Drosophila* female germ line transcriptional control elements. *Development* **114**:481–491.

Gilmour, D. S., Thomas, G. H. and Elgin, S. C. R. (1989). *Drosophila* nuclear proteins bind regions of alternating C and T residues in gene promoters. *Science* **245**:1487–1490.

Glaser, R. L., Wolfner, M. F. and Lis, J. T. (1986). Spatial and temporal pattern of *hsp26* expression during normal development. *EMBO J.* **5**:747–754.

Hoffman, E. and Corces, V. (1986). Sequences involved in temperature and ecdysterone-induced transcription are located in separate regions of a *Drosophila melanogaster* heat shock gene. *Mol. Cell. Biol.* **6**:663–673.

Holmgren, R., Corces, V., Morimoto, R., Blackman, R. and Meselson M. (1981). Sequence homologies in the 5′ regions of four *Drosophila* heat-shock genes. *Proc. Natl Acad. Sci.* (*USA*) **78**:3775–3778.

Hultmark, D., Klemenz, R. and Gehring, W. J. (1986). Translational and transcriptional control elements in the untranslated leader of the heat-shock gene *Hsp22. Cell* **44**:429–438.

Ingolia, T. D. and Craig, E. A. (1981). Primary sequence of the 5′ flanking regions of the *Drosophila* heat shock genes in chromosome subdivision 67B. *Nucl. Acids Res.* **9**:1627–1642.

Ingolia, T. D. and Craig, E. A. (1982). Four small *Drosophila* heat shock proteins are related to each other and to mammalian alpha-crystallin. *Proc. Natl Acad. Sci* (*USA*) **79**:2360–2364.

Klemenz, R. and Gehring, W. J. (1986). Sequence requirement for expression of the *Drosophila melanogaster* heat shock protein *hsp22* gene during heat shock and normal development. *Mol. Cell. Biol.* **6**:2011–2019.

Leicht, B. G. and Bonner, J. J. (1988). Genetic analysis of chromosomal region 67A–D of *Drosophila melanogaster*. *Genetics* **119**:579–594.

Lindquist, S. and Craig, E. A. (1988). The heat-shock proteins. *Ann. Rev. Genet.* **22**:631–677.

Mason, P. J., Hall, L. M. C. and Gausz, J. (1984). The expression of heat shock genes during normal development in *Drosophila melanogaster*. *Mol. Gen. Genet.* **194**:73–78.

Mestril, R., Sciller, P., Amin, J., Klapper, H., Ananthan, J. and Voellmy, R. (1986). Heat shock and ecdysterone activation of the *Drosophila melanogaster* hsp23 gene; a sequence element implied in developmental regulation. *EMBO J.* **5**:1667–1673.

Pauli, D., Spierer, A. and Tissières, A. (1986). Several hundred base pairs upstream of *Drosophila* hsp23 and 26 genes are required for their heat induction in transformed flies. *EMBO J.* **5**:755–761.

Pauli, D. and Tonka, C-H. (1987). A *Drosophila* heat shock gene from locus 67B is expressed during embryogenesis and pupation. *J. Mol. Biol.* **198**:235–240.

Pauli, D., Tonka, C-W. and Ayme-Southgate, A. (1988). An unusual split *Drosophila* heat shock gene expressed during embryogenesis, pupation and in testis. *J. Mol. Biol.* **200**:47–53.

Pelham, H. (1985). Activation of heat shock genes in eukaryotes. *Trends Genet.* **1**:31–35.

Riddihough, G. and Pelham, R. B. (1986). Activation of the *Drosophila* hsp27 promoter by heat shock and by ecdysone involves independent and remote regulatory sequences. *EMBO J.* **5**:1653–1658.

Simon, J. A. and Lis, J. T. (1987). A germline transformation analysis reveals flexibility in the organization of heat shock consensus elements. *Nucl. Acids Res.* **15**:2971–2988.

Sirotkin, K. and Davidson, K. (1982). Developmentally regulated transcription from *Drosophila melanogaster* chromosomal site 67B. *Dev. Biol.* **89**:196–210.

Southgate, R., Ayme, A. and Voellmy, R. (1983). Nucleotide sequence analysis of the *Drosophila* small heat shock gene cluster at locus 67B. *J. Mol. Biol.* **165**:35–57.

Thomas, G. H. and Elgin, S. C. R. (1988). Protein/DNA architecture of the DNase I hypersensitive region of the *Drosophila* hsp26 promoter. *EMBO J.* **7**:2191–2201.

Thomas, S. R. and Lengyel, J. A. (1986). Ecdysteroid-regulated heat-shock gene expression during *Drosophila melanogaster* development. *Develop. Biol.* **115**:434–438.

Vázquez, J. (1991). Response to heat shock of *gene1*, a *Drosophila melanogaster* small heat shock gene, is developmentally regulated. *Mol. Gen. Genet.* **226**:393–400.

Zimmerman, J. L., Petri, W. and Messelson, M. (1983). Accumulation of a specific subset of *D. melanogaster* heat shock mRNAs in normal development without heat shock. *Cell* **32**:1161–1170.

18

The *Hsp70* Gene Family: *Hsp70A7d*, *Hsp70A7p*, *Hsp70C1d1*, *Hsp70C1d2*, *Hsp70C1p*

Chromosomal Location:			**Map Position:**
Hsp70A7d, *Hsp70A7p*	3R,	87A7	3-[51]
Hsp70C1d1/2, *Hsp70C1p*	3R,	87C1	3-[51]

Products

Heat-shock proteins of 70 kD, HSP70s, the most abundant type of heat-shock proteins.

Structure

The sequence of *Drosophila* HSP70s is 70–80% identical to heat-shock proteins of groups as distant as vertebrates and vascular plants (Fig. 18.1), and some of the properties discussed below are from studies in other organisms. The different members of the *Drosophila* HSP70 family are no less than 97% identical. Two distinct regions have been identified in these proteins. The more highly conserved region is near the N-terminus; it contains an ATP-binding site and has weak ATPase activity. The other region, closer to the C-terminus, is more variable, and it has sites important for nucleolar localization. It has been suggested that a hydrophobic pocket on the protein surface is the site of HSP70 binding to hydrophobic residues of partly denatured proteins (Lindquist and Craig 1988; Schlesinger 1990 and references therein).

Function

Organisms subjected to mildly elevated temperatures become more tolerant of subsequent high-temperature exposure. It has been suggested that HSP70 may be capable of preventing the denaturation of cellular proteins; and, with the

```
Hsp70-C1 ...M                   Y       N Y                S    N EP          R              YD   KIAE          VS . G
Hsp70-A7 ...M                   Y                          S      EP          R              YD   KIAE          VS . G
     Pig MAKSV                  F                    S     T      DA       L  Q              FG   VVQG        R IN . D
 Petunia ..EG                   W  DR                  G   T      DA          I             RFS   SVQS I L      IPGP D
     CON ----PAIGID LGTTYSCVGV -QHGKVEIIA NDQGNRTTPS YVAFTD-ERL IG--AKNQVA MNP-NTVFDA KRLIGRK--D P----DMKHW PFKV--D-G-
         1                                                   50                                                   100

Hsp70-C1     G E     S R                  T XX     ES  TD                       H                   L      N L ..
Hsp70-A7     G E     S R                  T A      ES  TD                       H                   L      N L ..
     Pig  VQ   S     T G Y                I . G    HP VSN                       V                   I     RT G ..
 Petunia   M V T     E Q  A               I .      TT  KN  V                    V       M           I      K ASSA  K
     CON KPKI-V-YKG E-K-FAPEEI SSMVLTKMKE -A-EAYLG-- I--AVITVPA YFNDSQRQAT KDAG-IAGLN VLRIINEPTA AA-AYGLDK- -K--GERNVL
         101                                                150                                                   200

Hsp70-C1                     S L    RS                  T LAE       Y   LRS             A            E       A    Q
Hsp70-A7                     S L    RS                  T LAD       Y   LRS             A            E       A    Q
     Pig                   D . I    KA                  N FVE       H   YSQ  K  V       C            Q SL    S    I
 Petunia              L   E  . I    KA                M N FVQ       N   ISG             C          TAQT      S Y  I   S
     CON IFDLGGGTFD VSILTIDEG- -FEV--TAGD THLGGEDFDN RLV-H---EF KRK-KKD--- NPRALRRLRT A-ERAKRTLS SST-ATIEID -LFEG-DFYT
         201                                                250                                                   300

Hsp70-C1 KVS         AN    N  Q        N      G    I               S   E  H   N  L                        Q  G I  V  V
Hsp70-A7 KVS         A     N  Q        N      G    I               S      H   N  L                        Q  G I  V  V
     Pig SIT         S     S  E        R   L  A    L               K      N   RD K                     M  K  ENV  L  L
 Petunia TIT        NM     KCME    C   R      SSV  V               Q      N   E CK                   EGN  E V  L  L
     CON ---RARFEEL C-DLFR-TL- PVEKAL-DAK MDK-QIHD-V LVGGSTRIPK VQ-LLQDFF- GK-LN-SINP DEAVAYGAAV QAAILSGD-S -K-QD-LL-D
         301                                                350                                                   400
```

```
Hsp70-C1         I         K  E   CR  C    KT            S                A   T D                     L           KEM
Hsp70-A7         I         K  E   CR  C    KT    A       S                A   T D                     L           KEM
     Pig         L         A  K   ST  T    QI T          L            R   L   R E                     I         T TDK
 Petunia   T     L      G  V  P   TT  T KE QV            L          R     L   K E               T  C  I           EDKT
     CON  VAPLSLG-ET AGGVMT-LI- RN--IP-KQT --FSTYSDNQ PGV-IQVYEG ERAMTKDNN- LG-F-LSGIP PAPRGVPQIE VTFD-DANGI LNVSA---ST
          401                                                450                                                    500

Hsp70-C1     KN   K        QA   D   N      AD   KH Q  ITSR       VF V QS  Q AP.A   D     NSV    N  T R   S   T   E  D
Hsp70-A7     KN   K        QA   D   N      AD   KR Q   TSR     H VL V QA  Q AP.A   D     NSD    N DT R   S   T   E  D
     Pig     NK   T        KE   E   Q      KA   IQ E   GAK       AF M SV  D EGLK  IS     KKV    Q  V S   A   L   D  E
 Petunia   QKNK   T        KE   E   Q      KS   ELKKK  EAK    N  AY MRNTIKD DKINSQ SA    KRIE AID  A K   N Q L  AD  ED
     CON  GKA--ITI-N DKGRLS--EI -RMV-EAEKY --EDE--R-R V---NALESY --N-K--VE- ----GKL-EA DK---LDKC- E-I-WLD-NT -AEK-EF-HK
          501                                                550                                                    600

Hsp70-C1  ME  TRH S   MT  H Q    AAG .P  N CGQQAG FGG YS   V     *
Hsp70-A7  LE  TRH S   MT  H Q    AGA GP  N CGQQAG FGG YS R V     *
     Pig  RK  EQV N   ISGLY GAG  PGP GF  P DLKGGS S.. ..   I     .
 Petunia  MK  ESI N   IA  Y G    GATMDEDGP SVGGSA SQT GA  KI     *
     CON  --EL---C-P I--KM-Q-GA G---G--GA- ------G--- --GPT-EEVD -
          601                                                650
```

FIG. 18.1. Comparison of HSP70s from *Drosophila*, the pig (Accession, M69100) and *Petunia* (Accession, X13301). The CON(sensus) line indicates positions at which all four sequences agree. Where there is no such agreement, the residue occupying that position in each sequence is indicated. There is 89% overall identity between Hsp70A7 and the porcine sequence. Sequences aligned with the GCG *Pileup* program.

expenditure of ATP, it may be involved in the renaturation of denatured proteins and the dissociation of abnormal protein complexes. HSP70 is related to other non-heat-shock proteins known as "molecular chaperones". Molecular chaperones are involved in the translocation of proteins across membranes and also seem to have a role in controlling denaturation and renaturation of proteins (Schlesinger 1990; Gething and Sambrook 1992).

Tissue Distribution

HSP70 is present at low levels in untreated flies. During heat stress, *Hsp70* transcription increases, and HSP70 becomes prominent in the nucleus and the nucleolus where it forms insoluble complexes. After return to normal temperatures, HSP70 levels remain high for some time, but the protein returns to the cytoplasm (Velazquez and Lindquist 1984; Schlesinger 1990).

Organization and Expression of the Clusters

The two *Hsp70* genes at 87A7 are separated by a 1.6 kb spacer, and they are divergently transcribed. At 87C1, a centromere proximal gene, *Hsp70C1p*, is separated from two centromere-distal genes, *Hsp70C1d1* and *Hsp70C1d2*, by 40 kb of DNA. The distal genes are tandemly transcribed toward the telomere while the proximal gene is transcribed in the opposite direction. None of the *Hsp70* genes have introns (Fig. 18.2).

A large portion of the spacer between the proximal and distal copies at 87C1 is made up of simple sequences designated alpha, beta and gamma; these are arranged in various repeat patterns. The gamma element includes a copy of the *Hsp70* regulatory region and, in response to heat shock, it promotes transcription of the spacer sequences. As far as is known, the spacer transcripts have no coding capacity and area non-functional (Ish-Horowicz and Pinchin 1980; Hackett and Lis 1981).

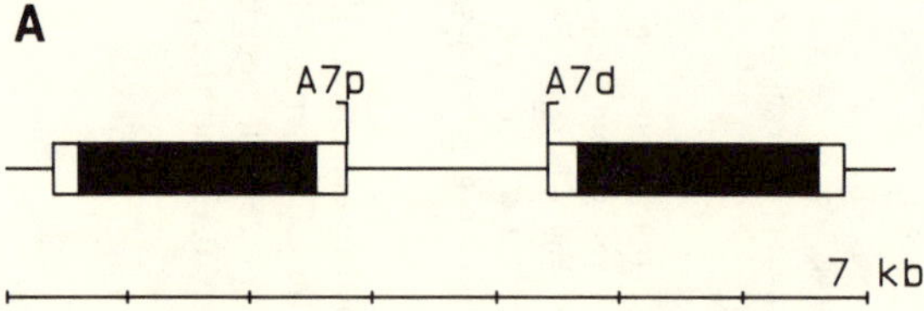

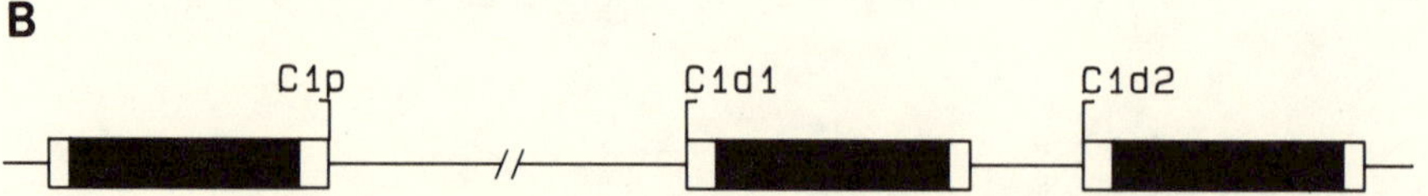

FIG. 18.2. Organization of the *Hsp70* clusters. (A) Cluster at 87A7. (B) Cluster at 87C1.

In *D. simulans* and *D. mauritiana*, there are only four *Hsp70* genes; that is, two divergently transcribed genes occur at each of two loci corresponding to 87C1 and 87A7. Thus, it appears that duplication of the distal gene at 87C1 and multiplication of the simple spacer sequences are recent events unique to *D. melanogaster* (Leigh-Brown and Ish-Horowicz 1981; see reviews in Schlesinger et al. 1982).

All copies of *Hsp70* are very similar, especially in the coding and 5′ regions. In a segment that extends from −610 to 1 (the first codon), the various genes present the following frequencies of base substitution, addition, or deletion relative to *Hsp70C1d1*: *Hsp70C1d2*, 0.5%; *Hsp70C1p*, 1.4%; *Hsp70A7d*, 6.5%. In contrast, at the 3′ end of the genes, *Hsp70C1d1* is much more similar to *Hsp70C1p* than to *Hsp70C1d2*. Within the segment −610/1 *Hsp70A7d* and *Hsp70A7p* differ by 3%, but further upstream the two sequences appear unrelated. Because sequence similarities and repeats occur in blocks having no apparent functional significance, it has been suggested that much of the sequence conservation of the *Hsp70* genes may be due to intergenic corrections rather than negative selection against deleterious mutations (Török et al. 1982).

Developmental Pattern and Promoter

Transcription of the *Hsp70* genes in *Drosophila* occurs only in response to heat and other stressful conditions usually associated with protein denaturation; i.e., no developmentally related expression occurs (Mason et al. 1984).

Most studies of transcription regulation were carried out with *Hsp70A7* promoter sequences. However, given the great deal of sequence conservation in the promoter regions, the available information about transcriptional regulation probably applies equally to all *Hsp70* genes. As is true for other *Hsp* genes, the heat-shock response seems controlled by the heat-shock element (hse) consensus sequence CTNGAANNTTCNAG that must be present in at least two adjacent copies (Pelham 1985; Bienz and Pelham 1987).

Germline transformations involving 5′ deletions demonstrated that the 97 bp upstream of the transcription initiation site (to coordinate −348), a segment that includes hse1 and hse2 (*Hsp70* Sequences), are sufficient for normal levels of heat-induced transcription (an approximately 100-fold increase as compared to the uninduced state) (Dudler and Travers 1984). Thus, these two hse's seem to be the main functional regulatory elements. Repositioning a 51-bp segment that includes the two hse's at various distances from the TATA box does not affect expression very much (Simon and Lis 1987). This flexibility contrasts with the sequence conservation noted earlier.

More detailed studies involving *in vitro* mutagenesis, germline transformation, and *in vitro* binding assays led to a reassessment of the sequence elements responsible for heat induction. The conclusion from those studies is that hse's possess alternating repeats of the 5-bp unit NGAAN and its reverse complement, NTTCN. There are three or four such units in the hse's of the *Hsp70* genes (Xiao and Lis 1988; Perisic et al. 1989).

Transcription is activated by a heat-shock transcription factor (HSF). HSF

Hsp70-C1d1 and *Hsp70-A7d*

```
             .         .         .         .         .         .         .         .         .
A7d    TGTCA AG TCCATAGGCC------------   C G   GACAAC  C   -       - C            C        T A C
C1d1  TCAGACATTTATTGGTTTAGAAGCGCAGTATTTTTTTTGCGA-----ATACGCATAACAAAGCGCTTCGATTATCTTTAACATAAGTTAT   -568
             .         .         .         .         .         .         .         .         .
                                             TATATATATAAATAAA                         C      G
-567  TTAAGCAGCCGTATTTATAAAGAAATTTCCAAAATAAAGC---------------GAATATTCTAGAATCCCAAAACAAA-CTGGTTATT   -478
                                                             ***  ***  ***hse4
                                                             |-         f4           -|

             .         .         .         .         .         .         .         .         .
       C CG   -          C                                      TC     ---------G     T A
-477  GTGGTAGGTCATTTGTTTGGCAGAAAGAAAACTCGAGAAATTTCTCTGGCCGTTATTCGTTATTCTCTCTTTTCTTTTTGGGTCTCTCCC   -388
                                      **  ***  ***hse3
                                    |-         f3           -|  |-       GAGA     -|    |-

             .         .         .         .         .         .         .         .         .
            T    T                 G              CT  AT        C
-387  TCTCTGCACTAATGCTCTCTCACTCTGTCACACAGTAAACGGCATACTGCTCTCGTTGGTTCGAGAGAGCGCGCCTCGAATGTTCGCGAA   -298
                                                        **  *    ***  ** hse2      ***  ***  ***hse1
         GAGA                  -|                    |-           f2         -|-      f1
                                                                |-    GAGA   -|

             .         .         .         .         .         .         .         .         .
                                        G        T  -->-251                     A    G    --
-297  AAGAGCGCCGGAGTATAAATAGAGGCGCTTCGTCTACGGAGCGACAATTCAATTCAAACAAGCAAAGTGAACACGTCGCTAAGCGAAAGC   -208
            -|     ------
             .         .         .         .         .         .         .         .         .
       G      C             -                     A                                      A
-207  TAAGCAAATAAACAAGCGCAGCTGAACAAGCTAAACAATCTGCAGTAAAGTGCAAGTTAAAGTGAATCAATTAAAAGTAACCAGCAACCA   -118
             .         .         .         .         .         .         .         .         .
           TTAAACT AA                   AA         C         G                               GTC
-117  AGTAA-------ATCAACTGCAACTACTGAAATCTGCCAAGAAGTAATTATTGAATACAAGAAGAGAACTCTGAATACTTTCAACAA---   -28
             .         .         .         .         .         .         .         .         .
               G     ---                                                                  A
-27   GTTACCGAGAAAGAAGAACTCACACACAATGCCTGCTATTGGAATCGATCTGGGCACCACCTACTCCTGCGTGGGTGTCTACCAGCATGG   62
                                  MetProAlaIleGlyIleAspLeuGlyThrThrTyrSerCysValGlyValTyrGlnHisGl  (21)
             .         .         .         .         .         .         .         .         .
            G       T    A C                                            T             TC  C
63    CAAGGTTGAGATTAACGCCTATGACCAGGGCAACCGCACCACGCCGTCCTACGTGGCTTTCACAGACTCGGAACGCCTCAATGGTGAACC   152
      yLysValGluIleAsnAlaTyrAspGlnGlyAsnArgThrThrProSerTyrValAlaPheThrAspSerGluArgLeuAsnGlyGluPr  (51)
                   Ile   Asn                                                         Ile
             .         .         .         .         .         .         .         .         .
                                                                                    C     G
153   GGCCAAGAACCAGGTGGCCATGAACCCCAGAAACACAGTGTTTGACGCCAAGCGACTCATCGGCCGAAAATACGACGATCCCAAAATCGC   242
      oAlaLysAsnGlnValAlaMetAsnProArgAsnThrValPheAspAlaLysArgLeuIleGlyArgLysTyrAspAspProLysIleAl  (81)
             .         .         .         .         .         .         .         .         .
                                 G  G        C
243   AGAGGACATGAAGCACTGGCCTTTCAAAGTTGTAAGCGATGGCGGAAAGCCCAAGATCGGGGTGGAGTATAAGGGTGAGTCCAAGAGATT   332
      aGluAspMetLysHisTrpProPheLysValValSerAspGlyGlyLysProLysIleGlyValGluTyrLysGlyGluSerLysArgPh  (111)
             .         .         .         .         .         .         .         .         .
         C              C                                 CGG     A                    A  C
333   TGCTCCCGAGGAGATCAGTTCGATGGTGCTGACCAAGATGAAGGAGACGGCGG---AGGCGTATCTGGGCGAGAGCATCACGGATGCAGT   422
      eAlaProGluGluIleSerSerMetValLeuThrLysMetLysGluThrAlaG---luAlaTyrLeuGlyGluSerIleThrAspAlaVa  (141)
                                                         AlaGlu
```

```
              .         .         .         .         .         .         .         .         .
                     C              C
  423 CATCACAGTTCCAGCTTACTTCAACGACTCTCAGCGCCAGGCTACCAAAGACGCCGGTCACATCGCCGGCCTGAATGTGCTCCGCATCAT   512
      lIleThrValProAlaTyrPheAsnAspSerGlnArgGlnAlaThrLysAspAlaGlyHisIleAlaGlyLeuAsnValLeuArgIleIl (171)
              .         .         .         .         .         .         .         .         .
                             C                     C
  513 CAATGAGCCCACGGCGGCAGCATTGGCCTACGGACTGGACAAGAATCTCAAGGGTGAGCGCAATGTGCTTATCTTCGACTTGGGCGGCGG   602
      eAsnGluProThrAlaAlaAlaLeuAlaTyrGlyLeuAspLysAsnLeuLysGlyGluArgAsnValLeuIlePheAspLeuGlyGlyGl (201)
              .         .         .         .         .         .         .         .         .
                                                                  C           G
  603 CACCTTCGATGTCTCCATCCTGACCATCGACGAGGGATCTCTGTTCGAGGTGCGCTCCACAGCCGGAGACACACACTTGGGCGGCGAGGA   692
      yThrPheAspValSerIleLeuThrIleAspGluGlySerLeuPheGluValArgSerThrAlaGlyAspThrHisLeuGlyGlyGluAs (231)
              .         .         .         .         .         .         .         .         .
                            T T        C
  693 CTTTGACAACCGGCTAGTCACCCACCTGGCGGAGGAGTTCAAGCGCAAGTACAAGAAGGATCTGCGCTCCAACCCTCGCGCCCTACGACG   782
      pPheAspAsnArgLeuValThrHisLeuAlaGluGluPheLysArgLysTyrLysLysAspLeuArgSerAsnProArgAlaLeuArgAr (261)
                                     Asp
              .         .         .         .         .         .         .         .         .
                                                C                       T      C
  783 CCTCAGAACAGCAGCTGAACGGGCCAAGCGCACACTCTCCTCTAGCACGGAGGCCACCATCGAGATCGACGCATTGTTTGAGGGCCAAGA   872
      gLeuArgThrAlaAlaGluArgAlaLysArgThrLeuSerSerSerThrGluAlaThrIleGluIleAspAlaLeuPheGluGlyGlnAs (291)
              .         .         .         .         .         .         .         .         .
                     G     C                        G
  873 CTTCTACACCAAAGTAAGCCGTGCCAGGTTTGAGGAGCTGTGCGCGAACCTCTTCCGCAACACCCTGCAGCCTGTGGAGAAGGCCCTCAA   962
      pPheTyrThrLysValSerArgAlaArgPheGluGluLeuCysAlaAsnLeuPheArgAsnThrLeuGlnProValGluLysAlaLeuAs (321)
                                                    Asp
              .         .         .         .         .         .         .         .         .
                     T
  963 CGATGCCAAGATGGACAAGGGTCAGATCCACGACATCGTGCTCGTCGGCGGATCCACTCGCATTCCCAAGGTGCAAAGTCTGCTGCAGGA  1052
      nAspAlaLysMetAspLysGlyGlnIleHisAspIleValLeuValGlyGlySerThrArgIleProLysValGlnSerLeuLeuGlnGl (351)
                                                                                              As
              .         .         .         .         .         .         .         .         .
      C                                                  T
 1053 GTTCTTCCACGGCAAGAACCTCAACCTATCCATCAACCCAGACGAGGCAGTGGCATACGGAGCTGCTGTGCAGGCCGCTATCCTCAGCGG  1142
      uPhePheHisGlyLysAsnLeuAsnLeuSerIleAsnProAspGluAlaValAlaTyrGlyAlaAlaValGlnAlaAlaIleLeuSerGl (381)
      p
              .         .         .         .         .         .         .         .         .

 1143 AGACCAGAGCGGCAAGATCCAGGACGTGCTGCTGGTGGACGTGGCCCCACTTTCATTGGGAATTGAGACCGCTGGAGGTGTAATGACCAA  1232
      yAspGlnSerGlyLysIleGlnAspValLeuLeuValAspValAlaProLeuSerLeuGlyIleGluThrAlaGlyGlyValMetThrLy (411)
              .         .         .         .         .         .         .         .         .
                        C                          A        A   G
 1233 GCTGATCGAGCGCAACTGTCGCATTCCGTGCAAGCAGACTAAGACGTTCTCCACGTACTCGGACAACCAGCCCGGAGTCTCCATCCAGGT  1322
      sLeuIleGluArgAsnCysArgIleProCysLysGlnThrLysThrPheSerThrTyrSerAspAsnGlnProGlyValSerIleGlnVa (441)
                                                                Ala
              .         .         .         .         .         .         .         .         .

 1323 GTATGAGGGCGAACGTGCGATGACGAAGGACAACAATGCATTGGGCACCTTCGATCTGTCCGGCATTCCACCTGCACCAAGGGGTGTGCC  1412
      lTyrGluGlyGluArgAlaMetThrLysAspAsnAsnAlaLeuGlyThrPheAspLeuSerGlyIleProProAlaProArgGlyValPr (471)
              .         .         .         .         .         .         .         .         .
                  T                                                     C
 1413 CCAGATAGAAGTAACCTTCGACTTGGACGCCAATGGAATCCTGAACGTCAGCGCCAAGGAGATGAGTACGGGCAAGGCCAAGAACATCAC  1502
      oGlnIleGluValThrPheAspLeuAspAlaAsnGlyIleLeuAsnValSerAlaLysGluMetSerThrGlyLysAlaLysAsnIleTh (501)
```

(continued)

```
             .         .         .         .         .         .         .         .         .
                                                                     A                    G    G
1503  GATCAAGAACGACAAGGGACGCCTCTCGCAGGCCGAGATTGATCGCATGGTGAACGAGGCTGAGAAGTACGCCGACGAGGACGAAAAGCA  1592
      rIleLysAsnAspLysGlyArgLeuSerGlnAlaGluIleAspArgMetValAsnGluAlaGluLysTyrAlaAspGluAspGluLysHi  (531)
                                                                                              Ar

             .         .         .         .         .         .         .         .         .
               AG                C         C    CC    T  G      G       A     A  T
1593  TCGCCAGCGCATAACCTCTAGAAATGCTCTGGAGAGCTACGTATTCAACGTAAAGCAGTCCGTGGAGCAGGCGCCCGCTGGCAAACTGGA  1682
      sArgGlnArgIleThrSerArgAsnAlaLeuGluSerTyrValPheAsnValLysGlnSerValGluGlnAlaProAlaGlyLysLeuAs  (561)
      g         Val                          His   Leu            Ala

             .         .         .         .         .         .         .         .         .
            T             A                  C     C  G                    T
1683  CGAGGCCGACAAGAACTCCGTCTTGGACAAGTGCAACGAAACTATTCGATGGCTGGACAGCAACACCACCGCCGAGAAGGAGGAGTTCGA  1772
      pGluAlaAspLysAsnSerValLeuAspLysCysAsnGluThrIleArgTrpLeuAspSerAsnThrThrAlaGluLysGluGluPheAs  (591)
                         Asp               Asp

             .         .         .         .         .         .         .         .         .
             C             C              C                           T        T GA CT   GGT
1773  CCACAAGATGGAGGAGCTCACTCGCCACTGCTCCCCTATCATGACCAAGATGCATCAGCAGGGAGCGGGAGCAGCTGGGGGT---CCGGG  1862
      pHisLysMetGluGluLeuThrArgHisCysSerProIleMetThrLysMetHisGlnGlnGlyAlaGlyAlaAlaGlyGly---ProGl  (621)
             Leu                                                                 GlyAla   Gly

             .         .         .         .         .         .         .         .         .
         A     C     G     G           A             G   G                  G      G  TCTAAT  TT
1863  AGCCAACTGTGGCCAACAGGCCGGAGGATTTGGCGGCTACTCTGGACCCACAGTCGAGGAGGTCGACTAAAGCCAAATAGAAATTATTCA  1952
      yAlaAsnCysGlyGlnGlnAlaGlyGlyPheGlyGlyTyrSerGlyProThrValGluGluValAspEnd                      (643)
                                                    Arg

             .         .         .         .         .         .         .         .         .
          ATCAA GG   A A C TA  GGT ATA    AA T    TTTA GTTTTTGAG CTG T AG A GT T GATCGA  A  CCA
1953  GTTCTGGCTTAAGTTTTTAAAAGTGATATTATTTATTTGGTTGTAACCAACCAAAAGAATGTAAATAACTAATACATAATTATGTTAGTT  2042
             .         .         .         .         .         .         .         .         .
      AG  CAACAAT GT T ACC AA TA C AG CTTAATT A CAA  ATGT  TTGCT   AG AAA  TA ATTA TTA   G AAT T
2043  TTAAGTTAGCAACAAATTGATTTTAGCTATATTAGCTACTTGGTTAATAAATAGAATATATTTATTTAAAGATAATTCGTTTTTATTGTC  2132
                                                   ------(87C1)          ------(87A7)

             .         .         .         .         .         .         .         .         .
       AA TCAACT
2133  AGGGAGTGAGTTTGCTTAAAAACTCGTTTAGATCTGTCCTCGAGAAATTATTTATTTAAATGCGATGGAGAGCCGGCGCCGAATCGAAAA  2222
               |(A)n (87A7)

             .         .         .         .         .         .         .         .         .
2223  CTTTACGCGCTTAAAAGCACGAGTTGGCATCCCTAGTAAACAGCTGTTCGTGAAGATATGCAGTGCAAACGAAAAACCCGCCTACAAATA  2312
             .         .         .         .         .         .         .         .         .
2313  TTGTTATTTTGATTAGATTACGGATTACAGAATGGAACCGCCGTTCGCCCCGCTAAGTGAGTCCTGCACCAAGGCGTGGGCGACAGGTGT  2402
             .         .         .         .         .         .         .         .         .
2403  ACGAGAAATGTAAGCTGGCCTCGCAGGAGATCCGTCATCCCAATTGGGAAATGTAATCTTTGCCAGAATGGTTACGGAGTTCAACAACAA  2492
             .         .         .         .         .         .         .         .         .
2493  AAACAGTCTATAGAAATAATAGCCTTTCCTTTCCTCATATGTATGTAAATATGTAAAATAAGTCACAACTAAATTCTAATACACTTCTCA  2582
             .         .         .         .         .         .         .         .         .
2583  GTCTTAAATTAATTTTATCGTATATTAAAACAGAAGAAAGTCCGTTAATCGTTGATTTCGTTAACTAAAAGTACAAAATAATCTTTAATC  2672
             .         .
        | Ca. coordinate -640 of Hsp70-C1d2
2673  TTTAGAAGCGCAGCAATGTT  2692
```

Hsp70 SEQUENCES. Accession, J01104, J01105 (DROHSP7D1) and J01103 (DROHSP7A2). The numbered line shows the sequence of *Hsp70C1d1*; where the sequence of *Hsp70A7d* differs, the changed bases are indicated above and the amino acid substitutions below the *Hsp70C1d1* sequence. Dashes represent gaps in one sequence relative to the other. Asterisks below the sequence mark positions that match the hse consensus.

has an apparent M_r of 110 kD and binds with high affinity (dissociation constant, 4×10^{-12}) to two contiguous segments designated f1 and f2 in the *Hsp70* Sequences; these binding sites extend from −315 to −290 and from −340 to −315, respectively (between 40 and 90 bp upstream of the transcription initiation site). Two additional binding sites occur farther upstream, at −440 to −415 (f3) and at −510 to −485 (f4). The binding of HSF to these secondary sites has a minor effect on *in vitro* transcription; it is not clear what their *in vivo* role might be. All the binding sites overlap hse's (Wu et al. 1987; Topol et al. 1985).

HSF seems to preexist as an unbound monomer in all cells. In response to heat, it is reversibly changed to the active form capable of specific DNA binding; that change includes the formation of oligomers (Westwood et al. 1991). Heat treatments as short as 30 s are sufficient to induce detectable binding of HSF to *Hsp70* promoter fragments (Zimarino and Wu 1987). Binding of HSF to hse is highly cooperative, and the cooperativity is itself temperature-dependent (Xiao et al. 1991).

At normal temperatures, RNA polymerase II binds to the region around the transcription initiation site of *Hsp70* (coordinates −186 to −263), and transcription is initiated but blocked. It is only after heat shock, and presumably after binding of HSF, that the transcription block is released and RNA polymerase II becomes detectable along the whole length of the gene (Gilmour and Lis 1985, 1986; Rougvie and Lis 1988).

Another factor, a 66 kD protein, seems to associate with the segments of alternating CT (or GA) sequence found between positions −415 and −360 and between −325 and −319. This same protein, the GAGA, factor binds sequences upstream of the histone genes *His3* and *His4*, the heat shock gene *Hsp26* and *Ultrabithorax* (Gilmour et al. 1989).

Hsp70 mRNA is very stable and efficiently translated at 36°, but has a half-life of only minutes at 25°. This insures that when *Drosophila* flies or cells are returned to 25° after a heat shock, HSP70s cease to be synthesized. AU-rich sequences in the 3′ untranslated region of the mRNA are responsible for the specificity of temperature-dependent degradation of *Hsp70* mRNA (Petersen and Lindquist 1989).

Hsp70A7d
(Distal gene at 87A7)

Gene Organization and Expression

Open reading frame, 643 amino acids, expected mRNA length, 2,389 bases. Primer extension and S1 mapping were used to define the 5′ end. The 3′ end was obtained by S1 mapping (*Hsp70* Sequences) (Karch et al. 1981; Török and Karch 1980; Török et al. 1982).

The transcription initiation site of *Hsp70A7p* (the proximal gene at 87A7) is approximately 1,630 bp upstream of the *Hsp70A7d* transcription initiation site. Only a few hundred bp at the 5′ and 3′ ends of *Hsp70A7p* have been sequenced. Assuming conservation of the intervening coding region, the segment of sequence similarity between *Hsp70A7d* and *Hsp70A7p* extends approximately from coordinates −600 to 2,130 (*Hsp70* Sequences and Fig. 18.1). This would leave, between the inverted repeats, a spacer of approximately 940 bp made up largely of blocks of simple sequence DNA (Mason et al. 1982). The two genes do not seem to share regulatory elements, i.e., each has its own *cis*-acting hse's.

Hsp70C1d1
(first distal gene at 87C1)

Gene Organization and Expression

Open reading frame, 641 amino acids; expected mRNA length, ca. 2,360 bases. Primer extension and S1 mapping were used to define the 5′ end. The 3′ end was not defined (*Hsp70* Sequences) (Ingolia et al. 1980; Karch et al. 1981).

The repeat containing *Hsp70C1d2* (the second distal gene at 87C1) begins 576 bp downstream of the *Hsp70C1d1* termination codon (ca. coordinate 2510). The two genes are part of a tandem duplication of approximatey 2,900 bp from coordinates −820 to 2,080 of *Hsp70C1d1*. Such alignment leaves a spacer of approximately 430 bp between the repeats (from 2,080 to 2,510) (*Hsp70* Sequences and Fig. 18.2); it is not clear whether this spacer originated at the time of the duplication or whether it arose by sequence divergence at one or both ends of the repeat. *Hsp70C1d2* has been only partially sequenced but it appears very similar to *Hsp70C1d1*. The most extensive sequence divergence occurs within the last 100 bp at the 3′ end, especially around the polyadenylation signals (Török et al. 1982).

The region of sequence similarity between *Hsp70C1d1* and *Hsp70C1p* (for which only the 5′ and 3′ end-sequences are available) starts near coordinate −600 and extends to coordinate 2,540. The overlap of *Hsp70A7* with *Hsp70C1p* sequences extends from −600 to 1,940.

Related Genes

Hsp68 at chromosomal location 95 D is a heat-shock gene related to the *Hsp70* family. Hybridization data indicate that *Hsp68* and *Hsp70* are 75–85% identical.

Seven other genes identified by cross-hybridization are the *heat shock cognate genes*, *Hsc1–Hsc7*. They are expressed very strongly during development but are not heat-inducible or clustered (Lindquist and Craig 1988).

References

Bienz, M. and Pelham, H. R. B. (1987). Mechanisms of heat-shock gene activation in higher eukaryotes. *Adv. Genet.* **24**:31–72.

Dudler, R. and Travers, A. A. (1984). Upstream elements necessary for optimal function of the *hsp70* promoter in transformed flies. *Cell* **38**:391–398.

Gething, M-J. and Sambrook, J. (1992). Protein folding in the cell. *Nature* **355**:33–45.

Gilmour, D. S. and Lis, J. T. (1985). *In vivo* interactions of RNA polymerase II with genes of *Drosophila melanogaster. Mol. Cell Biol.* **5**:2009:2018.

Gilmour, D. S. and Lis, J. T. (1986). RNA polymerase II interacts with the promoter region of the noninduced *hsp70* gene in *Drosophila. Mol. Cell Biol.* **6**:3984–3989.

Gilmour, D. S., Thomas, G. H. and Elgin, S. C. R. (1989). *Drosophila* nuclear proteins bind regions of alternating C and T residues in gene promoters. *Science* **245**:1487–1490.

Hackett, R. W. and Lis, J. T. (1981). DNA sequence analysis reveals extensive homologies of regions preceding hsp70 and $\alpha\beta$ heat shock genes in *Drosophila melanogaster. Proc. Natl Acad. Sci.* (*USA*) **78**:6196–6200.

Ingolia, T. D., Craig, E. A. and McCarthy, B. J. (1980). Sequence of three copies of the gene for the major *Drosophila* heat shock induced protein and their flanking regions. *Cell* **21**:669–679.

Ish-Horowicz, D. and Pinchin, S. M. (1980). Genomic organization of the 87A7 and 87C1 heat-induced loci of *Drosophila melanogaster. J. Mol. Biol.* **142**:231–245.

Karch, F., Török, I. and Tissières, A. (1981). Extensive regions of homology in front of the two *hsp70* heat shock variant genes in *Drosophila melanogaster. J. Mol. Biol.* **148**:219–230.

Leigh-Brown, A. J. and Ish-Horowicz, D. (1981). Evolution of the 87A and 87C heat shock loci in *Drosophila. Nature* **290**:677–682.

Lindquist, S. and Craig, E. A. (1988). The heat-shock proteins. *Ann. Rev. Genet.* **22**:631–677.

Mason, P. J., Hall, L. M. C. and Gausz, J. (1984). The expression of heat shock genes during normal development in *Drosophila melanogaster. Mol. Gen. Genet.* **194**:73–78.

Mason, P. J., Török, I., Kiss, I., Karch, F. and Udvardy, A. (1982). Evolutionary implications of a complex pattern of DNA sequence homology extending far upstream of the *hsp70* genes at loci 87A7 and 87C1 in *Drosophila melanogaster. J. Mol. Biol.* **156**:21–35.

Pelham, H. (1985). Activation of heat shock genes in eukaryotes. *Trends Genet.* **1**:31–35.

Perisic, O., Xiao, H. and Lis, J. T. (1989). Stable binding of *Drosophila* heat shock factor to head-to-head and tail-to-tail repeats of a conserved 5 bp recognition unit. *Cell* **59**:797–806.

Petersen, R. B. and Lindquist, S. (1989). Regulation of HSP70 synthesis by messenger RNA degradation. *Cell Regulation* **1**:135–149.

Rougvie, A. E. and Lis, J. T. (1988). The RNA polymerase II molecule at the 5′ end of the uninduced *hsp70* gene of *D. melanogaster* is transcriptionally engaged. *Cell* **54**:795–804.

Schlesinger, M. J., Ashburner, M. and Tissières, A. (eds) (1982). *Heat Shock from Bacteria to Man.* Cold Spring Harbor, NY: Cold Spring Harbor Laboratory.

Schlesinger, M. J. (1990). Heat shock proteins. *J. Biol. Chem.* **265**:12111–12114.

Simon, J. A. and Lis, J. T. (1987). A germline transformation analysis reveals flexibility in the organization of heat shock consensus elements. *Nucl. Acids Res.* **15**:2971–2988.

Török, L. and Karch, F. (1980). Nucleotide sequences of heat shock activated genes in *Drosophila melanogaster*. I. *Nucl. Acids Res.* **8**:3105–3123.

Török, I., Mason, P. J., Karch, F., Kiss, I. and Udvardy, A. (1982). Extensive regions of homology associated with heat-induced genes at loci 87A7 and 87C1 in *Drosophila melanogaster*. In *Heat shock from Bacteria to Man*, eds. M. J. Schlesinger, M. Ashburner and A. Tissieres. Cold Spring Harbor, NY: Cold Spring Harbor Laboratory, pp. 19–25.

Topol, J., Ruden, D. M. and Parker, C. S. (1985). Sequences required for *in vitro* transcriptional activation of a *Drosophila Hsp70* gene. *Cell* **42**:527–537.

Velázquez, J. M. and Lindquist, S. (1984). hsp70: nuclear concentration during environmental stress and cytoplasmic storage during recovery. *Cell* **36**:655–662.

Westwood, J. T., Clos, J. and Wu, C. (1991). Stress-induced oligomerization and chromosomal relocalization of heat-shock factor. *Nature* **353**:822–827.

Wu, C. S., Wilson, S., Walker, B., Dawid, I., Paisley, T. and Zimarino, V. (1987). Purification and properties of *Drosophila* heat shock activator protein. *Science* **238**:1247–1253.

Xiao, H. and Lis, J. T. (1988). Germline transformation used to define key features of heat-shock response elements. *Science* **239**:1139–1142.

Xiao, H., Perisic, O. and Lis, J. T. (1991). Cooperative binding of *Drosophila* heat shock factor to arrays of a conserved 5 bp unit. *Cell* **64**:585–593.

Zimarino, V. and Wu, C. (1987). Induction of sequence specific binding of *Drosophila* heat shock activator protein without protein synthesis. *Nature* **327**:727–730.

19

janus: janA, janB

Chromosomal Location:
3R, 99D4–8

Map Position:
3-[101]

Products

The properties and functions of *janus* products are unknown. Allowing for nine gaps, there is 37% sequence identity between JANA and JANB.

Organization of the *janus* Cluster

janus is a small but complex locus that includes two partly overlapping transcription units: *janA* is upstream and its 3′ untranslated region contains the transcription initiation site of *janB* (*jan* Sequences and The *Serendipity* Gene Cluster Fig. 28.1) genes probably originated by duplication, judging from their sequence similarities and the comparable positions of two of the three introns in each gene (Yanicostas et al. 1989).

janA

Gene Organization and Expression

The structure of all the transcription products is not yet clear; a final description will be possible only after more cDNA sequences become available. There are two initiation sites, 18 bp apart, two polyadenylation sites, 5 bp apart, and a facultatively spliced intron that spans part of the leader and part of the coding region. The open reading frames are 119 and 135 amino acids long, and the expected mRNA length is between 638 and 731 bases depending on which 5′ and 3′ ends occur and whether or not the leader intron is spliced. These sizes are in agreement with a 0.8 kb band observed in RNA gels (but see below). Primer extension and a cDNA sequence were used to define the upstream 5′

janA and *janB*

```
                                        <-- Sry-beta
-351  ATTCGGCTTAAACAATTTAATTTGTGTATATTTTGTTGTGAACGCCAGAGCTGTGCCGATAGTGCCGATAGTATCGACTGCGTGCTGTCG  -262

-261  GCGTAATCGATAAATTTGCTGTCACTGATAACAACGTTTTCTTTTTGAGTTTATTAATTATTACTAAAATATACTGAGTAGTACAAAAAA  -172
                                                                    -------

-171  CGTTTTCCCAAATGTACTAAAGAAATAACTGAATATTATAATTTTAATAGTATCGATACATAAGGTGAACGAGAATAAAAGTATCTGGTC  -82
          -----                                          _         -------
      -->-81 (janA)   -->-63                             |
 -81  ACATTGCTGGACTAAAGCAGCGTTTTTGGAAAATTTGCCGGTTGGTAAGACATTAAATTCTGTTTTCAAACACTTTTCCACAATGAATCG  8
                                                                                     MetAsnAr  (3)
                          _                              _|
                          |
   9  CCTCCAACTGCTTTCCAAAGGACTACGACTGATTCACAAAATGTCCGAGGAAGCACTTGCCGGCGTGCCACTGGTGCACATCAGTCCAGA  98
      gLeuGlnLeuLeuSerLysGlyLeuArgLeuIleHisLysMetSerGluGluAlaLeuAlaGlyValProLeuValHisIleSerProGl  (33)
                          |_

  99  GGGCATCTTCAAGTATGTCATGATCAATGTCTTCGATGGAGGAGATGCTTCAAAGGCGGTGATCCGCGGATTTGCGGACTGCACATGGCA  188
      uGlyIlePheLysTyrValMetIleAsnValPheAspGlyGlyAspAlaSerLysAlaValIleArgGlyPheAlaAspCysThrTrpHi  (63)

 189  TGGTAAGTCGGATCCTCATCACCCATCAAGTGCCCACTTAGCTTGGTTACTGTCCCACAGCCGACATCTTCGAGCGCGAGGAGGAGGTCT  278
      sA                                                          laAspIlePheGluArgGluGluGluValP  (74)

 279  TTAAAAAACTGGGGCTGCGGGCCGAGTGTCCTGGCGGCGGTCGCATTGAACACAATCCCGAGAAGAAGTACTTGAAGGTCTACGGATACT  368
      heLysLysLeuGlyLeuArgAlaGluCysProGlyGlyGlyArgIleGluHisAsnProGluLysLysTyrLeuLysValTyrGlyTyrS  (104)

 369  CGCAGGTGGGTCTATTCCTTGAGTAAAGGGGTCGCTGGGCAGTGGATGGACTGATGTATCTAACTACTTGAAAAATTTTCTGGAGTTCTA  458
      erGln                                                                                       (105)

 459  ATCAAGTCTTTCTATTTAAGGGCTTTGGAAAAGCTGATCACGCGCAGACCAAACGCATCCTGGCCACCAAATACCCGGACTACACGATCG  548
                          GlyPheGlyLysAlaAspHisAlaGlnThrLysArgIleLeuAlaThrLysTyrProAspTyrThrIleG  (129)

 549  AAATCTCCGATGAGGGATATTAGCTGCAATCAACGAGAGAAGACTCCACATAAGCACACTGACATGAATTTATACCATTGGCTTCGATCC  638
      luIleSerAspGluGlyTyrEnd                                                                     (135)
                                                                                 ----
                                                                -->696 (janB)
 639  TGTGTGCCATGATTTTATTGGAAATGGCATTTAAAATTGAGAAATACTCTGAAAGGCAGTTAGTCTGTAGCTTTGCAACTGCTCGCACTA  728
                           ------

 729  AACCTTTTCGGATCTAAATTAATCAGTTTGTACACAAATTTCGTTTCTTTTCCTTTGGTTAAATAAAATGAAAATGTTCAAGTCATTGCG  818
                                                                          MetLysMetPheLysSerLeuAr  (8)
                                                                               |(A)n|(A)n (janA)

 819  TCTGCTTCCTCATATTGTTTCTCCGTTTCGTAAGGCTTAGGAATATTCAATATTAAGATTTACAAGCCCTAATATACTTGGTTTTAGAAA  908
      gLeuLeuProHisIleValSerProPheG                                                           lnL  (19)

 909  AATGTTACTCAACCGATTTGATAAGTTTGGTAGGCGTTCCCCGGGTCAAGATAACCAAGGGTCAGAATCGTTATTTGTTGGTGAATATTC  998
      ysCysTyrSerThrAspLeuIleSerLeuValGlyValProArgValLysIleThrLysGlyGlnAsnArgTyrLeuLeuValAsnIleH  (49)

 999  ATACGCATGGCTTCACGAAGTATGGAAGAGTTATTGTCCGTGGCGCCGATGTTGACAATCACTGTGAGTTTCCACTGCTGGACGCTTAAC  1088
      isThrHisGlyPheThrLysTyrGlyArgValIleValArgGlyAlaAspValAspAsnHisL                             (70)

1089  CTTGAGCAGTCTTACAAATCCTTCTTTCAGTGGCGGTCTTCGACTCGATTTTGGAGGAGCTGGAACCCGAGGGCATATGTGCCAAAATCC  1178
                                  euAlaValPheAspSerIleLeuGluGluLeuGluProGluGlyIleCysAlaLysIleL  (90)
```

```
1179 TCGGTGGTGGAAGGATTCTCAACGAGGCAGAAAATAAAAAAATTAAGATCTATGGCACCTCCAGGGTAAGTAGAGGATCCTTGGTCCTTG 1268
     euGlyGlyGlyArgIleLeuAsnGluAlaGluAsnLysLysIleLysIleTyrGlyThrSerArg                         (111)

1269 AAGCACCGGCTAATGGTTCTTGATGGGTCTCCCTAGACTTTCGGCGGTGCTGATCACACAAGGACAAGGAATATACTTCAAGCGTGGACC 1358
                                                ThrPheGlyGlyAlaAspHisThrArgThrArgAsnIleLeuGlnAlaTrpThr (129)

1359 ACTTATAAGGACTTTAAGATAACCGTTAAACAATAAAGTTGCATAAATTTCGAAAATGGAAATTCAGTACTAATAAAAAGAAAATAGAAT 1448
     ThrTyrLysAspPheLysIleThrValLysGlnEnd                                  ------               (140)

1449 ATAAAACTAGCGCTCTTTCAATATTATTAAGGGGTAATCGACAGGCGATTGTAATTTGGCTTCGATCCTGTGTGCCATCATTTTATTGGA 1538
      |(A)n (janB)
```

jan SEQUENCES. Strain, *Canton S*. Accession, M27033. The TATA box and transcription initiation site of *Sryβ* are indicated (near −200).

end; and primer extension was used to define the downstream 5′ end. The 3′ ends were obtained from two cDNA sequences. There is a leader intron starting at −41, with an acceptor site at +29. One cDNA in which this intron was spliced out, and one in which it was not, were sequenced. If this intron is spliced out, translation might start with Met-17 at +49. There are also introns in the Ala-64 and after the Gln-105 codons (*jan* Sequences) (Yanicostas et al. 1989).

Developmental Pattern

The 0.8 kb *janA* transcript is present at all developmental stages in both sexes; it is particularly high in 0–12 h embryos and in the ovaries of adult females. In addition, there is a 0.95 kb transcript that differs from the 0.8 kb transcript only in the length of its poly(A) tail. The 0.95 kb transcript is sex-specific, occurring only in males from the third larval instar onward; the highest levels are in the adult male, where it is found in the gonads (Yanicostas et al. 1989).

Promoter

The gene *Sryβ* is upstream of *janA* and transcribed in the opposite direction (see *Sry*, Fig. 28.1). Less than 100 bp separate the putative TATA boxes of *Sryβ* and *janA*; and since they are both expressed at high level in ovaries, it is likely that the two genes share regulatory sequences (Yanicostas et al. 1989).

janB

Gene Organization and Expression

Open reading frame, 140 amino acids; expected mRNA length, 579 bases. Primer extension, S1 mapping and a cDNA sequence defined the 5′ end. The 3′ end was obtained from S1 mapping and a cDNA sequence. There are introns in the Gln-18 and Leu-70 codons and after the Arg-111 codon (*jan* Sequences) (Yanicostas et al. 1989).

Developmental Pattern

janB transcripts are present only in males from the third larval through the adult stages; the highest levels occur in adults. Expression appears to be restricted to the gonads. The leader region of *janB* has striking sequence similarity with the leader element of *mst(3)gl–9*, a gene that is thought to mediate spermatid-specific translation (Yanicostas et al. 1989).

Promoter

Accurate and tissue-specific transcription requires no more than 175 bp upstream of the transcription initiation site of *janB* (Yanicostas et al. 1989; Yanicostas and Lepesant 1990). When there is active transcription of *janA*, there is a reduced accumulation of RNA from the *janB* transcription initiation site; this is probably a case of transcription interference similar to that observed in *Adh* (Yanicostas and Lepesant 1990).

References

Yanicostas, C. and Lepesant, J-A. (1990). Transcriptional and translational *cis*- regulatory sequences of the spermatocyte-specific *Drosophila janusB* gene are located in the 3′ exonic region of the overlapping *janusA* gene. *Mol. Gen. Genet.* **224**:450–458.

Yanicostas, C., Vincent, A. and Lepesant, J-A. (1989). Transcriptional and posttranscriptional regulation contributes to the sex-regulated expression of two sequence-related genes at the *Janus* locus of *Drosophila melanogaster*. *Molec. Cell. Biol.* **9**:2526–2535.

20

knirps and Related Genes: *kni, knrl, egon*

Chromosomal Location:		**Map Position:**
kni	3L, 77E1–2	3-[46]
knrl	3L, 77E1–2	3-[46]
egon	3L, 79B	3-[47]

Products

DNA-binding regulatory proteins of the steroid/thyroid hormone receptor superfamily that includes receptors for vitamin D and retinoic acid in vertebrates.

Structure

Throughout this superfamily of proteins, the region extending from Cys-5 to Arg-81 is conserved, with 20 amino acids being identical in all of the related proteins and 40 others common to several of them (*kni* Sequence). In the *kni*-group proteins, as in other proteins in the superfamily, the conserved region is divided into two putative finger domains: one with four Cys (C_4) and one with five (C_5) (Evans 1988; Evans and Hollenberg 1988; Nauber et al. 1988; Harrison 1991). The three proteins encoded by the *kni*-group genes are more than 80% identical in the finger regions and identical for a group of 19 amino acids adjacent to the fingers (KNI box), but they are completely divergent in other regions. *kni* and *knrl* have several segments of short repeats (Fig. 20.1) (Rothe et al. 1989).

It should be noted that the classification of KNI-group proteins with the hormone receptor superfamily is based on similarities in the DNA-binding

kni

```
              .         .         .         .         .         .         .         .         .
-2577  GAATTCCTCTGCCTGATGCAACAAATGAAAGTCAAATGGAAAATCTTCTGGGAAGTCAGCTAACGAGTTTTTGTTAAGAGTATACCTTAG  -24
              .         .         .         .         .         .         .         .         .
-2487  ACATGGTTTAGTACATCGGTTGAAGTTTTATATTTTATAATACTAGCCACACTTCGGAGTGAAAAAGTCAAGGTTCCTGTCTTTGGGTCT  -23
              .         .         .         .         .         .         .         .         .
-2397  GAACAACCCTTTTGGTACAATGCGCGCCCATAAAAGGGTTAAGCACATCGGTTAGCGGCATAAAAGGGTTAAACAGGTAGCTCCTTCTTT  -23
                                             ---------->kr2                  ---------->kr1

              .         .         .         .         .         .         .         .         .
-2307  CTTTTTGGCTTTGAGCAAACAACAATAAATATTCATAAAAAGAGCTTAAGTGCCGCCATAAGGCTCCTTGTTTACACAAAGGAGAAATTA  -22
              .         .         .         .         .         .         .         .         .
-2217  TGTTGGAAGTTGACTTTTAAAAGGGTTACAATTAAATTCGATTGATATTTGTATTTTATTGAGTATAATGATGGTGAAGGTGTGGATAAG  -21
              .         .         .         .         .         .         .         .         .
-2127  AAAGTTTTATAATATTTAAGAATAATATAATTTCATGATTTATTTGCGAAATATTACTCTACTAAAAGTGTAATATTAATAAAATTATTA  -20
              .         .         .         .         .         .         .         .         .
-2037  AAATAATTATAATTATATTCTATTCATATTGAACTTGTATGGTTTAAACCTATTTTTGTATGCTATTTTAGAACCAGCTTGCAAATCAAC  -19
              .         .         .         .         .         .         .         .         .
-1947  TACTTTAATATGAATCATTCTGAATCCGGGTAATAGCCCGTCTAAATAGTATTTTTTATAACTTTTCGGACGCAATTACATACTCAATAA  -18
              .         .         .         .         .         .         .         .         .
-1857  TACTCAACTATCGTTTTTTTGCTATGAATCAATGCAGATCTCTTATTGATTAACTTCTAATTAAAGCGTTTCAATTTATTGCCAAGTCGC  -17
              .         .         .         .         .         .         .         .         .
-1767  GGTTATGCAAATTTTAACACATTTCATGAAATCTTGAGAATCAGTTTGTGAATCACACAGAAAGTGGGAATATTTCCCGCGGAAAAAGGT  -16
              .         .         .         .         .         .         .         .         .
-1677  TTTGAAAATCAAACTAGGTGTTAGGCATACAGGCAACTCTAAATGTACCCAAAAACCGGCGGACTTTGAAAAGAAAACCCCAAGCGAATT  -15
              .         .         .         .         .         .         .         .         .
-1587  GGCCTCCAACCATTTCGATTTCGAGCAGCCAAAACCGTCCGCCCATGCCAAAAAAATGAGCAGCTGTTAAAAATGAAGTCAATAGCTTAG  -14
              .         .         .         .         .         .         .         .         .
-1497  TCAATGTGGTGTGTGTGTGTGTGTGTGGTGTTCGAGTGAGTGAGAAATCCAGCCGCCCTTAGCACGCGAGTATCTTTAATAAATAAACGA  -14
              .         .         .         .         .         .         .         .         .
-1407  ATAACGAATAATATCAGGGCCATGCAAATAGCCTGATTACAGGGAACTCAAAATCGAGAGAGAGAGAGAGACTGAGCGTGAGATCTGAAT  -13
              .         .         .         .         .         .         .         .         .
-1317  GAGTGAGTGAGTGTTTTCTATTCATTCAACAACAGAGCGTTAACATTCTGCTAACATTTCGCTCGAGTGGGGTTCGAACTCAATGCGCAT  -12
              .         .         .         .         .         .         .         .         .
-1227  GTGTGCGTGCTCGATCGCTCTCTCACTCGATCCGAGTCTTAAAGGTGGTGGTTTCAGCCGTGATTTATCAGAGAGCTGGGGCTGAAAACT  -11
              .         .         .         .         .         .         .         .         .
-1137  GGTAAGTTTGCTTTGGTGTGAGTGCGAGTACATAAGCCAAAGAGTTCGGCCAGTAGGCGAAACACAGAAACGCTTCTTGCCAGTCGAACC  -10

                                                      !
              .         .         .         .         .         .         .         .         .
-1047  TCTCAGCCAGAAATCAGTCGTCAGCTAGTAGCCGAAGTGTCAGTTACACACATCGAGGATTCCCAAACCGCGTTGTTTCGGCGATCAAAA  -95
              .         .         .         .         .         .         .         .         .
 -957  ACCATACATCACATCAAATCGTATCCAAAATTATAATCAAAGTGTTGAAAACCTGAATCACTCAGACTGATCGAAAAGTGCTTGCAAACC  -86
              .         .         .         .         .         .         .         .         .
 -867  GAAAACAAAACAAAAAAAAAAAAAATAACAGACACAAACTGCCAAAGTGAGAAAACGTGCAGCAACGATCGCGAATCGCTGCAATTAAAA  -77
                                                     _
              .         .         .         .      | .         .         .         .         .
 -777  AGTCGTCATAAACCTAGTGCCCATCCAACAAAAAAAAGCAAAAAGGTAAGTTCTGAGCGAGGGAGACAACATGAGATCAAAGTCAATTCA  -68
                                                    _|

              .         .         .         .         .         .         .         .         .
 -687  CGGTTCTTGTTTTTGGGTCGCGGCCGTCGGCAAATTTTACGATCCCTTGTGCACTACTTTTATTTTTTACTGTTTCACGCGAAAAACTAA  -59
              .         .         .         .         .         .         .         .         .
 -597  CGGCCACATTTCCATCTTTTCCTTATTTTTTTGCGTCCCGAGGCAGCGGGAAAAAAATCACAATTTTCATAAGCCGAATTTTCTATTTTT  -50
              .         .         .         .         .         .         .         .         .
 -507  TGTTTGACTCGGGAAAAATCGGCGTGAGTTTTTATGGCCGAACGTCAAGGTCTGGACTGCTGCTGCATTTTTTTAGGGAACACTATTTTC  -41
              .         .         .         .         .         .         .         .         .
 -417  GCAGCACTCGTAATGACTTCGAATAAAAAAGAAAATCTACCTGAGTTTTATGACTGGACAGCGCGAAAAAAATGAAAATGAACGCATAGG  -32
```

```
             .         .         .         .         .         .         .         .         .
 -327 GTTGCATAATCCGGCAGCTTAAGTTTTTTGGCCTGTTGTGATCATAAAAAACGCCCATCCTGTCTAGTTTTTCCCAGTTTCCTATATATA   -238
             .         .         .         .         .         .         .         .         .
 -237 TCCTGGCCCGCCTAGAGCTCAGCATCAGTTGCTCAGCAGCATTCCAAGCGAACAGATCATACAGCAGATCCTCACAGCGATCGTGAGAAA   -148
             .         .         .         .         .         .         .         .         .
 -147 AGCATTCAAAATTCCAACAAATATCATTCCAAAATGGTGTTCAACTTGGTTAGTGTCCAGAGCCGTTCGCCTACTTGTGGATTACACATA   -58

                                                                -
             .         .         .         .         .         .|        .         .         .
  -57 TACCTCTCGACCAGTGGATTAAACCCTTACTAACGCGGATTTCTTTACAATCTTCCAGATGAACCAGACATGCAAAGTGTGCGGTGAGCC   32
                                                                MetAsnGlnThrCysLysValCysGlyGluPr   (11)
                                                                |_          ***   ---***


             .         .         .         .         .         .         .         .         .
   33 GGCGGCGGGCTTCCATTTTGGCGCCTTCACCTGCGAGGGCTGCAAGGTAAGTTGTGTCTCAAGAAATCCATTGACAAATAAATTAGCAGA   122
      oAlaAlaGlyPheHisPheGlyAlaPheThrCysGluGlyCysLys                                               (26)
       ---   ---   ---   ---      ---***------***---
             .         .         .         .         .         .         .         .         .
  123 ACGTAACCCCCAAGGGGTTAGTTTTAGAAATGTTCGAGGAACAGGCCATCGGCGATTCAAATCGTTGTACCATTGGCCTTAAGTTCTTGA   212
             .         .         .         .         .         .         .         .         .
  213 ATGAATTTCTGCTCTTCTTTTGCTAATCAGTTGCATACCACATAACTAAGCCACATTCGTCCTTCCCTTCGCCATTGCAGTCCTTCTTTG   302
                                                                                      SerPhePheG   (30)
                                                                                         ------

             .         .         .         .         .         .         .         .         .
  303 GCCGCTCTTACAACAACATCAGCACCATCAGCGAGTGCAAGAACGAGGGCAAGTGCATCATCGACAAGAAGAACCGCACCACCTGCAAGG   392
      lyArgSerTyrAsnAsnIleSerThrIleSerGluCysLysAsnGluGlyLysCysIleIleAspLysLysAsnArgThrThrCysLysA   (60)
        ------                           ***               ***   ---------      ---      ***   -
             .         .         .         .         .         .         .         .         .
  393 CGTGCCGCTTGAGGAAGTGCTACAACGTGGGCATGTCGAAGGGGGGATCCCGCTACGGACGTCGCTCCAACTGGTTCAAGATCCATTGTC   482
      laCysArgLeuArgLysCysTyrAsnValGlyMetSerLysGlyGlySerArgTyrGlyArgArgSerAsnTrpPheLysIleHisCysL   (90)
      --***------------***      ---------   ---------   ---      ------
             .         .         .         .         .         .         .         .         .
  483 TGCTGCAGGAGCACGAACAGGCCGCCGCAGCGGCGGGCAAGGCGCC1CCATTAGCGGGTGGCGTATCGGTGGGTGGTGCCCCGTCGGCCT   572
      euLeuGlnGluHisGluGlnAlaAlaAlaAlaAlaGlyLysAlaProProLeuAlaGlyGlyValSerValGlyGlyAlaProSerAlaS   (120)
             .         .         .         .         .         .         .         .         .
  573 CTTCCCCGGTGGGCTCGCCACACACTCCCGGATTTGGGGACATGGCCGCCCATTTGCACCACCATCATCAGCAGCAGCAGCAGCAGCAGG   662
      erSerProValGlySerProHisThrProGlyPheGlyAspMetAlaAlaHisLeuHisHisHisHisGlnGlnGlnGlnGlnGlnGlnV   (150)
             .         .         .         .         .         .         .         .         .
  663 TGCCGCGTCATCCACATATGCCTCTGCTGGGCTATCCCAGCTATCTGTCCGACCCATCCGCCGCCCTGCCCTTCTTCAGCATGATGGGCG   752
      alProArgHisProHisMetProLeuLeuGlyTyrProSerTyrLeuSerAspProSerAlaAlaLeuProPhePheSerMetMetGlyG   (180)
             .         .         .         .         .         .         .         .         .
  753 GTGTACCGCACCAGTCGCCCTTCCAGCTGCCCCCACACCTCCTCTTCCCAGGCTACCATGCAAGTGCTGCCGCTGCAGCGGCTTCTGCTG   842
      lyValProHisGlnSerProPheGlnLeuProProHisLeuLeuPheProGlyTyrHisAlaSerAlaAlaAlaAlaAlaAlaSerAlaA   (210)
             .         .         .         .         .         .         .         .         .
  843 CCGATGCCGCTTACCGGCAGGAGATGTACAAGCACCGCCAGAGCGTGGATTCCGTTGAGTCGCAGAACCGCTTTAGTCCCGCCAGCCAGC   932
      laAspAlaAlaTyrArgGlnGluMetTyrLysHisArgGlnSerValAspSerValGluSerGlnAsnArgPheSerProAlaSerGlnP   (240)
             .         .         .         .         .         .         .         .         .
  933 CACCAGTGGTGCAGCCCACCTCCTCGGCCCGCCAGTCGCCCATCGATGTCTGCCTGGAGGAGGATGTTCACTCCGTGCACAGCCATCAGT   1022
      roProValValGlnProThrSerSerAlaArgGlnSerProIleAspValCysLeuGluGluAspValHisSerValHisSerHisGlnS   (270)
             .         .         .         .         .         .         .         .         .
 1023 CGTCCGCAAGCCTCCTGCATCCCATTGCCATCCGAGCCACGCCAACCACTCCGACTAGCAGCAGCCCGCTGAGTTTTGCGGCCAAGATGC   1112
      erSerAlaSerLeuLeuHisProIleAlaIleArgAlaThrProThrThrProThrSerSerSerProLeuSerPheAlaAlaLysMetG   (300)
             .         .         .         .         .         .         .         .         .
 1113 AGAGCTTGTCGCCCGTTTCGGTTTGCTCCATTGGCGGCGAAACCACCAGCGTTGTACCAGTGCATCCTCCCACCGTTTCCGCTCAAGAAG   1202
      lnSerLeuSerProValSerValCysSerIleGlyGlyGluThrThrSerValValProValHisProProThrValSerAlaGlnGluG   (330)
```

(*continued*)

```
1203 GACCCATGGATCTGAGCATGAAGACCTCGCGGAGCTCCGTGCACAGCTTCAACGACAGCGGCTCCGAGGATCAAGAAGTGGAGGTGGCTC 129
     lyProMetAspLeuSerMetLysThrSerArgSerSerValHisSerPheAsnAspSerGlySerGluAspGlnGluValGluValAlaP (36

1293 CGCGCCGGAAGTTCTACCAACTGGAGGCCGAGTGCCTGACCACCAGCAGCAGCAGTTCCTCCCACTCCGCCGCCCACTCACCGAACACCA 138
     roArgArgLysPheTyrGlnLeuGluAlaGluCysLeuThrThrSerSerSerSerSerSerHisSerAlaAlaHisSerProAsnThrT (39

1383 CCACCGCCCATGCGGAAGTCAAGCGGCAGAAGCTAGGTGGTGCAGAGGCCACCCACTTCGGTGGCTTCGCGGTGGCCCACAATGCGGCTA 147
     hrThrAlaHisAlaGluValLysArgGlnLysLeuGlyGlyAlaGluAlaThrHisPheGlyGlyPheAlaValAlaHisAsnAlaAlaS (42

1473 GTGCCATGAGGGGAATATTCGTGTGTGTCTAAGTACACGGCGAAAAAACCAAGTGGGAGGAGTCGCCCCAAAAACCCTCGTTGTTTATTT 156
     erAlaMetArgGlyIlePheValCysValEnd                                                           (42

1563 TTTGTTACTTAAAGAAAATGTAAATTTATTCGTGTGCTCGCTCACACTTAGGGAAGTGGAAAGAGATAGGGACAGACAGGTTTTGCTGGA 165

1653 AAGAGACAGCCTGACCAGTTAGTTGCATTGCACTCGCACACATACACCTATATACCACCACACACACACTCACACTCACCTATTGAGCTC 174

1743 GGATCCAAAAATTATTTTTTATGAAAACGTTAAAATTGTAAATATATCTTTGAGCTTGTTTGCAATTGTATTTTAAAGTTAGCCGGCGGA 183

1833 AGAGCCGTAGAAGTAGTAATCATTCCCACCCTCAAATGCTATTGTACATACAAATTGTTAAGTCTAAGAATGATCTTTATTGTCTCTAAG 192

1923 TATTTTATTCTATTATAGTCCTAGTTATGGTATGTCTAAAGATTGGCATTTAGGTTTTATACAAAGAAAAATAAAAACTATTAAAAATTA 201
                                                                                 ------      |(A)n

2013 AACTTTTGTCGTTTCCAATGCTTTTCGGTGTATTTCAGAATACACAAATTCATATTTGAAGTTTTTGCTTATGGATAATTGAACTAACTT 210

2103 ATTTACAATGTCGTCTTGAAGCTCATTAATTCCACGGCACGTTTACTTCGGTGTTGCTTTTATTGATTTAATTTTAGTTGTGACATCATA 219

2193 GAAAAGTGTATTTAATTACAAAACAAACACTTTAAGAAAATTATTTAAAAATACTCATACAACGTATTCGTTGTACCTTAAAGTTAACGA 228

2283 ACTCTTCTGATTTGTTTAAGCACATTATTATGGACTATATGTCTGGTGCAAACTATCTTTCGGATGTATCTGCTGGATGTATCAACGAAG 237

2373 TTTGTCGGCTACCGCACATTCCTATAGGATCAAAGCCAATAAATTATTTACGTATTCAGCTACCGCTCTTGTTTCAAAATCAGTTCTGTT 246

2463 CAAACATGCGAGCATATATCCATATACATATATCTGATCGGCGGTGTCTTTGGCTCGATGTTCGTTAACCACGGGCCAAATGGCGTGGCC 255

2553 TGTCAATGGCAACACCAAAAAGAGACGGAAAACAAATGCTTTGGCATAAATTCAATCAACATTCGGTTGCAAGTCGATCGGCGATGGCCG 264

2643 ATAATAAAACCGATATAGCAGCCGTTAAGTGCTTTGCTGTGCCTGCCCATTGCACTCGCGATTGCTGTAAGGCAGTTGTGTAGTAAATTA 273

2733 AAAATGCCACAAATGTTACGCACAGAAATTCGATGCAACCCCCC 2776
```

kni SEQUENCE. Accession, X14153 (DROKNR1). kr1 and kr2 are two KR binding sites. An exclamation sign at −1,003 marks the 5′ end of a cDNA. Dashes under the amino-acid sequence mark conserved positions in the C_4/C_5 finger regions, and asterisks, the relevant Cys residues.

regions only; the C-terminal regions of KNI-type proteins bear no resemblance to the C-terminal regions of the mammalian receptors to which hormones bind. Further, there is no evidence that function of the KNI-type proteins in *Drosophila* requires the presence of a ligand, as is the case for the steroid/thyroid hormone receptors.

kni

Product

Functions

KNI plays an important role in the early stages of embryonic pattern determination in the posterior region of the embryo. The consensus binding site of KNI is AA/TCTAA/GATC (Hoch et al. 1992).

1. KNI is one of the regulators of the embryonic "zebra" pattern of expression of the pair-rule gene *hairy* (*h*): two of the functions of KNI appear to be the activation of stripe 6 of *h* and the repression of anterior expansion of stripe 7. Strong binding of KNI to the *h* promoter in the stripe-7 regulatory element and weak binding to that of stripe 6, has been observed (Pankratz et al. 1990).

2. KNI has a binding site in cd1, a *cis*-acting regulatory region of *Kr*. This binding site partly overlaps a *bicoid* protein (BCD) binding site and the two regulatory proteins compete for binding: excess KNI prevents BCD from activating *Kr* (Hoch et al. 1992).

Tissue Distribution

At blastoderm stage, the KNI protein is localized in a band that extends approximately between 43% and 27% egg length (Appendix, Fig. A.2).

Mutant Phenotypes

This is one of the gap genes. In embryos homozygous for a null allele, abdominal segments A1–A7 are fused and replaced by a single segment with a broad band of ventral denticles (embryonic lethal) (Nüsslein-Volhard and Wieschaus 1980; Ingham 1988).

Gene Organization and Expression

Open reading frame, 429 amino acids; expected mRNA length, ca. 2,068 bases. A cDNA sequence provides the only information on the 5′ and 3′ ends. There are two introns at −732/0 and after the Lys-26 codon. These parameters agree with an RNA of 2.2 kb detected in northerns. A second RNA of 2.5 kb has been reported; it is not clear if this is generated by alternative splicing or by alternative initiation or termination. *kni* is transcribed toward the telomere (*kni* Sequence) (Nauber et al. 1988).

Developmental Pattern

Accumulation of *kni* RNA is first evident in 2–4 h embryos and reaches a maximum by 4–6 h. After 8 h the RNA level is very low and it becomes

```
      1                                                  50                                                  100
EGON  .........M NQLCKVCGEP AAGFHFGAFT CEGCKSFFGR TYNNIAAIAG CKHNGDCVIN KKNRTACKAC RLRKCLLVGM SKSGSRYGRR SNWFKIHCLL
KNRL  MMNQDNPYAM NQTCKVCGEP AAGFHFGAFT CEGCKSFFGR SYNNLSSISD CKNNGECIIN KKNRTACKAC RLKKCLMVGM SKSGSRYGRR SNWFKIHCLL
KNI   .........M NQTCKVCGEP AAGFHFGAFT CEGCKSFFGR SYNNISTISE CKNEGKCIID KKNRTTCKAC RLRKCYNVGM SKGGSRYGRR SNWFKIHCLL
CON   ---------M NQ-CKVCGEP AAGFHFGAFT CEGCKSFFGR -YNN---I-- CK--G-C-I- KKNRT-CKAC RL-KC--VGM SK-GSRYGRR SNWFKIHCLL
                    *  *                *  *                *     *           *  *     *         |____  _________

      101                                                150                                                 200
EGON  QEQQ...... .....TTSGL GGGSSVGSGS GGGVSSASLE QLARLQQASN QARQTYQDKT NPC...IKSA TATTSPRIEG AAVGTGIGGG ..........
KNRL  QEQQQQAVAA MAAHHNSQQA GGGSSGGSGG GQGMPNGVKG MSGVPPPAAA AAALGMLGHP GGYPGLYAVA NAGGSSRSKE ELMMLGLDGS VEYGSHKHPV
KNI   QEHEQAAAAA ....GKAPPL AGGVSVGGAP SASSPVGSPH TPGFGDMAAH LHHHHQQQQQ QQVPRHPHMP LLGYPSYLSD .......... ........PS
CON   QE-------- ---------- -GG-S-G--- ---------- -------A-- ---------- ---------- ---------- ---------- ----------
      _| KNI BOX

      201                                                250                                                 300
EGON  .ASPSFLQAA KLHHQRQLKL DSRLSN.... ..TPSDSGAS SAGD...... PNEDGVTSVL GGQIATPSST NATSLPKLDL RHPNFPATSE PDA.DMQRQR
KNRL  VASPSVSSPD SHNSDSSVEV SSVRGNPLLH LGGKSNSGGS SSGA...... DGSHSGGGGG GGGGVTPGRP PQM...RKDL S.PFLPLPFP GLA.SMPVMP
KNI   AALPFFSMMG GVPHQSPFQL PPHLLFPGYH ASAAAAAASA ADAAYRQEMY KHRQSVDSVE SQNRFSPASQ PPVVQPTSSA RQSPIDVCLE EDVHSVHSHQ
CON   -A-P------ ---------- ---------- ---------- ---------- ---------- ------P--- ---------- ---------- ----------

      301                                                350                                                 400
EGON  HQELLE.... IFRSHSEPLY SSFAPFSHLP PVLLAAGVPQ LPI...FKDQ FKAELLFPTT SSPELEEPID LSFRSRADHA SPMAHNSNSP SLSEPAAASH
KNRL  PPAFLPPSHL LFPGYHPALY SHHQGLLKPT PEQQQAAVAA AAVQHLFNSS GAGQRFAPGT SPFANHQQHH KEEDQPAPAR SPSTHANNNH LLTNGGAADE
KNI   SSASLLHPIA IRATPTTPTS SSPLSFAAKM QSLSPVSVCS IG........ .......GET TSVVPVHPPT VSAQEGPMDL SMKTSRSSVH SFNDSGSEDQ
CON   ----L----- ---------- S--------- -------V-- ---------- ---------T ---------- ---------- S--------- ----------
```

```
     401                                               450                                                500
EGON .....CLGES TNFVRKSTPL DLTLVR.... ....SQTLTG *......... .......... .......... .......... .......... ..........
KNRL LTKRFYLDAV LKSQQQSPPP TTKLPPHSKQ DYSISALVTP NSESGRERVK SRQNEEDDEA RADGIIDGAE HDDEEEDLVV SMTPPHSPAQ QEERTPAGED
KNI  .......EVE VAPRRKFYQL EAECLTTSSS SSSHSAAHSP NTTTAHAEVK RQKLGGAEAT HFGGFAVAHN AASAMRGIFV CV*....... ..........
CON  ---------- ---------- ---------- ----S----- ---------- ---------- ---------- ---------- ---------- ----------

     501                                               550                                                600
EGON .......... .......... .......... .......... .......... .......... .......... .......... .......... ..........
KNRL PRPSPGQDNP IDLSMKTTGS SLSSKSSSPE IEPETEISSD VEKNDTDDDD EDLKVTPEEE ISVRETADPE IEEDHSSTTE TAKTSIENTH NNNNSISNNN
KNI  .......... .......... .......... .......... .......... .......... .......... .......... .......... ..........
CON  ---------- ---------- ---------- ---------- ---------- ---------- ---------- ---------- ---------- ----------

     601                                               650
EGON .......... .......... .......... .......... .......... .........
KNRL NNNNNNNNSI LSDSEASETI KRKLDELIEA SSENGKRLRL EAPVKVATSN ALDLTTKV*
KNI  .......... .......... .......... .......... .......... .........
CON  ---------- ---------- ---------- ---------- ---------- ---------
```

FIG. 20.1. Alignment of the KNI-related polypeptides by the GCG program *Pileup*. Asterisks mark Cys in the C_4/C_5 finger domains. The KNI box is underlined. The CON(sensus) sequence identifies residues identical in the three sequences.

undetectectable in larval stages. The 2.2 kb transcript is present transiently during the blastoderm stage while the 2.5 kb transcript predominates in the later stages (Nauber et al. 1988). RNA is first detectable, by *in situ* hybridization, after the 11th round of embryonic nuclear division when it appears in a broad band centered at 40–35% egg length (Appendix, Figs A.1–A.3). Soon thereafter, RNA appears at the anterior tip; and, still later, during blastoderm cellularization, a third zone of expression becomes evident as a narrow stripe at 75–70% egg length. Expression in the posterior domain diminishes during gastrulation and eventually ceases altogether. In the anterior tip, on the other hand, expression persists through gastrulation when it exhibits a complex pattern. In yet older embryos, *kni* transcription is limited to distinct areas of the epidermis and gut (Rothe et al. 1989).

Promoter

The expression of *kni* is stimulated by the *Krüppel* protein (KR) either directly or indirectly, and there are two KR binding sites between −2,300 and −2,400 (Pankratz et al. 1989; Capovilla et al. 1992). Anteriorly, transcription of *kni* seems mainly regulated by the product of *hunchback* (HB) (and perhaps the product of *bicoid*), being repressed at intermediate and high concentrations, and stimulated at low concentrations. Similarly, *kni* is repressed by the *tailless* product (TLL) which is present at the posterior end of the embryo. It is proposed that these interactions explain the expression of *kni* in a broad band immediately posterior to the band of *Kr* expression in the mid-section of the embryo (Appendix, Figs A.2 and A.4) (Hülskamp et al. 1990).

A 4.4 kb fragment upstream of the transcription initiation site is sufficient for normal *kni* expression. Deletion mapping of this DNA segment indicates the presence of several sub-regions in whose absence *kni* expression in embryos expands either anteriorly or posteriorly. The presence of HB and TLL binding sites in those sub-regions led to the suggestion that *kni* expression is activated throughout the embryo and that the broad band of *kni* transcription in the posterior half of the embryo is achieved through repression by HB (anteriorly) and TLL (posteriorly) (Pankratz et al. 1992).

knrl
(*knirps-related*)

Product

Unknown. No mutations are known in this gene (Fig. 20.1).

Gene Organization and Expression

Open reading frame, 647 amino acids. One cDNA of 3,505 bases was sequenced; a single band of 3.8 kb is detected by northern analysis. That cDNA sequence provides the only information on the 5′ and 3′ ends (Oro et al. 1988).

Developmental Pattern

A low level of maternal *knrl* RNA is uniformly distributed throughout pre-blastoderm embryos. After the 12th nuclear division a posterior band forms as is also the case for *kni* RNA; and afterwards expression of the two genes is almost the same. However, *knrl* transcription never ceases altogether; a low level of expression is maintained in all stages (Oro et al. 1988; Rothe et al. 1989).

egon
(embryonic gonad)

Product

Unknown. No mutations are known in this gene (Fig. 20.1).

Gene Organization and Expression

Open reading frame, 373 amino acids. There is one intron after Lys-26.

Developmental Pattern

Transcripts are restricted to late embryogenesis and they are 10-fold less abundant than for *kni* or *knrl*. After germ band shortening, transcripts appear only in the gonadal primordia that form in abdominal segment 5, as demonstrated by *in situ* hybridizations (Rothe et al. 1989).

References

Capovilla, M., Eldon, E. D. and Pirrotta, V. (1992). The *giant* gene of *Drosophila* encodes a b-ZIP DNA-binding protein that regulates the expression of other segmentation gap genes. *Development* **114**:99–112.

Evans, R. M. (1988). The steroid and thyroid hormone receptor superfamily. *Science* **240**:889–895.

Evans, R. M. and Hollenberg, S. M. (1988). Zinc fingers: gilt by association. *Cell* **52**:1–3.

Harrison, S. C. (1991). A structural taxonomy of DNA-binding domains. *Nature* **353**:715–719.

Hoch, M., Gerwin, N., Taubert, H. and Jäckle, H. (1992). Competition for overlapping sites in the regulatory region of the *Drosophila* gene *Krüppel*. Science **256**:94–97.

Hülskamp, M., Pfeifle C. and Tautz, D. (1990). A morphogenetic gradient of *hunchback* protein organizes the expression of the gap genes *Krüppel* and *knirps* in the early *Drosophila* embryo. *Nature* **346**:577–580.

Ingham, P. W. (1988). The molecular genetics of embryo pattern formation in *Drosophila*. *Nature* **335**:25–34.

Nauber, U., Pankratz, M. J., Kienlin, A., Seifert, E., Klemm, U. and Jäckle, H. (1988). Abdominal segmentation of the Drosophila embryo requires a hormone receptor-like protein encoded by the gap gene *knirps*. *Nature* **336**:489–492.

Nüsslein-Volhard, C. and Wieschaus, E. (1980). Mutations affecting segment number and polarity in *Drosophila*. *Nature* **287**:795–801.

Oro, A. E., Ong, E. S., Margolis, J. S., Posakony, J. W., McKeown, M. and Evans, R. M. (1988). The *Drosophila* gene *knirps-related* is a member of the steroid-receptor gene superfamily. *Nature* **36**:493–496.

Pankratz, M. J., Busch, M., Hoch, M., Seifert, E. and Jäckle, H. (1992). Spatial control of the gap gene *knirps* in the *Drosophila* embryo by posterior morphogen system. *Science* **255**:986–989.

Pankratz, M. J., Hoch, M., Seifert, E. and Jäckle, H. (1989). *Krüppel* requirement for *knirps* enhancement reflects overlapping gap gene activities in the *Drosophila* embryo. *Nature* **341**:337–340.

Pankratz, M. J., Seifert, E., Gerwin, N., Billi, B., Nauber, U. and Jäckle, H. (1990). Gradients of *Krüppel* and *knirps* gene products direct pair-rule gene stripe patterning in the posterior region of the *Drosophila* embryo. *Cell* **61**:309–317.

Rothe, M., Nauber, U. and Jäckle, H. (1989). Three hormone receptor-like *Drosophila* genes encode an identical DNA-binding finger. *EMBO J.* **8**:3087–3094.

21

Krüppel: Kr

Chromosomal Location:
2R, 60F3

Map Position:
2-107.6

Product

DNA-binding regulatory protein of the Zn-finger type that plays a central role in the early stages of embryonic pattern determination in the mid-section of the embryo.

Structure

The protein sequence can be divided into three regions. Two of the regions, the amino-terminal and carboxy-terminal segments (221 and 108 amino acids, respectively) are >30% Ala, Ser, and Pro. The third region, at the middle of the protein, is made up of four and a half repeats of a 28-amino-acid segment; these segments have the characteristics of C_2H_2 Zn-fingers. A potential glycosylation site is found near the C-terminus. Short segments near the N-terminus show similarities with the *hunchback* (*hb*) protein (Rosenberg et al. 1986; Evans and Hollenberg 1988; Harrison 1991).

Function

Distinct DNA-binding and repressor domains have been identified in KR. KR finger domains bind to AANGGGTTAA decamers, sequences that are known to occur in the promoters of several genes controlled by KR (Pankratz et al. 1989; Stanojevic et al. 1989; Treisman and Desplan 1989). Transcriptional repression is effected through an Ala-rich region of the protein included in the segment from amino acids 26–110 (Licht et al. 1990).

KR acts as a repressor of the anterior gap gene *hb* (*hb*), the pair-rule gene *eve-skipped* (*eve*) (Licht et al. 1990) and probably *giant* (*gt*) (Kraut and Levine 1991). In the posterior regions of the embryo, it interacts with the pair-rule gene *hairy* (*h*) and with the gap gene *knirps* (*kni*), which it activates, perhaps indirectly, by repressing the expression of *gt*, a repressor of *kni* (Capovilla et al. 1992). KR-binding sites exist in several of the stripe-specific regulatory elements in the promoters of *eve* and *h*, as well as in the promoters of *hb* and *kni*. These interactions play important roles in the periodic expression of the primary pair-rule genes *eve* and *h* and consequently, in the future segmentation pattern along the antero-posterior axis of the embryo (Pankratz et al. 1989, 1990; Small et al. 1991; Stanjojevic et al. 1989, 1991; Treisman and Desplan 1989).

KR may be involved in developmental processes other than segmentation. After gastrulation, the protein can be seen in the nuclei of some neuroblasts, Malpighian tubule anlagen, amnion serosa and other cells; in some sites, KR persists until the end of embryogenesis (Gaul et al. 1987).

Tissue Distribution

The *Kr* protein is localized in nuclei with a pattern of accumulation that agrees roughly with what would be expected from the regions and stages of embryonic *Kr* transcription (see below). The correspondence is not exact, however; the protein begins to appear 30 min later than mRNA, during the 13th nuclear division, suggesting that there exists a mechanism of post-transcriptional control. During the blastoderm stage, when metameric determination is taking place, KR accumulates in a bell-shaped concentration profile. That is, the protein is detectable between approximately 60 and 33% egg length (Appendix, Figs A.2 and A.3) with the concentration being maximum in the middle of the embryo and declining in steep exponential fashion toward each pole (Knipple et al. 1985; Gaul et al. 1987).

Mutant Phenotypes

This is one of the gap genes. Embryos amorphic for *Kr* lack the three thoracic and first five abdominal segments; *Kr* is an embryonic lethal without maternal effect (Nüsslein-Volhard and Wieschaus 1980; Ingham 1988).

Gene Organization and Expression

Open reading frame, 466 amino acids. Primer extension and cDNA sequencing were used to define the 5′ end. Two cDNAs having different 3′ ends (368 bp apart) were sequenced. Spliced and unspliced RNAs are abundant. Thus it might be expected that RNAs of several sizes ranging between 1,851 and 2,591 bases would be observed. However, only two bands of approximately 2.5 kb are detectable; one has an intron, the other does not. The intron is in the Thr-13 codon, and there are several short open reading frames within the intron that might serve for translational control (*Kr* Sequence) (Gaul et al. 1987).

When a genomic library was searched extensively with a *Kr* probe consisting of only the finger domains, eight cross-hybridizing clones were identified. One

Kr

```
        Kr730 element
        ______BamH1
-3267   GGATCCTAAGTTAACTATAATCCAGGCTTAATCACTGGATCAATAACTAAGTAGCATTTTCCGGGATGGAAATATGAAGTTACCTGCATA   -3178

-3177   TGACCTACCGATCCTGAAAACTGCTTTAACTTAATCGACATGCATGATCATAAAAAGCAATTTGCTACAATTTATATTTTTTTGCTTTTC   -3088
                            ----------------tll1   ///////////////////////////////////////////////

-3087   CTTCTTTTAAGCATCTGGGATCTGGATCAGAAAAGAAAAAGTGTAACGCCTACCTTCAGAAACGGATTAAATTTTTTCAGACAAATAATC   -2998
        ///////hb1-2              ////////////////////hb3                ----------tll2       -------
                                                             ||||||||||||||||||||||||bcd1 |||||||
                                                                      /////////////hb4

-2997   CAGCCTTAAGCATGGTGATTAAGCTTGATCCCCTACCAAGGGGCGTAATATTGACGGATTTTCCTTAAATCCGTCTGTTAATCTCCGGCT   -2908
        -tll3                                        ---------tll4                  -----------tll5
        ||||||||||||||||||||||||||bcd2-3                     ||||||||||||||||||||||||||||||||||||||

-2907   TAGAGCGCGACGCGTTTTTTCGCGACTCCGCCTGCATTGTTTTTTTTTCAGTTTCTTCAATTCGCAAGAAGGCAGGCCTATGGACCGAAT   -2818
        |||||bcd4-5 ///////////hb5          //////////////////hb6

-2817   GAGGATCATAATTATGGAATTCCTAAATAAACTAAGAAGGGCAGTCGGCATAGTATTGATCTACCTGTAAGCGTGGGTTCTATCTTTGCC   -2728

-2727   CCTCGCATTCGAGACTCTCTAGTCACAGGTAGACTGTATACCAGCCTTGAGTTCGTCGGCAATTAAGAAGTCAAATTTCTCTTAAAAACA   -2638
                                                                  ------------------tll6
                                                                                     ////////////

-2637   ACAAAAAATGTCAAAGTAAAAACAATGCAAAAAAGATGTGTAACTGAACTAAATCCGGCTTAGGATTCTTGCGTCATAAACGTGACTAGG   -2548
        ////////////////////////hb7-10            ----------------kni1 ---------tll7
                                                      ||||||||||||||||||||||||bcd6
                                                                 \\\\\\\\\\\\\\\\gt

-2547   TAGCC -2543

                                                                                           -->-184
 -267   AATATAATCGAATGAAATTTCAACTACCTCATTTTGCTAAGTCNGTAGACTTTTATAAAAGACAATTTTTGTGAAATCTCTCTACCTCAA   -178
                                                             ------

 -177   AGTACAAAAGTGTGTACAAAAATTATTCATATCCCTGAAAGTGCACAAAATTCTCAAATGAAATTTTGTTGTCTAAAAAACTAAGCTCCA   -88

  -87   AAATCACTAAGGCGAATATTATAGGTGTTTTCTGTGTGCGGGAAAACATTGCGCGACACAAAATTAGGAGCACAAGAAGAATTTGTTGAT   2
                                                                                                Me   (1)

    3   GTCCATATCAATGCTTCAAGACGCACAAACGCGAAGTAAGTATAGACCAAATTAAAATATTCCCCAAGAAAGTAAACTATCTAGAACTTC   92
        tSerIleSerMetLeuGlnAspAlaGlnThrArgT                                                        (13)

   93   TAGTGTCCCCGATCACTTTCTCATTATTAAACAGTCCGATGTCTTTAGGATAGAAAATACAAATGTAATGTAATTGCAGCACATACCGAT   182

  183   TAGTTGAATTTGTTTACATGTTTGGACAGGAACCGGCACTTAACTCGTTATCGACCAAAACAAAAACTAGTTAGACGAAAATAGAGAGCT   272

  273   GCGAAAACACTAAGAGTTCGCTCCGTACGAAACTTTCTCTCACACATGAATCATATGTAAAATTTTTTTCTCTTTTAAGCCGTTGCTCTT   362

  363   AAGACATTTCCAAATGAAAACATACTAACTTATGATTTTTTTTTTAGCCTTAGCTGCTGCATTAGCTGGCATAAAACAAGAGGACGTTCA   452
                                                       hrLeuAlaAlaAlaLeuAlaGlyIleLysGlnGluAspValHi   (27)

  453   TCTAGACCGTTCCATGTCGCTATCGCCCCCCATGTCGGCCAACACATCAGCTACAAGCGCCGCTGCGATTTATCCAGCTATGGGTCTCCA   542
        sLeuAspArgSerMetSerLeuSerProProMetSerAlaAsnThrSerAlaThrSerAlaAlaAlaIleTyrProAlaMetGlyLeuGl   (57)
```

(continued)

```
             .         .         .         .         .         .         .         .         .
543   ACAGGCGGCCGCTGCCTCAGCTTTTGGAATGCTATCACCTACCCAACTTCTGGCTGCAAACCGTCAAGCTGCCGCATTCATGGCCCAACT   632
      nGlnAlaAlaAlaAlaSerAlaPheGlyMetLeuSerProThrGlnLeuLeuAlaAlaAsnArgGlnAlaAlaAlaPheMetAlaGlnLe   (87)

             .         .         .         .         .         .         .         .         .
633   GCCCATGAGCACATTGGCCAACACTCTCTTTCCACACAATCCGGCGGCTTTGTTTGGGGCTTGGGCTGCCCAACAGTCGCTCCCGCCCCA   722
      uProMetSerThrLeuAlaAsnThrLeuPheProHisAsnProAlaAlaLeuPheGlyAlaTrpAlaAlaGlnGlnSerLeuProProGl   (117

             .         .         .         .         .         .         .         .         .
723   GGGTACGCATTTACATTCGCCGCCAGCCAGCCCGCACTCGCCGCTGTCCACTCCTTTAGGTAGTGGCAAGCACCCATTAAATTCCCCCAA   812
      nGlyThrHisLeuHisSerProProAlaSerProHisSerProLeuSerThrProLeuGlySerGlyLysHisProLeuAsnSerProAs   (147

             .         .         .         .         .         .         .         .         .
813   CAGCACTCCCCAGCACCATGAGCCAGCGAAGAAGGCTCGAAAGTTATCGGTTAAGAAGGAGTTTCAGACCGAGATCAGCATGAGTGTAAA   902
      nSerThrProGlnHisHisGluProAlaLysLysAlaArgLysLeuSerValLysLysGluPheGlnThrGluIleSerMetSerValAs   (177

             .         .         .         .         .         .         .         .         .
903   CGATATGTACCTATCATCGGGAGGCCCAATATCTCCGCCTTCCAGTGGCAGCTCTCCTAATTCAACGCACGACGGAGCGGGTGGAAATGC   992
      nAspMetTyrLeuSerSerGlyGlyProIleSerProProSerSerGlySerSerProAsnSerThrHisAspGlyAlaGlyGlyAsnAl   (207

             .         .         .         .         .         .         .         .         .
993   TGGATGTGTCGGTGTCTCCAAGGATCCATCTCGCGACAAAAGCTTCACCTGTAAAATCTGCTCACGCAGCTTTGGCTATAAGCACGTGCT   1082
      aGlyCysValGlyValSerLysAspProSerArgAspLysSerPheThrCysLysIleCysSerArgSerPheGlyTyrLysHisValLe   (237
                                                       ---      ---

             .         .         .         .         .         .         .         .         .
1083  TCAGAACCACGAACGCACCCACACCGGTGAGAAGCCTTTCGAATGTCCGGAGTGCGACAAGCGGTTTACTCGGGACCATCACTTAAAAAC   1172
      uGlnAsnHisGluArgThrHisThrGlyGluLysProPheGluCysProGluCysAspLysArgPheThrArgAspHisHisLeuLysTh   (267
             ---         ---                     ---      ---

             .         .         .         .         .         .         .         .         .
1173  CCACATGCGTTTGCATACTGGAGAAAAACCATATCATTGCTCGCACTGCGATCGTCAATTCGTTCAGGTGGCCAATCTTAGACGACATTT   1262
      rHisMetArgLeuHisThrGlyGluLysProTyrHisCysSerHisCysAspArgGlnPheValGlnValAlaAsnLeuArgArgHisLe   (297
       ---         ---                     ---      ---                                    ---

             .         .         .         .         .         .         .         .         .
1263  GCGAGTCCACACTGGAGAGCGTCCCTATACTTGTGAAATCTGCGATGGCAAATTCAGTGACTCCAATCAGCTTAAGTCCCACATGCTGGT   1352
      uArgValHisThrGlyGluArgProTyrThrCysGluIleCysAspGlyLysPheSerAspSerAsnGlnLeuLysSerHisMetLeuVa   (327
             ---                     ---      ---                                    ---

             .         .         .         .         .         .         .         .         .
1353  ACACACCGGTGAAAAGCCGTTCGAGTGCGAACGGTGTCACATGAAGTTCCGACGGCGGCACCATCTGATGAATCACAAGTGTGGCATCCA   1442
      lHisThrGlyGluLysProPheGluCysGluArgCysHisMetLysPheArgArgArgHisHisLeuMetAsnHisLysCysGlyIleGl   (357
       ---                     ---      ---                                    ---

             .         .         .         .         .         .         .         .         .
1443  GTCGCCGCCTACTCCCGCGCTTTCACCGGCCATGAGTGGAGATTACCCCGTGGCAATCTCCGCAATTGCTATCGAGGCATCCACGAATAG   1532
      nSerProProThrProAlaLeuSerProAlaMetSerGlyAspTyrProValAlaIleSerAlaIleAlaIleGluAlaSerThrAsnAr   (387

             .         .         .         .         .         .         .         .         .
1533  ATTTGCGGCAATGTGTGCCACCTACGGAAGTTCGAATGAGTCGGTCGACATGGAAAAAGCGACACCGGAGACGATGGTCCATTGGATTTG   1622
      gPheAlaAlaMetCysAlaThrTyrGlySerSerAsnGluSerValAspMetGluLysAlaThrProGluThrMetValHisTrpIleCy   (417
                                        ---------

             .         .         .         .         .         .         .         .         .
1623  TCTGAAGATGGAGCCAGCTCTGTGGATGGCCATTACAGCAACATCGCACGGCGCAAGGCACAGGACATTCGTCGGGTTTTCCGGCTGCCT   1712
      sLeuLysMetGluProAlaLeuTrpMetAlaIleThrAlaThrSerHisGlyAlaArgHisArgThrPheValGlyPheSerGlyCysLe   (447

             .         .         .         .         .         .         .         .         .
1713  CCACCGCAAATCCCTCACGTACCCAGTGATATGCCTGAGCAAACCGAGCCAGAGGATTTGAGCATGCATTCTCCTCGTTCTATCGGATCT   1802
      uHisArgLysSerLeuThrTyrProValIleCysLeuSerLysProSerGlnArgIleEnd                               (466

             .         .         .         .         .         .         .         .         .
1803  CACGAGCAAACCGATGATATTGACTTGTATGATTTAGATGATGCCCCGGCTTCTTATATGGGCCATCAACAACATTAGGCCACAACCAGT   1892

             .         .         .         .         .         .         .         .         .
1893  CCGAATTGTACATAGCCCTAATCAGTTTTCATTTGATGAAATTGACTGGCATTTATTAACACAAAATTGAAAATTTTGCTATTTCAAAGT   1982
```

```
                .         .         .         .         .         .         .         .         .
1983  GGAAAGTAAAAATTGTTGCAACAGGAATATAATGATAAGTACAAGTTTAAAAAAATAACATACAAAAAGTCGAAATTGTACAAAGTAAGC  2072
                                ------                       |(A)n
                .         .         .         .         .         .         .         .         .
2073  CATACGTATGCTTGTTACGCCAAACCCACCAAATCAAATCGAAAATGTCGTGCCATTCTTTACCTTAAATTTAAGTTATATTCTTAGGTT  2162
                .         .         .         .         .         .         .         .         .
2163  CGGAATCTTAAATTGTACATATTCAGCTTACACAGCTGCCAATTGTAAAGTAATCGGCGCTCTAAACATGCTTGTTGCAGAAAAATAAAA  2252
                .         .         .         .         .         .         .         .         .
2253  GACACAAAGGTTTAATTAGGAAATCTATAACTAATTTTATTTAATTTATTACGCTTAATTTTTTTATAATTTAATCAAATTCTTTAAGAA  2342
                .         .         .         .         .         .         .         .         .
2343  AACAATCGCAATAATCTCAAACAAAACTAACTTCAAGTTAAATAATAAAAAACATTTGTTTGATAATTGTTCTGTTTGCATTCTCTATTT  2432
                                                  ------            |(A)n
                .         .         .         .         .
2433  AAAACTATTATTAAATATAAAAATTTAGTTAATCCTGTTTTTTTAAAGATC  2483
```

Kr SEQUENCE. The segment from -267 to 2,483 is from GenBank, Accession, X03414 (DROKR). His and Cys of the Zn-finger repeats are underlined, as is a potential glycosylation site (Asn-399). The segment from $-3{,}267$ to $-2{,}543$ (Kr730) is from Hoch et al. (1992) and is numbered arbitrarily. This regulatory region starts at the *Bam*H1 site approximately 3 kb upstream of *Kr*. Symbols under the sequence indicate various footprints: ---tll, for TLL; |||bcd for BCD, ///hb for HB (Hoch et al. 1991, 1992) and \\\gt for GT (Capovilla et al. 1992).

of these was characterized in some detail. It was localized to the left arm of chromosome two in region 26A–B. Sequence analysis identified three finger domains of the *Kr* type; greatest similarity was found in the seven amino acids that separate adjacent fingers (the "H/C-link"; Schuh et al. 1986).

Developmental Pattern

Both *Kr* transcripts are present primarily in 2–5 h embryos, blastoderm to gastrula stages (Rosenberg et al. 1986).

Kr transcripts are first detected in syncytial blastoderm embryos, after the 11th nuclear division (Appendix, Fig. A.1). RNA occurs in the peripheral cytoplasm confined to a band 8–10 nuclei wide in the mid-embryo (55–45% egg length; Appendix, Figs A.2 and A.4). By the cellular blastoderm stage (3.5 h of development), the level of transcript has greatly increased; the RNA appears in a band about 12–14 cells wide as well as in the cytoplasm of yolk cells. During this stage, *Kr* RNA also accumulates in a posterior cap; the cap is 10 cells wide and does not include the pole cells. Early in gastrulation, a third zone of gene expression develops in the anterior portion of the embryo; and, as gastrulation progresses, expression becomes yet more widespread. By the end of germ-band extension (6 h), *Kr* RNA occurs throughout the embryo, from the posterior edge of the cephalic furrow and through the thoracic and abdominal anlagen. The transcripts then begin to diminish; and, by the beginning of germ-band shortening (8 h), they reach near background level (Knipple et al. 1985).

Promoter

An upstream segment of DNA 18-kb long is necessary for normal *Kr* expression. Within this region, there are at least seven independent *cis*-acting elements that, alone or in various combinations control *Kr* expression at each of the ten identified embryonic sites where *Kr* product is found.

Two of the *cis*-acting elements (cd1 and cd2), located from 1 to 3 kb upstream of the transcription initiation site, are primarily responsible for expression in the central domain of the embryo (Hoch et al. 1990). During the blastoderm stage, the central region of expression is, at least in part, defined by the gradients of *bicoid* (*bcd*) and *hb* gene products; *Kr* transcription appears to be stimulated by low concentrations and repressed by high concentrations of those proteins (Hülskamp et al. 1990). A 400-bp segment in cd1 is essential for expression of a reporter gene in the central region of the embryo. The *cis*-acting function of cd1 depends on the presence of wild-type alleles of *hb* (repressing *Kr* transcription) and *bcd* (activating transcription). Clustered in 730 bp of cd1 (the Kr730 element) are 10 HB and 6 BCD binding sites (Hoch et al. 1991). Seven binding sites for the product of *tll* (TLL) are also found in the Kr730 element. The TLL sites partly overlap BCD binding sites, and there is competition for occupancy such that the activating function of BCD can only occur if TLL concentration is low enough. Similar competition occurs between BCD and the *kni* product (KNI); but there is only on KNI binding site, so its effect does not appear to be so significant as TLL's (*Kr* Sequence) (Hoch et al. 1992).

The repressive action of BCD may be effected directly or through its activation of *gt*, which in turn would interact with HB to repress *Kr* (Kraut and Levine 1991). The repressive action of *gt* on *Kr*, if it occurs, would be mediated by *gt* protein binding sites in the regulatory regions cd1 (*Kr* Sequence) and cd2 (Capovilla et al. 1992).

References

Capovilla, M., Eldon, E. D. and Pirrotta, V. (1992). The *giant* gene of *Drosophila* encodes a b-ZIP DNA-binding protein that regulates the expression of other segmentation gap genes. *Development* **114**:99–112.

Evans, R. M. and Hollenberg, S. M. (1988). Zinc fingers: gilt by association. *Cell* **52**:1–3.

Gaul, U., Seifert, E., Schuh, R. and Jäckle, H. (1987). Analysis of *Krüppel* protein distribution during early *Drosophila* development reveals posttranscriptional regulation. *Cell* **50**:639–647.

Hülskamp, M., Pfeifle, C. and Tautz, D. (1990). A morphogenetic gradient of *hunchback* protein organizes the expression of the gap genes *Krüppel* and *knirps* in the early *Drosophila* embryo. *Nature* **346**:577–580.

Harrison, S. C. (1991). A structural taxonomy of DNA-binding domains. *Nature* **353**:715–719.

Hoch, M., Schröder, C., Seifert, E. and Jäckle, H. (1990). Cis-acting control elements for *Krüppel* expression in the *Drosophila* embryo. *EMBO J.* **9**:2587–2595.

Hoch, M., Seifert, E. and Jäckle, H. (1991). Gene expression mediated by *cis*-acting sequences of the *Krüppel* gene in response to the *Drosophila* morphogenes *bicoid* and *hunchback*. *EMBO J.* **10**:2267–2278.

Hoch, M., Gerwin, N., Taubert, H. and Jäckle, H. (1992). Competition for overlapping sites in the regulatory region of the *Drosophila* gene *Krüppel*. *Science* **256**:94–97.

Ingham, P. W. (1988). The molecular genetics of embryo pattern formation in *Drosophila*. *Nature* **335**:25–34.

Knipple, D. C., Seifert, E., Rosenberg, U. B., Preiss, A. and Jäckle, H. (1985). Spatial and temporal patterns of *Krüppel* gene expression in early *Drosophila* embryos. *Nature* **317**:40–44.

Kraut, R. and Levine, M. (1991). Mutually repressive interactions between the gap genes *giant* and *Krüppel* define middle body regions of the *Drosophila* embryo. *Development* **111**:611–621.

Licht, J. D., Grossel, M. J., Figge, J. and Hansen, U. M. (1990). *Drosophila Krüppel* protein is a transcriptional repressor. *Nature* **346**:76–79.

Nüsslein-Volhard, C. and Wieschaus, E. (1980). Mutations affecting segment number and polarity in *Drosophila*. *Nature* **287**:795–801.

Pankratz, M. J., Hoch, M., Seifert, E. and Jäckle, H. (1989). *Krüppel* requirement for *knirps* enhancement reflects overlapping gap gene activities in the *Drosophila* embryo. *Nature* **341**:337–340.

Pankratz, M. J., Seifert, E., Gerwin, N., Billi, B., Nauber, U. and Jäckle, H. (1990). Gradients of *Krüppel* and *knirps* gene products direct pair-rule gene stripe patterning in the posterior region of the *Drosophila* embryo. *Cell* **61**:309–317.

Rosenberg, U. B., Schröder, C., Preiss, A., Kienlin, A., Côté, S., Riede, I. and Jäckle, H. (1986). Structural homology of the product of the *Drosophila* Krüppel gene with *Xenopus* transcription factor IIIA. *Nature* **319**:336–339.

Schuh, R., Aicher, W., Gaul, U., Côté, S., Preiss, A., Maier, D., Seifert, E., Nauber, U., Schröder, C., Kemler, R. and Jäckle, H. (1986). A conserved family of nuclear proteins containing structural elements of the finger protein encoded by Krüppel, a *Drosophila* segmentation gene. *Cell* **47**:1025–1032.

Small, S., Kraut, R., Warrior, R. and Levine, M. (1991). Transcriptional regulation of a pair-rule stripe in *Drosophila*. *Genes Dev.* **5**:827–839.

Stanojevic, D., Hoey, T. and Levine, M. (1989). Sequence-specific DNA-binding activities of the gap proteins encoded by *hunchback* and *Krüppel* in *Drosophila*. *Nature* **341**:331–335.

Stanojevic, D., Small, S. and Levine, M. (1991). Regulation of a segmentation stripe by overlapping activators and repressors in the *Drosophila* embryo. *Science* **254**:1385–1387.

Treisman, J. and Desplan, C. (1989). The products of the *Drosophila* gap genes *hunchback* and *Krüppel* bind to the *hunchback* promoters. *Nature* **341**:335–337.

22

The Metallothionein Genes: *Mtn, Mto*

Chromosomal Location:		**Map Position:**
Mtn	3R, 85E10–15	3-48.8
Mto	3R, 92	3-[68]

Products

Small, Cys-rich cadmium- and copper-binding proteins.

Structure

MTN and MTO share properties with the metallothioneins (MT) of other invertebrate and vertebrate species: they are small, they lack aromatic amino acids and Cys residues constitute 25% or more of the protein (Lastowski-Perry et al. 1985; Mokdad et al. 1987). One striking feature of MTN is the arrangement of its 10 Cys residues in Cys-X-Cys groups that are distributed almost identically to the Cys-X-Cys groups in the N-terminal half of mammalian MT (Lastowski-Perry et al. 1985; Maroni 1990). Otherwise, sequence identity beteen MTN and MTO, or between either one of the *Drosophila* MTs and a mammalian MT is not extensive, being only 20–25% in all pairwise combinations.

Cu-MTs may be precursors of the copper- and sulfur-rich concretions that are detectable in the middle mid-gut of larvae fed on Cu^{++}-containing food (Tapp and Hockaday 1977; Maroni et al. 1986b; Lauverjat et al. 1989).

MTO has been purified and partially sequenced (Silar et al. 1990); but MTN has proven surprisingly intractable in this respect, and purification of the protein has not been achieved (Silar et al. 1990; G. Maroni, unpublished observations).

Function

MTs are involved in metal tolerance as evidenced by the fact that flies with duplications for *Mtn* have increased tolerance to Cu^{++} and Cd^{++} in the medium. Such duplication-carrying flies have been obtained from many natural populations where it is thought that elevated Cu^{++} level has acted as a selective agent (Otto et al. 1986; Maroni et al. 1987; Theodore et al. 1991). Also, cells in culture that had been selected for increased tolerance to Cd^{++} showed higher levels of MT (probably MTO) accumulation (Debec et al. 1985; Mokdad et al. 1987). Whether these proteins also serve a role in metal homeostasis is not known; null mutations are not available.

Tissue Distribution

Synthesis of MT is stimulated by the presence of Cd^{++} or Cu^{++} in the food and the proteins accumulate primarily in the midgut of individuals so treated (Maroni and Watson 1985).

Mtn

Gene Organization and Expression

Open reading frame, 40 amino acids. There are two common alleles: *Mtn*$^{.3}$, thought to be closer to the ancestral allele, is expected to make an mRNA 387 bases long; *Mtn*^{1} has lost 49 bases of the 3′ untranslated region (*Mtn* Sequence) and is expected to make an mRNA 338 bases long. These estimates are in agreement with RNA bands of 0.4 and 0.5 kb detected by northern analysis. Primer extension and cDNA sequencing were used to define the 5′ end. The 3′ end was obtained from a cDNA sequence that included a poly(A) tail. There is an intron in the Gly-8 codon (*Mtn* Sequence) (Lastowski-Perry et al. 1985; Maroni et al. 1986a; Theodore et al. 1991).

Duplications occur in natural populations and in laboratory strains; they always involve the *Mtn*^{1} allele. The two copies are in direct tandem repeats at a distance of 1–5 kb of each other (Otto et al. 1986; Maroni et al. 1987; Lange et al. 1990).

Flies carrying the allele *Mtn*$^{.3}$, an allele that is present primarily in African populations, accumulate approximately 30% as much mRNA as those carrying *Mtn*^{1}; the extra 49 bases in the 3′ untranslated region of *Mtn*$^{.3}$ may increase its mRNA turnover rate (Theodore et al. 1991).

Developmental Pattern

Cadmium, copper, mercury, silver and zinc induce transcription of *Mtn* in larval and adult mid-guts, zinc being the least effective of these metals. Treatment with high metal concentrations leads to expression in the fat bodies and other tissues as well. *Mtn* RNA is not detectable early in embryogenesis, but it is clearly present, even in the absence of a metal supplement, in 18–24 h embryos, larvae, and adults (Lastowski-Perry et al. 1985; Silar et al. 1990).

Mtn

```
       ______EcoRI     .         .        Begin Mtn-.3|        .         .         .         .
 -496  GAATTCGTTGCAGGACAGGATGTGGTGCCCGATGTGACTAGCTCTTTGCTGCAGGCCGTCCTATCCTCTGGTTCCGATAAGAGACCCAGA  -407
             .         .         .         .         .         .         .         .         .
 -406  ACTCCGGCCCCCCACCGCCCACCGCCACCCCCATACATATGTGGTACGCAAGTAAGAGTGCCTGCGCATGCCCCATGTGCCCCACCAAGA  -317

        C    .         .         .         .         .         .         .         .         .
 -316  GTTTTGCATCCCATACAAGTCCCCAAAGTGGAGAACCGAACCAATTCTTCGCGGGCAGAACAAAAGCTTCTGCACACGTCTCCACTCGAA  -227
                                                                            ------->            =
             .         .         .         .         .         .         .     G   .         .
 -226  TTTGGAGCCGGCCGGCGTGTGCAAAAGAGGTGAATCGAACGAAAGACCCGTGTGTAAAGCCGCGTTTCCAAAATGTATAAAACCGAGAGC  -137
       ----           <-------                        <-------                   -----
             .      -->-123   .         .         .         .         .         .         .
 -136  ATCTGGCCAATGTGCATCAGTTGTGGTCAGCAGCAAAATCAAGTGAATCATCTCAGTGCAACTAAAGGCCTAAATAGCCCATACCTACCT  -47
             .         .         .         .         .         .         .         .         .
  -46  TTTTTGTAAACAAGTGAACAAGTTCGAGGAAATACAACTCAATCAAGATGCCTTGCCCATGCGGAAGCGGTAAGTTCGCAGTCTGGTGTG  43
                                                      MetProCysProCysGlySerG                       (8)
             .         .         .         .         .         .         .         .         .
   44  ATCCTTTAGGATATCACAGATCTTTCAGAGAAATGGTATTATACTAGTATAAAAATTCAATGGTGATTCAATAGTATAAAAATTCAAGGC  133
             .         .         .         .         .         .         .         .         .
  134  TGAAACTATCTGCAAAGTGAAATCTCTGAGTTCGTCTCTCTAAGAAAAGAAGTTCTTCAACTGCGTTTTATAAAATGGAACACTAATGTT  223
             .         .         .         .         .         .         .         .         .
  224  ATATGGCTTATGGATTACAGGATGTACCAGCATGTACTAATTTTTAAATTCTACTTCTTTCCAGGATGCAAATGCGCCAGCCAGGCCACC  313
                                                                       lyCysLysCysAlaSerGlnAlaThr  (16)

             .         .         .         .         .         .         .  A      .         .
  314  AAGGGATCCTGCAACTGCGGATCTGACTGCAAGTGCGGCGGCGACAAGAAATCCGCCTGCGGCTGCTCCGAGTGAGCTTTCCCCCAAAAA  403
       LysGlySerCysAsnCysGlySerAspCysLysCysGlyGlyAspLysLysSerAlaCysGlyCysSerGluEnd                 (40)
                                                                            Lys

             .         .         .         .         .         .         .         .         .
                                       CGAACTGATTTCTGTATAACTCCCAATACTAAAACGACATGTTTTCTCA T
  404  AGATCTGGAGTAGAGGCGCTGCATCTTGTCTC.................................................TCTACACAC  493
             .         .         .         .         .         .         .         .         .
  494  CCTGCAATAAATGTCCAATTAAAGTAATTGATGCCTAACTGCGTCTTTTCGGGTTGCATAATCAATTGGTCTGCGGCATTCTAGGTTAGA  583
            ------                       |(A)n
             .         .         . |End Mtn-.3       .         .         .         .         .
  584  TTCGCTTTTATTGGAGGTAGCTTCTAGCTACGTGGTCGGCAATATGCGTCGTGGAAATGGGATGGTCAAGTGTTTTCCACAATGTGCATA  673
             .         .         .         .         .         .         .         .         .
  674  TACATATGTACATAACACTAAAGTCAGTTGAGCAATATGGTAATCTGAGATGACTACTTCTGAAGCGACTGAGGGATGAGTTCAAACACA  763
             .         .         .         .         .         .         .         .         .
  764  CGGCTGACCATGACTGTAGATAAAAATACAGTTCGGCGTTAGAATATAGCCGCTATCGAATGGATAATATTAAAGAATACTAGCTTTAGA  853
             .         .         .         .         .         .         .         .         .
  854  AATAATAAAAATATATTACCCTATCAAATTTAAAACGATTTTAGGCATAACAACGAAATGGGTAATGAAAGTTCATATTTAAATCGGCTT  943
             .         .         .         .         .         .         .         .         .
  944  CCATTATTTTATAGGTGATTCATAGAAATATATGATTGTAGACTTATTATTGCTCAGTCTGTTTTGTGAAATGCCTCGTTTATAGCGCAA  1033
             .         .         .         .         .         .
 1034  AAGTGCCATATAGTTTTAGATGTAATATGATCGCGCAATTAACATGAAAGTGTAAGAACCCG  1095
```

Mtn SEQUENCE. Accession, M12964 (DROMETG). The numbered lines represent the sequence of the *Mtn*[1] allele, above it are the four base substitutions and the extra 49 bases present in the allele *Mtn*·[3]. Between positions −250 and −170, the 8-bp cores of putative metal regulatory elements are underlined.

Mto

```
      ______SspI
-1072 AATATTGAGTTCTACAGGAATGTTCCCAGGACTACACGGAGAAAAATCGAAGGACACTTTGGGGATGAGAGGATATTCATGCAATTTGTG   -983

-982  GTAAGGAACTGAAGTCATACTCTAACTGAACGGTGCTGGCTGGTCAAGTTATATATGTTTATGTGATGTTAAATATATACCTTGTGGTCA   -893

-892  ACACAAGCACAAGAATCAATTATATATTCATTATACCCGTTTAAGATATAGTAAGGTAAATAAAATGTAGAAGGATCAGGATAACTAGAC   -803

-802  GACTTAAAGACTGCAGACAACTTATCTAAGTCATTCTTTCGTTGCAGGTACACCTACCAAAAAACTATTTCTATATTTGTTTTCGAAAAC   -713

-712  TTTTTTTTTTACTAAAAGTCATAAATATATATAAAGTTGTTCCGGGTGTTTGGTTTTCCGTGCAACGAACTGTTTTCGTAGCTCCCGCAG   -623

-622  AGCTTATAGTTTTTGCCTAATTTGCAGCGCGTTTTTTCCTCTATTAATTTTTAGTTAGCTTTCCACATGTGATATTTTTATGGCATTTAC   -533

-532  GCTGGGTTTTTTTTGAAAAGAGTTTAGTCGTAAAGCGTTTTTGCAGCCAATATGAGCATTTAAATTTGTTTTACTACAGGAAAGTCTTTT   -443

-442  ATTTATTGTGAAAAACCCGCTGGGTAGCTGCCTGCGCTTTTCATGCTTTTTATTGTGTGCTTCTGGGCTGTGGGCTGAGTCACGATACGC   -353

-352  GGCGTATACGCAACGTATACGCAACGTGGGCAGCTGATAAGCTGATGAGGAGTTCGTGTGCACCGAGTTGGCGAGCAATCGCGTGCGCAA   -263
                                                                         <-------                     <-------

-262  AAAGAATTGCCTGGCCTATCGTCTGATAAATTGCGAACCACTCGCCCCAGGCTTGCACACGACGTGATAAGTTGGGTCAAACAAACAAAT   -173
                                 ----                                    ------->
                                 -->-143
-172  TTGTTTTGGATTTGTGCAATTTTGCACTCGTTCGAGTTCGAGGCAATCGAAGTGGGTATAAAAGTGGGGGAGTTGCCGGACTGGGTCATC   -83
                             ------->

-82   AGTTGAATAGCCAAGCAACAAGCAAACAAGTGAATATCAGTTCGCCTCAGCCAAGTGAAAGTCGAGAAATAGATACATACAAGATGGTTT   7
                                                                                               MetValC   (3)

8     GCAAGGGTTGTGGAACAAGTAAGTGGTACAACGCAGCAGCAAGCTGTATAATTGACAATCGTTCTCGATTCCTCGACAGACTGCCAGTGC   97
      ysLysGlyCysGlyThrA                                                                   snCysGlnCys   (12)

98    TCGGCCCAAAAGTGCGGGGACAACTGCGCCTGCAACAAGGATTGCCAGTGCGTTTGCAAGAATGGGCCCAAGGACCAGTGCTGCAGCAAC   187
      SerAlaGlnLysCysGlyAspAsnCysAlaCysAsnLysAspCysGlnCysValCysLysAsnGlyProLysAspGlnCysCysSerAsn   (42)

188   AAATAAGCGGGCCAACTATATAACTAACTGTTTAACTTCTAAACTGGAGCTTAACTCCCAACGAGTTGGCCGCAATAAATAAAGTTTATA   277
      LysEnd                                                                      -----/----     (43)

278   AAGATTTTGAGCATTTAAAAGTTTCTGCCGTTAACTTTTTGTTACTGGGCGGTCGGTCATCTTACCAAGCGATAATTATATTTTCGGCTT   367
                     |(A)n

368   TTGGCAGCTAAAACCAATTATGGTAAAATAATAAACGTGAGCTGGCATTCAGTTAAGCAAACCGCAAAATAGAATTACATGAAAAATAAG   457

458   CAAACGCAATGCGACAATTTGGGCGGGATTTGCAAATATTTGTATGTTCGCGGACAGCTGCACCGGAATTAAAATCCAATCCATCAGCCG   547
                                                            ____TaqI
548   TGATTTCGGTAGAAAACTCACCGAAAGTCCATTGAATTGTGCGCAAAACGGAACATAAATCGA                              610
```

Mto SEQUENCE. Strain, *Oregon R*. Accession, X52098 (DROMTOG). Between positions −300 and −140, the 8-bp cores of putative metal regulatory elements are underlined.

Promoter

A fragment that extends from 373 bp upstream to 54 bp downstream of the transcription initiation site is sufficient for apparently full metal response and for control of the expression of reporter genes. The addition of 3,500 bp farther upstream does not seem to increase the metal-induced response. Within the 373-bp segment that precedes the transcription initiation site, there are several copies of a 12-bp sequence that is related to the mammalian metal regulatory element (*Mtn* Sequence). The *Drosophila Mtn* promoter is capable of supporting metal-regulated expression of a reporter gene transfected into baby hamster kidney cells (Maroni et al. 1986a; Otto et al. 1987).

Mto

Gene Organization and Expression

Open reading frame, 43 amino acids; expected mRNA length, 376 bases, in agreement with RNA detected in northern blots. Primer extension was used to define the 5′ end. The 3′ end was obtained from a cDNA sequence that included a poly(A) tail. There is an intron in the Asn-9 codon (*Mto* Sequence) (Mokdad et al. 1987; Silar et al. 1990).

Developmental Pattern

Cadmium, copper, zinc, mercury and silver induce transcription of *Mto* in larvae and adults, zinc being the least effective inducer (Silar et al. 1990). RNA accumulations reach levels that are only 30–50% of the levels reached by *Mtn* when the same metals are used (G. Maroni and J. E. Young, unpublished observations). During embryonic and larval development, in the absence of a metal supplement, *Mto* RNA is present at approximately constant levels; in adult females it is barely detectable; and it is absent from males (Silar et al. 1990).

Promoter

There is no canonical TATA box upstream of the transcription initiation site. As in the *Mtn* promoter, there are several short sequences related to the metal regulatory elements found in mammalian metallothionein promoters (Silar et al. 1990).

References

Debec, A., Mokdad, R. and Wegnez, M. (1985). Metallothioneins and resistance to cadmium poisoning in Drosophila cells. *Biochem. Biophys. Res. Comm.* **127**:143–152.

Lange, B. W., Langley, C. H. and Stephen, W. (1990). Molecular evolution of Drosophila metallothionein genes. *Genetics* **126**:921–932.

Lastowski-Perry, D., Otto, E. and Maroni, G. (1985). Nucleotide sequence and expression of a *Drosophila* metallothionein. *J. Biol. Chem.* **260**:1527–1530.

Lauverjat, S., Ballan-Dufrancais, C. and Wegnez, M. (1989). Detoxification of cadmium. Ultrastructural study and electron-probe microanalysis of the midgut in a cadmium-resistant strain of *Drosophila melanogaster. Biol. Metals* **2**:97–107.

Maroni, G. (1990). Animal metallothioneins. In *Heavy Metal Tolerance in Plants*, ed. A. J. Shaw (Boca Raton, Florida: CRC Press), pp. 215–232.

Maroni, G., Lastowski-Perry, D., Otto, E. and Watson, D. (1986b). Effects of heavy metals on Drosophila larvae and a metallothionein cDNA. *Environm. Health Persp.* **65**:107–116.

Maroni, G., Otto, E. and Lastowski-Perry, D. (1986a). Molecular and cytogenetic characterization of a metallothionein gene of Drosophila. *Genetics* **112**:493–504.

Maroni, G. and Watson, D. (1985). Uptake and binding of cadmium, copper and zinc by *Drosophila melanogaster* larvae. *Insect Biochem.* **15**:55–63.

Maroni, G., Wise, J. and Otto, E. (1987). Metallothionein gene duplications and metal tolerance in natural populations of *Drosophila melanogaster. Genetics* **117**:739–744.

Mokdad, R., Debec, A. and Wegnez, M. (1987). Metallothionein genes in *Drosophila melanogaster* constitute a dual system. *Proc. Natl Acad. Sci.* (*USA*) **84**:2658–2662.

Otto, E., Allen, J. M., Young, J. E., Palmiter, R. D. and Maroni, G. (1987). A DNA segment controlling metal-regulated expression of the *Drosophila melanogaster* metallothionein gene *Mtn. Mol. Cell. Biol.* **7**:1710–1715.

Otto, E., Young, J. E. and Maroni, G. (1986). Structure and expression of a tandem duplication of the *Drosophila* metallothionein gene. *Proc. Natl Acad. Sci.* (*USA*) **83**:6025–6029.

Silar, P., Theodore, L., Mokdad, R., Errais, N-E., Cadic, A. and Wegnez, M. (1990). Metallothionein *Mto* gene of *Drosophila melanogaster*: structure and regulation. *J. Mol. Biol.* **215**:217–224.

Tapp, R. L. and Hockaday, A. (1977). Combined X-ray and microanalytical studies on the copper accumulating granules in the midgut of larval *Drosophila. J. Cell Sci.* **26**:201–215.

Theodore, L., Ho, A-S. and Maroni, G. (1991). Recent evolutionary history of the metallothionein gene *Mtn* in Drosophila. *Genet. Res.* **58**:203–210.

23

ovarian tumor: otu

Synonym: Transcription unit K of the chorion gene cluster on the X

Chromosomal Location:	**Map Position:**
X, 7F1	X-23.2

Products

Proteins of 98 and 104 kD of uncertain function.

Structure

In each case, the apparent M_r is slightly larger than predicted from the sequence, probably due to the skewed amino-acid composition. OTU proteins are largely hydrophilic and rich in Pro (approximately 10%) (Steinhauer et al. 1989; Steinhauer and Kalfayan 1992).

Tissue Distribution

OTU proteins are localized in ovaries. The 104 kD form predominates in pupal stages when advanced stages of oocyte maturation are absent. The 98 kD form is the more abundant one in adult females, when most of the ovarian mass comprises egg chambers at more advanced stages (G. L. Sass and L. L. Searles, personal communication).

Mutant Phenotypes

Mutations in *otu* lead to female-sterility; they have no effect on viability in either sex or male fertility. Null alleles of *otu* (the QUI alleles) result in the total absence of germ cell proliferation. Severely deficient alleles (the ONC alleles), seem to result in germ cell proliferation with little or no differentiation while more subtle mutations (DIF alleles) produce ovarioles with mixtures of egg chambers that have reached various degrees of differentiation (King et al. 1986; Steinhauer and Kalfayan 1992; Sass et al. 1993.

otu

```
          .         .         .         .         .         .         .         .         .
-1331  GAATTCATAGTCGTTGCGTTTTGCACACTCGCAAGATAACCAACTAACGACATTTACTAACAATAAACAAAAACATAACTTTACACGAGA   -1242
          .         .         .         .         .         .         .         .         .
-1241  ACACAAAAAACACAAAAAAAAAACAGGAAAACAAAAGGCACACACAGTCACACACTCACATCTCTTCCAGACAACTTTTGTCGCGGTAAC   -1152
          .         .         .         .         .         .         .         .         .
-1151  AGCGCGAACTGAAAGTTTGCTCCTGGCTTCATTGACTCGCAATTTCGAACTGAGTCTGATGAACAAGAACAACAGTGCGCCGTGTGGAAA   -1062
          .         .         .         .         .         .         .         .         .
-1061  GCGGCATTTTCCACCCCCTAAAAAGCGGCCAGCAACAACAGCAACGACAGTAACAAGAACAATTTGAAGGTAACAGAAACTTTTGGGGAT   -972
          .         .         .         .         .         .         .         .         .
 -971  GACACGGAACAGATGATGCCGCTATCGGTGTCATCGATAGACGGCGATAACAGGAGTTTTTAACCGCTCAGCAATATATTTCAAGTATA   -882
          .         .         .         .         .         .         .         .         .
 -881  TCATACACTTGTGTATTTCATTTAGAAAGTATTCAACAAGATCAGATATATTTATTTTGTTGATAAAATCACGAACCAACTCCATTGATT   -792
                                                                                        ----
          .         .         .         .         .         .         .         .  -->    . -->
 -791  CATTTCCGCACATCACTATTGCCCAATTTCGTTTGTCGGCATCCTTCCAGGCACTGGAAGTTCGTTCTTATACTTTTCGTTCGCATTCTA   -702
                    ----
                              ||=P3
          .         . !    ||=P1,P2,P4  . -->-668   --->-659/658    .         .         .
 -701  GTTCGCGGGTTCTCTGAAAGGCTAGATCGCGCCATTCGCTTCAATTCTTCGTGTAACGGTGCTAGGTGCGGATGCCAGTGTTATTTTTAA   -612

                                                                              _
          .         .         .         .         .         .         .     |.         .
 -611  TTGTTAATTTAATTGTTAACTATTTATAAAAATAGAATTTGTACAACAGAAGACGAACAGCAGAACACCGGTAATATCTCGATTCGATTT   -522
                                                                             _|
          .         .         .         .         .         .         .         .         .
 -521  TAACTGTATTAGTTGAAACATTTATAGTAACGGTAATTTGTCAAGTGACGAAATTAACTAATTAAGCGCAGCATGAGAGGCTTTTAAATC   -432
          .         .         .         .         .         .         .         .         .
 -431  ATTAAATTTTAAACAAATATTTAATTTTCATCAGCTTCATCACATTTAATTTTGCTCTTTTGCTTCATTTGCCTTTCTACTGCGCCATCT   -342
          .         .         .         .         .         .         .         .         .
 -341  TGAATTCGCAGGTGCATATTGTCATCTCGCTCTGAAGCCCGGCTTGTATGGAGTCGGTTAATAATTGGAATATATTTGTATTGCAGCAAA   -252
          .         .         .         .         .         .         .         .         .
 -251  TTTGCTTTAAAACTATTAAAGTTAAAAAAACTATACAATAGTTAACATAAAATAAGTAATAAAGCTTAGTATGCGCACTTCTTAGTGAAA   -162
          .         .         .         .         .         .         .         .         .
 -161  CGACAATAGATAGCAGTTGAAAAGTGATTGTGAAGGTCAAATAGATCGAGGTCAGGGCCCTCTTCTAACTGTTAATTGTGCAATACTTGT   -72

                                                                         _
          .         .         .         .         .         .         .|         .         .
  -71  ATTTCAAAGGGAAAACATGACAAAAAAAAAAATGAAATGAATAAAATTTAAGTTTCTCGATTCCAGAGTCGCCATGGACATGCAAGTGCAG   18
                                                                         |_          MetAspMetGlnValGln   (6)
          .         .         .         .         .         .         .         .         .
   19  CGCCCCATTACGTCAGGCAGCCGGCAGGCCCCGGATCCGTATGATCAGTATCTGGAGAGCCGTGGACTCTACCGTAAGCACACGGCCCGG   108
       ArgProIleThrSerGlySerArgGlnAlaProAspProTyrAspGlnTyrLeuGluSerArgGlyLeuTyrArgLysHisThrAlaArg   (36)
          .         .         .         .         .         .         .         .         .
  109  GACGCCTCCAGTTTGTTCCGTGTGATCGCCGAGCAGATGTACGACACCCAGATGCTGCACTACGAGATTCGGCTAGAGTGCGTCCGCTTC   198
       AspAlaSerSerLeuPheArgValIleAlaGluGlnMetTyrAspThrGlnMetLeuHisTyrGluIleArgLeuGluCysValArgPhe   (66)
          .         .         .         .         .         .         .         .         .
  199  ATGACCCTAAAACGACGCATCTTTGAGAAGGTAGGCCTCTAACAATCACACATTTTGTAAAAAAAAAAGAAATAATTTTATTTATATCCC   288
       MetThrLeuLysArgArgIlePheGluLys                                                              (76)
          .         .         .         .         .         .         .         .         .
  289  AGGAAATTCCTGGCGATTTCGATAGCTACATGCAGGACATGTCCAAGCCCAAGACATATGGAACCATGACAGAACTACGCGCTATGTCCT   378
         GluIleProGlyAspPheAspSerTyrMetGlnAspMetSerLysProLysThrTyrGlyThrMetThrGluLeuArgAlaMetSerC   (106)
          .         .         .         .         .         .         .         .         .
  379  GCCTATATCGGTAATTAATCCTTAGTTACTATTTTCTATTAAACTACAAATATATATGATTTCTGTACGACTTCCAGCCGCAATGTTATC   468
       vsLeuTyrAr                                                                      gArgAsnValIle   (113)
```

(*continued*)

```
 469 CTGTATGAGCCCTACAACATGGGCACCAGCGTCGTTTTTAATCGTCGCTATGCGGAAAACTTCCGTGTCTTCTTCAACAATGAGAATCAC  558
     LeuTyrGluProTyrAsnMetGlyThrSerValValPheAsnArgArgTyrAlaGluAsnPheArgValPhePheAsnAsnGluAsnHis (143)

 559 TTTGATTCGGTTTATGACGTTGAATATATAGAAAGAGCCGCCATTTGTCAATGTACGTAGCCTATTAATATATCCAATTTTGCTTTTTGT  648
     PheAspSerValTyrAspValGluTyrIleGluArgAlaAlaIleCysGlnS                                       (161)

 649 ATATGTACGTTGCTTTCAGCAATCGCCTTTAAGTTGCTGTACCAGAAGCTTTTCAAATTGCCTGACGTATCCTTTGCTGTGGAGATTATG  738
                       erIleAlaPheLysLeuLeuTyrGlnLysLeuPheLysLeuProAspValSerPheAlaValGluIleMet (184)

 739 TTGCATCCACACACCTTCAATTGGGATCGCTTCAATGTGGAGTTCGATGACAAGGGCTATATGGTTCGCATTCATTGCACCGATGGACGA  828
     LeuHisProHisThrPheAsnTrpAspArgPheAsnValGluPheAspAspLysGlyTyrMetValArgIleHisCysThrAspGlyArg (214)

 829 GTTTTTAAGCTTGATCTGCCAGGGGACACAAACTGCATACTGGAAAACTATAAGCTGTGCAATTTCCATAGCACCAATGGAAATCAGAGC  918
     ValPheLysLeuAspLeuProGlyAspThrAsnCysIleLeuGluAsnTyrLysLeuCysAsnPheHisSerThrAsnGlyAsnGlnSer (244)

 919 ATTAATGCTCGAAAGGGAGGCCGGCTGGAGATTAAAAACCAGGAGGAGCGAAAGGCATCCGGCAGCAGTGGCCACGAACCAAACGATCTG 1008
     IleAsnAlaArgLysGlyGlyArgLeuGluIleLysAsnGlnGluGluArgLysAlaSerGlySerSerGlyHisGluProAsnAspLeu (274)

1009 TTGCCCATGTGTCCAAACCGATTGGAGTCCTGTGTCCGCCAGCTGCTAGATGATGGTCAGTAGAGGTGGTTTCAAACATCAAATGCTTAC 1098
     LeuProMetCysProAsnArgLeuGluSerCysValArgGlnLeuLeuAspAspG                                    (293)

1099 ATAATACTCTCTTTTTAGGTATCTCTCCGTTTCCCTACAAAGTGGCCAAGTCCATGGACCCCTATATGTATCGTAATATAGAATTTGATT 1188
                       lyIleSerProPheProTyrLysValAlaLysSerMetAspProTyrMetTyrArgAsnIleGluPheAspC (317)

1189 GCTGGAACGATATGCGCAAGGAGGCCAAGCTTTATAATGTCTACATAAATGACTATAACTTTAAGGTAAACTGTGCAGAACATTGGATTA 1278
     ysTrpAsnAspMetArgLysGluAlaLysLeuTyrAsnValTyrIleAsnAspTyrAsnPheLys                          (338)

1279 TCGTTAGCACACATACACACGCACACCAACACACGTTTCATGTCAACCACCCATCCAAATTAACACCCTTTCATTTTGATCTATACACTG 1368
                                                                                        A=13
1369 GATACACCTTATACTTTACTATACATGTATGTCTTGCCTTATCCTTCCTCGTCTCGTCGCCGTGTTATTTGTTTTCCAGGTGGGCGCCAA 1458
                                                                                     ValGlyAlaLy (342)
     .A=11
1459 GTGCAAGGTGGAATTGCCGAACGAAACGGAGATGTACACGTGCCACGTTCAAAATATCTCCAAAGATAAGAATTACTGCCACGTCTTTGT 1548
     sCysLysValGluLeuProAsnGluThrGluMetTyrThrCysHisValGlnAsnIleSerLysAspLysAsnTyrCysHisValPheVa (372)
      Tyr

1549 TGAGAGGATTGGCAAAGAGATAGTGGTACCTCTTCTTTTTATCTGATTTTCTAGACCCTTGCAGAGAAATGCAAAAATTTCGATTAGAAA 1638
     lGluArgIleGlyLysGluIleVal                                                                  (380)

1639 CGATTATCATATTTAACAATTAGTTAAATTTGTTAAAGTTTAGTTAAAAGTATATTAATTGTGGCCCAATGAACTGGTATATAAGTCTAT 1728

1729 AAAATAATTGATCTGCAAGGGCTAAAAATGTTCGGTATCCGAAGCTAATTGTAACTATTTCGCTTTAATAGAGAGCTTACTAATATACAA 1818

1819 ACATATCTGTTGGCTTAGGTCCCGTATGAATCGCTCCATCCCCTGCCGCCAGATGAGTACCGCCCATGGTCGTTGCCATTCCGCTATCAT 1908
                       ValProTyrGluSerLeuHisProLeuProProAspGluTyrArgProTrpSerLeuProPheArgTyrHis (404)

1909 CGCCAGATGCCTCGCTTGCCGTTGCCCAAGTATGCCGGTAAGGCCAACAAGTCTTCCAAATGGAAGAAGAACAAGCTGTTCGAAATGGAC 1998
     ArgGlnMetProArgLeuProLeuProLysTyrAlaGlyLysAlaAsnLysSerSerLysTrpLysLysAsnLysLeuPheGluMetAsp (434)

1999 CAGTATTTTGAGCACAGCAAGTGTGATTTGATGCCCTACATGCCCGTGGACAATTGCTATCAGGGTGTGCACATTCAGGACGATGAGCAG 2088
     GlnTyrPheGluHisSerLysCysAspLeuMetProTyrMetProValAspAsnCysTyrGlnGlyValHisIleGlnAspAspGluGln (464)

2089 CGGGATCATAATGATCCTGAACAAAATGACCAGAACCCGACTACGGAGCAGCGGGATCGTGAAGAACCGCAGGCACAGAAGCAACACCAG 2178
     ArgAspHisAsnAspProGluGlnAsnAspGlnAsnProThrThrGluGlnArgAspArgGluGluProGlnAlaGlnLysGlnHisGln (494)
```

```
2179  CGCACGAAGGCATCAAGGGTTCAGCCGCAGAACTCGAGTTCCAGCCAAAACCAGGAGGTTTCGGGTTCGGCTGCCCCGCCACCCACTCAG  2268
      ArgThrLysAlaSerArgValGlnProGlnAsnSerSerSerSerGlnAsnGlnGluValSerGlySerAlaAlaProProProThrGln  (524)

2269  TATATGAATTACGTGCCAATGATACCGAGTCGTCCTGGGCATTTACCGCCACCTTGGCCTGCATCTCCGATGGCTATTGCCGAGGAGTTT  2358
      TyrMetAsnTyrValProMetIleProSerArgProGlyHisLeuProProProTrpProAlaSerProMetAlaIleAlaGluGluPhe  (554)

2359  CCGTTCCCCATTTCAGGAACCCCGCATCCACCGCCAACCGAAGGTTGTGTATACATGCCATTCGGTGGTTATGGTCCACCACCACCGGGA  2448
      ProPheProIleSerGlyThrProHisProProProThrGluGlyCysValTyrMetProPheGlyGlyTyrGlyProProProProGly  (584)

2449  GCTGTTGCTTTATCGGGACCGCATCCATTTATGCCGCTTCCTTCTCCACCGCTAAATGTTACCGGAATTGGCGAGCCACGTCGTTCTCTA  2538
      AlaValAlaLeuSerGlyProHisProPheMetProLeuProSerProProLeuAsnValThrGlyIleGlyGluProArgArgSerLeu  (614)

2539  CACCCAAACGGTGAAGATTTGCCCGTGGATATGGTGACTTTGAGATACTTCTACAACATGGGCGTGGATTTGCATTGGCGCATGTCGCAC  2628
      HisProAsnGlyGluAspLeuProValAspMetValThrLeuArgTyrPheTyrAsnMetGlyValAspLeuHisTrpArgMetSerHis  (644)

                                                                      T=5
2629  CACACGCCGCCTGATGAACTAGGAATGTTTGGATACCATCAGCAGAACAACACTGATCAACAGGCAGGACGGACTGTAGTCATTGGCGCC  2718
      HisThrProProAspGluLeuGlyMetPheGlyTyrHisGlnGlnAsnAsnThrAspGlnGlnAlaGlyArgThrValValIleGlyAla  (674)
                                                                      End
                                                                                   T=14
2719  ACAGAGGACAATTTGACTGCCGTGGAGTCAACACCACCACCTTCGCCAGAGGTGGCAAATGCCACAGAGCAGTCACCGCTTGAGAAAAGT  2808
      ThrGluAspAsnLeuThrAlaValGluSerThrProProProSerProGluValAlaAsnAlaThrGluGlnSerProLeuGluLysSer  (704)
                                                                                   End

2809  GCCTACGCCAAGCGCAATTTGAATTCGGTTAAGGTGCGCGGCAAACGTCCGGAGCAGCTGCAAGATATTAAGGATTCGCTGGGGCCAGCG  2898
      AlaTyrAlaLysArgAsnLeuAsnSerValLysValArgGlyLysArgProGluGlnLeuGlnAspIleLysAspSerLeuGlyProAla  (734)

2899  GCATTTTTGCCCACTCCAACGCCATCGCCAAGCTCGAATGGCAGTCAGTTTAGTTTCTATACTACTCCATCGCCGCATCATCACCTGATA  2988
      AlaPheLeuProThrProThrProSerProSerSerAsnGlySerGlnPheSerPheTyrThrThrProSerProHisHisHisLeuIle  (764)

2989  ACACCGCCGAGGTTGCTCCAACCGCCGCCACCGCCACCGATATTCTACCACAAGGCGGGACCACCACAGCTAGGGGGAGCAGCTCAAGGA  3078
      ThrProProArgLeuLeuGlnProProProProProProIlePheTyrHisLysAlaGlyProProGlnLeuGlyGlyAlaAlaGlnGly  (794)

3079  CAGGTAGGAGTGATACATGCACTAACAAATTCAAAATATTCTATAGGCAATCGACACTCGACCATTTTTAGACTCCCTACGCCTGGGGCA  3168
      Gln                                                                          ThrProTyrAlaTrpGlyM  (802)

3169  TGCCAGCTCCGGTGGTGTCCCCCTATGAGGTGATCAACAACTATAACATGGACCCGTCGGCTCAGCCACAACAACAGCAGCCAGCCCCCT  3258
      etProAlaProValValSerProTyrGluValIleAsnAsnTyrAsnMetAspProSerAlaGlnProGlnGlnGlnGlnProAlaProL  (832)

3259  TGCAACCAGCTCCCTTATCTGTCCAATCTCAGCCGGCAGCTGTCTATGCTGCAACGCGTCATCACTAAACAAAGAAAGAGAAAAAAAAGG  3348
      euGlnProAlaProLeuSerValGlnSerGlnProAlaAlaValTyrAlaAlaThrArgHisHisEnd  (853)

3349  GAGCGGGGGCAAAAAACAGATCACTTGAAAGAGAGAGGCCATACAGATCGAAGGCACTACATTCCATTGCAATTAACGGCTTTTAAAATT  3438

3439  TAATCTCACTTTTAAATTTGTAGTTAACTTTTTATAGGCCATAAGCGTTGGCGCTCTATCATAAACCATTCAGCTTCTGTACAACAATCG  3528

3529  ATTGCATAACCTAACGCAAATGTCAACCCAACTTCATTTTAAAAATGTAATTTAACGTAATTTTATGCGAATTTTTTTAAAGTTAGCCGT  3618

3619  CACGAAATCAAAGAACCACCTATTTATATGATTTATTTAAAACCCTTCTAACCAAAAATATCTACATACTATCTACTATATATATACATA  3708

3709  TATATATATATATATATTTATGTGCTCGCTGTTCGGCTAGAGACTCACCTATGTAAAGTGTACCATCAAAAATTAACCATAAATAAAACA  3798
                                                                                         ------
3799  AGATTCAACTGCAG  3812
                  |(A)n
```

By *in vitro* mutagenesis of *otu*, two constructions were prepared, one that could produce only the 104 kD protein and another that could produce only the 98 kD protein. When introduced into QUI mutants, the 104 kD protein restores fertility. The 98 kD protein is unable to rescue the QUI mutant phenotype but does restore some fertility to ONC or DIF type alleles. Thus, it would appear that the 104 kD protein is capable of carrying out all *otu* functions while the 98 kD protein can perform some of the late oocyte maturation functions but is unable to carry out early oocyte maturation functions or those required for controlled cell proliferation (A. R. Comer and L. L. Searles, personal communication).

Gene Organization and Expression

Open reading frame, 811 (98 kD protein) or 853 (104 kD protein) amino acids depending on splicing; mRNA, 3,045–3,230 bases, depending on the start site and splicing. The most common RNA is approximately 3.2 kb, but other cross-hybridizing RNAs occur. The 5′ end was defined by S1 mapping, primer extension, and the sequencing of two cDNA clones. Several sites are used for transcription initiation, the main ones being those at positions −668, −659 and −658 (*otu* Sequence). There is no TATA box associated with any of the 5′ ends. The 3′ end was defined from a cDNA sequence that contained a poly-A tail. There are eight introns: one is in the leader between positions −541 and −7, the others are after the Lys-76 codon, in the Arg-109, Ser-161 and Gly-293 codons, and after the Lys-338, Val-380 and Gln-795 codons. The 126-base exon starting with the Val-339 codon is often spliced out to produce mRNA that codes for the 98 kD protein (*otu* Sequence, Fig. 23.1) (Champe and Laird 1989; Steinhauer et al. 1989; Comer et al. 1992; Steinhauer and Kalfayan 1992).

---

otu SEQUENCE (*previous pages*). Strain, *Canton S.* Accession, M30825 (DROOTUA) and X13693 (DROOTU). Arrows above the sequence, between −720 and −658, indicate possible sites of transcription initiation; the exclamation mark at −688 marks the 5′ end of two independently obtained cDNAs. Several mutations are indicated in the sequence: otu^{5} and otu^{14} cause premature termination, and homozygotes accumulate smaller proteins (both alleles belong to the DIF class); otu^{13} is unable to produce the 104 kD protein because it has a disabled acceptor site in exon 7; and otu^{11} has an amino-acid substitution in exon 7 (both otu^{11} and otu^{13} affect the 104 kD protein but not the 98 kD protein and both are ONC alleles) (Steinhauer and Kalfayan 1992). The four *P* element insertions near the 5′ end seem to affect transcription, and the severity of their phenotypes is generally proportional to the size of the insertion: otu^{P1} (2.9 kb) is a QUI allele, otu^{P2} (2.0 kb) is an ONC allele, and otu^{P3} (0.6 kb) and otu^{P2} (0.5 kb) are DIF alleles (Sass et al. 1993).

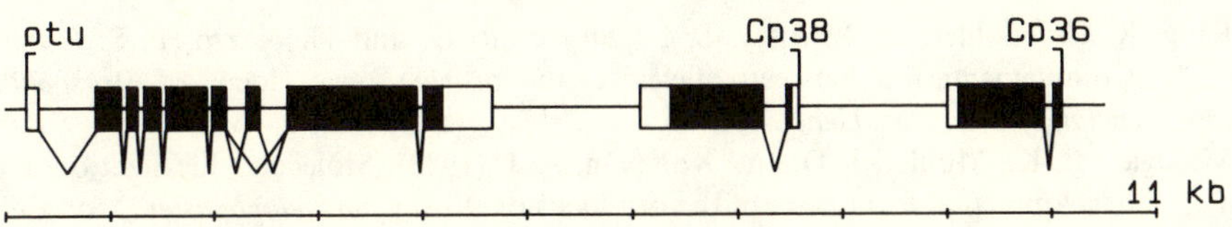

FIG. 23.1. *otu* and neighboring genes *Cp36* and *Cp38*

The *otu* gene is 0.06 map units away from the chorion protein gene *Cp38*, closer to the centromere, and transcribed convergently with *Cp38*, toward the telomere; the two 3′ ends are approximately 1.4 kb apart (Fig. 23.1). *otu* is amplified, together with the chorion genes, in follicular cells, but it is not expressed in those cells (Parks and Spradling 1987; see also Chorion Protein Genes).

Developmental Pattern

The predominant 3.2 kb transcript is present mainly in female pupae and adults. It occurs in nurse cells and oocytes, and the peak of expression is egg chambers between stages 8 and 10. This transcript is found at much lower levels in female heads and thoraxes and in male testes along with other cross-hybridizing transcripts. Given that null mutations have no effect other than on female sterility, it is likely that the non-ovarian transcripts lack any function (Mulligan et al. 1988).

Promoter

Studies of a reporter gene under the control of an *otu* fragment that extends from 452 bp upstream of the transcription initiation site to the end of the first exon, showed expression, in ovaries, in nurse cells and oocytes as well as in the germarium. In males, expression was detected in the anterior tip of the testes, in the region of stem cells and primary spermatocytes (Comer et al. 1992).

Constructions with 310 bp of upstream sequence and the complete transcribed region produced apparently normal levels of 3.2 kb RNA and rescued *otu* mutations. Similar constructions with only 190 bp of the promoter region, however, were unable to support gene expression (Comer et al. 1992).

References

Champe, M. A. and Laird, C. D. (1989). Nucleotide sequence of a cDNA from the putative *ovarian tumor* locus of *Drosophila melanogaster*. *Nucl. Acids Res.* **17**:3304.

Comer, A. R., Searles, L. L. and Kalfayan, L. (1992). Identification of a genomic DNA fragment containing the *Drosophila melanogaster ovarian tumor* (*otu*) gene and localization of regions governing its expression. *Gene.* **118**:171–179.

King, R. C., Mohler, J. D., Riley, S. F., Storto, P. D. and Nicolazzo, P. S. (1986). Complementation between alleles at the *ovarian tumor* locus of *Drosophila melanogaster*. *Dev. Genet.* **7**:1–20.

Mulligan, P. K., Mohler, J. D. and Kalfayan, L. J. (1988). Molecular localization and developmental expression of the otu locus of *Drosophila melanogaster*. *Mol. Cell. Biol.* **8**:1481–1488.

Parks, S. and Spradling, A. (1987). Spatially regulated expression of chorion genes during *Drosophila* oogenesis. *Genes Dev.* **1**:497–509.

Sass, G. L., Mohler, J. D., Walsh, R. C., Kalfayan, L. J. and Searles, L. L. (1993). Structure and expression of hybrid dysgenesis-induced alleles of the *ovarian tumor* (*otu*) gene in *Drosophila melanogaster*. *Genetics* **133**:253–263.

Steinhauer, W. R. and Kalfayan, L. J. (1992). A specific *ovarian tumor* protein isoform is required for efficient differentiation of germ cells in *Drosophila* oogenesis. *Genes Dev.* **6**:233–243.

Steinhauer, W. R., Walsh, R. C. and Kalfayan, L. J. (1989). Sequence and structure of the *Drosophila melanogaster* ovarian tumor gene and generation of an antibody specific for the ovarian tumor protein. *Mol. Cell. Biol.* **9**:5726–5732.

24

6-Phosphogluconate Dehydrogenase Gene: *Pgd*

Chromosomal Location:
X, 2D4-6

Map Position:
1-0.6

Product

6-Phosphogluconate dehydrogenase (6-PGD) (E.C. 1.1.1.44), a member of the pentose shunt.

Structure

The sequence of *Drosophila* 6-PGD is 50% identical to prokaryotic 6-PGD and 60–70% identical to the porcine and ovine enzymes (Fig. 24.1) (Scott and Lucchesi 1991). 6-PGD is a homodimer; the monomer has a M_r of approximately 53 kD (Williamson et al. 1980).

Function

6-PGD is responsible for the oxidative decarboxylation of 6-phosphogluconate (6-PG) to yield ribulose-5-phosphate and reduced nicotinamide adenine dinucleotide phosphate (NADPH); these two products are important for the biosynthesis of ribose and lipids, respectively (Wood 1985).

Tissue Distribution

The specific activity of the enzyme increases during the larval stages to reach a maximum early in the third instar. Activity diminishes late in the third instar and early pupal stages, then climbs again in late pupae and adults (Williamson et al. 1980). In larvae, highest activity is observed in fat bodies and actively dividing imaginal cells (Gutierrez et al. 1989; Scott and Lucchesi 1991).

```
            1                                                     50                                                    100
Dm pgd  MSGQADIALI GLAVMGQNLI LNMDEKGFVV CAYNRTVAKV KEFLANEAKD TKVIGADSLE DMVSKLKSPR KVMLLVKAGS AVDDFIQQLV PLLSAGDVII
 Ovine  .MAQADIALI GLAVMGQNLI LNMNDHGFVV CAFNRTVSKV DDFLANEAKG TKVLGAHSLE EMVSKLKKPR RIILLVKAGQ AVDNFIEKLV PLLDIGDIII
   CON  ---QADIALI GLAVMGQNLI LNM---GFVV CA-NRTV-KV --FLANEAK- TKV-GA-SLE -MVSKLK-PR ---LLVKAG- AVD-FI--LV PLL--GD-II

        101                                                  150                                                    200
Dm pgd  DGGNSEYQDT SRRCDELAKL GLLFVGSGVS GGEEGARHGP SLMPGGHEAA WPLIQPIFQA ICAK.ADGEP CCEWVGDGGA GHFVKMVHNG IEYGDMQLIC
 Ovine  DGGNSEYRDT MRRCRDLKDK GILFVGSGVS GGEDGARYGP SLMPGGNKEA WPHIKAIFQG IAAKVGTGEP CCDWVGDDGA GHFVKMVHNG IEYGDMQLIC
   CON  DGGNSEY-DT -RRC--L--- G-LFVGSGVS GGE-GAR-GP SLMPGG---A WP-I--IFQ- I-AK---GEP CC-WVGD-GA GHFVKMVHNG IEYGDMQLIC

        201                                                  250                                                    300
Dm pgd  EAYHIMKS.L GLSADQMADE FGKWNSAELD SFLIEITRDI LKYKDGKG.Y LLERIRDTAG QKGTGKWTAI AALQYGVPVT LIGEAVFSRC LSALKDERVQ
 Ovine  EAYHLMKDVL GLGHKEMAKA FEEWNKTELD SFLIEITASI LKFQDADGKH LLPKIRDSAG QKGTGKWTAI SALEYGVPVT LIGEAVFARC LSSLKDERIQ
   CON  EAYH-MK--L GL----MA-- F--WN--ELD SFLIEIT--I LK--D--G-- LL--IRD-AG QKGTGKWTAI -AL-YGVPVT LIGEAVF-RC LS-LKDER-Q

        301                                                  350                                                    400
Dm pgd  ASSVLKGPST KAQVANLTKF LDDIKHALYC AKIVSYAQGF MLMREAAREN KWRLNYGGIA LMWRGGCIIR SVFLGNIKDA YTSQPELSNL LLDDFFKKAI
 Ovine  ASKKLKGPQN IPFEGDKKSF LEDIRKALYA SKIISYAQGF MLLRQAATEF GWTLNYGGIA LMWRGGCIIR SVFLGKIKDA FDRNPGLQNL LLDDFFKSAV
   CON  AS--LKGP-- ---------F L-DI--ALY- -KI-SYAQGF ML-R-AA-E- -W-LNYGGIA LMWRGGCIIR SVFLG-IKDA ----P-L-NL LLDDFFK-A-

        401                                                  450                                    485
Dm pgd  ERGQDSWREV VANAFRWGIP VPALSTALSF YDGYRTAKLP ANLLQAQRDY FGAHTYELLG QEGQFHHTNW TGTGGNVSAS TYQA*
 Ovine  ENCQDSWRRA ISTGVQAGIP MPCFTTALSF YDGYRHAMLP ANLIQAQRDY FGAHTYELLA KPGQFIHTNW TGHGGSVSSS SYNA*
   CON  E--QDSWR-- -------GIP -P---TALSF YDGYR-A-LP ANL-QAQRDY FGAHTYELL- --GQF-HTNW TG-GG-VS-S -Y-A-
```

FIG. 24.1. Comparison of the sheep (Accession, 60195) and *Drosophila* (Dm) sequences. There is 72% overall identity between the proteins. Sequences aligned with the GCG *Pileup* program.

Pgd

```
       ______XhoI
-1206  CTCGAGCAGTTCAAGTTCCTGAAGTGAGTTGCGCCACCTTTGTCTTCTCTGAGCGTTACCAATCCTGTTCACAAACTTATTTCCCATAGC  -1117

-1116  TCCCCCATTTCGGGATTTCCCTTCTACATGCTCATCGAGACCTCGGGCAGCAACGGTGACCACGACGAGGAGAAGATCAACCAGTTCATT  -1027

-1026  GGGGACGGTATGGAGCGTGGCGAGATCCAGGATGGCACCGTAACCGGTGATCCCGGCAAGGTGCAGGAGATCTGGAAGATCGCGAAATGG  -937

 -936  TGCCGCTGGGTCTGATCGAGAAGAGCTTCTGCTTCAAGTACGACATCTCGCTGCCTCTGCGGGACTTCTACAACATTGTGGACGTGATGC  -847

 -846  GAGAGAGGTGCGGTCCCTTGGCCACAGTTGTCTGCGGATACGGCCATCTGGGGGACTCTAATCTGCACCTGAACGTCTCCTGCGAGGAGT  -757

 -756  TTAACGACGAGATCTACAAGCGGGTCGAACCCTTCGTCTACGATACACCTCCAAGCTGAAGGGCAGCATTATGGCGGAGCACGGCATTGG  -667

 -666  CTTCCTGAAGAAGGACTACCTGCACTACTCCAAGGACCCGGTGGCCATTGGCTACATGCGCGAGATGAAGAAGCTGCTGGACCCCAACAG  -577

 -576  CATCCTCAATCCCTACAAGGTGCTTAACTGAAGGCTTCTACCTAATAGATTCTATTTTTTTTGTTTGTGTGTAATTTTCATAACCTTATA  -487

 -486  ATACAGAAATGGCATTAGAAGTGAATTTTGTTAACTTGTGAAGTTAAAAAGGACCATCATATTTGGCACGAAACCAATGGGCAAAACTTA  -397

 -396  CTTATAAAATAGTCCGAAAAAATAGTATATACCAGTTTTTACAGTACCACATTATAGGTACTCGGAGGTAATAATAGAAAAAACACTATC  -307

 -306  TTTGCATTTACTGTTACACTACGAAGCACTATATTTAGTAGCAGTACTCATTAGAGTCCACTCACAAAATTAGCACCAACCGGCAGTAAT  -217

 -216  TGGTCAAGGATCGGCGATAGCTTCAAACTCCGAAGTTCAAAGTCAAACTGCCGCCCTGCGAAAGCTTCGCGAGTGGAGCTTTTCTGCACT  -127

 -126  TATCGATAGCTAACATTGTGGCGCGACTATCGATCGACGAGCTGCCGCTTAACAGTGCCATATATAGATTGTAACATTAGGAGCTCAAAT  -37
                   ----                                                 -----

         -->-34
  -36  CATTGTTGGAACACAAACCACAAAGAACACACGAAACATGAGCGGGTGAGTAGAGGGAAATTCTCTTTTCCCCGGAGTTTTCCGCGATCC  53
                                             MetSerGl                                               (3)

   54  TAACGTCGCCCATTTCCGGATTTCTTCCAGACAAGCGGATATTGCCCTCATCGGCCTGGCCGTCATGGGCCAAAACCTGATACTCAACAT  143
                                   yGlnAlaAspIleAlaLeuIleGlyLeuAlaValMetGlyGlnAsnLeuIleLeuAsnMe  (23)

  144  GGACGAGAAGGGATTCGTGGTGTGCGCCTACAACCGCACGGTGGCCAAGGTCAAGGAGTTCCTCGCCAATGAGGCTAAGGACACCAAAGT  233
       tAspGluLysGlyPheValValCysAlaTyrAsnArgThrValAlaLysValLysGluPheLeuAlaAsnGluAlaLysAspThrLysVa  (53)

  234  GATTGGAGCCGACTCGCTCGAGGACATGGTCTCCAAGCTGAAGAGCCCCCGGAAGGTCATGCTGCTGGTCAAGGGTGAGTTGCATATCCA  323
       lIleGlyAlaAspSerLeuGluAspMetValSerLysLeuLysSerProArgLysValMetLeuLeuValLysA                    (78)

  324  AATTCAGCGGCTGGGTAGCGCAGAGCATCGAAAACCCATTGAAACCTGCTGCAAGCGATCGCTGTGTTGGTGACTCAACTTACATGTGTG  413

  414  CGCGCGTGCTTGTGAATTGGTGAAAAAGTCGAAGCAAAGTCATCATGATGACGATTTTTGCGGCTCATATTCCAATGTGCAAAGGGGAAC  503

  504  GATAGGATAAGCAGGTGAGCTCAATGCTTAAGTTTCGAATCCTATAAAGAGCTTTGAATTCTGTCTAGTTTTCAAGTCAAAACTATCGCA  593

  594  TACAAAACCTACGAAATGCCATCCCTATCATTTGTACAAAAAGAACTCCTAACCCAGACTTAGTGGTTAAGGCCGCAGCTCAATGATCTC  683

  684  TAAACAGTTGTTTTTTGTGTTTACTCCACCCCCTCACCGTTTTCTCGCGCTCCCTCCTCTTCCTACTTCCTTTTAAAACCGCACTTCTGA  773

  774  TAAAAGGTTTATAAATGGATCAGTCCCATTTCGAAAACCGTAACCACAAGTGTGGCGTGAGTTTTGTCTAATCACATAGTTGTGGTAAGC  863

  864  TGCCTCCACTTACCTAAACCATCGAGCGAACCCATCAGGTGATTTCCAGGTCACTCACGCGTTCGTCTACCACTCTCGCGTGTCCGAAAC  953
```

(continued)

954 TCTGCTCACCTCTAGATCGGCGTGCCCGGCTTATCTGTTCGTGCGAAAGCAACAACAACGCGGCGCAGAGAGAAATCTTTGACATTCATA 1043

1044 ATAGGTCACACAAAATGGGCGATTTTCAGGTGGATTTACTCGGATTTGACCAGCCGAAAAACCTACATATTCCTCTTCTGCGAGTTGCCA 1133

1134 GGCCAGTGAGTCATTTCGTCTGGAGACTGCTCCTTAGAAGAATACAGTGCGGGTCAATAACATATGTACATAGCTCTGGAGGTTTTTGTG 1223

1224 CTGAACATATGTAGATTTGAAAGTTGCGTGACAGGTTGTGCGAATTCCCACATTCACAGGGTGGGTGGGAGTAAGGATGACGACACAAAA 1313

1314 AGCTAGTTGGTCATTGAACAGAGCGAGTCCAACAATCTTGACCGCTAGTGTGCCCCACAAACCACCACCAACGACCGCTAGATAGATAGA 1403

1404 TCAATGGTAGTATCGCCACGACTCGTTGGCCTTATCTGGGTCCACTGCGCTGGAGAACTGCTCACCCGGCGCTAGGGGAATTCCTCATCG 1493

1494 GGGTTCTCAAAAGCTCAACTATCGTAGACTCATTTTCCAAAGCGTTCTTAGCGAGCGCCAGTTCTTTTAACGTAAAGAAATCTTCGATTT 1583

1584 AGCCAGAAAGTAGAGCGTGCGATTGGACAAGGTCGGTTGGTTGCTTTTGGAAAGTCACTGTTTTGGAGGTCACCCTGGTGGCGAGGCGTG 1673

1674 ATCTGCTTTAATCGACTTTACGCTAATCAGATGTAAACTCGATACAATTTCAGCTGGAAGTGCAGTCGACGACTTCATCCAGCAGCTGGT 1763
laGlySerAlaValAspAspPheIleGlnGlnLeuVa (90)

1764 GCCGCTGCTTTCCGCCGGCGATGTGATCATCGATGGTGGCAACTCGGAGTATCAGGACACATCTCGCCGCTGCGACGAGTTAGCCAAACT 1853
lProLeuLeuSerAlaGlyAspValIleIleAspGlyGlyAsnSerGluTyrGlnAspThrSerArgArgCysAspGluLeuAlaLysLe (120)

1854 TGGCCTGCTCTTCGTCGGATCCGGCGTGAGCGGTGGCGAGGAGGGCGCCCGCCACGGACCCTCGCTGATGCCCGGCGGACACGAGGCCGC 1943
uGlyLeuLeuPheValGlySerGlyValSerGlyGlyGluGluGlyAlaArgHisGlyProSerLeuMetProGlyGlyHisGluAlaAl (150)

1944 GTGGCCCCTTATCCAACCCATCTTCCAGGCGATCTGCGCCAAGGCCGACGGTGAACCCTGCTGCGAGTGGGTGGGCGATGGAGGCGCCGG 2033
aTrpProLeuIleGlnProIlePheGlnAlaIleCysAlaLysAlaAspGlyGluProCysCysGluTrpValGlyAspGlyGlyAlaGl (180)

2034 TCACTTCGTCAAGATGGTGCACAACGGCATCGAATACGGTGACATGCAGCTGATCTGCGAGGCGTACCACATCATGAAGAGCCTGGGACT 2123
yHisPheValLysMetValHisAsnGlyIleGluTyrGlyAspMetGlnLeuIleCysGluAlaTyrHisIleMetLysSerLeuGlyLe (210)

2124 GTCGGCTGACCAGATGGCAGACGAGTTCGGCAAGTGGAACTCGGCCGAACTGGACTCCTTCCTCATTGAAATCACGCGTGATATTCTTAA 2213
uSerAlaAspGlnMetAlaAspGluPheGlyLysTrpAsnSerAlaGluLeuAspSerPheLeuIleGluIleThrArgAspIleLeuLy (240)

2214 GTACAAGGACGGCAAAGGTTATCTGCTGGAGCGGATTCGCGATACCGCCGGCCAGAAGGGCACGGGCAAGTGGACGGCAATCGCTGCTCT 2303
sTyrLysAspGlyLysGlyTyrLeuLeuGluArgIleArgAspThrAlaGlyGlnLysGlyThrGlyLysTrpThrAlaIleAlaAlaLe (270)

2304 GCAGTATGGAGTGCCTGTGACGCTAATTGGCGAGGCGGTCTTCTCGCGATGCCTGTCTGCCCTGAAGGACGAGCGCGTCCAGGCCAGCAG 2393
uGlnTyrGlyValProValThrLeuIleGlyGluAlaValPheSerArgCysLeuSerAlaLeuLysAspGluArgValGlnAlaSerSe (300)

2394 CGTGCTGAAGGGACCCTCGACCAAGGCGCAAGTGGCCAACCTCACCAAGTTCCTCGACGACATCAAGCACGCTCTCTACTGCGCCAAGAT 2483
rValLeuLysGlyProSerThrLysAlaGlnValAlaAsnLeuThrLysPheLeuAspAspIleLysHisAlaLeuTyrCysAlaLysIl (330)

2484 CGTGTCCTACGCCCAGGGATTCATGCTCATGCGAGAGGCGGCCAGGGAGAACAAGTGGAGACTTAATTACGGCGGCATTGCGCTGATGTG 2573
eValSerTyrAlaGlnGlyPheMetLeuMetArgGluAlaAlaArgGluAsnLysTrpArgLeuAsnTyrGlyGlyIleAlaLeuMetTr (360)

2574 GCGTGGCGGCTGCATCATCCGCAGCGTCTTTCTGGGCAACATTAAGGACGCGTATACGTCGCAGCCGGAGCTGTCTAATCTGCTGCTGGA 2663
pArgGlyGlyCysIleIleArgSerValPheLeuGlyAsnIleLysAspAlaTyrThrSerGlnProGluLeuSerAsnLeuLeuLeuAs (390)

2664 TGACTTCTTCAAGAAGGCCATCGAGCGCGGCCAGGACTCGTGGCGCGAGGTGGTGGCCAATGCCTTCCGCTGGGGCATTCCCGTGCCGGC 2753
pAspPhePheLysLysAlaIleGluArgGlyGlnAspSerTrpArgGluValValAlaAsnAlaPheArgTrpGlyIleProValProAl (420)

2754 CCTGTCTACCGCCCTAAGCTTCTACGACGGCTACCGCACGGCCAAGCTGCCAGCCAACTTGCTGCAGGCCCAGAGGGATTACTTCGGCGC 2843
aLeuSerThrAlaLeuSerPheTyrAspGlyTyrArgThrAlaLysLeuProAlaAsnLeuLeuGlnAlaGlnArgAspTyrPheGlyAl (450)

```
2844 CCACACCTATGAGCTGCTGGGCCAGGAGGGTCAGTTCCACCACACGAACTGGACAGGCACCGGCGGCAATGTGTCCGCCAGCACTTACCA  2933
     aHisThrTyrGluLeuLeuGlyGlnGluGlyGlnPheHisHisThrAsnTrpThrGlyThrGlyGlyAsnValSerAlaSerThrTyrGl  (480)

2934 GGCGTAGGTTCCACCTGCTCCACTTTCCCGTTCACACATTCCATGTCATTGGCGCCGGTGTCTTAGATGTTTCTTTTTTTTCTGGAGTAC  3023
     nAlaEnd                                                                                     (481)

3024 TTTAGTACTTATTTATACCATTAATATATATGTATGTATATAGAATTTCATAATTGTTGTTAAACATAACATTAAATTGGTGTTTTTTTG  3113
                                                                                  ------

3114 CTAGCAAATGATTTTGATTCCTTAGTTTCATGAATGCAAGTGCCATTTAAAATCAACAATGCGTGTGGTTTGGTGTGTTGTGTTGTGTGT  3203
         |(A)n

3204 GGGTCGAGTCTTTCGAGTTGTGTCTTCATCTGGAGACGCCTCCTGCTCCTTCTACCGCTCCTTCCCTGCTATTGTACTCTCTTCAGCTAG  3293

3294 CGCGCTTTTTTCGCTCCGTATTTCCCTTAGTCGTCCGAGGGCTTCAGGGTCTTCTTGTTCTCTATAACCAGTTTGTCAGCGGAATACAGG  3383

3384 TGGCCGATGATTACCTGTGGACATTCAAAGGTTAATAAACTCAACCGGCTGATAAGCGAAAAAGGGGCAAAATGGTTACTTTCGATTTCT  3473
                                                                                         ______SspI
3474 AATAGGATGGTAATTGAGTTTTCCATTCCCCATATTTGCAAAATCAGATATATATGATAAAATCTACTTTAAATATACATTAATATT  3560
```

Pgd SEQUENCE. Strain, *Canton S*. Accession, M80598 (DROPGD).

Phenotype of Mutations

Two electrophoretic variants (A and B) have been described (Kazazian et al. 1965). *Pgd* null mutations are lethal due to the accumulation of 6-PG; viability can be improved by dietary manipulations that reduce 6-PG synthesis or by the introduction of a null mutation on a second gene in the pentose shunt, *Zwischenferment* (*Zw*). *Zw* is the structural gene for glucose-6-phosphate dehydrogenase, G-6-PD, the enzyme that precedes 6-PGD in the pentose biosynthetic pathway, and it is required for the synthesis of 6-PG (Hughes and Lucchesi 1977, 1978).

Gene Organization and Expression

Open reading frame, 481 amino acids; mRNA length, 1,659 bases, in agreement with an RNA of 1.7 kb observed in gels. Primer extension and S1 mapping were used to define the major 5′ end (there seem to be several minor transcription initiation sites as well). The 3′ end was obtained from a cDNA sequence. There is a short intron in the Gly-3 codon and a long one in the Ala-78 codon (*Pgd* Sequence) (Scott and Lucchesi 1991).

Promoter

In transgenic animals, a 4.7 kb fragment that extends 1,172 bp upstream of the transcription initiation site and 442 bp downstream of the poly(A) site is sufficient for apparently normal expression of *Pgd* in larvae. Removal of the small first intron does not significantly affect expression, but removal of the larger second intron leads to a 10-fold reduction in enzyme levels. The second intron is specifically required for expression in the fat body, but apparently not necessary for expression in actively dividing imaginal cells. Expression in imaginal cells requires only a 421-bp segment immediately upstream of the transcription initiation site (Scott and Lucchesi 1991).

References

Gutierrez, A. G., Christensen, A. C., Manning, J. E. and Lucchesi, J. C. (1989). Cloning and dosage compensation of the *6-phosphogluconate dehydrogenase* gene (Pgd^+) of *Drosophila melanogaster*. *Dev. Genet.* **10**:155–161.

Hughes, M. B. and Lucchesi, J. C. (1977). Genetic rescue of a lethal "null" activity allele of *6-phosphogluconate dehydrogenase* in *Drosophila melanogaster*. *Science* **196**:1114–1115.

Hughes, M. B. and Lucchesi, J. C. (1978). Dietary rescue of a lethal "null" activity allele of *6-phosphogluconate dehydrogenase* in *Drosophila melanogaster*. *Biochem. Genet.* **16**:469–475.

Kazazian, H. H., Young, W. J. and Childs, B. (1965). X-linked *6-phosphogluconate dehydrogenase* in *Drosophila*: Subunit association. *Science* **150**:1601–1602.

Scott, M. J. and Lucchesi, J. C. (1991). Structure and expression of the *Drosophila melanogaster* 6-phosphogluconate dehydrogenase gene. *Gene* **109**:177–183.

Williamson, J. H., Krochko, D. and Geer, B. W. (1980) *6-Phosphogluconate dehydrogenase* from *Drosophila melanogaster*, I. Purification and properties of the A isozyme. *Biochem. Genet.* **18**:87–101.

Wood, T. (1985). *The Pentose Phosphate Pathway* (New York: Academic Press).

25

paired: *prd*

Chromosomal Location:
2L, 33C1-2

Map Position:
2-45

Product

A DNA-binding regulatory protein of the homeodomain type important in establishing the segmentation pattern in early embryos.

Structure

The following potentially important sequence features occur.

1. The segment between residues 27 and 154 has great similarity to regions in both *gooseberry* genes and has been designated the "paired domain" (Bopp et al. 1986). This is a DNA-binding region (Treisman et al. 1991).

2. A homeodomain occurs between Gln-213 and His-272 (Frigerio et al. 1986; Harrison 1991). The sequence similarities between *prd* and *gooseberry* extends 18 amino acids upstream of the homeodomain (Bopp et al. 1986). A Ser in position 9 of the recognition helix (H3 in *prd* Sequence) differentiates the binding specificity of PRD from that of the products of *bicoid* (*bcd*), and *fushi tarazu* (*ftz*), which have Lys and Gln, respectively, in that position. *In vitro*, the PRD H3 does not bind sequences derived from the "standard" homeodomain binding site (TAAT). It is able to bind to the sequence TTTGACGT but only if the C-terminal region of the protein is removed. *In vivo*, the latter may be a regulatory region that is moved out of the way by interactions with other molecules (Treisman et al. 1989).

3. The C-terminus of PRD is characterized by a high proportion of His and Pro residues called the "PRD repeat". Using a DNA fragment from the PRD repeat, 11 other cross-hybridizing sequences were identified, one of which was *bcd* (Frigerio et al. 1986).

prd

```
-495  AGCTGAGACGCCCCCTGGGCGCGACGCGAGACGGTTGCTAAATGGGTCGAGTCGAGCCAGAGCGAGATGCCGTTGTGGAGAGCGCTGCGA   -406

-405  TTGGTCCGCGTAGTGGTTACCTGCCAAGTGACTGTGGGATATGGCCGACGTCTGGGCCGTGGCTTCACAGAAAGGCAACGATCTTGGCCG   -316

                                                                                  !-244
-315  ACGTTCGGATGGTGAAGTCAGTCAGGCACAGACTGCGCAGCGAGCCACACCGCATCTCGTCCTCGTTCTCGTCTTCGCCTTCGCCTCCGT   -226

-225  TTCATCTTTCCCATCGAGATTGCGAACTCACAGATACTTAGATATTCGAAGTGCAACTAATCGGTTAATCAATACCTCGCAACGCTTACT   -136

-135  TATGACTTTGACAAAGTGTCCAGACATTGTCCAAAACTAAAGTGATATAATCAAGTGATACACGAACTTCGAGACTGAGTTAACACCGGT    -46

 -45  TTTGTGCCGGGACAAGCTTACGCATCTTGGAGCTCCTCCAGAAACTATGACCGTAACCGCCTTTGCTGCCGCAATGCACAGACCCTTCTT     44
                                                      MetThrValThrAlaPheAlaAlaAlaMetHisArgProPhePh   (15)

  45  CAATGGATATTCTACGATGCAAGGTGAGTGTCTATCGATCTTATAGAACATCCAGCAAAAGTCACTTTCACAATTTACTTACTAAATATC    134
      eAsnGlyTyrSerThrMetGlnA                                                                        (23)

 135  AAAGCCTAGTTGATCATTTCCATATATCTCCATTTCTAAACCTACTACCCAAGATCCCGCTAAAGATCTCAGTTTGGGCCAAGGCGTCGG    224

 225  CTACTCTCTAATGGCCATTAGTTGCCCGGCGGGAGAGTCGCGCGCCTCTGACCTTCGACCTTAGCTCCGAGTTTCCCGTCTTCCCGGGAA    314

 315  GTCAACTCCGGTCGAAGGTGTCGTAAATCAAGTGACACGCGCTCCGCTCTACCTAGCTAGTATTGGAAAAGCCTCTAAAATTTCCATTTT    404

 405  CTCATCTTCCTCATTCCAGACATGAACAGCGGCCAGGGGCGCGTCAATCAACTAGGTGGAGTTTTCATCAACGGTCGTCCTTTGCCCAAC    494
                        spMetAsnSerGlyGlnGlyArgValAsnGlnLeuGlyGlyValPheIleAsnGlyArgProLeuProAsn  (46)
                                      |-

                                                                                  |---|2.45.17
 495  AATATTCGTCTTAAAATCGTCGAGATGGCCGCCGATGGCATTCGGCCCTGTGTGATCTCCAGACAGCTACGTGTATCCCATGGCTGCGTA    584
      AsnIleArgLeuLysIleValGluMetAlaAlaAspGlyIleArgProCysValIleSerArgGlnLeuArgValSerHisGlyCysVal     (76)

 585  TCGAAGATCCTGAATCGCTACCAGGAGACTGGCTCCATTAGACCAGGTGTGATCGGTGGCTCCAAGCCGAGGATAGCCACGCCCGAAATC    674
      SerLysIleLeuAsnArgTyrGlnGluThrGlySerIleArgProGlyValIleGlyGlySerLysProArgIleAlaThrProGluIle    (106)

 675  GAAAACCGAATTGAGGAGTACAAGCGCAGTAGCCCGGGCATGTTCTCGTGGGAGATCAGGGAGAAGCTGATCCGCGAGGGTGTCTGCGAC    764
      GluAsnArgIleGluGluTyrLysArgSerSerProGlyMetPheSerTrpGluIleArgGluLysLeuIleArgGluGlyValCysAsp    (136)

                                                                                      A
 765  AGGAGCACAGCACCATCTGTGTCCGCCATATCGCGCCTGGTGCGCGGCCGAGATGCTCCATTGGACAATGATATGTCTTCTGCCTCTGGA    854
      ArgSerThrAlaProSerValSerAlaIleSerArgLeuValArgGlyArgAspAlaProLeuAspAsnAspMetSerSerAlaSerGly    (166)
                                                    -|PRD DOMAIN                       Thr

 855  TCTCCGGCGGGTGATGGCACCAAAGCATCGAGTTCCTGTGGCTCCGATGTCTCCGGCGGCCATCACAACAACGGCAAGCCCTCCGATGAG    944
      SerProAlaGlyAspGlyThrLysAlaSerSerSerCysGlySerAspValSerGlyGlyHisHisAsnAsnGlyLysProSerAspGlu    (196)

                                                                           A
 945  GACATCTCAGACTGCGAAAGTGAGCCGGGAATCGCCTTGAAGCGCAAACAGCGCCGCTGCAGGACCACCTTTTCCGCTTCCCAGTTGGAC   1034
      AspIleSerAspCysGluSerGluProGlyIleAlaLeuLysArgLysGlnArgArgCysArgThrThrPheSerAlaSerGlnLeuAsp    (226)
                                                    |-    *     *      Ile*  ------*--------

1035  GAACTGGAACGCGCCTTCGAGCGCACCCAATACCCTGATATCTACACCCGTGAGGAGCTGGCCCAGCGCACCAATCTCACGGAGGCACGC   1124
      GluLeuGluArgAlaPheGluArgThrGlnTyrProAspIleTyrThrArgGluGluLeuAlaGlnArgThrAsnLeuThrGluAlaArg    (256)
      ---*-----------------H1     *          ------------------*--------------H2 *   ---------
```

```
1125  ATCCAGGTGTGGTTCAGCAACCGGCGTGCTCGTCTCCGCAAGCAGCACACCTCGGTCTCAGGCGGAGCACCTGGCGGAGCAGCTGCCTCA  1214
      IleGlnValTrpPheSerAsnArgArgAlaArgLeuArgLysGlnHisThrSerValSerGlyGlyAlaProGlyGlyAlaAlaAlaSer  (286)
      *-----*--*--*-----*--*--*H3   *     *           -|HOMEODOMAIN

1215  GTAAGCCATGTCGCCGCGTCCAGCTCTCTTCCCAGTGTGGTATCAAGTGTGCCCAGCATGGCTCCGCTGGCCATGATGCCGGGATCCCTG  1304
      ValSerHisValAlaAlaSerSerSerLeuProSerValValSerSerValProSerMetAlaProLeuAlaMetMetProGlySerLeu  (316)

1305  GATCCAGCCACTGTGTACCAGCAGCAATACGATTTCTACGGCAGTCACGCCAACATTTCCGTATCCGCCGCAGCTCCAATGGCCAGTAGT  1394
      AspProAlaThrValTyrGlnGlnGlnTyrAspPheTyrGlySerHisAlaAsnIleSerValSerAlaAlaAlaProMetAlaSerSer  (346)

1395  AATCTATCGCCCGGAATTACAACCACGCCACCGCACCACCATCAGTTCTACAATCCCAGCGCTAACACAGCCAGCTACATAATGCCGGGT  1484
      AsnLeuSerProGlyIleThrThrThrProProHisHisHisGlnPheTyrAsnProSerAlaAsnThrAlaSerTyrIleMetProGly  (376)

1485  GAGAATGGCAACACCACACCCACCGGGAACATCATCGTCTCCAGCTATGAGACTCAGTTGGGTTCAGTTTACGGCACCGAAACGGAAACC  1574
      GluAsnGlyAsnThrThrProThrGlyAsnIleIleValSerSerTyrGluThrGlnLeuGlySerValTyrGlyThrGluThrGluThr  (406)

1575  CACCAGACTATGCCACGCAACGAGAGCCCCAACGAGTCCGTGTCCTCCGCCTTCGGGCAACTGCCACCCACACCCAACAGCCTTTCCGCG  1664
      HisGlnThrMetProArgAsnGluSerProAsnGluSerValSerSerAlaPheGlyGlnLeuProProThrProAsnSerLeuSerAla  (436)

1665  GTGGTGAGTGGAGCTGGTGTGACCTCCTCCAGTGGGGCCAACTCGGGAGCCGATCCCTCGCAGTCGCTGGCCAATGCCAGTGCTGGAAGT  1754
      ValValSerGlyAlaGlyValThrSerSerSerGlyAlaAsnSerGlyAlaAspProSerGlnSerLeuAlaAsnAlaSerAlaGlySer  (466)

1755  GAGGAGCTATCGGCTGCCCTGAAAGTGGAATCGGTGGACCTGATCGCGGCCAGTCAGTCGCAGTTGTACGGCGGATGGAGCTCCATGCAG  1844
      GluGluLeuSerAlaAlaLeuLysValGluSerValAspLeuIleAlaAlaSerGlnSerGlnLeuTyrGlyGlyTrpSerSerMetGln  (496)

1845  GCACTGCGCCCCAATGCGCCACTTTCGCCGGAGGACTCGCTGAACTCCACCAGCTCGACCAGCCAGGCTCTGGATGTCACCGCCCACCAG  1934
      AlaLeuArgProAsnAlaProLeuSerProGluAspSerLeuAsnSerThrSerSerThrSerGlnAlaLeuAspValThrAlaHisGln  (526)

1935  ATGTTCCATCCGTATCAGCATACGCCGCAGTATGCATCCTATCCGGCACCAGGCCACGCCCATTCGCATCACGGACATCCCCATGCGCCG  2024
      MetPheHisProTyrGlnHisThrProGlnTyrAlaSerTyrProAlaProGlyHisAlaHisSerHisHisGlyHisProHisAlaPro  (556)
                                                                                         |-

2025  CATCCGCACGCACATCCGCATCCGCAGTACGCAGGCGCACATCCGCACTATCCGCCGCCCAGTTCGTCGGCGCACTTCATGCCGCAGAAC  2114
      HisProHisAlaHisProHisProGlnTyrAlaGlyAlaHisProHisTyrProProProSerSerSerAlaHisPheMetProGlnAsn  (586)
                                                    -|PRD REPEAT

2115  TTCAATGCCGCCGCCTTTCCTTCGCCCTCGAAGGTCAACTACACAACGATGCCGCCACAGCCGTTCTATCCCTCCTGGTACTAGAATCAA  2204
      PheAsnAlaAlaAlaPheProSerProSerLysValAsnTyrThrThrMetProProGlnProPheTyrProSerTrpTyrEnd        (613)

2205  AGAGACACGGATCCACCACCTACTCCTCCAGGAGCAGGAGCAGTGTCACCAGATCCATGGTACAAGTCGCCAAAGATGTACATACCCATA  2294

2295  GAGCAGGGGACGAAAATATAAATAACATTTTATTTGTGGTGGAGCAGTACAGACATTTTCCGTTTGAGAAAACCGCTGACAGACTCGCTC  2384

2385  CCAAACAATAAACATATGTATTAGTTCCAATTCGTAGATGTAAGCCTAGAAAATAGTACCGACTTAGGATTAGAGTTTAAGATGATTAGC  2474

2475  CTAAGTAGCAAGTGCTCTTAAATAAAAAAATATATCTATGCTAATTTACAACGTACTCCAATGATCTTTCAC  2546
                        ------                           |(A)n
```

prd SEQUENCE. Accession No. M14548 (DROPRD). An exclamation mark at −244 marks the 5′ end of the longest cDNA. Allele *prd*$^{2.45.17}$ is an insertion of 1.1 kb following position 569, with a concomitant 5-bp deletion of positions 569–573. In cDNA sequences, two natural variants were detected; these involve changes in the amino-acid sequence at codons 164 and 220. A homeodomain spanning Gln-213 to His-272 is delimited by vertical bars and conserved residues are marked with asterisks; within the domain, the three putative helices, H1, H2 and H3, are

(*continued*)

Function

Treisman et al. (1989, 1991) have demonstrated direct binding of PRD to element e5 of the *even-skipped* (*eve*) promoter. The homeodomain and the paired domain bind to different sub-regions of e5.

Mutant Phenotype

prd is one of the pair-rule genes. Null mutants are embryonic lethals with only half the correct number of segments. The missing elements correspond mainly to odd-numbered parasegments (Appendix, Fig. A.3); i.e., posterior region of T2 and the adjacent boundary to T3, the posterior of A1 and the adjacent boundary to A2, etc. (every other segment boundary and neighboring areas are missing). The pattern is similar to that affected by *eve*, but the position of the missing elements is shifted anteriorly by a fraction of a parasegment in *prd* as compared to *eve* (Nüsslein-Volhard and Wieschaus 1980; Nüsslein-Volhard et al. 1985). It would appear that, in mutants, the regions of the segmented embryo that are lacking are those in which *prd* is maximally expressed in normal embryos (see below).

Gene Organization and Expression

Open reading frame, 613 amino acids; expected mRNA length, 2,417+ bases; in agreement with a 2.5 kb band detected by northern analysis); information on the 5′ and 3′ ends is from a cDNA sequence. There is an intron in the Asp-23 codon (*prd* Sequence) (Frigerio et al. 1986).

Developmental Pattern

The *prd* transcript is absent from oocytes and barely detectable in 0–2 h embryos; it peaks in 2–4 h embryos and disappears soon afterward. The transcript is first detectable by *in situ* hybridization during nuclear cycle 12 (syncytial blastoderm) in the primordial cephalic region (77–63% egg length; Appendix, Figs A.1–A.3). By nuclear cycle 14 (late syncytial blastoderm), expression is localized in seven bands covering the area from the cephalic region to the eighth abdominal segment (75–20% egg length). These bands are more intense on the dorsal than on the ventral side of the embryo. In general terms, the seven bands of *prd* expression have a two-segment periodicity similar to that of other pair-rule genes such as *eve*, *ftz* and *hairy* (*h*). The *prd* bands,

---

(*continued*) underlined; these were identified based upon their similarity to *Antennapedia* helical regions. The PRD repeat, spanning His-552 to His-572, and PRD domain, spanning Gly-27 to Asp-154, are also delimited by vertical bars.

however, are broader with the area covered by each band corresponding to more than one segment, i.e., they extend posteriorly from the middle of one segment to the posterior boundary of the next segment. The intensity of expression increases posteriorly within each band so that the regions of highest *prd* expression correspond to the posterior compartments of the mandibular, labial, T2, A1, A3, A5 and A7 segments. At this time expression starts in a new domain in the anterior pole of the embryo, at 93–87% egg length, but in the dorsal region only (Kilchherr et al. 1986; Akam 1987; Baumgartner et al. 1987).

Around the time of blastoderm cellularization, an eighth band appears posteriorly (at 13% egg length), and bands 2–7 of the original seven become double because transcripts disappear from the central portion of each band. Thus, in the segmented germ band region, there are 14 stripes; 13 of them are two-cell-wide bands that appear to correspond to the two most posterior cells of each segment in the region from the mandibular segment to the A7 segment. The 14th band is wider and includes A8 and A9. This banded pattern persists until the beginning of gastrulation but disappears soon thereafter (Kilchherr et al. 1986; Baumgartner et al. 1987). In later stages, expression is restricted to the head region and central nervous system (Gutjahr et al. 1993).

References

Akam, M. E. (1987). The molecular basis for metameric development in the *Drosophila* embryo. *Development* **101**:1–22.

Baumgartner, S., Bopp, D., Burri, M. and Noll, M. (1987). Structure of two genes at the *gooseberry* locus related to the *paired* gene and their spatial expression during *Drosophila* embryogenesis. *Genes Dev.* **1**:1247–1267.

Bopp, D., Burri, M., Baumgartner, S., Frigerio, G. and Noll, M. (1986). Conservation of a large protein domain in the segmentation gene *paired* and in functionally related genes of *Drosophila. Cell* **47**:1033–1040.

Frigerio, G., Burri, M., Bopp, D., Baumgartner, S. and Noll, M. (1986). Structure of the segmentation gene *paired* and the *Drosophila* PRD gene set as part of a gene network. *Cell* **47**:735–746.

Gutjahr, T., Frei, E. and Noll, M. (1993). Complex regulation of early paired expression: Initial activation by gap genes and pattern modulation by pair-rule genes. *Development.* (In press.)

Harrison, S. C. (1991). A structural taxonomy of DNA-binding domains. *Nature* **353**:715–719.

Kilchherr, F., Baumgartner, S., Bopp, D., Frei, E. and Noll, M. (1986). Isolation of the *paired* gene of *Drosophila* and its spatial expression during early embryogenesis. *Nature* **321**:493–499.

Nüsslein-Volhard, C., Kluding, H. and Jürgens, G. (1985). Genes affecting the segmental subdivision of the *Drosophila* embryo. *Cold Spring Harbor Symp. Quant. Biol.* **50**:145–154.

Nüsslein-Volhard, C. and Wieschaus, E. (1980). Mutations affecting segment number and polarity in *Drosophila. Nature* **287**:795–801.

Treisman, J., Gonczy, P., Vashishtha, M., Harris, E. and Desplan, C. (1989). A single amino acid can determine the DNA binding specificity of homeodomain proteins. *Cell* **59**:553–562.

Treisman, J., Harris, E. and Desplan, C. (1991). The paired box encodes a second DNA-binding domain in the Paired homeo domain protein. *Genes Dev.* **5**:594–604.

26

Ribosomal protein 49: Rp49

Chromosomal Location: 3R, 99D4-8
Synonym: *M(3)99D*

Map Position: 3-[101]

Product

Protein 49 of the large ribosomal subunit (Vaslet et al. 1980; O'Connell and Rosbash 1984). The syntheses of ribosomal proteins are coordinately regulated and at least part of that regulation occurs at the level of translation. For instance, while almost all *Rp49* mRNA is translated during oogenesis, only a small fraction is associated with polysomes early in embryogenesis (Al-Atia et al. 1985).

Mutant Phenotype

Heterozygotes for a deletion show a strong *Minute* phenotype (Kongsuwan et al. 1986).

Gene Organization and Expression

Open reading frame, 133 amino acids; mRNA length, ca. 520 bases in agreement with a 0.6 kb band from RNA blots. S1 mapping was used to define the 5′ and 3′ ends; the 3′ end is near position 570. There is no apparent TATA box. There is an intron after Ser-31 (*Rp49* Sequence) (O'Connell and Rosbash 1984). *Rp49* is in the *Serendipity* cluster (Chapter 28, Fig. 28.1); it is transcribed convergently with *Sryδ*, the 3′ ends of these genes being approximately 300 bp apart.

Rp49

```
               .         .         .         .         .         .         .         .         .
-418 ACGACGTTCGATGTTTAACCACAGCTTTCTTTCGCTTCTGTTTCCGGCAAGGTATGTGCCGTGATTTTGGGCCCACGTGTATGTCCATTA  -329
               .         .         .         .         .         .         .         .         .
-328 ATTTTAAGCCGTAATGTCGTTTTTGCGTTTCGAGTTGAACTGCGTTAGTCCTCGGGCTAGTGAACTAGTTAGCAAGTAGTTGCGGCTAGT  -239
               .         .         .         .         .         .         .         .         .
-238 ATTTCAGACCATTCTTGATTCCTGTGAGCAGTTACTGCCGAATGGCTTCTGTGTTTGCTGAATTCGGTATTCGATGTTCGACATCACGGT  -149
               .         .         .         .         .         .         .         .         .
-148 ACTGTCAATGGATACTGCCCAAGCAGCTAGCCCAACCTGGTTGAATTATGCATTAGTGGGACACCTTGTGTGTTATTAGCTTGATAAGTG  -59

               .         .         .         .         . -->-8   .         .         .         .
 -58 ATATTTCCAGTGGGTCAGTGCACTAATGGCTACACTTGTTGTGTCCTACCAGCTTCAAGATGACCATCCGCCCAGCATACAGGCCCAAGA  31
                                                               MetThrIleArgProAlaTyrArgProLysI  (11)
               .         .         .         .         .         .         .         .         .
  32 TCGTGAAGAAGCGCACCAAGGACTTCATCCGCCACCAGTCGGATCGATATGCTAAGCTGTCGGTGAGTGCCACGGATTGTGCCAAATTGT  121
     leValLysLysArgThrLysAspPheIleArgHisGlnSerAspArgTyrAlaLysLeuSer                             (31)
               .         .         .         .         .         .         .         .         .
 122 ACCCGTGTTTAATCAACATGTCTCCTTGCAGCACAAATGGCGCAAGCCCAAGGGTATCGACAACAGAGTCGGTCGCCGCTTCAAGGGACA  211
                                        HisLysTrpArgLysProLysGlyIleAspAsnArgValGlyArgArgPheLysGlyGl  (51)
               .         .         .         .         .         .         .         .         .
 212 GTATCTGATGCCCAACATCGGTTACGGATCGAACAAGCGCACCCGCCACATGCTGCCCACCGGATTCAAGAAGTTCCTGGTGCACAACGT  301
     nTyrLeuMetProAsnIleGlyTyrGlySerAsnLysArgThrArgHisMetLeuProThrGlyPheLysLysPheLeuValHisAsnVa  (81)
               .         .         .         .         .         .         .         .         .
 302 GCGCGAGCTGGAGGTCCTGCTCATGCAGAACCCGCGTTTACTGCGCGAGATGCCCACGGCGTCTCCTCCAAGAAGCAAGGAGATTATCGA  391
     lArgGluLeuGluValLeuLeuMetGlnAsnProArgLeuLeuArgGluMetProThrAlaSerProProArgSerLysGluIleIleGl  (111)
               .         .         .         .         .         .         .         .         .
 392 GCGCGCCAAGCAGCTGTCGCTCCGCTCACCAACCCCAACGGTCGCCTGCGTCTCAAGAAGAACGAGGTAAGCTTAAGATTCTTGAGAGTT  481
     uArgAlaLysGlnLeuSerLeuArgSerProThrProThrValAlaCysValSerArgArgThrArgEnd                     (133)
               .         .         .         .         .         .         .         .         .
 482 CTTGTAACGTGGTCGGAATACACATTTGTAAACGTTAATATACCGGACTTTTAGTTAAAAAATGATGTGCCAGTGCCGAGTTCAATTGTC  571
                                                   ------
               .         .         .         .         .         .         .         .         .
 572 ATTTCTGAGATCGGGATAGCAGCACCATCGATAACATGTGCATTATCTGGATGGATATCAGTTAATCCAGACCATTGCGGTCTTTCTTTC  661
               .         .         .         .         .         .         .         .         .
 662 TGATAGCAACTGCCTCGAGATATTAGACCAATATAAATTCTTGACGTGCCAAAACTAGACAGCATCAATCCTTATCAGGGAATTTTGTTA  751
               .         .         .         .         .         .         .         .         .
 752 TATATTTTACATTTTTCCCCCTTAGTATTCAAAGAGGTTGTTTATATGAAATCATATATATATTCGCAATTATTTTTACAGAACAGTGTA  841
```

Rp49 SEQUENCE. Accession X00848 (DRORP49).

References

Al Atia, G. R., Fruscoloni, P. and Jacobs-Lorena, M. (1985). Translational regulation of mRNAs for ribosomal proteins during early *Drosophila* development. *Biochemistry* **24**:5798–5803.

Kongsuwan, K., Dellavalle, R. and Merriam, J. R. (1986). Deficiency analysis of the tip of chromosome 3R in *Drosophila melanogaster*. *Genetics* **112**:539–550.

O'Connell, P. and Rosbash, M. (1984). Sequence, structure, and codon preference of the *Drosophila* ribosomal protein 49 gene. *Nucl. Acids Res.* **12**:5495–5513.

Vaslet, C. A., O'Connell, P., Izquierdo, M. and Rosbash, M. (1980). Isolation and mapping of a cloned ribosomal protein gene of *Drosophila melanogaster*. *Nature* **285**:674–676.

27

Salivary Gland Secretion Protein Genes: Sgs3, Sgs5, Sgs7, Sgs8

Chromosomal Location:			**Map Position:**
Sgs3, Sgs7, Sgs8	3L,	68C3-5	3-35.0
Sgs5	3R,	90B3-8	3-[60]

Products

Glue proteins in the salivary gland secretion of third-instar larvae.

Structure and Function

Genes for seven proteins have been identified, SGS1 and SGS3-8. Proteins are numbered in order of increasing electrophoretic mobility except for SGS6 which is slightly slower than SGS3 (Velissariou and Ashburner, 1981). Partial sequences for SGS3, SGS7 and SGS8 confirmed the primary structure derived from nucleotide sequences and the existence of a 23-amino-acid signal peptide (Fig. 27.1) (Crowley et al. 1983). These proteins attach larvae to a solid substratum prior to pupariation. Cys residues and glycosylation appear to play a role in the function of SGS.

Tissue Distribution

The glue proteins are synthesized in the larval salivary glands between 106 h and 120 h after fertilization, during the second half of the third instar (Beckendorf and Kafatos 1976; for reviews, see Berendes and Ashburner 1978; Ashburner and Berendes 1978).

Evolutionary Relationships

The evolutionary relationships among the *Sgs* genes are not entirely obvious. It is clear from amino-acid sequence similarities, intron position, sequence of

```
       1                                                      50                                                   100
Sgs5   MFNIKLLLLL LAVSWFHHGQ AVQET..... .......... .......... .......... .......... .......... .......... ..........
Sgs3   MKLTIATALA SILLIGSANV ANCCDCGCPT TTTTCAPRTT QPPCTTTTTT TTTTCAPPTQ QSTTQPPCTT SKPTTPKQTT TQLPCTTPTT TKATTTKPTT
Sgs7   MKLIAVTIIA CILLIGFSDL ALGGA..... .......... .......... .......... .......... .......... .......... ..........
Sgs8   MKLLVVAVIA CIMLIGFADP ASGCK..... .......... .......... .......... .......... .......... .......... ..........
 CON   MKL------A -I-LIG---- A--------- ---------- ---------- ---------- ---------- ---------- ---------- ----------
                ^                     |

       101                                                    150                                                  200
Sgs5   .......... .......... .......... .......... .......... .......... .........K IEEKPVSEPE IESEIKNSTS VPSKCNIYYR
Sgs3   TKATTTKATT TKPTTTKQTT TQLPCTTPTT TKQTTTQLPC TTPTTTKPTT TKPTTTKPTT TKPTTTKPTT TKPTTTKPTT TKPTTTKPTT TKPTTTKPTT
Sgs7   .......... .......... .......... .......... .......... .......... .......... .......... .......... ..........
Sgs8   .......... .......... .......... .......... .......... .......... .......... .......... .......... ..........
 CON   ---------- ---------- ---------- ---------- ---------- ---------- ---------- ---------- ---------- ----------

       201                                                    250                                                  300
Sgs5   NYQWALQDCV CRCFQNECLM QIESDQRKKE GRSPFVPVTE ELCRSFICKK CSVGFPVVAE FPIPAPCGCN RKPGSIATER FYSLCHLLKF SAENSKPFLT YSYCWPF*
Sgs3   TKPTTTKPTT TKPTTTKPTT TKPTTTKPTT TKPTTTKPTT TKPTTTKPTT TKPTTPKPCG CKSCGPGGEP CNGCAKRDAL CQDLNGVLRN LERKIRQCVC GEPQWLL*
Sgs7   .......... .......... .......... .......... .......... ........CE CQPCGPGGKA CTGCPEKPQL CQQLISDIRN LQQKIRKCVC GEPQWMI*
Sgs8   .......... .......... .......... .......... .......... .......DCS CVICGPGGEP CPGCSARVPV CKDLINIMEG LERQVRQCAC GEQVWLF*
 CON   ---------- ---------- ---------- ---------- ---------- --------C- C--CGPGG-- C-GC------ C--L------ L----R-C-C GE--W---
                                         ^                                                                        ^
```

FIG. 27.1. Comparison of SGS3, SGS7, SGS8 and SGS5 amino-acid sequences. Only positions in which three of the four sequences agree are represented in the CON(sensus). The vertical line at position 23 marks the last residue in the signal peptides of SGS3, SGS7 and SGS8 (Crowley et al. 1983). The caret at position 10 marks the intron in *Sgs3*, *Sgs7* and *Sgs8*, and at positions 234 and 297, the introns in *Sgs5*.

regulatory elements and clustering of the genes, that the three genes at 68C have a common ancestor (Martin and Meyerowitz 1988). On the other hand, *Sgs5* is similar to the other genes in the group only with respect to protein function and possibly some regulatory sequences (Fig. 27.1) (Shore and Guild 1986; Todo et al. 1990). It is likely that at least some SGS proteins are functionally equivalent since natural variants causing a deficiency in SGS5 (Shore and Guild 1987), SGS4 or SGS6 (Velissariou and Ashburner 1981) have no obviously deleterious effect.

Gene Expression and Developmental Pattern

The *Sgs* genes are expressed in salivary glands during the third larval instar (Meyerowitz and Hogness 1982). Transcription starts approximately 96–98 h after fertilization, reaches a plateau by approximately 112 h, and becomes undetectable by 120 h, the time of pupariation (Hansson and Lambertsson 1983; Georgel et al. 1991). An increase in ecdysterone level is necessary for the start of transcription in the middle third instar (Hansson and Lambertsson 1983). Subsequently, however, in late third instar larvae, high levels of this hormone repress transcription (Crowley and Meyerowitz 1984). Chromosomal puffing accompanies transcriptional activity, but the two processes seem to be somewhat independent of each other (Crowley et al. 1984; Hansson and Lambertsson 1983). There is considerable information on the expression of *Sgs4*; the complete sequence, however, is not available.

Promoters

The consensus sequence TNTTTGN_XTCCAT(T/A), in which N_X represents a variable number of nucleotides (values between 18 and 39 have been observed), was identified as a tissue-specific, *cis*-acting regulatory element of *Sgs3*; such sequences were also found upstream of *Sgs5*, *Sgs7* and *Sgs8* (Todo et al. 1990; Hofmann et al. 1991).

Sgs Gene Cluster at 68C: *Sgs3*, *Sgs7* and *Sgs8*

Organization and Expression of the Cluster

The three genes are contained in less than 5 kb of DNA. The arrangement of the genes is shown in Fig. 27.2; *Sgs8* is centromere distal (Garfinkel et al. 1983). *Sgs7* and *Sgs8* are separated by 475 bp and they are transcribed divergently. The developmental expression of *Sgs7* and *Sgs8* seems to be controlled by common enhancer elements (Todo et al. 1990; Hofmann et al. 1991). The levels of RNA accumulation are comparable for *Sgs3* and *Sgs7* and an order of magnitude lower for *Sgs8* (Crowley and Meyerowitz 1984).

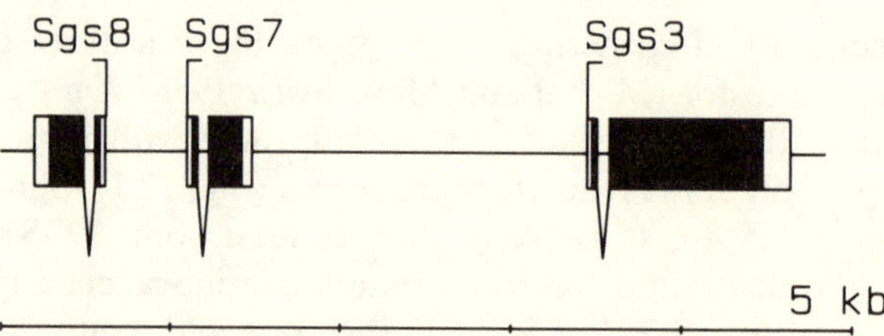

FIG. 27.2. Organization of the 68C cluster.

Sgs3

Product

SGS3 is heavily glycosylated (Beckendorf and Kafatos 1976) and very rich in the likely target of glycosylation, Thr residues (45% in the mature peptide). SGS3 is most similar to SGS7 and SGS8 in the amino-terminal 20–25 residues and the carboxy-terminal 50 residues. In particular, the position of eight Cys is conserved among the three sequences. The middle segment of SGS3 is not represented in SGS7 or SGS8; this segment is 235 amino acids long and contains most of the Thr residues: the first 50 amino acids constitute a Thr- and Cys-rich region (residues 23 to 72), and the last 185 amino-acid segment (from 73 to 257) is composed of 37 repeats of the peptide Pro Thr Thr Thr Lys, and variants thereof (Garfinkel et al. 1983). Twenty of the repeats are lacking from a natural variant found in the strain *Formosa* (*Sgs3* Sequence) (Mettling et al. 1985).

Gene Organization and Expression

Open reading frame, 307 amino acids; expected mRNA length, 1,117 bases. The strain *Formosa* makes an mRNA that is 300 bases shorter due to an internal deletion (Mettling et al. 1985). Primer extension was used to define the 5′ end. The 3′ end was obtained from a cDNA sequence. There is an intron in the Ala-10 codon (Garfinkel et al. 1983).

Promoter

Two *cis*-acting regulatory elements were identified by *in vitro* mutagenesis and analysis of DNase-hypersensitive sites. Either element is sufficient for correct developmental regulation of transcription, albeit at reduced level; when both elements are present, the transcription level increases 20-fold (Martin et al. 1989a, 1989b; Meyerowitz et al. 1987; Roark et al. 1990). Mutational analysis of the proximal element established that it is bipartite: the critical sequences being TGTTTG (pa, at −120, in *Sgs3* Sequence) and TCCATT (pb at −96). Sequences related to these two hexanucleotides are also found in the promoter

Sgs3

```
         .         .         .         .         .         .         .         .         .
-782  TCGTTGAATCAATGTCAAATTGCCTGTCAAAGTGCAAACGAAGCCCAAAATGTCTATCCTAATTCGAACCTAAAAATATATATTTTTTGA  -693
         .         .         .         .         .         .         .         .         .
-692  ATATGCAATACTATAAGATAATTGAATAGTTTTATGGGGCTTATTTGTAAAGCTAAATTAAGCTAAATTTAACTGTCCTTATTTATATTA  -603
                                              ----->da                          ----->db

         .         .         .         .         .         .         .         .         .
-602  TTATATTTACTCAGCCTATATTAAAGACCTATTATTTATAGAATTTAACGCAGTTTGTCTGCAAAACATCTCTACACCTTTTTCTACCCG  -513
         .         .         .         .         .         .         .         .         .
-512  TTACTCGTAGAGTAAAAGGGTATACTCGTTTCGCTGAGAAGTAACAGGCAGAATATAAAGCATATATATTCTTGATTAGGGTCAATAGCC  -423
         .         .         .         .         .         .         .         .         .
-422  GAGTCGATCTGGCCATGTCCGTCTGATTCTGTTTGCCACTCCCACATTTTTGAAAAATGTTTTATAATTTTTTCATATTTTTATTATCTA  -333
         .         .         .         .         .         .         .         .         .
-332  AATCTATCCCTTCCACACCTTAGAGCATTAAATTTAATTTCTTTCCCCCAATTTTTACCGATATTCGTGAAAAATGTTATACATTTTCCA  -243
         .         .         .         .         .         .         .         .         .
-242  TTTCACTTGAACTAGCTAAGTAACGGGTATCTGTTAGTCTCGTTAGCGTTCTCTCTTGTTTTAAAATAAAGTCTAGGCGATCGAGTCGAC  -153
         .         .         .         .         .         .         .         .         .
-152  CCAAAAGTATCAAACAAAGGGGAGAAGGCTTGTGTTTGCATAATCGAAATACTGACTCCATTTTTAGAATTGCAGTTTCAGTGAAAGCGT  -63
                                     ----->pa                ----->pb    ----

         .         .         .         . --> -28 .         .         .         .
 -62  ACCTATAAAAAGGTGAGGTATCCGCAAGAAAAGTATCAGTTTGTGGAGAATTAAGTAAAAAACATGAAGCTGACCATTGCTACCGCCCTA  27
         -----                                                          MetLysLeuThrIleAlaThrAlaLeu  (9)
         .         .         .         .         .         .         .         .         .
  28  GGTAGGTTTCACCGAATGCTCTTGTTTTCGGTATTTGAGCCACTGATATATTCATCCGTTTGCCTTCTCCACAGCGAGCATCCTGCTTAT  117
      A                                                                         laSerIleLeuLeuIl  (15)
         .         .         .         .         .         .         .         .         .
 118  TGGCTCCGCTAATGTTGCCAACTGTTGCGATTGTGGATGCCCCACAACTACAACTACTTGTGCGCCACGTACCACGCAACCTCCGTGCAC  207
      eGlySerAlaAsnValAlaAsnCysCysAspCysGlyCysProThrThrThrThrThrCysAlaProArgThrThrGlnProProCysTh  (45)
         .         .         .         .         .         .         .         .         .
 208  AACTACGACAACAACAACCACAACTACTTGTGCGCCACCCACACAACAATCTACCACGCAACCTCCATGCACGACATCTAAGCCCACCAC  297
      rThrThrThrThrThrThrThrThrThrCysAlaProProThrGlnGlnSerThrThrGlnProProCysThrThrSerLysProThrTh  (75)

         .         .         .         .         .         |-       .         .         .
 298  ACCTAAGCAAACTACCACGCAACTTCCGTGCACAACACCCACCACCACTAAGGCCACCACCACGAAGCCCACCACCACTAAAGCCACCAC  387
      rProLysGlnThrThrThrGlnLeuProCysThrThrProThrThrThrLysAlaThrThrThrLysProThrThrThrLysAlaThrTh  (105)
         .         .         .         .         .         .         .         .         .
 388  CACTAAGGCCACCACCACTAAGCCCACCACCACTAAGCAAACTACCACGCAACTTCCGTGCACAACACCCACCACCACTAAGCAAACTAC  477
      rThrLysAlaThrThrThrLysProThrThrThrLysGlnThrThrThrGlnLeuProCysThrThrProThrThrThrLysGlnThrTh  (135)

         .         .         .         . -| Deleted in Formosa       .         .         .
 478  CACGCAACTTCCGTGCACAACACCCACCACCACTAAGCCCACCACCACGAAGCCCACCACCACGAAGCCCACCACCACTAAGCCCACCAC  567
      rThrGlnLeuProCysThrThrProThrThrThrLysProThrThrThrLysProThrThrThrLysProThrThrThrLysProThrTh  (165)

         .         .         .         .         .         .         .         .        |-
 568  CACGAAGCCCACCACCACCAAGCCCACCACCACGAAGCCCACCACCACTAAGCCCACCACCACGAAGCCCACCACCACTAAGCCCACCAC  657
      rThrLysProThrThrThrLysProThrThrThrLysProThrThrThrLysProThrThrThrLysProThrThrThrLysProThrTh  (195)
         .         .         .         .         .         .         .         .         .
 658  CACGAAGCCCACCACCACGAAGCCCACCACCACTAAGCCCACCACCACGAAGCCCACCACCACTAAGCCCACCACCACGAAGCCCACCAC  747
      rThrLysProThrThrThrLysProThrThrThrLysProThrThrThrLysProThrThrThrLysProThrThrThrLysProThrTh  (225)

         .         .         .         . -| Deleted in Formosa       .         .         .
 748  CACTAAGCCCACCACCACGAAGCCCACCACCACTAAGCCCACCACCACGAAGCCCACCACCACGAAGCCCACCACCACTAAGCCCACCAC  837
      rThrLysProThrThrThrLysProThrThrThrLysProThrThrThrLysProThrThrThrLysProThrThrThrLysProThrTh  (255)
```

(continued)

```
838  ACCTAAGCCGTGCGGTTGCAAGAGCTGCGGTCCTGGAGGAGAGCCATGCAATGGATGTGCTAAGAGGGATGCACTGTGCCAGGATCTTAA   927
     rProLysProCysGlyCysLysSerCysGlyProGlyGlyGluProCysAsnGlyCysAlaLysArgAspAlaLeuCysGlnAspLeuAs  (285)

928  CGGCGTACTCCGCAATCTGGAGCGCAAGATCCGTCAATGCGTCTGCGGTGAACCGCAATGGTTGCTGTGAAGCGTCGAAGGAGCGTCTAA   1017
     nGlyValLeuArgAsnLeuGluArgLysIleArgGlnCysValCysGlyGluProGlnTrpLeuLeuEnd                      (307)

1018 TCCACTCCCGTACTGATCGATGTGACTGCACCCCTGCGAAATATATTCTGTGGGGGAGCTCGGCCAGGACTTTGACTACGCTTTGTTTTT   1107

1108 GTTATCATCAATTGATTTTACGTGTAAGAATTAATAAAATTAGTTAGACTGCATAAATTTTAAAAGCATTTATTATTATTTTACTTGTAT   1197
                                 ------                    |(A)n

1198 TATTTATGACAAATTATTATTTATCTGTTGGGTTTTCGAAAATGTTGGTTCTAAATTAAGTTTGGCCATCATTTGATCGACTTTTTCGAA   1287

1288 TGTATCTGTTACTTTTACCAATGCGTTGGCTTTGGCTCCTAGTTCTATGCGAAGTCTTAACTATCCGAGCTCTTATGACTTGGTCAACTT   1377

1378 GTCTCAGCTAACTACTGTTGG  1398
```

Sgs3 SEQUENCE. Strain, *Oregon R*. Accession, X01918 (DROSGS378). Arrows labeled da, db, pa and pb underline the a and b parts of the distal and proximal promoter elements. The *Formosa* strain deletions that occur in the repetitive middle portion of the coding region are delimited by vertical bars (Mettling et al. 1985).

region of *Sgs7* and *Sgs8* and within the distal element of *Sgs3*, at −651 (da) and −617 (db) (Todo et al. 1990). A DNase-hypersensitive site near −630 occurs only in the chromatin of salivary glands of third instar larvae, and DNase protection experiments identified two footprints overlapping da and db. There are three other hypersensitive sites near the 5′ end of *Sgs3*, including one around −100; but these are not restricted to the tissue in which *Sgs3* is expressed (Georgel et al. 1991).

A 115-kD protein that binds specifically to the distal promoter element, the Glue Enhancer-Binding Factor, GEBF1, was isolated from nuclear extracts. It appears that GEBF1 binds to both parts of the distal promoter element (da and db). The amount of GEBF1 found in extracts rises in parallel with the transcriptional activity of the salivary gland secretion genes during the third instar. GEBF1 is absent from extracts of the *Broad Complex* mutant $l(1)t^{435}$ (located in region 2B5, the site of an early, ecdysone-irreducible, puff) (Georgel et al. 1991). This allele also reduces or eliminates expression of the glue genes (Crowley et al. 1984). These observations suggest that (a) a gene in 2B5 is, directly or indirectly, responsible for the synthesis of GEBF1; and that (b) ecdysone induces the glue genes indirectly, by inducing the appearance of a regulatory factor (Georgel et al. 1991).

Sgs7

Product

SGS7 is not glycosylated (Beckendorf and Kafatos 1976); it contains only 4% Ser/Thr (Shore and Guild 1986).

Sgs7

```
      .         .         .         .<-- -507 Sgs8      .         .         .         .
-540  TGGTTGTTGCTTTAACAAATTAACTTTACCAGATGGTAACCGTTTATGAACACCCTACCCCTTTTATAGCAAAACAAATGTGTTATAGGA  -451
                                                                    ------

      .         .         .         .         .         .         .         .         .
-450  TCAATGGAAATTTCATTGAATTCATCCAAAAATAAAATATATAACCATTTGTGCTTAAGCAAATAGAAACACGATATTAAACTTCGCCCT  -361

      .         .         .         .         .         .         .         .         .
-360  TTGTTCTCACCATTTTCTGTGTCATCGTTCATACTAATATAATATAACATTTTACATGCCCTTTTTACTAAAGAAAGTATTACTCATAAA  -271

      .         .         .         .         .         .         .         .         .
-270  ATGAAATCTAAATTATATCTGAGTAACAAATATATTAAATTAATAAGTATCTATAAAAAGTTAATTCTATAAATAAAGCGCCTGCCGTAT  -181

      .         .         .         .         .         .         .         .         .
-180  AAAAAGCCAAGTGTTTGGTGTTTTATTTATTTTAATACAATTGGTTTGTCCAGTACTTTTTATTTTTGGATGTGCTCACTGAAATTTTCC  -91
                 ----->1a                     2a----->----->1b     -------                   ---

      .         .         .         .         .         .       --> -32   .         .
-90   ATTGATCCAGCTAACTTTTTGCGCTATATAAAGGTGTTGCTTTCCTTGAGTTGGTACCATCTGGTAAAGTAGTCTCAATCTAGATAGAAC  -1
      -->2b                     -----

      .         .         .         .         .         .         .         .         .
   0  CATGAAACTGATCGCAGTCACCATCATCGGTAACTACATAATAAGATCTTTAATCCACAACCAACTTCAATATCTCGCATCCTCAATATC  89
       MetLysLeuIleAlaValThrIleIleA                                                               (10)

      .         .         .         .         .         .         .         .         .
  90  CCCAGCTTGCATCCTGCTCATTGGATTCTCCGATCTAGCCCTGGGTGGTGCCTGTGAGTGCCAACCGTGTGGTCCTGGTGGAAAGGCCTG  179
           laCysIleLeuLeuIleGlyPheSerAspLeuAlaLeuGlyGlyAlaCysGluCysGlnProCysGlyProGlyGlyLysAlaCy  (38)

      .         .         .         .         .         .         .    T=Formosa      .
 180  CACGGGCTGTCCCGAAAAGCCCCAACTTTGTCAGCAGCTCATTAGCGATATTCGCAATCTCCAGCAGAAGATCCGGAAATGCGTCTGCGG  269
      sThrGlyCysProGluLysProGlnLeuCysGlnGlnLeuIleSerAspIleArgAsnLeuGlnGlnLysIleArgLysCysValCysGl  (68)
                                                                      Leu

      .         .         .         .         .         .         .         .         .
 270  AGAACCACAATGGATGATTTAGACACCAATCACTTTTAAAGATCACAAAAATTCTTCCTTAATAAAATTGTTATTACTGCTTCAAAAAAA  359
      yGluProGlnTrpMetIleEnd                                      ------                |(A)n     (74)

      .         .         .         .         .         .         .         .         .
 360  AAAAAAAAAAATGTTTGAGTTCTTTTTTATCATTTATTTCAGTATATATCGTCCAGAAAAGAACAAAACTAGTTTTTCCTGTGGGTCACA  449
      .         .         .         .         .         .         .         .         .
 450  ATCACGATGGTCTTGTGGCACCTCTTGGGATTCTTGCACTTCCGCTTGGGAATGCGGGTGTGGCACCAGTTGTTGCCCCTCCAACAAAGC  539
      .         .         .         .         .         .         .         .         .
 540  TTTTTGTGAGTGGAAGCGGCCGTTGTTGTTGTTGTGGTTGATGCTGCAGTGGTGGTTGTTGTTGTGGTGGTGGCATCAGTTGTTGTTGTG  629
      .         .         .         .         .         .         .         .         .
 630  TTACTAGCAGAGGAAGCCATCTGGATGGCCAGGGCAATAAGGGCAACCACGAAAAGGTATTTCATTTTGAAATTTGATGGAATTTATCTA  719
      .         .         .         .         .         .         .         .         .
 720  AGAAGTCCGCAGTGAAATAATCGAATTTGCTAGATGCTGTGTTCTGATTTTCTGGAGTTGCAATTAAGTCTTTTATAGTGGAATTTCTCT  809
      .         .         .         .         .         .         .         .         .
 810  TCTGTTTAGTTCCTCGTTTTGTGCTATCGAGTACATTTGCCAAATAATAATTCCACAATGATTTCCTTCCTGCAGACAAAATAAGTCTCA  899
      .         .         .         .         .         .         .         .         .
 900  TGAACTATATTAAATATTTGCTATCAATAAACGCCGATCCATTGGGTTACCGACGACACTAAGACAGCTGTATAAAGGTTTATGATATTC  989
      .         .         .         .         .         .         .         .         .
 990  ATAGCAATGTACCAAATCAAACATGATAGGAAAAATAAGCCGAGATCACAAATAAAATTGATAAAAAATAGCTTAAGTATTTATGTTCGG  1079
      .         .         .         .
1080  ATTAGATTTTTTGTTCTACTTTTATTATATTCATATTTG  1118
```

Sgs7 Sequence. Strain, *Oregon R.* Accession, X01918 (DROSGS378). Arrows labeled 1a, 1b, 2a and 2b, underline the a and b parts of the two putative promoter elements responsible for the regulation of *Sgs7* and *Sgs8*. Element 2a partially overlaps a putative CAAT box. The 5′ end of *Sgs8* (on the opposite strand) is marked by <-- at

(continued)

Gene Organization and Expression

Open reading frame, 74 amino acids; expected mRNA length, 319 bases. S1 mapping was used to define the 5′ end. The 3′ end was obtained from a cDNA sequence. There is an intron in the Ala-10 codon (*Sgs*7 Sequence) (Garfinkel et al. 1983).

Promoter

A region between −243 and −75 is necessary for transcription of *Sgs7* and *Sgs8*. Within this region, the segment between −165 and −80 enhances transcription from the *Sgs3* promoter in promoter fusion experiments. This 85-bp segment contains two copies of the bipartite regulatory element defined experimentally for *Sgs3*: 1a, 1b, 2a and 2b (*Sgs7* Sequence) (Todo et al. 1990; Hofmann et al. 1991).

Sgs8

Product

SGS8 is not glycosylated (Beckendorf and Kafatos 1976); it contains only 4% Ser/Thr (Shore and Guild 1986).

Gene Organization and Expression

Open reading frame, 75 amino acids; expected mRNA length, 353 bases. S1 mapping was used to define the 5′ end. The 3′ end was obtained from a cDNA sequence. There is an intron at Ala-10 (*Sgs8* Sequence) (Garfinkel et al. 1983).

Promoter

The putative regulatory elements, between positions −452 and −370, are the same as those described above for *Sgs7* (Todo et al. 1990).

(*continued*) −507 and the TATA box at −478 is double underlined. A base substitution (at 245) found in the strain *Formosa* is indicated (Mettling et al. 1985).

Sgs8

```
-270  ATTTTATGAGTAATACTTTCTTTAGTAAAAAGGGCATGTAAAATGTTATATTATATTAGTATGAACGATGACACAGAAAATGGTGAGAAC  -181

-180  AAAGGGCGAAGTTTAATATCGTGTTTCTATTTGCTTAAGCACAAATGGTTATATATTTTATTTTTGGATGAATTCAATGAAATTTCCATT  -91
                                                                                           ---
                                                                     --> -32
-90   GATCCTATAACACATTTGTTTTGCTATAAAAGGGGTAGGGTGTTCATAAACGGTTACCATCTGGTAAAGTTAATTTGTTAAAGCAACAAC  -1
      -                -----

0     CATGAAGCTGCTCGTTGTCGCCGTCATTGGTAAGTGCCAAAAAGTACTATTTTTTATGTGACCCAAATCCACTTAGCCATCCGTTCATTC  89
       MetLysLeuLeuValValAlaValIleA                                                              (10)

90    TGACCCAGCGTGCATCATGCTCATCGGATTCGCCGATCCTGCCTCGGGCTGCAAGGATTGTTCATGCGTGATTTGTGGACCTGGTGGCGA  179
              laCysIleMetLeuIleGlyPheAlaAspProAlaSerGlyCysLysAspCysSerCysValIleCysGlyProGlyGlyGl  (37)

180   GCCGTGTCCTGGGTGTTCCGCACGGGTTCCCGTCTGCAAAGATCTGATCAACATTATGGAGGGTCTTGAGCGGCAGGTGCGTCAGTGCGC  269
      uProCysProGlyCysSerAlaArgValProValCysLysAspLeuIleAsnIleMetGluGlyLeuGluArgGlnValArgGlnCysAl  (67)

270   CTGCGGAGAGCAGGTTTGGCTGTTCTAGAGATGTGCCCTCAACCTAATCGGCACTGACCTTTTATCTGCTGGCATTTAAAACTGCTGTCT  359
      aCysGlyGluGlnValTrpLeuPheEnd                                                               (75)

360   AATAAAACTATTATCATTCCTGCACGACCCAAACTCCTTTTCTTTGTTTTTTAATTATTTATTTTCAGATGTATTGCTTAAAAAGTGTCA  449
                         |(A)n

450   GAACTAGTCTTTTCTGTGTGTCACAATCACGATGGTCTTGTGGCACCTCTTTGGATTCTTGCACTTCCGCTTGGGAATGCGGGTGTGGCA  539

540   CCAATTGTTGCCCCTCCAACAAAGCTTTTTGTGAGTGGAAGCGGCCGTTGTTGTTGTTGTGGTTGATGCTGCAGTAGTGGTTGTTGTTGT  629

630   GGTGGTGGCATCCGTGGTTGTTGTGGTACTAGCAGATGACGCAACCTGAATGGCCAGGGCAATAAGGGCAACCACGAAAAGATACTTCAT  719

720   TTTGAAATAAGATTAGATTTTTCGATACGACTGGAATTGAACGATCAGGTGTTGTGATTAATTAAAATCATACCCACTGCTTTTATAGCA  809

810   AAACAAGCAGATTTCCGCATTCGCTTTACTATGTTTTTGCTTCCCATAACGCATAAGCACATAAAAAGCGAGTACAATAGCAAAAGCATT  899

900   TAATAATCAAATGTTTGAACAGTAAGCAAAGACGGTTTTGTTGACATATTTGTAATATCAACAATTAAATGGGTTACTATTCCTAAAAAA  989

990   ATTCCCTAAAAAGTATGCAATAATGTTTACCCACGACGATTGTATTTCAATGTCAAAACACTGCAACAGAAATAAAAAATATTTCAAAAT  1079

1080  ATTCTAGAAGCTTTTGGAAGAATATTACCCAGAAGAAAAAAACACATTAAATTTGTTCACATTT  1143
```

Sgs8 SEQUENCE. Strain, *Oregon R.* Accession, X01918 (DROSGS378).

Sgs5

Product

SGS5 is lightly glycosylated (Beckendorf and Kafatos 1976; it contains 12% Ser/Thr distributed throughout the sequence; Cys (6.75%) is also distributed without any apparent pattern (Shore and Guild 1986).

Sgs5

```
          .         .         .         .         .         .         .         .         .   A
-205 AAGCTTTTTTTTGGAGTGGAAAATTTATGGCTGTGTTTTTTTGGCCAGTCAAGGTTGTTTGCGTACGTTCTGCAAACATTTTACTTTCAG   -116
                                                            ----->                        --->
          .         .  A      .         .         .         .         .         .G      ->>>> -32
-115 ATGCACTAAGTCAATAAAGCGCTTTGCCACAACTGCTAAAACAGTGGAGTGTATTCAATATAAATAGCCAAATGAGATATTATGGGGACA   -26
                ----                                           -----
     >    .         .         .         .         .         G         .         .A        .
 -25 GTTATATTCTTAGCCACTTTTACGACATGTTCAATATTAAATTGCTGCTTTTGTTATTGGCCGTTTCGTGGTTCCACCATGGACAAGCCG   64
                               MetPheAsnIleLysLeuLeuLeuLeuLeuLeuAlaValSerTrpPheHisHisGlyGlnAlaV   (22)
                                                                               Gln
          .         .         .         .         .         .G        .         . T       .
  65 TCCAGGAGACGAAAATCGAAGAAAAACCAGTATCAGAGCCTGAAATTGAATCCGAAATAAAGAACTCTACGAGCGTCCCAAGTAAATGCA   154
     alGlnGluThrLysIleGluGluLysProValSerGluProGluIleGluSerGluIleLysAsnSerThrSerValProSerLysCysA   (52)
                                                             Val                  Ser
          .         .         .         .         .         .         .         .         .
 155 ATATTTACTATAGGAACTACCAATGGGCTCTTCAGGATTGTGTCTGCCGTTGTTTCCAAAACGAATGCCTTATGCAAATCGAGAGCGACC   244
     snIleTyrTyrArgAsnTyrGlnTrpAlaLeuGlnAspCysValCysArgCysPheGlnAsnGluCysLeuMetGlnIleGluSerAspG   (82)
          .         .         .         .         .         .         .         .         .
 245 AGCGCAAAAAGGAGGGTAGATCCCGTAAGTAAATTAACCAGTTAAGCAAAATGTATTTTATTTATAACTTGTAAATACAGCATTTGTGCC   334
     lnArgLysLysGluGlyArgSerP                                                        roPheValPr   (93)
          .         .         .         .         .         .         .         .         .
 335 CGTTACGGAGGAACTCTGCCGTTCCTTCATCTGCAAAAAGTGCAGCGTGGGTTTCCCCGTGGTTGCTGAATTCCCCATTCCGGCTCCCTG   424
     oValThrGluGluLeuCysArgSerPheIleCysLysLysCysSerValGlyPheProValValAlaGluPheProIleProAlaProCy   (123)
          .         .         .         .         .         .         .         .         .
 425 TGGATGCAATCGAAAGCCAGGATCAATTGCCACAGAGAGATTCTACAGTTTGTGCCACCTGCTGAAATTCTCAGCGGAGAACAGCAAGCG   514
     sGlyCysAsnArgLysProGlySerIleAlaThrGluArgPheTyrSerLeuCysHisLeuLeuLysPheSerAlaGluAsnSerLysP   (153)
          .         .         .         .         .         .         .         .         .
 515 TAAGTCCAAAGAATTGGTTCCAAATTATCGGTAATATATACATTTTGTATCTTTACAGCATTCCTGACTTATTCCTATTGTTGGCCCTTC   604
                                                               roPheLeuThrTyrSerTyrCysTrpProPhe   (163)
          .         .         .         .         .         .         .         .         .
 605 TAAGTGAGGTGGATTCAGTTGGATCACGTTACTAATATCTTTGTTTGTTTGTTTTATTATTTTGTTGATTTGTTCATTTAAAGGGAGATG   694
     End
          .         .         .         .         .         .         .         .         .
 695 GATTACAAATAATAAAGAAATATATTCAATGACGAGTGCAATAAATTTTTTTGAATATGAAAATCTTTTTTAGACTAAACAGCTATGCAT   784
              ------                         |(A)n
          .         .
 785 ATGTTTAAACATTGAAAAGCTT  806
```

Sgs5 SEQUENCE. Strain, *Oregon R*. Accession, X04269 (DROSGS5). The sequences with similarity to the *Sgs3* regulatory element are underlined by the arrows at −150 and −120. The natural variant $Sgs5^{n1}$, found in strain *CA-2* fails to express this gene. The base substitutions that distinguish *CA-2* from *Oregon R* are shown above the *Oregon R* sequence (Shore and Guild 1987).

Gene Organization and Expression

Open reading frame, 163 amino acids; expected mRNA length, 646–653 bases. The average polyadenylation tail is 100–150 bases long. Transcription appears to initiate with equal frequency at the first A or at any of the five Gs between −33 and −25. Nuclease protection was used to define the 5′ and 3′ ends. There are introns in the Pro-90 and Pro-153 codons (*Sgs5* Sequence) (Shore and Guild 1986).

Promoter

A DNA fragment that extends from −205 to 806 is capable of autonomous expression in a somatic transformation assay. A segment that extends from −151 to −93 contains *cis*-acting sequences necessary for expression (Shore and Guild 1987). The shorter interval includes sequences that resemble the bipartite regulatory elements of *Sgs3* (Todo et al. 1990).

References

Ashburner, M. and Berendes, H. D. (1978). Puffing of polytene chromosomes. In *The Genetics and Biology of Drosophila 2b*, eds. M. Ashburner and T. R. W. Wright (New York: Academic Press), pp. 315–395.

Beckendorf, S. K. and Kafatos, F. C. (1976) Differentiation in the salivary glands of *Drosophila melanogaster*: characterization of the glue proteins and their developmental appearance. *Cell* **9**:365–373.

Berendes, H. D. and Ashburner, M. (1978). The salivary glands. In *The Genetics and Biology of Drosophila 2b*, eds. M. Ashburner and T. R. W. Wright (New York: Academic Press), pp. 453–498.

Crowley, T. E. and Meyerowitz, E. M. (1984). Steroid regulation of RNAs transcribed from the Drosophila 68C polytene chromosome puff. *Dev. Biol.* **102**:110–121.

Crowley, T. E., Bond, M. W. and Meyerowitz, E. M. (1983). The structural genes for three *Drosophila* glue proteins reside at a single polytene puff locus. *Mol. Cell. Biol.* **3**:623–634.

Crowley, T. E., Mathers, P. H. and Meyerowitz, E. M. (1984). A *trans*-acting regulatory product necessary for expression of the *Drosophila melanogaster* 68C glue gene cluster. *Cell* **39**:149–156.

Garfinkel, M. D., Pruitt, R. E. and Meyerowitz, E. M. (1983). DNA sequence, gene regulation and modular protein evolution in the Drosophila 68C glue gene cluster. *J. Mol. Biol.* **168**:765–789.

Georgel, P., Ramain, P., Giangrande, A., Dretzen, G., Richards, G. and Bellard, M. (1991). *Sgs-3* chromatin structure and *trans*-activators: developmental and ecdysone induction of a glue enhancer-binding factor, GEBF-I, in Drosophila larvae. *Mol. Cell. Biol.* **11**:523–532.

Hansson, L. and Lambertsson, A. (1983). The role of the *su(f)* gene function and ecdysterone in transcription of glue polypeptide mRNAs in *Drosophila melanogaster*. *Mol. Gen. Genet.* **192**:395–401.

Hofmann, A., Garfinkel, M. D. and Meyerowitz, E. M. (1991). *cis*-acting sequences required for expression of the divergently transcribed *Drosophila melanogaster Sgs-7* and *Sgs-8* glue protein genes. *Mol. Cell. Biol.* **11**:2971–2979.

Martin, M., Giangrande, A., Ruiz, C. and Richards, G. (1989a). Induction and repression of the Drosophila *Sgs3* glue gene are mediated by distinct sequences in the proximal promoter. *EMBO J.* **8**:561–568.

Martin, M., Mettling, C., Giangrande, A., Ruiz, C. and Richards, G. (1989b). Regulatory elements and interactions in the Drosophila 68C glue gene cluster. *Dev. Genet.* **10**:189–197.

Martin, C. H. and Meyerowitz, E. M. (1988). Mosaic evolution in the *Drosophila* genome. *BioEssays* **9**:65–69.

Mettling, C., Bourouis, M. and Richards, G. (1985). Allelic variation at the nucleotide level in *Drosophila* glue genes. *Mol. Gen. Genet.* **201**:265–268.

Meyerowitz, E. M. and Hogness, D. S. (1982). Molecular organization of a Drosophila puff site that responds to ecdysone. *Cell* **28**:165–172.

Meyerowitz, E. M., Vijay Raghavan, K., Mathers, P. H. and Roark, M. (1987). How Drosophila larvae make glue: control of *Sgs3* gene expression. *Trends Genet.* **3**:288–293.

Roark, M., Vijay Radhaven, K., Todo, T., Mayeda, C. and Meyerowitz, E. M. (1990). Cooperative enhancement at the Drosophila *Sgs3* locus. *Dev. Biol.* **139**:121–133.

Shore, M. E. and Guild, G. M. (1986). Larval salivary gland secretion proteins in Drosophila. Structural analysis of the *sgs5* gene. *J. Mol. Biol.* **190**:149–158.

Shore, M. E. and Guild, G. M. (1987). Closely linked DNA elements control the expression of the *Sgs5* glue protein gene in Drosophila. *Genes Dev.* **1**:829–839.

Todo, T., Roark, M., Vijay Radhavan, K., Mayeda, C. and Meyerowitz, E. (1990). Fine-structure mutational analysis of a stage- and tissue-specific promoter element of the Drosophila glue gene *Sgs3. Mol. Cell. Biol.* **10**:5991–6002.

Velissariou, V. and Ashburner, M. (1981). Cytogenetics and genetic mapping of a salivary gland secretion protein in *Drosophila melanogaster. Chromosoma* **84**:173–185.

28

The *Serendipity* Gene Cluster: *Sryα, Sryβ, Sryδ*

Chromosomal Location:
3R, 99D4-8

Map Position:
3-[101]

Organization of the Cluster

Sryα, *Sryβ* and *Sryδ* are grouped in a dense cluster within an 8 kb segment that also includes *janA*, *janB* and the ribosomal protein gene *rp49* (Fig. 28.1). The three *Sry* genes are transcribed in the same direction; the distance between the poly-A signal in one gene and the TATA box of the next is a few hundred bp. In addition to the gene-specific transcripts, two other longer poly-A RNAs are detectable. These include sequences from neighboring genes: either β plus α or α plus δ are combined. These longer RNAs are thought to be the consequences of transcription starting normally in one gene but then proceeding to "read-through" to the end of the next gene downstream (Vincent et al. 1984, 1985).

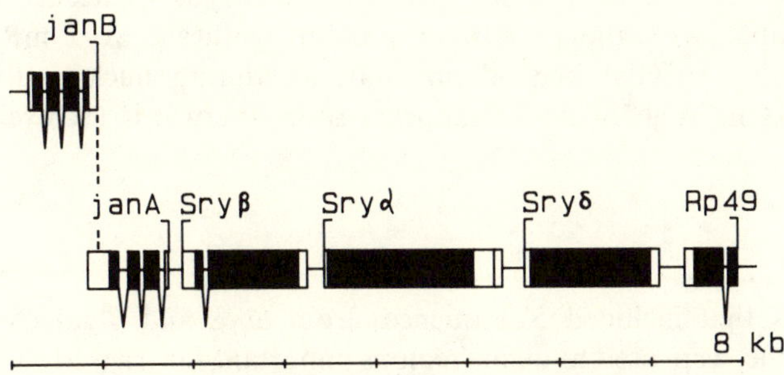

FIGURE 28.1. Organization of the *Sry* cluster. For the sake of clarity, *janB* was drawn on a separate line; it actually overlaps *janA*

Sryα

Product

A 58 kD protein without resemblance to other known proteins (Vincent et al. 1985).

Tissue Distribution

It is present very briefly in embryos undergoing blastoderm cellularization. The *Sryα* protein (SRYα) accumulates sharply during nuclear cycle 14 and disappears as gastrulation proceeds. As the embryonic syncytium becomes partitioned into individual cells, SRYα is concentrated at the leading edges of the invaginations of the plasma membrane; this intracellular distribution is very similar to that of actin filaments (Schweisguth et al. 1990).

Mutant Phenotypes

In embryos homozygous for a deletion of *Sryα*, the process of cellularization is severely disrupted, surface invaginations of the plasma membrane are very irregularly distributed, and they often encompass multiple nuclei. Such *Sryα* mutation is an embryonic lethal (Schweisguth et al. 1990).

Gene Organization and Expression

Open reading frame, 530 amino acids; expected mRNA length, 1,862 or 1,952 bases depending on whether the major or minor polyadenylation site is used. The 5′ end was determined by S1 mapping and primer extension; the 3′ end was defined by S1 mapping. There are no introns (*Sry* Sequences) (Vincent et al. 1985).

Developmental Pattern

Expression of *Sryα* is restricted to 2–4 h embryos (Vincent et al. 1985). In accordance with the pattern of protein synthesis, *Sryα* mRNA is first detectable in syncytial blastoderm embryos, during nuclear division cycle 11; it peaks in cycle 14 and disappears soon afterwards (Schweisguth et al. 1989).

Promoter

P elements that included 5′ sequences from *Sryα* and *β*-galactosidase as a reporter gene, were used to define regions important for transcription. A 248-bp segment that extends from 118 bp upstream of the transcription initiation site to 130 bp downstream of the transcription initiation site is sufficient for specific

Sryβ

```
-580  GGATCCGACTTACCATGCCATGTGCAGTCCGCAAATCCGCGGATCACCGCCTTTGAAGCATCTCCTCCATCGAAGACATTGATCATGACA  -491

-490  TACTTGAAGATGCCCTCTGGACTGATGTGCACCAGTGGCACGCCGGCAAGTGCTTCCTCGGACATTTTGTGAATCAGTCGTAGTCCTTTG  -401

-400  GAAAGCAGTTGGAGGCGATTCATTGTGGAAAAGTGTTTGAAAACAGAATTTAATGTCTTACCAACCGGCAAATTTTCCAAAAACGCTGCT  -311

                <-- janA.
-310  TTAGTCCAGCAATGTGACCAGATACTTTTATTCTCGTTCACCTTATGTATCGATACTATTAAAATTATAATATTCAGTTATTTCTTTAGT  -221
                                                -------

                                                                                      -->-144
-220  ACATTTGGGAAAACGTTTTTTGTACTACTCAGTATATTTTAGTAATAATTAATAAACTCAAAAAGAAAACGTTGTTATCAGTGACAGCAA  -131
        -----                                        -------

-130  ATTTATCGATTACGCCGACAGCACGCAGTCGATACTATCGGCACTATCGGCACAGCTCTGGCGTTCACAACAAAATATACACAAATTAAA  -41

 -40  TTGTTTAAGCCGAATTTTCGATTGGATTCCACGGCGACTAGATGAGCTCCACGCGTCCGTTTTGCTTCGTTTGCGGCAAGGAGAAGTCCG  49
                                              MetSerSerThrArgProPheCysPheValCysGlyLysGluLysSerV  (17)

  50  TGGGGGTGTTCCAGCTGATAGAAGGTAACGTTCGCTTACGCCGCACTCGAAAGTCCTGATAGCCGACTTTTTCACAGGCTGCATTGTGCC  139
      alGlyValPheGlnLeuIleGluG                                                       lyCysIleValPr  (29)

 140  AGGAACCTTTAAGCCCATCAAGGATATACTGAAATACTTCGAGAAGATCATAAACCAGCGGCTGGAGCTCCTGCCCAACTCGGCCGCCTG  229
      oGlyThrPheLysProIleLysAspIleLeuLysTyrPheGluLysIleIleAsnGlnArgLeuGluLeuLeuProAsnSerAlaAlaCy  (59)

 230  CCGGGACTGCCTGGAGTACCTCTTCAACTACGACAGGCTGGTGAGGAATCTCAGCCAAGTGCAGCGCCAGATTGCGGACGCACTGCTCGG  319
      sArgAspCysLeuGluTyrLeuPheAsnTyrAspArgLeuValArgAsnLeuSerGlnValGlnArgGlnIleAlaAspAlaLeuLeuGl  (89)

 320  CTGCAGGCAGGTGGAGGGCAAGGCGGAGACCAAGCAACAGGCGGCAAAGAGGGCCCGCGTCCAGGTGCCGGCCTTCAAGATCGTCCAGGC  409
      yCysArgGlnValGluGlyLysAlaGluThrLysGlnGlnAlaAlaLysArgAlaArgValGlnValProAlaPheLysIleValGlnAl  (119)

 410  CACCGCCCTCAAGGAGCCCGAAAGGCAGCCGGGCGAGGAGGATGAGTGCGAGGAATTCATGAAGGAGGAGATGCTGGACGAGGAGTTCCA  499
      aThrAlaLeuLysGluProGluArgGlnProGlyGluGluAspGluCysGluGluPheMetLysGluGluMetLeuAspGluGluPheGl  (149)

 500  GTTCAGCGAGCCGGACGACAGCATGCCGTCGTCGGAGGAGGAGTTCTTCACCGAGACCACCGAGATACCCTGCCATATCTGCGGCGAGAT  589
      nPheSerGluProAspAspSerMetProSerSerGluGluGluPhePheThrGluThrThrGluIleProCysHisIleCysGlyGluMe  (179)
                                                                                   ---     ---

 590  GTTTTCCAGCCAGGAGGTGCTCGAGCGGCACATCAAGGCGGACACCTGCCAGAAGAGCGAGCAGGCCACCTGCAACGTGTGTGGCTTGAA  679
      tPheSerSerGlnGluValLeuGluArgHisIleLysAlaAspThrCysGlnLysSerGluGlnAlaThrCysAsnValCysGlyLeuLy  (209)
                                 ---                                            ---     ---

 680  AGTGAAGGACGACGAGGTACTCGATCTGCATATGAACTTGCACGAGGGCAAAACAGAACTTGAATGCCGCTACTGCGACAAAAAGTTCTC  769
      sValLysAspAspGluValLeuAspLeuHisMetAsnLeuHisGluGlyLysThrGluLeuGluCysArgTyrCysAspLysLysPheSe  (239)
                                  ---      ---                          ---      ---

 770  GCACAAGCGGAACGTCCTGCGCCACATGGAGGTGCACTGGGACAAGAAGAAGTACCAGTGCGACAAGTGCGGCGAACGCTTCTCGCTCTC  859
      rHisLysArgAsnValLeuArgHisMetGluValHisTrpAspLysLysLysTyrGlnCysAspLysCysGlyGluArgPheSerLeuSe  (269)
                               ---        ---                       ---      ---

 860  CTGGCTGATGTACAACCATCTGATGCGCCACGACGCCGAGGAGAACGCCCTGATCTGCGAGGTGTGCCACCAGCAGTTCAAGACCAAGCG  949
      rTrpLeuMetTyrAsnHisLeuMetArgHisAspAlaGluGluAsnAlaLeuIleCysGluValCysHisGlnGlnPheLysThrLysAr  (299)
                        ---      ---                        ---      ---
```

(continued)

```
 950  CACCTACAAGCACCACTTGCGCACCCACCAGACGGACCGGCCGCGCTACCCCTGCCCCGACTGCGAGAAATCGTTCGTGGACAAGTACAC  1039
      gThrTyrLysHisHisLeuArgThrHisGlnThrAspArgProArgTyrProCysProAspCysGluLysSerPheValAspLysTyrTh  (329)
                ---         ---                                         ---      ---

1040  CCTGAAGGTGCACAAGCGGGTCCACCAGCCGGTCGAGAAGCCAGAGTCGGCGGAGGCCAAGGAAGCCACCGTCACGTTCTTTTAGGGTAG  1129
      rLeuLysValHisLysArgValHisGlnProValGluLysProGluSerAlaGluAlaLysGluAlaThrValThrPhePheEnd       (356)
             ---         ---

1130  TCCTTTGCTAGATTAATCTAAGAAGCCCAGCTCATGGGTGCATTAGCGCGCGTATGTATCATAAAATTAAATCTAAACAATTATTGACGG  1219
                                                                            =====

1220  AAAGTACCAGTTCTTTGCCGTTCTTCGCCCATTTTCCAGGAACCCCAGTAGGTAAAGTAGCGGATTTCGCGAATTTTCGCGGGTATGGCA  1309
         |(A)n
```

Sryα

```
1310  ATAAAACAGGCAGATGTTTTTTAATCCCCAAAATAGGTCCTTTCTACCTGTGCGCTTGGCAAAGTATATAAAGGTGTTGCGTCGTCCGCC  1399
                                ======                                      ======

            -->1407                                  1450
1400  AGAACTTAGTTGAACATTTCTGTTTCCCGGAGCACATCTGATAGAACAGCATGGAACAGCTATTGGCCCAATTACACACTTGCAGTGAGC  1489
                                                      MetGluGlnLeuLeuAlaGlnLeuHisThrCysSerGluL    (14)

1490  TGATTGCAGAGGGCTACAGCAGCACCGGCAACATTGGCTGGCTGAACGAGTTCTGCGCCACTTTCCTGGACTTCGCCAGCGATCTGAAGG  1579
      euIleAlaGluGlyTyrSerSerThrGlyAsnIleGlyTrpLeuAsnGluPheCysAlaThrPheLeuAspPheAlaSerAspLeuLysA    (44)

1580  CTAGGCTGCCGGAGGTGGCGCCCAGTGGCGCAAACCTTGATGTGGAGACCATCTTCCTGTGCCTCACCCAGGTGGTAACCTGCATCACCC  1669
      laArgLeuProGluValAlaProSerGlyAlaAsnLeuAspValGluThrIlePheLeuCysLeuThrGlnValValThrCysIleThrH    (74)

1670  ACCTAGAGCGGACCATCAGCATGGAGGCACCGCATATGACCAGGCAGCACTTCCTCGACCGCTTGGACTGGTGCTTGCGGCGACTGCTCG  1759
      isLeuGluArgThrIleSerMetGluAlaProHisMetThrArgGlnHisPheLeuAspArgLeuAspTrpCysLeuArgArgLeuLeuV   (104)

1760  TCTCCTTGACGCAACTGGAAGGCAACGTGACCCCAGTCAAGAACCTAGAGGATCACTCCTTCGTTGAGCTCATGGACCTGGCTCTGGACC  1849
      alSerLeuThrGlnLeuGluGlyAsnValThrProValLysAsnLeuGluAspHisSerPheValGluLeuMetAspLeuAlaLeuAspH   (134)

1850  ACTTGGATGACTACATGGAGAAGCTGGCCCAGCAGAGAAACAACTCCCTGCACATTCTAGAAGAGAGCTTCACGGAAGACACCTACCAGC  1939
      isLeuAspAspTyrMetGluLysLeuAlaGlnGlnArgAsnAsnSerLeuHisIleLeuGluGluSerPheThrGluAspThrTyrGlnL   (164)

1940  TGGCCAGCATAGTTAATCACATCGTTCGCCACGCCCTGGCCTTTGCCAATGTGGCCATTCATTCGGACAAGAAGGCTTTGACGGCTTTGT  2029
      euAlaSerIleValAsnHisIleValArgHisAlaLeuAlaPheAlaAsnValAlaIleHisSerAspLysLysAlaLeuThrAlaLeuC   (194)

2030  GCGAGACCTTGCTCGCCGAATGTGCCACTTTCCACGAGGAGGCGGGCGAGCCCAACAGTGGTCATCGGAAGCTGGAGGCCCTCTCCCTGG  2119
      ysGluThrLeuLeuAlaGluCysAlaThrPheHisGluGluAlaGlyGluProAsnSerGlyHisArgLysLeuGluAlaLeuSerLeuG   (224)

2120  AACGTGCCCTCTATGCCCTGGAATCCTTTCTCAATGAGGCGCTGCTGCACTTGCTGTTCGTCAGTCTGATAGATCTGGAAAACGCTTCGG  2209
      luArgAlaLeuTyrAlaLeuGluSerPheLeuAsnGluAlaLeuLeuHisLeuLeuPheValSerLeuIleAspLeuGluAsnAlaSerV   (254)

2210  TGGAGAAGCTAAAGGATGCACTGCAAAGGGATCCTGCGGGAGCTCAGGAGCTAATCTCCGCATTCGACACGAACATGGATCGCATTCAGC  2299
      alGluLysLeuLysAspAlaLeuGlnArgAspProAlaGlyAlaGlnGluLeuIleSerAlaPheAspThrAsnMetAspArgIleGlnG   (284)

2300  AGATTGGGGTTCTGGCCATAGCCTTCTCGCAGGACATCAAAACGAAGACGATTGTCAGGAGCTGCCTGGCCTCACTGGAATCCCTGGATG  2389
      lnIleGlyValLeuAlaIleAlaPheSerGlnAspIleLysThrLysThrIleValArgSerCysLeuAlaSerLeuGluSerLeuAspA   (314)

2390  CGTGCATTGTGCCCGCTCTCCAGCTGCCAGAGTCCACTTCATCCGCACACCACGCGGAGGTCTTGCAGGAGCATTTTAACCAGGAGCTGC  2479
      laCysIleValProAlaLeuGlnLeuProGluSerThrSerSerAlaHisHisAlaGluValLeuGlnGluHisPheAsnGlnGluLeuL   (344)
```

2480 TGATCTTTAGGAACGTCATCCACGAAATCATCGATAGCTGCTCCCTGATCAACAACTACCTGGACATGCTGGGCGAGAGGATCCACGTAC 2569
euIlePheArgAsnValIleHisGluIleIleAspSerCysSerLeuIleAsnAsnTyrLeuAspMetLeuGlyGluArgIleHisValG (374)

2570 AGGACAAAAGCCATCTGAAGCTGATTGTCCAGAGGGGCGGAGTGGTGGTGGATCACTTTCGGCTGCCCGTCAATTACTCGGGACTCAGTG 2659
lnAspLysSerHisLeuLysLeuIleValGlnArgGlyGlyValValValAspHisPheArgLeuProValAsnTyrSerGlyLeuSerG (404)

2660 AAGATGGCAAGCGGGTGCACAAGGACCTCATTCTGATCCTGCGCGAGTGCCAGGCCGTGGTCAACCTGGACGTCCCAGTGGATCCCAAGC 2749
luAspGlyLysArgValHisLysAspLeuIleLeuIleLeuArgGluCysGlnAlaValValAsnLeuAspValProValAspProLysA (434)

2750 GCATCGTGAAGCGCCTTAAGATACTGTACTCCGTGCTGGCCAAGCTGAGGGACTTGATATGCAGGGATAATCTGGAGCCCGATTCCTCAG 2839
rgIleValLysArgLeuLysIleLeuTyrSerValLeuAlaLysLeuArgAspLeuIleCysArgAspAsnLeuGluProAspSerSerV (464)

2840 TTGCTTCCGAAGCTCAAGTGCCCTCAAGTGCAACCCGAACTTTTGTGCGGAGCAGTCGATCCTTTGGCAAACGGCATCGATCCTTTGTAA 2929
alAlaSerGluAlaGlnValProSerSerAlaThrArgThrPheValArgSerSerArgSerPheGlyLysArgHisArgSerPheValL (494)

2930 AACAAACCGGAAATTGCTCAGTTTTCGGGCCACAGGACTCACTTGCTGAATCCGGACACAGCGAAAGCGATCTTATTAGTTTCCAAATCA 3019
ysGlnThrGlyAsnCysSerValPheGlyProGlnAspSerLeuAlaGluSerGlyHisSerGluSerAspLeuIleSerPheGlnIleT (524)

3020 CTGAGATTCTTAGGTTAGATTGAGTGGGGCGAGCCATATTCTAAATACGCGGGCTTATCTGTATGAGATTTTTTTAATACTTCATTGGC 3109
hrGluIleLeuArgLeuAspEnd (530)

3110 TTGAAGTGCTTAATTAAGATTAAACATATCCTTACAAAGATTCTAATTAGCTACCTAAGTCAATTGTGTTCTTTACACTTATGTAATTAC 3199

3200 TTCATTAAGTTGAAGCCATTCGCATAATTTATATAAAATACAATTAAAACATACCATTATAAAAAAATTATATAATTCAATTCAATTTTT 3289
------- |$(A)_n$ (major)

3290 TTCAGGGAATCAATAAATTAAATGCTACTCGTTTTCATAACTAAAGAAACCAACACCACCATAATAATCAACAACAAATTATGTATTTAT 3379
|$(A)_n$ (minor)

Sryδ

3380 GCAAATTGAATATCCGTTTGCAATATTGAGCAAACACATATTTTTATTATTCAACTCATATATTTCAGTTTTCACACCCTGTTCCATTCC 3469

3470 CACCGTTCCGTTCCTGGCATCAATCGGCATCGTTCGCCACGCCTGGTGGGCATTATGCCATGGTTGCATTTCACGCATTTTAGTATAGCT 3559

.-->3601

3560 TCCGATATTCATCATTTTGCCAACTCTATTAAATTTCATACACAATTTAAAAGATTGTAAACAAACGCCGCACGAAACGGAATCGCTAAA 3649

3668 A=12

3650 AGGACCATCGTCGGCGCAATGGATACTTGCTTCTTCTGCGGCGCCGTCGATCTGAGCGACACGGGCTCCTCCAGCTCCATGCGCTACGAG 3739
MetAspThrCysPhePheCysGlyAlaValAspLeuSerAspThrGlySerSerSerSerMetArgTyrGlu (24)
Tyr

3740 ACGCTGTCGGCCAAGGTGCCGTCGTCGCAGAAAACAGTGTCCCTGGTGCTCACCCACCTGGCCAACTGCATCCAGACGCAGCTGGACCTG 3829
ThrLeuSerAlaLysValProSerSerGlnLysThrValSerLeuValLeuThrHisLeuAlaAsnCysIleGlnThrGlnLeuAspLeu (54)

3830 AAGCCCGGCGCCCGGCTGTGTCCGCGCTGCTTTCAGGAGCTCTCCGACTACGACACGATCATGGTGAACCTGATGACCACCCAGAAGAGG 3919
LysProGlyAlaArgLeuCysProArgCysPheGlnGluLeuSerAspTyrAspThrIleMetValAsnLeuMetThrThrGlnLysArg (84)

3920 CTGACGACCCAGCTAAAGGGCGCTCTAAAGTCCGAGTTCGAGGTGCCGGAGTCCGGCGAGGACATACTCGTGGAGGAGGTGGAGATACCC 4009
LeuThrThrGlnLeuLysGlyAlaLeuLysSerGluPheGluValProGluSerGlyGluAspIleLeuValGluGluValGluIlePro (114)

(continued)

```
     .         .         .         .         .         .         .         .         .      G=SF1
4010 CAAAGCGATGTCGAGACAGACGCCGATGCCGAGGCGGACGCCCTGTTCGTGGAGCTGGTCAAGGATCAGGAGGAGTCCGACACGGAGATA   4099
     GlnSerAspValGluThrAspAlaAspAlaGluAlaAspAlaLeuPheValGluLeuValLysAspGlnGluGluSerAspThrGluIle   (144)
                                                                                            Val

     .         .         .         .         .         .         .         .         .
4100 AAGAGAGAGTTCGTGGACGAGGAGGAGGAGGAGGACGACGACGACGACGACGACGAGTTCATCTGCGAGGACGTGGATGTGGGCGACTCC   4189
     LysArgGluPheValAspGluGluGluGluGluAspAspAspAspAspAspAspGluPheIleCysGluAspValAspValGlyAspSer   (174)
     .         .         .         .         .         .         .         .         .
4190 GAGGCCCTGTATGGCAAGTCCTCCGATGGCGAGGACAGGCCGACGAAGAAGCGCGTCAAGCAGGAGTGCACTACCTGCGGCAAGGTGTAC   4279
     GluAlaLeuTyrGlyLysSerSerAspGlyGluAspArgProThrLysLysArgValLysGlnGluCysThrThrCysGlyLysValTyr   (204)
                                                                       ---      ---

     .         .         .         .         .         .         .         .         .
4280 AACTCCTGGTATCAACTGCAGAAGCACATCAGCGAGGAGCACTCCAAGCAGCCCAACCACATCTGCCCCATCTGCGGGGTGATCCGGCGC   4369
     AsnSerTrpTyrGlnLeuGlnLysHisIleSerGluGluHisSerLysGlnProAsnHisIleCysProIleCysGlyValIleArgArg   (234)
                             ---            ---                     ---      ---

     .         .         .         .         .       A=SF1       .         .         .   T=SF2
4370 GACGAGGAGTACTTGGAGCTGCACATGAATCTGCACGAGGGCAAGACGGAAAAGCAATGCCGCTACTGCCCCAAGAGCTTCTCGCGCCCG   4459
     AspGluGluTyrLeuGluLeuHisMetAsnLeuHisGluGlyLysThrGluLysGlnCysArgTyrCysProLysSerPheSerArgPro   (264)
                          ---         ---            Lys      ---      ---               Cys
     .         .         A=12      .         .         .         .         .         .
4460 GTGAACACCCTGCGCCACATGCGCATGCACTGGGACAAGAAGAAGTACCAGTGCGAGAAGTGCGGCCTGAGGTTCTCCCAGGACAACCTA   4549
     ValAsnThrLeuArgHisMetArgMetHisTrpAspLysLysLysTyrGlnCysGluLysCysGlyLeuArgPheSerGlnAspAsnLeu   (294)
                    ---Ile      ---                     ---      ---

     .         .         .         .         .         .         .         .         .
4550 CTCTACAACCACCGGCTGCGCCACGAGGCTGAGGAGAACCCCATCATATGCAGCATCTGCAATGTGTCGTTCAAGTCGCGCAAGACCTTC   4639
     LeuTyrAsnHisArgLeuArgHisGluAlaGluGluAsnProIleIleCysSerIleCysAsnValSerPheLysSerArgLysThrPhe   (324)
              ---         ---                        ---      ---

     .         .         .         .         .         .         .         .         .
4640 AACCATCACACGCTCATTCACAAGGAGAACCGCCCAAGACACTACTGCTCCGTCTGCCCCAAGTCCTTCACCGAGCGCTACACCCTCAAG   4729
     AsnHisHisThrLeuIleHisLysGluAsnArgProArgHisTyrCysSerValCysProLysSerPheThrGluArgTyrThrLeuLys   (354)
           ---         ---                        ---      ---

     .         .         .         .         .         .         .         .         .
4730 ATGCACATGAAGACCCACGAGGGCGACGTCGTTTACGGGGTTCGCGAGGAGGCGCCCGCCGACGAGCAGCAGGTGGTGGAGGAGCTGCAT   4819
     MetHisMetLysThrHisGluGlyAspValValTyrGlyValArgGluGluAlaProAlaAspGluGlnGlnValValGluGluLeuHis   (384)
        ---         ---

     .         .         .         .         .         .         .         .         .
4820 GTGGACGTCGACGAATCGGAGGCGGCCGTCACCGTCATCATGTCCGACAACGATGAGAACAGCGGCTTCTGTCTCATTTGCAATACCACC   4909
     ValAspValAspGluSerGluAlaAlaValThrValIleMetSerAspAsnAspGluAsnSerGlyPheCysLeuIleCysAsnThrThr   (414)
                                                                          ---      ---

     .         .         .         .         .         .         .         .         .
4910 TTCGAGAACAAGAAGGAGCTCGAACACCACTTGCAATTTGATCACGACGTGGTCTTGAAATAAGCTACATTGCCTACAATAAGTAATTGT   4999
     PheGluAsnLysLysGluLeuGluHisHisLeuGlnPheAspHisAspValValLeuLysEnd                              (434)
                                ---            ---

     .         .         .         .         .         .         .         .         .
5000 TTATCTTTCCCTAGTGTATTTCCTCCTCTTTGTACTTGATTATTGTAGATTCCTACAAAATATAATTTACTGGTATTTCAATTACTGCGT   5089
     .         .         .         .                           ------       |(A)n
5090 TTCATTTAGACAGAAGCATTTCCGATAATAATTGTAC  5126
```

Sry SEQUENCES. Accession X03121 (DRYOSRYG1). The Cys and His residues of the Zn-fingers are underlined. Four mutations in *Sryδ* are also indicated.

activation in blastoderm stage embryos, although at much reduced level. A segment extending between 311 and 118 bp upstream of the transcription initiation site is necessary to increase the level of transcript. The latter segment also seems to be responsible for repression of *Sryα* activity in the peripheral nervous system (Schweisguth et al. 1989).

Sryβ and *Sryδ*

Products

DNA-binding proteins of the Zn-finger type.

Structure

The amino-acid sequences of *Sryβ* and *Sryδ* proteins have some similarities to the *Xenopus* transcription factor TFIIIA and other Zn-finger proteins. There is a repeating unit of 28 or 29 amino acids characterized by Cys at positions 1 and 4 and His at positions 17 and 21/22 of the repeat (a C_2H_2 finger). A Phe at position 8 is also frequent. *Sryβ* has six such repeats and *Sryδ* seven, both are in the C-terminal half of the molecule (*Sry* Sequences). Although residues in other positions are not conserved from one repeat to the next, the C-terminal regions of SRY*β* and SRY*δ* and 50% identical; this suggests that the two genes were generated by a duplication. No sequence similarities are evident outside the coding regions (Vincent et al. 1985; Vincent 1986; Evans and Hollenberg 1988; Payre et al. 1990; Harrison 1991). An 18-amino-acid segment (residues 180–197) was identified as the nuclear localization signal of SRY*δ*; within that segment, the heptapeptide Pro-188/Lys-194 has strong similarity to the nuclear localization signals of SV40 large T antigen and c-*myc* (Noselli and Vincent 1991).

Function

The two proteins bind DNA, both in solution and in polytene chromosomes. SRY*β* binding sites include the consensus sequence YCAGAGATGCGCA and SRY*δ* binding sites the sequence YTAGAGATGGRAA (Payre et al. 1990; Payre and Vincent 1991).

Tissue Distribution

SRY*α* and SRY*δ* are maternally inherited and present in embryonic nuclei at the onset of zygotic transcription as well as in numerous cell types throughout development. Zygotic synthesis starts during the syncytial blastoderm stage (nuclear division cycles 12–13) for SRY*β* and during germ band extension (stage 10 embryos) for SRY*δ* (Payre et al. 1989, 1990).

Mutant Phenotypes

Four amino acid substitutions in *Sryδ* are lethal (*Sry* Sequences). These mutants can be rescued by germ line transformation with *Sryδ* but not by an extra copy of *Sryβ* sequences, an indication that the two genes have different functions (Crozatier et al. 1992).

Sryβ

Gene Organization and Expression

Open reading frame, 356 amino acids, expected mRNA length, 1,314 bases. The 5′ end was determined by S1 mapping and primer extension; the 3′ end was defined by S1 mapping. There is an intron within the Gly-25 codon (*Sry* Sequences) (Vincent et al. 1985; Payre et al. 1990).

Developmental Pattern

See *Sryδ*.

Sryδ

Gene Organization and Expression

Open reading frame, 434 amino acids; expected mRNA length, 1,476 bases. The 5′ end was determined by S1 mapping and primer extension; the 3′ end was defined by S1 mapping. There are no introns (*Sry* Sequences) (Vincent et al. 1985).

Developmental Pattern

Expression of *Sryδ* (and of *Sryβ*) is very high during oogenesis and early embryonic development; it remains significant, but lower, throughout the life cycle (Vincent et al. 1985; Payre et al. 1990).

Sryδ transcripts are abundant in nurse cells up to stage 10, at which time they begin to be transferred to the oocyte. Approximately 4 h after oviposition, transcripts from embryonic nuclei are added to the maternal complement. The total level of transcripts gradually decreases after germ band extension (Payre et al. 1989).

References

Crozatier, M., Kongsuwan, K., Ferrer, P., Merriam, J. R., Lengyel, J. A. and Vincent, A. (1992). Single amino acid exchanges in separate domains of the Drosophila serendipity s zinc finger protein cause embryonic and sex-biased lethality. *Genetics* **131**:905–916.

Evans, R. M. and Hollenberg, S. M. (1988). Zinc fingers: gilt by association. *Cell* **52**:1–3.

Harrison, S. C. (1991). A structural taxonomy of DNA-binding domains. *Nature* **353**:715–719.

Noselli, S. and Vincent, A. (1991). A *Drosophila* nuclear localization signal included in an 18 amino acid fragment from the *serendipity δ* zinc finger protein. *FEBS* **280**:167–170.

Payre, F., Noselli, S., Lefrère, V. and Vincent, A. (1990). The closely related *Drosophila Sryβ* and *Sryδ* zinc finger proteins show differential embryonic expression and distinct patterns of binding sites on polytene chromosomes. *Development* **110**:141–149.

Payre, F. and Vincent, A. (1991). Genomic targets of the *serendipity β* and *δ* zinc finger proteins and their respective DNA recognition sites. *EMBO J.* **10**:2533–2541.

Payre, F., Yanicostas, C. and Vincent, A. (1989). *Serendipity delta*: a *Drosophila* zinc-finger protein present in embryonic nuclei at the onset of zygotic gene transcription. *Dev. Biol.* **136**:469–480.

Schweisguth, F., Lepesant, J. A. and Vincent, A. (1990). The *Serendipity alpha* gene encodes a membrane-associated protein required for the cellularization of the *Drosophila* embryo. *Genes Dev.* **4**:922–931.

Schweisguth, F., Yanicostas, C., Payre, F., Lepesant, J. A. and Vincent, A. (1989). *cis*-regulatory elements of the *Drosophila* blastoderm specific *Serendipity alpha* gene: ectopic activation in the embryonic PNS promoted by the deletion of an upstream region. *Dev. Biol.* **136**:181–193.

Vincent, A., O'Connell, P., Gray, M. R. and Rosbash, M. (1984). *Drosophila* maternal and embryo mRNAs transcribed from a single transcription unit use alternate combination of exons. *EMBO J.* **3**:1003–1013.

Vincent, A., Colot, H. V. and Rosbash, M. (1985). Sequence and structure of the serendipity locus of *Drosophila melanogaster*. A densely transcribed region including a blastoderm-specific gene. *J. Mol. Biol.* **186**:149–166.

Vincent, A. (1986). TFIIIA and homologous genes. The "finger" proteins. *Nucl. Acids Res.* **14**:4385–4391.

29

Ultrabithorax: Ubx

Chromosomal Location:
3R, 89E1–2

Map Position:
3-58.8

Products

DNA-binding regulatory proteins of the homeodomain type involved in the determination of segmental identity in the mid-section of the embryo.

Structure

A family of at least five related polypeptides of approximately 40 kD, translated from alternatively spliced mRNAs (see Fig. 29.1B and discussion under Gene Organization and Expression). They all share the sequences encoded in exons at the 5′ and 3′ ends of the transcription unit; but they differ from each other with respect to whether they include one or more of three short internal segments, one nine amino acids long and the other two 27 amino acids each.

The homeodomain is near the C-terminus. Other sequence features include an alanine-rich segment near the homeodomain (see *eve*) and a glycine-rich segment between residues 111 and 129 (*Ubx* Sequence) (Weinzierl et al. 1987; O'Connor et al. 1988; Kornfeld et al. 1989). For a comparison of *Ubx* protein and DNA sequences in *D. melanogaster* and other species, see Wilde and Akam (1987).

All isoforms of UBX are multiply phosphorylated at Ser and Thr residues that occur between amino acids 39 and 183. Most of the phosphorylation is between residues 130 and 183 (Gavis and Hogness 1991).

Function

UBX helps to define segment identity (Lewis 1978) by acting as a transcriptional regulator. There is evidence that UBX acts on homeotic genes. In particular, it stimulates its own transcription while repressing transcription of *Antennapedia*

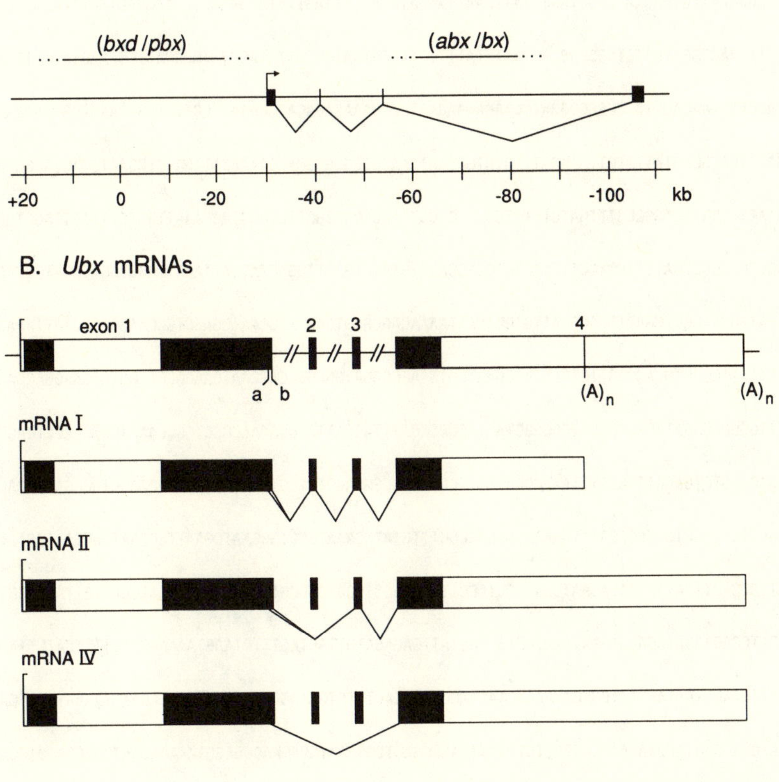

FIG. 29.1. Organization of the *Ubx* transcriptional unit: (A) based on Simon et al. (1990); and (B) based on O'Connor et al. (1988) and Kornfeld et al. (1989).

(*Antp*); this may serve to modulate the segmental distribution of the two products (Beachy et al. 1988; Biggin and Tjian 1989; Samson et al. 1989 and references therein). The identity of the UBX-controlled effector genes that are directly responsible for carrying out segmental differentiation is a subject of active current research.

UBX Ib, a member of the family that includes all three internal polypeptide segments, was produced in *E. coli* and cultured insect cells and tested for DNA-binding activity using DNA fragments from the neighborhoods of the *Antp* and *Ubx* genes. Five binding sites were found near *Antp*: A-1, A-2 and A-3 (approximately 6 kb upstream of the transcription initiation site P1), A-A and A-B (300–400 bp downstream of P1). Two binding sites, u-A and u-B, were detected near *Ubx*. These are 60 bp and 250 bp downstream of the transcription initiation site (*Ubx* Sequence). Multiple repeats of the trinucleotide TAA is a characteristic of all these binding sites (Beachy et al. 1988).

Ubx

|Hm

-4111 ATGACAGAAAAAAGTAAGAAACAGTTAAGTTATTCAATTAAAATGGATTATTAGTTTTAGGAAACTCCAAGCACTTGTTAAAATCGAATT -4022

-4021 TGTTCAATAACTGCATGATGTAGCAAGAACTAATGTATTTTTAAATATTATTGCCTTATAGCTATGGCCATTTTTAAGTATTTTTCCCCA -3932

-3931 GTGCACCATCTAACAGGTGCCGAGCCGCATCGAACAGAAGAATGCCGAAAGACACAGCCGAAATCCTTATAGACAATACGTAAACAAGTC -3842

-3841 GGAGAGTTCAGGCAGTATTTTGTTGAACATTTCTGTGTAAATAAACCTCGGGGCCACGAAAACTTTGATCTCGAATGGAGGAGGGCAACT -3752

-3751 AGTATACTCTGGTACTGGCCGTTTGATGTTTCTGGACTGGCGTCAGGCCGGCGCTTCCAGCTGCCAAATTGCTGCTTTATTAGCTGCGTA -3662

-3661 AGTGGCTCCCCCCTGATTTTCCTGCTTTCCACCTGGAGCAAATGTATCTGTTTTGGACTATGATTAGATTGGGTGCACCCATCGCACGCA -3572

-3571 TACGGATGGCATCGCTCGATTTGAGCGATTGTGGCCAATAAAACAGCGGGTGAGAAGGCAAACGAGCTGCCAAGGTGGCAATTAAACGGC -3482

-3481 TTGTCTAATTGCCCTGCACCAGTTCTCAACAGCGAATGGTGAACGGAGATGGAGGCCATCAATCAAGGCGTTGGTGTGAAGGAGGAGTCA -3392

-3391 GGTTTTGGCCACGGATTCGGCCGCCTCGGGGCTAATTGGCCACATTTAGCATTGTCCATATCCACTGGGCAACTGGTCAACCTCAGGCTA -3302

-3301 CTTGGACAGGTGTGAGCTTTGCATTTAATCCCCCTTTTCGCGAAACGGAAGCTCTCGTAAATTGCTGCAACAAGCTACCGATGACAGTGA -3212

-3211 AGCGGGGGCGCTGGTGGTGGCCATATGAAAATGAGATCGCTTTGTATGCAAATGCCTGGGAATCGAATTGCGAATCGGGAATCGTACAAT -3122

-3121 CTCATTGCGACTTTATGCCAAGACAATCGATGCCTCCCTTTTCGGGCTGCGTGGGCGTGGTGGTGGGGGCTTCCATTGTTAAAACGTGTT -3032

-3031 TACACATCCAGAAGAAAGAAATAAATAAATGCTGCTGCTATTGAGAGATGTTACTAGTTTCTAAGTAAAAAGCTCTTTTCATCTTAATCG -2942

-2941 TAATTTTCAAATTAATAGGATTGGTGAAAACTCAAAAACGTTTCCACTTTCTGAAAGAATTAGATTTCTCAGAACTAAAATACATCAACT -2852

-2851 CATAATCGAGCAGTAACTACAAACACTCCTCTTATTTCAGCCAACTCGGGAATGCAGAACGGAGGAAAAAACAATCATCGATGTCGAACA -2762

-2761 AAAACAAAACTTTCTGCAGGAGGAGGAGCTCCTGCAGGAGGTAACGAGAGCAACGAAAGGAGCAGAAGCCACAAAGGGGAACTGGCGCTG -2672

-2671 CCCCATTGCATACCCCACCATAACGTAGCAAGTTTGAATATACTCGCACCCGTAAGATTCCCGAGTATATTAGGTAGCAAAATTTTTACG -2582

-2581 AGCTCATTATGGCTCATTTCGGCGATTGTTGTGATCCTTTTATGCGCCTTCGAATGGCTTCAAATGGTTATGAGGCTATTTCTCTGTCAT -2492

-2491 CCCCGCGAGTCCTTCGCACTCGTGTCCTTCCCCTGGGTCCCAAATGCGGGATAGCGAGTTTCTGGGTCCTGGATTCCCGACTCACGATAT -2402

-2401 TTAGTTTGCAGTTGCGACTGCGATTTTTACTTTACTTTTGCTTTGCTCTGGGTCTTCTCGCCTTTTGGCTTCGCTTTTCCTTGGGCTTTT -2312

-2311 GGTCGCATATTTAGAAAATGTCGCCGTGTCTCCGAATGTGATTCAAGTGTTTGTCAGTGTGTGTGTATGTGGTGGTGCGTATCTGTGTTT -2222

-2221 TGTGCAAGTTTTTGTATCTTTCGCTTGTTGTTGATTTTTAAAACTTGGCACCGAAAATTTGGTCGGCGGAAAATGGCGCGAGCAGGGGTT -2132

-2131 GGAAAAGTAGAAAAGTAAATTTCTTCTTATAATAAGTTTATTTTGCAAACACTTGTGGCAAGCAACTTTGTTTGTTCGCCGTCGGTTTCT -2042

-2041 GGCTTAATCTACGAGCCGTTTGTACGAATGTGGAAACTTTGACAAATATTTGCGAAACTTTCGCGAATGGCGCGGCGTTCTCGGTCTGCA -1952

-1951 AGACCCGATCTATTTGCATTATGCATATGAATTTATTAGCATCAATTCCCACCATATTTACATTGACATTAATATCCCATTGAAAGTATT -1862

-1861 GGCGAACGGATTGAACATTTCGATTCGAATGCGAATTTGAATTTGAATAGGAAAAGGAAAACCGCAGCACACAAATTTTCCGGGTCCCTG -1772

-1771 GCTTCTATTACCGTACAATGCCGAGTTGGGGTTGACTTATTATCAATAAACTAATGCTAAACACGAGACTTTTAATAAATAAAAAATAAT -1682

```
                .         .         .         .         .         .         .         .         .
-1681  GTCTTCATTAATATATTATAATTCGTTTTTAAAGGCCTAATAAAGCCTTTAAGTGATAAGTATCTTTTTATGCCATACCAAATTGAATTA   -1592
                .         .         .         .         .         .         .         .         .
-1591  AGCTTATAGTTTTTGAGGTAAAGGTAAAAACGATTAATTTACTAATATTTTTCTAAATTGTAAAACAAATTATACGATATTTACTTCCAT   -1502
                .         .         .         .         .         .         .         .         .
-1501  AGAAATTAAATTAATAATTACAATATTCCGACTATTCTTTTAAAAATTGTTGTTTGCAAATTATTTCCCAAAATATTAAACAATTAAATA   -1412
                .         .         .         .         .         .         .         .         .
-1411  ATGACCAGCTTAAACTGAAAAGAAACAGACAAAGAATTTGCGTAAGCCGTATTCTAGCACAAAGATTGGGAACCGAAACTGTAGTCATGA   -1322
                .         .         .         .         .         .         .         .         .
-1321  GTGCGCTAAAACCGAGCAAATACGGGATGCGCATCGTTTGGCGTGCAACTGGCAACTGGCGGGCAGCCCGTCGAGCGTTTTTCGCCACTC   -1232
                .         .         .         .         .         .         .         .         .
-1231  AGTTGAAGGAAAATCAGCCCTCCTCCATGATGAATTTCCCGCGGCGAGCGCATTTTCCTTCCTCGCGAATGAATGAACGAGAGGCGCCCA   -1142
                                                                               //////////////>g4

                .         .         .         .         .         .         .         .         .
-1141  CCCCGATAAACTTAAACTGAACGAACACTCAAGAGAGAGCGCAAGAGCGCTCAAAAACAATCTGGTTTTGAGCGTTTCGCTGGCTCTCTG   -1052
                                     //////////////////>g3    -------------------fp4 .<//////////g2
                             \\\\\\\\\\\\\\\\z5   \\\\\\\\\\\\\\\z4      \\\\\\\\\\\\\\\z3

                                                                                             -->-->-966/-964
                .         .         .         .         .         .         .         .         .
-1051  TTTCTGTTTTCCACTCGTTTTTAGGCCGAGTCGAGTGAGTTGAGTCGGCAGAGCAAAGTCAAAACACTGGCAACTGCGATTTGGTGCCAC   -962
    g2/////                            /////////////////////>g1
             \\\\\\\\\\\\\\\\\\z2      \\\\\\\\\\\\\\\z1

                .         .         .         .         .         .         .         .         .
 -961  ATTCGTTCGATGGCAACGGATTGGATAACAGGCGCGCTTTGTTTTATTATCCACATTATCAGCGGCATTATTGTTATTATTGGCCCTC   -872
                MetAlaThrAspTrpIleThrGlyAlaArgPheValLeuLeuSerThrLeuSerAlaAlaLeuLeuLeuLeuLeuAlaLeu   (27)
                   ---------------------A        -------------------------------------- u-A

                .         .         .         .         .         .         .         .         .
 -871  AGCGCTTTACCGCTCGCCCACGCGTCCGCCCGTGAATGCCGCGCGGAAAAGTCGCTTTCCACTAGATTGGCGTCCAGATTCGAGGAAATC   -782
       SerAlaLeuProLeuAlaHisAlaSerAlaArgGluCysArgAlaGluLysSerLeuSerThrArgLeuAlaSerArgPheGluGluIle   (57)
                .         .         .         .         .         .         .         .         .
 -781  CGTCAGCAGACTCATTCGCGCCCGTTCGGTCAGCACTAAGGCTAATAATCGTTCAAATCGTTAAAACCATAAAAATAATAATAATTGCAA   -692
       ArgGlnGlnThrHisSerArgProPheGlyGlnHisEnd                                                      (69)
                                            -----------------------------------------------------

                .         .         .         .         .         .         .         .         .
 -691  TAACAATAAACATAGTAATAATCGTAACGCTTACGAGCCTTTGATAGTGCCAAGGCAAGCGCAATCCAAGTATTCAAATTCGAATTCAAT   -602
       --------------------------------- u-B
                .         .         .         .         .         .         .         .         .
 -601  TAACAGCAAAGTGCAATTGGCTAAAAACCGAAACCAAAACGCAACAAAGTATACGAAACACTTGTGAAACCGTACAAACAATTGTGGAAA   -512
                .         .         .         .         .         .         .         .         .
 -511  AAAAATTAAAAGATTATTAAGATTGAAGTCTCAATAAACATTAGTGCTTAAATAAATTTAAAACGACCCGCGTGGAGAGTGCAATAAAAA   -422
                .         .         .         .         .         .         .         .         .
 -421  GAATAACTTTTGAAATAAATATTTACCAAACAGAAAAATATTTTATAAATATTTAAATAAGTGAAAAACAAATTGGTTACTCTGAAACAA   -332
                .         .         .         .         .         .         .         .         .
 -331  AGAATATTCAAATTGGTGCTAAAACAAAGGAGAAAAAATTTCAAGAATTATTATACAAATAATAAGACATATTTAACTATATAAAACCAA   -242
                .         .         .         .         .         .         .         .         .
 -241  ACTTAATCAACAAAGACAAAGGAGTGAAAAAAATAAAAAAAAATTTTAAAAGAGTTAAAAAAAAATTGTTTATATCCAAAGGAGGCAAAG   -152
                .         .         .         .         .         .         .         .         .
 -151  GAACAGCACAGAAAGCGAGGAAACACTCAAATAAAATCCGCCAAAAATCGCAGATCCCTGGAAACCAATTCGTGTGAAATCGGTCAAGCC   -62
                .         .         .         .         .         .         .         .         .
  -61  CCCAACGACTTTTAGCCCGTCTCAGACGGAGCACCGCCAAGATTCTTACCGCCAGCAGCGCAATGAACTCGTACTTTGAACAGGCCTCCG   28
                                                                     MetAsnSerTyrPheGluGlnAlaSerG   (10)

                .         .         .         .         .         .|-       .         .         .-|Def 6.28
   29  GCTTTTATGGCCATCCGCACCAGGCCACCGGAATGGCAATGGGCAGCGGTGGCCACCACGACCAGACGGCCAGTGCAGCGGCGGCCGCGT   118
       lyPheTyrGlyHisProHisGlnAlaThrGlyMetAlaMetGlySerGlyGlyHisHisAspGlnThrAlaSerAlaAlaAlaAlaAlaT   (40)
```

(continued)

```
 119  ACAGAGGATTCCCTCTCTCGCTGGGCATGAGTCCCTATGCCAACCACCATCTGCAGCGCACCACCCAGGACTCGCCCTACGATGCCAGCA   208
      yrArgGlyPheProLeuSerLeuGlyMetSerProTyrAlaAsnHisHisLeuGlnArgThrThrGlnAspSerProTyrAspAlaSerI  (70)

 209  TCACGGCCGCCTGCAACAAGATATACGGCGATGGAGCCGGAGCCTACAAACAGGACTGCCTGAACATCAAGGCGGATGCGGTGAATGGCT   298
      leThrAlaAlaCysAsnLysIleTyrGlyAspGlyAlaGlyAlaTyrLysGlnAspCysLeuAsnIleLysAlaAspAlaValAsnGlyT  (100)

 299  ACAAAGACATTTGGAACACGGGCGGCTCGAATGGCGGCGGGGGTGGCGGCGGAGGCGGTGGTGGCGGCGGAGCGGGCGGAACAGGTGGAG   388
      yrLysAspIleTrpAsnThrGlyGlySerAsnGlyGlyGlyGlyGlyGlyGlyGlyGlyGlyGlyGlyGlyAlaGlyGlyThrGlyGlyA  (130)

 389  CCGGCAATGCCAATGGCGGTAATGCGGCCAATGCAAACGGACAGAACAATCCGGCGGGCGGTATGCCCGTTAGACCCTCCGCCTGCACCC   478
      laGlyAsnAlaAsnGlyGlyAsnAlaAlaAsnAlaAsnGlyGlnAsnAsnProAlaGlyGlyMetProValArgProSerAlaCysThrP  (160)

 479  CAGATTCCCGAGTGGGCGGCTACTTGGACACGTCGGGCGGCAGTCCCGTTAGCCATCGCGGCGGCAGTGCCGGCGGTAATGTGAGTGTCA   568
      roAspSerArgValGlyGlyTyrLeuAspThrSerGlyGlySerProValSerHisArgGlyGlySerAlaGlyGlyAsnValSerValS  (190)

 569  GCGGCGGCAACGGCAACGCCGGAGGCGTACAGAGCGGCGTGGGCGTGGCCGGAGCGGGCACTGCCTGGAATGCCAATTGCACCATCTCGG   658
      erGlyGlyAsnGlyAsnAlaGlyGlyValGlnSerGlyValGlyValAlaGlyAlaGlyThrAlaTrpAsnAlaAsnCysThrIleSerG  (220)

 659  GCGCCGCTGCCCAAACGGCGGCCGCCAGCAGTTTACACCAGGCCAGCAATCACACATTCTACCCCTGGATGGCTATCGCAGGTGAGTGTC   748
      lyAlaAlaAlaGlnThrAlaAlaAlaSerSerLeuHisGlnAlaSerAsnHisThrPheTyrProTrpMetAlaIleAlaGlyGluCysP  (250)

 749  CTGAAGATCCGACCAAAAGTGAGTGTCCACTGCAGCA*  INTRON 1 (10 KB)
      roGluAspProThrLysS                                                                         (256)

                                        *TTTCAGGTAAGATAAGATCTGATTTAACACAATACGGCGGCATATCAACAGA   838
                                             erLysIleArgSerAspLeuThrGlnTyrGlyGlyIleSerThrAs  (271)

 839  CATGGGTAAGAAAATTTCCACTTTTATTTCGTTACATTATTCGCTCTTAAGTTTTCCGAAAAATAGAGTATAAAGTGTAGAGCAGGTCCA   928
      pMetG                                                                                      (273)

 929  CTAACAAACCGTAGAGAACTAATCCCATTATGGTGTTGGTGGCTAAAATATTGTAGTATTCGTCTTTAAGGTGTGCAAAATTCATGAATC  1018

1019  AATGGGGCGGGTCTGTGGGTGGACCGGGAAAACCTGGGGGCCGCGTGTGGAAATGATTGATTCACTGTCCGCA* INTRON 2 (13 KB)

                                                                             *GACCTTTATAAACGTT  1108
                                                                      A=195|
1109  TTCTCTCTATTTTTTCCAGGTAAGAGATACTCAGAATCTCTTGCGGGCTCACTTCTACCAGACTGGCTAGGTAAGTCGAAGTTTTGTTAT  1198
                        lyLysArgTyrSerGluSerLeuAlaGlySerLeuLeuProAspTrpLeuG                      (290)
                                                                 End

1199  ATTTTTTGTAACCCC* INTRON 3 (50 KB)

                     *GGATCCTGTATTTTTGCTACCATTTCGTTAAGACTTTCTGAGAGATATGGCCGACAAATTGCCATAAACTGAC  1288

1289  GCATCGCAAATCTTGTGACCTGTCACTGGCCAATTTTCTGGCACATTAAATGGGTCTTTATAATTCTCGCAAGGCAGTTTAAAAATAAAG  1378

                                                                                        |- Def 9.
1379  CCACATTAAGGAAATTTATCAGCATGCATGAGCGAGTCCATATAGATGATAATTTCTTTGTTGCATCCTGGCCATTTTATTTCCATATCA  1468

1469  TTTTTATTGTCTATAAAATTTTTTCGCCATTTATTTCCACCACCCACACAATTGCAATCGCTACGTACGCTGCACATGTGTGTGCTTGTG  1558
```

```
1559  TGAGTGAAAGATAAATGCGTTTTCACTTTATGACTTTCGTGTCGGCATAAATTTGTTAATACCTTTAGGCCAAATTTATAACATAATAAA  1648

1649  TGCTCATAATATTTAACTTAACATTGTGCTCGGGCCCAGGAGAAAGACCTGGTCTCCAAAATGCCAAGTTAACATGGTCGAATGGGTGGG  1738

1739  TTGGTTGGTTGATATGGTGTGGTATGATTTGGTATGGGTTGATTTCGATAATATCAGACATTGTCTGGGCCTCTTCTTCGATGGGAGATG  1828

1829  GGCCAGAGACAGCTGCAGTGCATTTGCACACACACGAAATTGAGTTATTGCACTTGAAGGCAAATTAAACTTCATAAATATTTAAAATCA  1918

1919  GAGATTAAACACGGCATTGTTGCAACATGTTGATGCGACTTCTGGCTGCCCCGGCTCCCCTGGCTTCCCCCGGATTCCCCTGGATTCCCC  2008

2009  TGCTCCTCCTGCCCCATCTCGTCTCTCAGGTTGCCAATTAAACGGGCATTATCTGGCATAACTGCAATTTAAGTAGCCACATTCGCCATA  2098

2099  TCCCCAGTGCAATGCCACAACCGAGTGCTCGCACGTTTCTCCTTTTCATTTTAATGTGGCTGCATCTGCGGATCTGTGTATCTTTGTATC  2188

2189  TGAGGAACTGTGGAACTGCGAATCTGGATGCAATGACAGCACGGCAGCAACATTGGCGGTGCAGCGGCAAACGATCAATTTAAAGTAACG  2278

2279  ATCGCGCCGCAGAAACAAAAACCGCAACTGCAAACTGGCAAACTGGCAAATACTCGGCGATACTCGTAAAGATGAAATGTATTTTTTGCG  2368

2369  CTGAGATCCCCTTTCCTTTCGATTTGGGCCCCTTTGCAGGCAATTGCGGCCTACGTTCGAGCTGCTTGATGATCGCTGGCAAAAAGGAGA  2458

2459  ATTTATATTTACGACTTGGCCAAATAACAACGGCGAACAGCAAACAAATGTTGAGTTGCTCGTTGCAATTTTTCAATTTAATTTGCCCGA  2548

2549  TGAAAGGCCAAAATATAAATACCCGAAAACACTCTGTCACTGCTGCTCAATATGACTCAAATTTTGATGTCCTCATGTTCTCCTAAACGT  2638

2639  TAATATAACCAATTAAATCACTTTTGTGGCGATTTATATAAATAATTAGGCCAATAAATCGATAAAGATATATATCTACTTAGTCACTTT  2728

2729  GTCAATGTTTTCCTAACACATATCTGCATTTTGTAGCTGCTGTTATGAGACACATATTTTTGATTGCAAAATGAAATGTATGTATTTCGT  2818

2819  CGATGCAGGTCCAAAATGAATAATATTAAAAGTTTAATAATCTGGTTACTTACAAATGCTTTCCCATTCCGATCTACAGGTACAAATGGT  2908
                                                                                   lyThrAsnGly  (293)

2909  CTGCGAAGACGCGGCCGACAGACATACACCCGCTACCAGACGCTCGAGCTGGAGAAGGAGTTCCACACGAATCATTATCTGACCCGCAGA  2998
      LeuArgArgArgGlyArgGlnThrTyrThrArgTyrGlnThrLeuGluLeuGluLysGluPheHisThrAsnHisTyrLeuThrArgArg  (323)
        |-    *    *        *  -----*-----------*-----------------H1       *     ------
                                                  -|AAATTT=Def 9.22
2999  CGGAGAATCGAGATGGCGCACGCGCTATGCCTGACGGAGCGGCAGATCAAGATCTGGTTCCAGAACCGGCGAATGAAGCTGAAGAAGGAG  3088
      ArgArgIleGluMetAlaHisAlaLeuCysLeuThrGluArgGlnIleLysIleTrpPheGlnAsnArgArgMetLysLeuLysLysGlu  (353)
      ------------*--------------H2 *     ---------*-----*--*--*-----*--*H3*    *    *

3089  ATCCAGGCGATCAAGGAGCTGAACGAACAGGAGAAGCAGGCGCAGGCCCAGAAGGCGGCGGCGGCAGCGGCTGCGGCGGCGGCGGTCCAA  3178
      IleGlnAlaIleLysGluLeuAsnGluGlnGluLysGlnAlaGlnAlaGlnLysAlaAlaAlaAlaAlaAlaAlaAlaAlaAlaValGln  (383)
        | HOMEODOMAIN

3179  GGTGGACACTTAGATCAGTAGATCCTTAGATCCTTAGATCCTTAGATCCGTAGGGTGTATGTGGGATTGGGCGAAATGACGCGGAGACAG  3268
      GlyGlyHisLeuAspGlnEnd                                                                       (389)

3269  ATACAAAGCAACTATATTGTAACAAATGAACTATTTACTTAAATGAATAATATTTAAATATTTTGATGGTACTTGTGCGAATACGAAACT  3358

3359  TAACCTAAATCGAACCTAATGGAATTATTTCAAGCGTTTGAGCAGCAACCGAAAATACGTAAATGAAACAAAACTACAAACTAATTAACT  3448

3449  AGGCTAAGTAAATAAAAGTAGTGGAAGGAGCGCAGATTATAAACCTACTTAGAATTAAATGAGCAAAACAAACAATTTTAATTTAGTTCC  3538

3539  AAACGAAAAAAAAATTCAAGAGGATTCGCTCGAAATGGAAACCTCTGTCCTGCCCCTTTGTTGCTTACTGCTATGTTTAAATTAATTTCG  3628
```

(*continued*)

```
3629 CGAAAAATACTCAAAAATTGAAACACAAAGAAAACAAAAAATGAAAGTATACCATTATAATGTTGAATGCGAGCAAAATTCTGTTGATAT 3718

3719 GAATTTTTGGTAAAAACATGTTCTAAACCAATTTAAGATACGTAACGAAGGATGCAAAAACAAAATGAAAACTATTAAACTTTAACTTAA 3808

3809 ATATAAATAGAATTTGTTAGCCAAGTAAACATATTACGACACGAAGAACAAACGTTTTCGGGAGTATCGAATATTTGAATGTGTATAGTT 3898

3899 TGTGCTTATTAAATAAAATAATGCAATTTTAGTTAACTCTGTTTATTTGTAAACGAATTTGTTTAGTTCTCGCCCAAACGACTAGAGTGA 3988

3989 AGCTGTTTCTTTAAGTAATGTGTAGTGTGTTTACTTTTTAAATTAAATTAATGCCTAATTTTATTATTATTATGTTTAGTTTAATGACAA 4078

4079 GCGTTTATGAGATTATCCGACAGAAGCGGCGAGAAGAGGAGTGCGACAAACCGTTTGCCCCGGCAAACGCAAATAAATTATTGGTTTTGA 4168
                                                                                 ------

4169 AAAAATCTAAAGAAAACAAAAAAAAAAAACAATGAGAAATCGAATCCGATTGTTGTGTTATTATTTAGTTCTGCCATTGCGATTTTCCG 4258
                  |(A)n proximal

4259 TTCTCCCAGTGTAATTAGAGCCTGAGTTGTTTGAGAGAGTCTTCGGGCTACCCGCTTGCATGCGAAATTGCTTTTGATCTCGTTTTGAGC 4348

4349 CGTTAATTGATCGTGAGTTGTACGCTCTATAGAGATACCCATACCGATTAGCTATAACGATACCATACCGATACCAATACCATATATATA 4438

4439 GTTTAGTGGATCC 4451
```

Ubx SEQUENCE. Accession, Y00206, X05723 (DROUBX1), X05724 (DROUBX2), X05725 (DROUBX3), X05727 (DROUBX5), X05427 (DROUBXG5). Discontinuities in the sequence at 785, 1,091 and 1,213 correspond to introns that have not been completely sequenced; those gaps are not reflected in the numbering system. Position 739 is the alternative donor site of the first exon. Underlining in the interval between −1,160 and −600 marks protein-binding sites. The various sites are associated with the following proteins: g1–g4, GAGA protein; z1–z5, ZESTE; u-A and u-B, UBX; fp4, a protein that also binds to the *Ddc* promoter; and A, an unidentified protein (Biggin and Tjian 1988). Marks above the sequence indicate the following mutations: a vertical bar (at −4,036), the breakpoint of translocation Hm, a regulatory mutation of the *bxd/pbx* type (Bienz et al. 1988); |- -|Def. 6.28 (between position 81 and 112) and |- -|Def 9.22 (between positions 1,468 and 3,046), two *Ubx* deletions; and an A-for-G base substitution (at position 1,173), a nonsense mutation; all but Hm are null mutations (Weinzierl et al. 1987). The limits of the homeodomain are indicated by vertical lines below Arg-295 and Ala-356; helices 1, 2 and 3 (H1, H2, H3) are underlined and asterisks mark conserved positions. Helix 4, is seven amino acids long and follows immediately after H3 by analogy to the ANTP homeodomain (Qian et al. 1989).

The homeodomain controls specificity of DNA binding (Gehring 1987; Hayashi and Scott 1990; Harrison 1991), while other region(s) of the protein act as effectors, either stimulating or repressing transcription (Kuziora and McGinnis 1989). The Gln at position 9 of helix 3 (H3), characterizes UBX as an *Antp* class homeodomain (*bicoid* class homeoproteins have a Lys in that position) (Hanes and Brent 1991). Amino acids within the homeodomain but outside of H3 must distinguish the DNA-binding specificities of UBX and the product of *Deformed*, another homeotic of the *Antp* class, since both proteins are identical in H3 but interact with different genes (Kuziora and McGinnis 1989).

The optimal *in vitro* binding sequence was identified by the following procedure: an affinity matrix containing the UBX homeodomain was used to select random-sequence oligonucleotides capable of binding. The bound oligonucleotides were eluted and amplified by the polymerase chain reaction. The process was repeated several times. The sequence of the selected oligonucleotide, TTAATGG is found near the *decapentaplegic* gene in seven near-perfect copies of that consensus; and these sequences are afforded protection from DNase I digestion by a 70-amino-acid polypeptide that includes the UBX homeodomain (residues 295–365) (Ekker et al. 1991).

Tissue Distribution

Antibodies against an epitope common to all of the *Ubx* products were used to detect gene expression. UBX is first detectable in early stage 9 embryos (approximately 3 h 45 m of development) as a single band that occupies the posterior portion of parasegment 6 (anterior compartment of the first abdominal segment, A1a) (Appendix, Fig. A.3). Next UBX appears in parasegments 8, 10 and 12 and soon afterward in all parasegments between 5 and 13. In parasegment 7–12 UBX forms a repeating pattern wherein, in each parasegment, expression is weaker in the anterior portion and stronger in the posterior portion (Irvine et al. 1991). During the rest of embryogenesis UBX appears in a complex pattern that includes the nervous system; in larvae, UBX is found in imaginal discs. Highest antigen levels are observed in T3p and A1a structures (parasegment 6), in T2p and in the anterior compartment of A2–A7. UBX is localized in nuclei (Beachy et al. 1985).

The tissue distribution of UBX is in general agreement with the sites of gene transcription (see below) and with the sites of gene activity deduced from the effects of *Ubx* mutations.

Mutant Phenotypes

Ubx is a homeotic gene. Null mutations transform structures of parasegment 5 and parasegment 6 origin to parasegment 4 type differentiation, and they also cause minor abnormalities of the abdominal segments. (Lewis 1978; Sánchez-Herrero et al. 1985; Duncan 1987; Akam 1987).

Organization of the Complex

Ubx is part of the bithorax complex (BXC), a three-gene, 300-kb cluster. Approximately 60 kb upstream of *Ubx* is the 3′ end of *abdominal A*, which extends for 25 kb, and 90 kb further upstream is *Abdominal B*. All three genes are transcribed toward the centromere (reviewed by Duncan 1987; Peifer et al. 1987). *Ubx* itself is spread over 77 kb of DNA, and not all of it has been sequenced.

Gene Organization and Expression

The published *Ubx* sequence includes four exons and small sections of the neighboring introns. The open reading frames of several alternative splicing products vary between 346 and 389 amino acids. The expected size of mRNAs are 3,096 and 3,123 bases (Fig. 29.1B), forms Ia and Ib with polyadenylation at the proximal site) and approximately 4,100–4,200 (forms II and IV with polyadenylation at the distal site, see below). These sizes are in agreement with the occurrence of two main poly(A)+ RNA bands of 3.2 and 4.3 kb detected by northern analysis. There are introns within the Ser-256, Gly-273 and Gly-290 codons (Saari and Bienz 1987; Weinzierl et al. 1987; O'Connor et al. 1988; Kornfeld et al. 1989). Primer extension, S1 mapping and a cDNA sequence were used to define the 5′ end at −966/−964. There is no discernible TATA box appropriately positioned upstream of the *Ubx* transcription initiation site (Saari and Bienz 1987; O'Connor et al. 1988; Kornfeld et al. 1989). S1 protection was used to localize the two 3′ ends 1.1 kb apart; the proximal 3′ end was also identified by a cDNA sequence (Kornfeld et al. 1989).

Data from two studies (O'Connor et al. 1988; Kornfeld et al. 1989) on a total of 78 embryonic cDNAs indicates the following forms of splicing (see *Ubx* Sequence and Fig. 29.1B): Exon 1 has two donor sites, a and b; site a is used 80% of the time. Splicing can occur so that all four exons are included (forms Ia and Ib, 75%), or so that exon 2 is spliced out (forms IIa and IIb, 21%), or so that both exons 2 and 3 are spliced out (form IVa, 3%); IVb has not been observed. These alternative splicings introduce only small differences in the size of the mRNA: the two donor sites in exon 1 are only 27 bp apart while exons 2 and 3 (the two "micro" exons) are only 51 bp long. Thus, the main differences in the expected sizes of mRNAs depends on which polyadenylation site is used. As already mentioned, the proximal poly(A) site is used predominantly in form I RNA (the form that carries two micro exons) and the distal one in forms II and IV (one micro exon and no exon, respectively).

The unusually long leader region of this gene (1,066 bp) includes a potentially functional second open reading frame of 69 codons. The first 23 residues of this putative protein resemble a signal peptide; it has been suggested that translation of the leader peptide may be involved in regulating translation of the UBX protein (*Ubx* Sequence) (Saari and Bienz 1987).

In addition to the RNAs described above, there are other minor transcripts of uncertain function (O'Connor et al. 1988; Kornfeld et al. 1989).

Developmental Pattern

Ubx expression is undetectable before fertilization; transcripts are first detected at the end of the syncytial blastoderm stage, immediately after the 13th nuclear division at approximately 2 h 30 m (Akam and Martínez-Arias 1985). There is a 60–75 min lag between the time of appearance of *Ubx* RNA and the time when protein is first detected (Irvine et al. 1991). This delay has been ascribed to the enormous size of the 77 kb transcript, and Kornfeld et al. (1989) proposed that size may serve a regulatory function to insure the correct timing of UBX protein accumulation. *Ubx* expression increases dramatically between 3 h and 6 h of embryonic development, reaches a plateau by 9 h and remains at a high level until 15 h. The level of transcripts then decreases and remains relatively constant and low through to the adult stage (O'Connor et al. 1988).

The choice of splicing and polyadenylation sites are also developmentally regulated. Form I transcripts predominate (70–80% of *Ubx* transcripts) early in embryogenesis (3–8 h of development); they decrease during middle and late embryogenesis to approximately 30% and then rise once again to 50–60% during larval and adult stages. Form II rises from very low levels early in embryogenesis to 30–40% after 10 h of development and stays in that range. Form IV peaks late in embryogenesis and disappears after the second instar (O'Connor et al. 1988; Kornfeld et al. 1989).

Late in the cellular blastoderm stage (4 h), transcripts are detectable extending from 50% to 20% egg length (Appendix, Fig. A.3). The concentration of transcript is significantly higher in a zone that probably corresponds to parasegment 6 (between 50% and 45% egg length). With the onset of gastrulation, the distribution of transcripts becomes more complex. During the extended germ band stage (6–8 h) transcripts seem to accumulate in ectodermal and mesodermal derivatives of regions that correspond to parasegments 6–12. In parasegments 5 and 13, transcripts are more localized to ectodermal derivatives (Akam and Martínez-Arias 1985). In older embryos and larvae, *Ubx* expression is evident in ectodermal and many mesodermal (but not endodermal) derivatives. In 12–20 h embryos, strongest expression is in the nervous system. Expression is not uniform in all segments: in third instar larvae, expression in muscle extends primarily between A1 and A6; in the nervous system, highest RNA levels are detected in T3 and A1 (Akam 1983).

The mRNAs also display tissue specificity: form I predominates in embryonic myoblasts, while forms II and IV predominate in neuroblasts (O'Connor et al. 1988).

The pattern of *Ubx* expression early in development is determined by the action of maternal and segmentation genes. After the end of the germ band extension period, that pattern seems to be maintained through the rest of development by the products of genes of the *Polycomb* (*Pc*) group. It has been proposed that the *Pc* protein acts by modification of chromatin organization to prevent ectopic activation of *Ubx* (Paro and Hogness 1991 and references therein).

Promoter

P-Element-mediated transformation experiments showed that a segment extending from 1.7 kb upstream of the transcription initiation site to the first codon, when attached to a reporter gene, supports transcription in embryonic ectoderm. The expression is evident along the entire length of the embryo in a segmented pattern and is called the "basal pattern" of expression. The intensity of the "basal pattern" depends on sequences within 626 bp of the transcription initiation site while the segmented nature of the expression is dependent on regions of the *Ubx* leader that seem to coincide with homeoproteins binding sites (Bienz et al. 1988; *Ubx* Sequence).

In vitro transcription experiments defined a minimal promoter region that responds to nuclear extracts of staged embryos: a segment starting 154 bp upstream of the transcription initiation site and extending 41 bp into the leader is capable of supporting transcription in the presence of extracts from 8–12 h embryos (but not with extracts from 0–4 h embryos, where *Ubx* is not normally expressed). Proteins that bind 5′ upstream sequences include the GAGA protein, the *zeste* product, and a factor that also binds to a promoter element of *Dopa decarboxylase* (Biggin and Tjian 1988; Biggin et al. 1988; *Ubx* Sequence). At least one element downstream of the transcription initiation site is also required for *in vitro* transcription (designated A, in the *Ubx* Sequence). Just beyond this element are segments u-A and u-B to which UBX binds specifically and which are thought to be important in transcriptional regulation (Beachy et al. 1988; Kuziora and McGinnis 1989). Experiments using cultured cells demonstrated that ANTP and FTZ (the product of *fushi tarazu*) require element u-B to stimulate *Ubx* transcription (Winslow et al. 1989).

In addition to the proximal DNA elements responsible for the "basal pattern", there are at least two more distal regions that play a role (Fig. 29.1A). The *bxd/pbx* region, extending from 3 kb to >30 kb upstream of the transcription initiation site is thought to be involved in the regulation of *Ubx* expression in parasegments 5, 6, and perhaps also in the abdominal segments. A segment of DNA that extends from 35.4 kb upstream of the transcription initiation site to the eighth codon of *Ubx* can drive the expression of *lacZ* in an embryonic pattern identical to that of *Ubx*. A reporter gene construction in which the regulatory region extends 22.2 kb upstream of the transcription initiation site shows some deviations from normal *Ubx* expression; and, when only 5 kb of upstream DNA are included, *lacZ* is expressed in the "basal pattern" described above (Irvine et al. 1991).

The *abx/bx* regulatory region, found within the last intron of *Ubx* (Cabrera et al. 1985; White and Wilcox 1985; Peifer and Bender 1986) contains a 2–3-kb segment (approximately between −77 and −80 in Fig. 29.1A) that behaves as an enhancer and appears to be responsible for defining parasegment 5 as the anterior boundary of *Ubx* expression (Simon et al. 1990).

In a separate set of experiments, Qian et al. (1991) identified a 500-bp segment of the *bx* region (near coordinate −63 kb in Fig. 29.1A) containing an enhancer (called bre) that activates the minimal promoter to strong expression in parasegments 6, 8, 10 and 12 and represses its expression in the anterior half of the embryo. The *hunchback* product binds to three sites in the bre and this binding is necessary for repression of *Ubx* transcription in the anterior half of the embryo.

Other Transcripts

The *bxd* region produces a 27-kb transcript early in embryogenesis, between 3 h and 6 h of development. This transcript includes at least 11 exons that are spliced in different combinations to give rise to numerous distinct polyadenylated RNAs. It is doubtful, however, that these are functional mRNAs because their coding capacity is very poor as judged from the length of open reading frames and codon usage. Another *bxd* transcript is synthesized later, from the third larval instar onward. In contrast to the early transcripts, this is a simple, unspliced poly(A) + RNA with a 110-amino-acid open reading frame and good codon usage. It is not clear what role these upstream transcription units might play in the control of *Ubx* expression. It has been suggested that *bxd* transcripts are completely incidental, resulting because the strong *Ubx* enhancers can activate cryptic promoters (Lipschitz et al. 1987; Saari and Bienz 1987).

References

Akam, M. E. (1983). The location of *Ultrabithorax* transcripts in *Drosophila* tissue sections. *EMBO J.* **2**:2075–2084.

Akam, M. E. (1987). The molecular basis for metameric development in the *Drosophila* embryo. *Development* **101**:1–22.

Akam, M. E. and Martínez-Arias, A. (1985). The distribution of *Ultrabithorax* transcripts in *Drosophila* embryos. *EMBO J.* **4**:1689–1700.

Beachy, P. A., Helfand, S. L. and Hogness, D. S. (1985). Segmental distribution of bithorax complex proteins during *Drosophila* development. *Nature* **313**:545–551.

Beachy, P. A., Krasnow, M. A., Gavis, E. R. and Hogness, D. S. (1988). An *Ultrabithorax* protein binds sequences near its own and the *Antennapedia* P1 promoters. *Cell* **55**:1069–1081.

Bienz, M., Saari, G., Tremml, G., Müller, J., Züst, B. and Lawrence, P. A. (1988). Differential regulation of *Ultrabithorax* in two germ layers of *Drosophila. Cell* **53**:567–576.

Biggin, M. D., Bickel, S., Benson, M., Pirrotta, V. and Tjian, R. (1988). *Zeste* encodes a sequence-specific transcription factor that activates the *Ultrabithorax* promoter *in vitro. Cell* **53**:713–722.

Biggin, M. D. and Tjian, R. (1988). Transcription factors that activate the *Ultrabithorax* promoter in developmentally staged extracts. *Cell* **53**:699–711.

Biggin, M. D. and Tjian, R. (1989). A purified *Drosophila* homeodomain protein represses transcription *in vitro. Cell* **58**:433–440.

Cabrera, C. V., Botas, J. and Garcia-Bellido, A. (1985). Distribution of *Ultrabithorax* proteins in mutants of the *Drosophila* bithorax complex and its transregulatory genes. *Nature* **318**:569–571.

Duncan, I. (1987). The bithorax complex. *Ann. Rev. Genet.* **21**:285–319.

Ekker, S. C., Young, K. E., von Kessler, D. P. and Beachy, P. A. (1991). Optimal DNA sequence recognition by the Ultrabithorax homeodomain of *Drosophila. EMBO J.* **10**:1179–1186.

Gavis, E. R. and Hogness, D. S. (1991). Phosphorylation, expression and function of the *Ultrabithorax* protein family in *Drosophila melanogaster. Development* **112**:1077–1094.

Gehring, W. J. (1987). Homeo boxes in the study of development. *Science* **236**:1245–1252.

Hanes, S. D. and Brent, R. (1991). A genetic model for interaction of the homeodomain recognition helix with DNA. *Science* **251**:426–430.

Harrison, S. C. (1991). A structural taxonomy of DNA-binding domains. *Nature* **353**:715–719.

Hayashi, S. and Scott, M. (1990). What determines the specificity of action of *Drosophila* homeodomain proteins? *Cell* **63**:883–894.

Irvine, K. D., Helfand, S. L. and Hogness, D. S. (1991). The large upstream control region of the *Drosophila* homeotic gene *Ultrabithorax*. *Development* **111**:407–424.

Kornfeld, K., Saint, R. B., Beachy, P. A., Harte, P. J., Peattie, D. A. and Hogness, D. S. (1989). Structure and expression of a family of *Ultrabithorax* mRNAs generated by alternative splicing and polyadenylation in Drosophila. *Genes Dev.* **3**:243–258.

Kuziora, M. A. and McGinnis, W. (1989). A homeodomain substitution changes the regulatory specificity of the *Deformed* protein in *Drosophila* embryos. *Cell* **59**:563–571.

Lewis, E. B. (1978). A gene complex controlling segmentation in *Drosophila*. *Nature* **276**:565–570.

Lipschitz, H. D., Peattie, D. A. and Hogness, D. S. (1987). Novel transcripts from the *Ultrabithorax* domain of the bithorax complex. *Genes Dev.* **1**:307–322.

O'Connor, M. B., Binari, R., Perkins, L. A. and Bender, W. (1988). Alternative RNA products from the *Ultrabithorax* domain of the bithorax complex. *EMBO J.* **7**:435–445.

Paro, R. and Hogness, D. S. (1991). The Polycomb protein shares a homologous domain with a heterochromatin-associated protein of *Drosophila*. *Proc. Natl Acad. Sci.* (*USA*) **88**:263–267.

Peifer, M. and Bender, W. (1986). The *anterobithorax* and *bithorax* mutations of the bithorax complex. *EMBO J.* **5**:2293–2303.

Peifer, M., Karch, F. and Bender, W. (1987). The bithorax complex: control of segmental identity. *Genes Dev.* **1**:891–898.

Qian, Y. Q., Billeter, M., Otting, G., Müller, M., Gehring, W. J. and Wüthrich, K. (1989). The structure of the *Antennapedia* homeodomain determined by NMR spectroscopy in solution: comparison with prokaryotic repressors. *Cell* **59**:573–580.

Qian, S., Capovilla, M. and Pirrotta, V. (1991). The *bx* region enhancer, a distant *cis*-control element of the *Drosophila Ubx* gene and its regulation by *hunchback* and other segmentation genes. *EMBO J.* **10**:1415–1425.

Sánchez-Herrero, E., Vernos, I., Marco, R. and Morata, G. (1985). Genetic organization of the *Drosophila* bithorax complex. *Nature* **313**:108–113.

Saari, G. and Bienz, M. (1987). The structure of the *Ultrabithorax* promoter of *Drosophila melanogaster*. *EMBO J.* **6**:1775–1779.

Samson, M-L., Jackson-Grusby, L. and Brent, R. (1989). Gene activation and DNA binding by *Drosophila Ubx* and *abd-A* proteins. *Cell* **57**:1045–1052.

Simon, J., Peifer, M. and Bender, W. (1990). Regulatory elements of the bithorax complex that control expression along the anterio-posterior axis. *EMBO J.* **9**:3945–3956.

Weinzierl, R., Axton, M., Ghysen, A. and Akam, M. (1987). *Ultrabithorax* mutations in constant and variable regions of the protein coding sequence. *Genes Dev.* **1**:386–397.

White, R. A. H. and Wilcox, M. (1985). Regulation of the distribution of *Ultrabithorax* proteins in *Drosophila*. *Nature* **318**:563–567.

Wilde, C. D. and Akam, M. (1987). Conserved sequence elements in the 5′ region of the *Ultrabithorax* transcription unit. *EMBO J.* **6**:1393–1401.

Winslow, G. M., Hayashi, S., Krasnow, M., Hogness, D. and Scott, M. P. (1989). Transcriptional activation by the *Antennapedia* and *fushi tarazu* proteins in cultured Drosophila cells. *Cell* **57**:1017–1030.

30

vermilion: v

Chromosomal Location:
X, 10A1-2

Map Position:
1-33.0

Product

Tryptophan oxygenase, TO (EC 1.13.1.12), an enzyme involved in the biosynthesis of brown eye pigment.

Structure

A 150 kD protein, it requires a hematin cofactor for activity (see review by Phillips and Forrest 1980).

Function

TO converts tryptophan to N-formylkynurenine, the first step in the synthesis of xanthomatin from tryptophan. This is the major pathway for utilization of non-protein tryptophan in higher insects; and xanthomatin is the only brown eye pigment in *Drosophila* (Phillips and Forrest 1980 and references therein). There is considerable similarity between *Drosophila* and mammalian TO (Fig. 30.1).

Mutant Phenotypes

Null alleles such as v^{36f} and v^{48a} have no enzymatic activity, do not accumulate xanthomatin and display bright red eyes when present alone, or pure white when combined with *brown* (*bw*), a mutation that blocks synthesis of the red pigment. Severe hypomorphic alleles such as v^{1} have a few percent of normal enzyme activity, accumulate a small amount of xanthomatin and develop a slightly off-white eye color when in combination with *bw*. Mutations in another gene, *suppressor of sable* (*su*(*s*)), cause v^{1} homozygotes to accumulate 20% of normal TO level and to develop normal eye pigmentation. *v* mutations are not cell-autonomous (Phillips and Forrest 1980 and references therein) (see below).

```
     1                                                50                                                    100
Rat  MSGCPFSGNS VGYTLKNLSM EDNEEDGAQT GVNRASKGGL IYGDYLQLEK ILNAQELQSE IKGNKIHDEH LFIITHQAYE LWFKQILWEL DSVREIFQNG
Dm   .MSCPYAGNG .......... ....NDHDDS AVPLTTEVGK IYGEYLMLDK LLDAQCMLSE EDKRPVHDEH LFIITHQAYE LWFKQIIFEF DSIRDML.DA
CON  ---CP--GN- ---------- -----D---- -V------G- IYG-YL-L-K -L-AQ---SE ------HDEH LFIITHQAYE LWFKQI--E- DS-R------

     101                                              150                                                   200
Rat  HVRDERNMLK VMTRMHRVVV IFKLLVQQFS VLETMTALDF NDFREYLSPA SGFQSLQFRL LENKIGVLQS LRVPYNRKHY RDNFEGDYNE LLLKSEQEQT
Dm   EVIDETKTLE IVKRLNRVVL ILKLLVDQVP ILETMTPLDF MDFRKYLAPA SGFQSLQFRL IENKLGVLTE QRVRYNQKYS DVFSDEEARN SIRNSEKDPS
CON  -V-DE---L- ---R--RVV- I-KLLV-Q-- -LETMT-LDF -DFR-YL-PA SGFQSLQFRL -ENK-GVL-- -RV-YN-K-- ---------- ----SE----

     201                                              250                                                   300
Rat  LLQLVEAWLE RTPGLEPHGF NFWGKFEKNI LKGLEEEFLK IQAKKDSEEK EEQMAEFRKQ KEVLLCLFDE KRHDYLLSKG ERRLSYRALQ GALMIYFYRE
Dm   LLELVQRWLE RTPGLEESGF NFWAKFQESV DRFLEAQVQS AMEEPVEKAK NYRLMDIEKR REVYRSIFDP AVHDALVRRG DRRFSHRALQ GAIMITFYRD
CON  LL-LV--WLE RTPGLE--GF NFW-KF---- ---LE----- ---------K --------K- -EV----FD- --HD-L---G -RR-S-RALQ GA-MI-FYR-

     301                                              350                                                   400
Rat  EPRFQVPFQL LTSLMDIDTL MTKWRYNHVC MVHRMLGSKA .GTGGSSGYY YLRSTVSDRY KVFVDLFNLS SYLVPRHWIP KMNPIIHKFL YTAEYSDSSY
Dm   EPRFSQPHQL LTLLMDIDSL ITKWRYNHVI MVQRMIGSQQ LGTGGSSGYQ YLRSTLSDRY KVFLDLFNLS TFLIPREAIP PLDETIRKKL INKSV*....
CON  EPRF--P-QL LT-LMDID-L -TKWRYNHV- MV-RM-GS-- -GTGGSSGY- YLRST-SDRY KVF-DLFNLS --L-PR--IP -----I-K-L ----------

     401
Rat  FSSDESD*
Dm   ........
CON  --------
```

FIG. 30.1. Comparison of the rat (M55167) and *Drosophila* (Dm) sequences. There is 50% overall identity between the proteins. Sequences were aligned with the GCG *Pileup* program.

v

```
      ------EcoR1   .         .         .         .         .         .         .
-1170 GAATTCCAAGCACATTGCAAGAATCCCAAATCAAAAAATCGCATGAAATTGCCCCCGTACCTTTTGCGTTTTACTCCCAGATGTAACTCA   -1081
      .         .         .         .         .         .         .         .         .
-1080 ATTTTTTCTATGCAAAAGTAGTTGAAAATTATATAATAAAAACCGATTAGAAAAACAAAACATACATATATATATATATAAATATATATA   -991
      .         .         .         .         .         .         .         .         .
-990  TATATTTAGACACACATCGACAGTATCCTATTCAATTGATTTCTTTGAGAACTTTGATTTTGCGATTTTGGATATGCAGCAAGAAAAGTA   -901
      .         .         .         .         .         .         .         .         .
-900  AAACCAACAACAGAAAATGTGTAAGAAATAGTATAAATAAGGTCGGATATTAATGCCCCGACATTACGCTATATGTATGTGCTGGGGCAA   -811
      .         .         .         .         .         .         .         .         .
-810  ACAGCAACAAATCCAATAAAACAAAGTAATTAACAACAAACAAAACGGATGAAACAACCAAGATTATGTAAGCGATGGATCGAAGTAAGA   -721
      .         .         .         .         .         .         .         .         .
-720  ATTGATGCGAACGGCACAAGTATATAACAAATTTCAACAAGTATATGACTGAGCCAATGACTCAAAAATACATTTTAAATAAAAGGGAAA   -631
      .         .         .         .         .         .         .         .         .
-630  ACCAGAAATATATGAAAAATATAAAAACGATAAGCAAGTGAATGAAAGCTCTTTTTCTTTTTTGGGTTTTGGGAACTGCAAATTATTGAT   -541
      .         .         .         .         .         .         .         .         .
-540  TGTATGGAAATGTTTGTTTACCTATTTTTGCATATGGTGCGATTGTATCAAAACAAGTTTTGAATTATCAAAATTGGTTCCATTTATTTT   -451
      .         .         .         .         .         .         .         .         .
-450  TATACAACCTTGACCTATTTTCAAGGACCAATAAGATTGGACCCCACATTAACTTAGAAAACAATACTTGCCATGTTCAATTTTATTCCT   -361
      .         .         .         .         .         .         .         .         .
-360  ACGCAGGGTTTATTTATTATTATACTATGTTAATCAAAAAATTAAAATGTTAATTTCTCAGTTATTTAACTACACCTTAGGTAACTCTGA   -271
      .         .         .         .         .         .         .         .         .
-270  TTTGGCATTTCTCACTGAACTGTACTACTGTAGACTACCTTCCATTCAGGAAAATATTTGTGTGCGCCGCACTTTCACCTCAAGTGATTG   -181
      .         .         .         .         .         .         .         .         .
-180  ATAATTCCCAGCCTATCTGGCAGTGCCCATCGCCCAGATCACCGACTGTGCAATCAGTCGGAACTGGAGCTCTCTCGCTCTGTTATCGGT   -91
                                                                            ----
      .         .         .         .   -->-56.         .         .         ||=1=2=k  .
-90   TCGCTGGGGTCTCATCTCCGGTCCGCTGGCGGAGATCAGTTCGCCAGCATCCGCCGCTCGAGGAGTCACGATCTGATCTGAGCTGTGCAC   -1
      .         .         .         .         .         .         .         .         .
0     CATGAGCTGTCCCTATGCAGGAAACGGGTGAGCACCAGCACGTGCTGTCCAGGAATGCCAATCGATCTTCAGTTCTGCGATTCAATTCAA   89
       MetSerCysProTyrAlaGlyAsnGl                                                              (9)
      .         .         .         .         .         .         .     . T=203   .     G=kLTR
90    ACCCATACAGAAACGATCACGATGATTCGGCGGTGCCATTAACCACGGAAGTGGGCAAAATCTATGGAGAGTATCTGATGCTGGACAAAC   179
                yAsnAspHisAspAspSerAlaValProLeuThrThrGluValGlyLysIleTyrGlyGluTyrLeuMetLeuAspLysL   (36)
                                                                      Phe            Val
      . A       |- Def217.   -|    .         .         .    G=kLTR     |-|=Def208  .
180   TGCTGGATGCCCAGTGTATGCTGTCCGAGGAGGACAAGCGACCCGTGCACGATGAGCATCTGTTCATCATCACGCACCAGGGTGAGTAGG   269
      euLeuAspAlaGlnCysMetLeuSerGluGluAspLysArgProValHisAspGluHisLeuPheIleIleThrHisGlnA            (63)
       Gln                                                   Arg          ---
      .         .         .         .         .         |-Def251-|.  A=220 G=233   T=291
270   TTTACAACTTTGATGACAACACTCAATGGCATTTAAGTACCTTCGCCACAGCCTACGAGCTTTGGTTCAAGCAGATCATCTTTGAGTTCG   359
                                                            laTyrGluLeuTrpPheLysGlnIleIlePheGluPheA  (76)
                                                                      End    Arg      Phe        V
      T         .         .         .         .         .         . Def48a|-.    C=257.     T=225
360   ACTCCATACGAGACATGTTGGATGCAGAGGTCATCGATGAAACCAAGACGCTGGAGATTGTCAAGCGACTGAACCGAGTGGTTCTGATTC   449
      spSerIleArgAspMetLeuAspAlaGluValIleAspGluThrLysThrLeuGluIleValLysArgLeuAsnArgValValLeuIleL   (106)
      al                                                                     Pro           Phe
      .         .         .         .         .         .         .         .G=207    .
450   TAAAAGTGAGTGCTTTCTGAATCTCTTACCAAAATCCGTTTATAACTTCCTTTGTACAGCTCCTGGTGGACCAAGTGCCCATTCTGGAGA   539
      euLys                                                       LeuLeuValAspGlnValProIleLeuGluT   (118)
                                                                                 Glu
      .         .         .         .   Def226=|-   -|  C=270   234=A 252=|-.-| 253=G .A=245
540   CCATGACCCCGCTAGACTTCATGGACTTCCGCAAGTACCTGGCACCCGCATCTGGTTTTCAGTCGCTGCAGTTCCGTTTGATCGAGAACA   629
      hrMetThrProLeuAspPheMetAspPheArgLysTyrLeuAlaProAlaSerGlyPheGlnSerLeuGlnPheArgLeuIleGluAsnL   (148)
                                              A------laPro          Thr            TrpAsn
```

(*continued*)

```
      -|=Def48a .      T=218        .         .         .         .         .         .
 630  AGCTGGGAGTTCTGACAGAGCAGCGGGTGAGATACAACCAGAAGTACTCGGATGTCTTTAGCGACGAGGAGGCGCGGAATTCGATTCGCA  719
      ysLeuGlyValLeuThrGluGlnArgValArgTyrAsnGlnLysTyrSerAspValPheSerAspGluGluAlaArgAsnSerIleArgA  (178)
                       End

      .         .         .         .         .         .        T=223     .         .
 720  ACTCGGAGAAAGATCCCTCGCTACTGGAGCTAGTGCAGCGATGGCTGGAGAGGACGCCCGGACTGGAGGAGAGTGGCTTCAACTTCTGGG  809
      snSerGluLysAspProSerLeuLeuGluLeuValGlnArgTrpLeuGluArgThrProGlyLeuGluGluSerGlyPheAsnPheTrpA  (208)
                                                                 End

      .         .         .         .         .         .         .         .         .
 810  CCAAGTTTCAGGAGAGCGTCGATCGATTCCTGGAGGCGCAGGTACAGAGCGCCATGGAGGAGCCCGTGGAGAAGGCGAAAAACTACCGCC  899
      laLysPheGlnGluSerValAspArgPheLeuGluAlaGlnValGlnSerAlaMetGluGluProValGluLysAlaLysAsnTyrArgL  (238)

      .         .         .         .    201=A=214 A=219.         .   |-    .Def210 -|.
 900  TCATGGACATTGAGAAGCGACGCGAGGTGTATCGCTCCATCTTTGATCCGGCAGTGCACGATGCACTGGTGCGTCGTGGGGATCGCCGGT  989
      euMetAspIleGluLysArgArgGluValTyrArgSerIlePheAspProAlaValHisAspAlaLeuValArgArgGlyAspArgArgP  (268)
                                            Asn   Gly
      .         .         .         A=244     .         .         .         .         .
 990  TTAGCCATCGTGCCCTTCAGGGAGCCATCATGATCACCTTCTATAGGGATGAACCCAGGTTCAGCCAACCACACCAGTTGCTCACCCTGC  1079
      heSerHisArgAlaLeuGlnGlyAlaIleMetIleThrPheTyrArgAspGluProArgPheSerGlnProHisGlnLeuLeuThrLeuL  (298)
                                   Lys
      .         . T=227   .         .   T=266 .      ||=36f        .         .         .
1080  TCATGGACATCGACTCGTTAATAACCAAGTGGAGATGTAAGTATTGCATTCTTTGATACTCTTTTATAAATATATCTTATGTTTAAGACT  1169
      euMetAspIleAspSerLeuIleThrLysTrpArgT                                                        (310)
                 Val                    SerE
      .         .         .         .   250=A . T=246   .         .     T   .         .
1170  GGTTTTCCTAACCAAATACTTTCTATTCCCGCCGCAGACAATCACGTGATCATGGTGCAACGCATGATTGGATCCCAACAGTTGGGCACT  1259
                                           yrAsnHisValIleMetValGlnArgMetIleGlySerGlnGlnLeuGlyThr  (327)
                                           nd  Leu                     Phe
      .         |-Def237  .         .   -|    .         .         .         .         .
1260  GGTGGCTCGTCTGGATATCAATATCTGCGCTCCACTCTCAGGTGATCATCGCAGATGTGATTATATCGGGGATCAATGAACTCAAACTGT  1349
      GlyGlySerSerGlyTyrGlnTyrLeuArgSerThrLeuSe                                                   (341)
      .         .         . 242=AACAN  A=267 .         .         .         .         .     ||=Def281
1350  TCTCCCTTTGTTTTTTTTTGGTTTCAGTGATCGGTACAAGGTGTTTCTGGATCTGTTCAATCTGTCCACTTTTCTGATTCCCCGCGAGGC  1439
                                 rAspArgTyrLysValPheLeuAspLeuPheAsnLeuSerThrPheLeuIleProArgGluAl  (362)
      .         .    ||=H2a             .         .         .         .         .         .
1440  GATTCCACCGCTGGACGAGACCATTCGCAAGAAACTGATCAACAAAAGTGTCTGACAATCGGCAGGGTATCCAATTGGTCAATGTTTGGC  1529
      aIleProProLeuAspGluThrIleArgLysLysLeuIleAsnLysSerValEnd                                     (379)
      .         .         .         .         .         .         .         .         .
1530  TATGCGTTGTTTGTTCTGCCTACTGTTTTGTCGTTTTGGTGTAATAAAATTACTTGTTTAGTCTTTGTTATCACATTTGATGTGTTCCTT  1619
                                               ------                          |(A)n
      .         .         .         .         .         .         .         .         .
1620  TTCTTTATGTCTGACATATAATACATATAACATAACAAAATAAATATTCATATTTCAGACATAAACAAATTCTATGGGAATGTGTGAGTC  1709
      .         .         .         .         .         .         .         .         .
1710  AGCAGCCTGAAAGTAGACCATATATATTCTGGTTGTCTTTCTCGCTCGTTTCTATTAGTTCGTTAGCAAATTAAATTCCATAATATTGTG  1799
      .         .         .         .         .         .         .      ------HindIII
1800  TGGCAATACTTGTCAAAATAATAATGGTATAAGTGAATTTTAATTACAAAATACCGATTTAAACAAAAAGCTTG  1873
```

v SEQUENCE. Accession, M34147 (DROVERM). Mutations v^1, v^2, v^k, v^{H2a} and v^{48a} are indicated as well as a mutation produced *in vitro*, v^{kLTR}. v^{kLTR}, in which Met-32 and His-55 are replaced, codes for an inactive enzyme (*v* Sequence) (R. A. Fridell and L. L. Searles, personal communication). Allele v^{48a}, in which 50 amino acids are missing, accumulates normal levels of RNA, but produces no detectable enzymatic activity. The mutation v^{36f} is caused by insertion of a the transposable element *B104* and leads to a null phenotype. The mutation v^{H2a} is caused by insertion of a *P* element (Searles et al. 1990). Numerous mutations of *v* sequenced by Nivard et al. (1992), and designated by numbers between 200 and 299, are also shown.

Gene Organization and Expression

Open reading frame, 379 amino acids; expected mRNA length, 1,306 bases, in agreement with an RNA of 1.4 kb detected by northern analysis. Primer extension and S1 mapping were used to define the major 5′ end. The two longest cDNA clones identified extend 60–80 bp upstream of the major 5′ end; these may represent minor transcription initiation sites. There are no correctly positioned TATA boxes. The 3′ end was obtained from a cDNA sequence that included a poly(A) tail. There are 5 introns: in the Gly-9 and Ala-63 codons, after the Lys-107 codon and in the Tyr-310 and Ser-341 codons (*v* Sequence) (Searles et al. 1990).

Mutations v^1, v^2 and v^k are all the result of insertion of the transposable element *412* in the leader region. Homozygotes for these mutations accumulate trace amounts of a *v* RNA of almost normal size. This apparently functional RNA (its coding region is unaltered) is produced because transcription from the *v* promoter is normal, and because rare splicing events using cryptic splice sites near the ends of *412* remove most of the *412* sequences from the *v* transcript. Mutations in *suppressor of sable* (*su*(*s*)) lead to increased accumulation of these spliced RNAs and thus to suppression of the mutant phenotype (Fridell et al. 1990; Pret and Searles 1991). A similar mechanism of suppression is found in some *y* mutations.

Developmental Pattern

v RNA begins to accumulate in 12–24 h embryos, it remains at a constant level between the first larval instar and the beginning of the third larval instar, becomes very low during the pupal stages and rises again in adults. Using a chimeric *v-lacZ* construction that included 1.1 kb upstream of the transcription initiation site and the *v* leader, it was determined that larval expression is restricted to the fat body (Fridell and Searles 1992).

Promoter

Analysis of deletions of upstream and leader segments showed that sequences upstream of the 5′ end plus a segment of the leader are necessary and sufficient for normal expression in transgenic animals. The upstream elements are located in the intervals −550 to −350 and −210 to −110 and the leader element is between positions −38 and −12 (Fridell and Searles 1992).

References

Fridell, R. A., Pret, A-M. and Searles, L. L. (1990). A retrotransposon 412 insertion within an exon of the *Drosophila melanogaster vermilion* gene is spliced from the precursor RNA. *Genes Dev.* **4**:559–566.

Fridell, Y-W. C. and Searles, L. L. (1992). *In vivo* transcriptional analysis of the TATA-less promoter of the *Drosophila melanogaster vermilion* gene. *Mol. Cell. Biol.* **12**:4571–4577.

Nivard, M. J. M., Pastink, A. and Vogel, E. W. (1992). Molecular analysis of mutations induced in the *vermilion* gene of *Drosophila melanogaster* by methyl methanesulfonate. *Genetics* **131**:673–682.

Phillips, J. P. and Forrest, H. S. (1980). Ommochromes and pteridines. In *The Genetics and Biology of Drosophila 2d*, eds M. Ashburner and T. R. W. Wright (New York: Academic Press), pp. 541–623.

Pret, A-M. and Searles, L. L. (1991). Splicing of retrotransposon insertions from transcripts of the *Drosophila melanogaster vermilion* gene in a revertant. *Genetics* **129**:1137–1145.

Searles, L. L., Ruth, R. S., Pret, A-M., Fridell, R. A. and Ali, A. J. (1990). Structure and transcription of the *Drosophila melanogaster* vermilion gene and several mutant alleles. *Mol. Cell. Biol.* **10**:1423–1431.

31

Vitelline Membrane Protein Genes: *Vm26Aa, Vm26Ab, Vm32E, Vm34C, Fcp3C*

Chromosomal Location:			**Map Position:**
Vm26Aa, Vm26Ab	2L,	26A	2-[20]
Vm32E	2L,	32E	2-[44]
Vm34C	2L,	34C	2-[47]
Fcp3C	X,	3C	1-[3]

Products

The vitelline membrane is made up of 6–10 proteins that range in size from 10 to 100 kD; these proteins are secreted by the follicle cells that surround the developing oocyte.

Structure

The complete sequences of four genes for vitelline membrane proteins (*Vm26Aa, Vm26Ab, Vm32E* and *Vm34C*) are available: all of these genes are in the left arm of the second chromosome. The predicted amino-acid sequences for the four proteins include a common 38-amino-acid segment: within this segment, the sequences of *Vm26Aa* and *Vm34C* are identical to each other and to the consensus sequence; the *Vm26Ab* sequence differs from the consensus in 10% of the positions and the *Dm32E* sequence differs by 24%. Outside of this region, the protein sequences are quite different, but putative signal peptides have been identified. *Vm26Ab* has 6–7 repeats of the octapeptide Tyr-Ser-Ala-Pro-Ala-Ala-Pro-Ala, a sequence that occurs only once in *Vm32E* and *Vm34C* (Fig. 31.1). These predicted sequences indicate the proteins are rich in Ala (10–27%) and Pro (9–16%) (Popodi et al. 1988; Scherer et al. 1988).

```
        1                                            50                                                100
Vm34C   MKCIAIVSTI CLLAAFVAAD KEDKMLGSSY G......... .........G GYGK.PAAA. .......PAP SYSAPAAASP GLRAPAAPSY AAAPV.....
Vm26A1  MKSFVCIALV AFAAAALASP TNVASATGST GSSVTTQDGE LEGVTGQGFG DLTRLRKSAY GGSSGGYGGS .......... .......... ..........
Vm26A2  MAFNFGHLLI AGLVALSAVS SETIQLQPTQ GILIPAPLAE NIRVSRAAYG GYGAAPAAPS YSAPAAPAAQ AYSAPAAPAY SAPAAPAYSA PAAPAYSAPA
Vm32E   MQI.VALTLV AFVAIA.... .......... .......... .......... .......... .......... .......... .......... .GASCPYAAP
CON     M-------L- A--AA--A-- ---------- G--------- ---------G ---------- ---------- ---------- ---------- --A-------

        101                                          150                                              197
Vm34C   .......... ......SIPA PPCPKNYLFS CQPNLAPVPC SAPAPSYGSA GAYSQYAPVY APQPIQW*.. .......... .......... .......
Vm26A1  .......... ......SIPA PPCPKNYLFS CQPNLAPVPC SAPAPSYGSA GAYSSPVATY VAPNYGVPQH QQQLYSAYVP QTYGYQY*.. .......
Vm26A2  APAYSAPAAP AYSAPASIPS PPCPKNYLFS CQPSLQPVPL SAPAQSYGSA GAYSQYVPQY AVPFVREL*. .......... .......... .......
Vm32E   APAYSAPAA. ....SSGYPA PPCPTNYLFS CQPNLAPAPC AQEAPAYGSA GAYTEQVPTT WTSPNREQLQ QFHQRIGMAA LMEELRGLGQ GIQGQQY*
CON     ---------- ------SIPA PPCPKNYLFS CQPNLAPVPC SAPAPSYGSA GAYS--VP-Y ---------- ---------- ---------- -------
```

FIG. 31.1. Amino-acid sequence comparison of four vitelline membrane proteins. Gaps were introduced to highlight sequence features present in more than one protein. The CON(sensus) line indicates positions at which three of the four sequences agree.

Tissue Distribution

Synthesis takes place during egg-chamber stages 8–11, i.e., immediately before the synthesis of the chorion proteins that will form the outer eggshell (Petri et al. 1976; Fargnoli and Waring 1982; Mindrinos et al. 1985).

This chapter describes genes that are expressed exclusively in follicle cells at the time of vitelline membrane synthesis and, in addition to vitelline membrane proteins, includes the gene *Follicle cell protein at 3C*.

Follicle Cell Gene Cluster at 26A

Organization and Expression of the Cluster

The cluster consists of four transcriptional units (TU) contained in a little over 7 kb of DNA (Fig. 31.2). TU2 and TU4 (*Vm26Aa* and *Vm26Ab*, respectively) have been sequenced. Their *in vitro* translation products comigrate with identified vitelline membrane proteins. The other two transcription units are expressed at much lower levels: TU1 produces a 1.3-kb transcript; TU3 produces a 0.7-kb transcript which may be translated *in vitro* into a 20-kD protein. All four genes in the cluster are expressed exclusively in the follicle cells of egg chambers during the period of vitelline membrane deposition (Popodi et al. 1988).

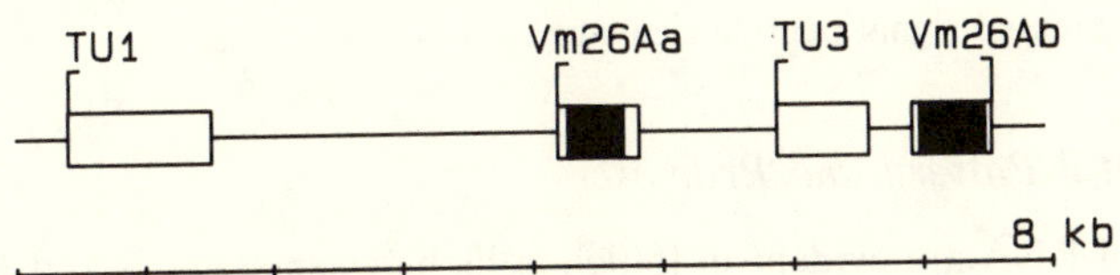

FIG. 31.2. Follicle cell gene cluster at 26A.

Vm26Aa

Product

Vitelline membrane protein Sv17.5.

Gene Organization and Expression

Open reading frame, 141 amino acids; expected mRNA length, 629 bases, in agreement with the results of northern analysis. S1 mapping and sequence features were used to define the 5′ end. The 3′ end was obtained from a cDNA sequence. There are no introns (*Vm26Aa* Sequence) (Burke et al. 1987).

Vm26Aa

```
            .         .         .         .    -->-80 .         .         .         .
-122   GGAGAGCTATAAAAGATGGGAGGCCAATTGAATGGTATTGGCATCAGTCACCTTTGGTAACTACCAGCAGCCCAACCAGCTCCCATCCGC   -33
            ------

            .         .         .         .         .         .         .         .
 -32   CTCCAGCTCAATCTTCAACCACCAACAACCAAGATGAAATCCTTCGTGTGCATCGCTCTGGTCGCCTTCGCCGCCGCCGCTCTGGCTTCG   57
                                        MetLysSerPheValCysIleAlaLeuValAlaPheAlaAlaAlaAlaLeuAlaSer   (19)
            .         .         .         .         .         .         .         .
  58   CCCACCAACGTGGCTTCGGCCACCGGCTCCACTGGCTCCTCGGTGACCACCCAGGACGGAGAGCTGGAGGGAGTGACCGGACAGGGATTC   147
       ProThrAsnValAlaSerAlaThrGlySerThrGlySerSerValThrThrGlnAspGlyGluLeuGluGlyValThrGlyGlnGlyPhe   (49)
                 |     |

            .         .         .         .         .         .         .         .
 148   GGTGACCTGACCCGTCTCCGTAAGTCTGCCTACGGCGGCAGCTCCGGCGGCTATGGCGGCTCCAGCATCCCAGCTCCTCCCTGCCCCAAG   237
       GlyAspLeuThrArgLeuArgLysSerAlaTyrGlyGlySerSerGlyGlyTyrGlyGlySerSerIleProAlaProProCysProLys   (79)
            .         .         .         .         .         .         .         .
 238   AACTACCTGTTCAGCTGCCAGCCCAACCTTGCCCCCGTGCCATGCAGCGCTCCAGCTCCCAGCTACGGATCCGCCGGCGCCTACTCCTCC   327
       AsnTyrLeuPheSerCysGlnProAsnLeuAlaProValProCysSerAlaProAlaProSerTyrGlySerAlaGlyAlaTyrSerSer   (109)
            .         .         .         .         .         .         .         .
 328   CCGGTGGCCACCTACGTCGCCCCCAACTACGGCGTGCCCCAGCACCAGCAGCAGCTGTACAGCGCCTACGTGCCCCAGACCTATGGCTAC   417
       ProValAlaThrTyrValAlaProAsnTyrGlyValProGlnHisGlnGlnGlnLeuTyrSerAlaTyrValProGlnThrTyrGlyTyr   (139)
            .         .         .         .         .         .         .         .
 418   CAGTACTAAGCACCTGCTCCGACTGCGACTCGATCATCGCCCAAGGACCACGAACCGACTGCCGAGAAACATAAGCTTTGATGGATTTGA   507
       GlnTyrEnd                                                                                    (141)
            .         .         .         .
 508   CAAAAAATATACCCAAAAATATGTACTGCAATTAAATCACT   548
                                               |(A)n
```

Vm26Aa SEQUENCE. Accession, M18280 (DROVITA). The vertical bars at Val-23 and Ser-25 mark potential signal peptide cleavage sites.

Developmental Pattern and Promoter

High levels of RNA are evident in follicle cells between stages 8 and 11 (Burke et al. 1987). A 170-bp segment upstream of the site of transcription initiation controls developmental specificity (Jin and Petri, personal communication).

Vm26Ab

Product

Vitelline membrane protein Sv23 (Popodi et al. 1988). The female sterile mutation *fs(2)QJ42* is rescued by transformation with *Vm26Ab* DNA (Savant and Waring 1989).

Gene Organization and Expression

Open reading frame, 168 amino acids; expected mRNA length, ca. 625 bases, in agreement with the results of northern analysis. Primer extension was used

Vm26Ab

```
                                                                        -207 GTCGACTGGCGGTTGCAGGTG   -187

-186  GTCAGCAGATTTCGAGCCGGGGTGCTTCCATTTGCATTTTTTTCGGAACGCTGTCGTTCTACTCCGTCAGTGCGATCAGCGTTTTCCGAG   -97
                                   -----
                                          -->-61
 -96  TGGGCTATAAAGTGGATTGGCTGGGAGGCTACAATCAACAGTCAGCCTTCGTTCGTCACTTCAGCAGCAAGTAGAGACAGCTCAAGAACC   -7
       ------

  -6  ATCCGCAATGGCATTCAACTTTGGTCACCTCCTCATCGCCGGCCTCGTGGCCTTGTCCGCCGTGTCCTCGGAGACCATCCAGCTGCAGCC   83
             MetAlaPheAsnPheGlyHisLeuLeuIleAlaGlyLeuValAlaLeuSerAlaValSerSerGluThrIleGlnLeuGlnPr   (28)
                                                                              |

  84  CACTCAGGGCATCCTCATCCCCGCCCCGCTGGCCGAGAACATCCGTGTGTCGCGTGCCGCCTACGGAGGATACGGCGCTGCCCCAGCCGC   173
      oThrGlnGlyIleLeuIleProAlaProLeuAlaGluAsnIleArgValSerArgAlaAlaTyrGlyGlyTyrGlyAlaAlaProAlaAl   (58)

 174  CCCATCGTACTCCGCCCCAGCCGCTCCCGCTGCCCAGGCCTACTCTGCTCCCGCTGCCCCAGCCTACTCCGCACCCGCTGCTCCCGCCTA   263
      aProSerTyrSerAlaProAlaAlaProAlaAlaGlnAlaTyrSerAlaProAlaAlaProAlaTyrSerAlaProAlaAlaProAlaTy   (88)

 264  CTCCGCACCCGCTGCTCCTGCCTACTCTGCTCCCGCTGCCCCAGCTTACTCTGCCCCAGCCGCACCAGCTTACTCCGCACCCGCCTCCAT   353
      rSerAlaProAlaAlaProAlaTyrSerAlaProAlaAlaProAlaTyrSerAlaProAlaAlaProAlaTyrSerAlaProAlaSerIl   (118)
                                                                                               __
 354  TCCGTCGCCGCCGTGCCCCAAGAACTACCTGTTCAGCTGCCAGCCCTCCCTGCAGCCCGTGCCCCTGTCCGCCCCAGCTCAGTCCTACGG   443
      eProSerProProCysProLysAsnTyrLeuPheSerCysGlnProSerLeuGlnProValProLeuSerAlaProAlaGlnSerTyrGl   (148)
BamHI ____
 444  ATCCGCCGGTGCCTACTCCCAGTACGTGCCCCAGTACGCCGTGCCCTTCGTCCGCGAACTTTAAGGATCGAACCGAATCTGACTTGACAT   533
      ySerAlaGlyAlaTyrSerGlnTyrValProGlnTyrAlaValProPheValArgGluLeuEnd                             (168)

 534  CTGAACCTAAGAATAAAGTAATGCTTTCATAAAA   567
                ------          |(A)n
```

Vm26Ab SEQUENCE. −96 to 567, from Popodi et al. (1988); the segment from −207 to −97 was kindly supplied by Gail L. Waring. The vertical bar at Thr-23 marks a potential signal peptide cleavage site.

to define the 5′ end, and S1 mapping gave the approximate position of the 3′ end. There are no introns (*Vm26Ab* Sequence) (Popodi et al. 1988).

Developmental Pattern and Promoter

High levels of RNA are present in follicle cells of stage 8–10 egg chambers. *Vm26Ab* RNA is approximately half as abundant as *Vm26Aa* RNA, but it is 20–40 times more abundant than TU1 or TU3 transcripts (Popodi et al. 1988). One hundred and forty-seven bp upstream of the transcription initiation site seem sufficient for correct gene expression (Savant and Waring 1989).

Vm32E

Product

Vitelline membrane protein of approximately 12 kD (Gigliotti et al. 1989).

Vm32E

```
                                                         . -->-28 .      .
-94  AAAAGTGCCGAGTTTTGTTATTAAAGTCAACGCATGAATGCTATAAGAATGCCACCATTGGTCACTAAATCGACAGTGTAAATCATTAGT   -5
                                              -----

-4   TCATCATGCAGATCGTTGCTCTCACCCTCGTTGCGTTTGTGGCCATTGCCGGTGCCTCCTGCCCGTATGCAGCTCCAGCTCCAGCTTATT   85
         MetGlnIleValAlaLeuThrLeuValAlaPheValAlaIleAlaGlyAlaSerCysProTyrAlaAlaProAlaProAlaTyrS   (29)

86   CAGCGCCCGCTGCTTCTTCTGGCTACCCGGCTCCACCATGCCCCACCAACTACCTGTTCAGCTGCCAGCCCAATTTGGCCCCAGCTCCTT  175
     erAlaProAlaAlaSerSerGlyTyrProAlaProProCysProThrAsnTyrLeuPheSerCysGlnProAsnLeuAlaProAlaProC   (59)

176  GTGCCCAGGAGGCCCCAGCCTATGGATCCGCCGGCGCCTACACAGAACAGGTGCCCACTACGTGGACAAGTCCCAACCGAGAGCAGTTGC  265
     ysAlaGlnGluAlaProAlaTyrGlySerAlaGlyAlaTyrThrGluGlnValProThrThrTrpThrSerProAsnArgGluGlnLeuG   (89)

266  AGCAATTTCACCAGCGCATTGGAATGGCGGCTTTGATGGAGGAACTGCGCGGCTTGGGCCAAGGAATCCAGGGTCAACAGTACTAGTGGC  355
     lnGlnPheHisGlnArgIleGlyMetAlaAlaLeuMetGluGluLeuArgGlyLeuGlyGlnGlyIleGlnGlyGlnGlnTyrEnd      (116)

356  AAAAAAAATTCATGTGAAGAATGTTTTCGAATTAAATCCGTCTATGCTTTAATTGGACTTTATACTATGGAACAAAAAAAAAATTGGGAT  445
                         ------                   |(A)n

446  TGGAGATAAGGAAAACTGGTAAAAAAAATAGGAGTTAAACTTATTTTGTTGTTTTGTGCCTCTGGCCTCCGATTCCTTTCGAAAGCCATA  535

536  AAGAACATTGTCCGTCTGTATTTATATATTCTAAC  570
```

Vm32E SEQUENCE. Strain, *Oregon R*. Accession, M27647 (DROVMP).

Gene Organization and Expression

Open reading frame, 116 amino acids; expected mRNA length, 434 bases, in agreement with a 0.46-kb RNA detected by northern analysis. Primer extension and S1 mapping were used to define the 5′ end. The 3′ end was obtained from the sequence of several cDNA clones. There are no introns (*Vm32E* Sequence) (Gigliotti et al. 1989).

Developmental Pattern

Transcription seems to be restricted to follicle cells in stage 10 egg chambers (Gigliotti et al. 1989).

Vm34C

Product

Vitelline membrane protein of approximately 10–11 kD (Mindrinos et al. 1985).

Gene Organization and Expression

Open reading frame, 119 amino acids. Northern analysis revealed an RNA of approximately 0.6 kb. Primer extension was used to define the 5′ end.

Vm34C

```
-523      GTTGCTAGGCAAAACTATAAACGAATATTTTTCCAATGACCGCATATTCGGCACGCGATTACAAATTCTTGTGGAAAATTAAG   -440

-439 CTCATTGAACTAAATAAATATTTTAGATATAAATAATTTATACACATATAATATTTATTTTAATACATTTATTCCAATTGTTCAGTAAAA   -350

-349 TAATGTAGCTCAATGCAAAGCTAAGTACATTCAATTCTTGGTGCTTCAACAATTTTTAGTTCCGTTACTTCATTAATTTACATTTTTGGC   -260

-259 ATGCGACAAATTGTTTACTCAACAAGTTCAGTGGCCCAAAAAAAGTAGAGGAAATGTTTGTTCTTTTCACTTTCTGTTGGCCGTGCAAAA   -170

-169 AAAGCGCCACTCACGTCGACTTCGAGGGGTCGTTGGGTAAACTGAAAACTTGGTCAGTGCTTGCATCTGCACTTTTGATGGCATTGCATC   -80
                                      -->
-79  GGGTATATAAACCTCAAGTGTCGAAGCCAGAAGCATCGCAGTCTGCTACCAACAGTCTAAGAAATCATCAACCAATCAACATGAAGTGCA   10
       ------                                                                            MetLysCysI   (4)

11   TCGCCATCGTCTCCACCATCTGCCTGCTGGCCGCTTTCGTTGCCGCCGATAAGGAGGATAAGATGCTCGGCTCCTCCTACGGTGGTGGCT   100
     leAlaIleValSerThrIleCysLeuLeuAlaAlaPheValAlaAlaAspLysGluAspLysMetLeuGlySerSerTyrGlyGlyGlyT   (34)
                                                     |

101  ACGGCAAGCCCGCCGCTGCTCCGGCTCCATCCTACTCCGCTCCGGCTGCCGCTTCCCCAGGCCTACGCGCCCCAGCTGCTCCATCCTACG   190
     yrGlyLysProAlaAlaAlaProAlaProSerTyrSerAlaProAlaAlaAlaSerProGlyLeuArgAlaProAlaAlaProSerTyrA   (64)

191  CCGCCGCTCCGGTCTCGATCCCGGCTCCTCCTTGCCCCAAGAACTACCTGTTCAGCTGCCAGCCCAACCTGGCCCCAGTGCCATGCAGCG   280
     laAlaAlaProValSerIleProAlaProProCysProLysAsnTyrLeuPheSerCysGlnProAsnLeuAlaProValProCysSerA   (94)

                  ______BamHI
281  CCCCAGCTCCCAGCTATGGATCCGCCGGTGCCTACTCGCAGTACGCCCCCGTCTACGCTCCTCAGCCCATCCAGTGGTAGGATGATCCAC   370
     laProAlaProSerTyrGlySerAlaGlyAlaTyrSerGlnTyrAlaProValTyrAlaProGlnProIleGlnTrpEnd              (119)

371  AGACTTCACTAACCCCTGATCAACGACAAAAGCAATGCAATAAAAAAATAAAAGAAAAATATTTATGTTTAATCATAAAAATTCATATGT   460
                                             ------

461  TTCAATTTGGGGGATAATAGCGTGCCTAATAGCTGAACTAAAAACATTAATAATTAATTGATCGAAGCTCGTCGTTATTCAAAGATTTTG   550

551  AAAAAAATTATTGTTTTATTGATTCATACTTAAATTCATAATTTTTAGAAATTTAACAACTTTTTAGATAATTCTGGTAAGTTCCTCTTT   640

641  AATTGTCGAC  650
```

Vm34C SEQUENCE. Kindly supplied by W. H. Petri and L. J. Sherer and from Mindrinos et al. (1985). The vertical bar at Ala-19 marks a potential signal peptide cleavage site. Also indicated are a *Bam*HI site present near the 3′ end of all *Vm* genes and a potential poly(A) signal.

The 3′ end was not determined. There are no introns (*Vm34C* Sequence) (W. H. Petri and L. J. Scherer, personal communication; Mindrinos et al. 1985).

Developmental Pattern

High levels of RNA are present in follicle cells of stage 8–10 egg chambers (Mindrinos et al. 1985).

Fcp3C
(*Follicle cell protein at 3C*)

Product

Unknown. The predicted amino-acid composition is relatively rich in Ser and Thr (11% each). The sequence shows no obvious similarity to other proteins (Burke et al. 1987).

Gene Organization and Expression

Open reading frame, 217 amino acids; expected mRNA length, 770/786 bases: two sites, 16 bp apart, were indicated to be the likely position of the 5′ end by S1

Fcp3C

```
-211 AAAAGTAATATTAGCTAAAGAACACATTTCATATCGTATATATTTCATATATCAGGCGCCTTTAAAAATTCCCTGCTGCTGCTGACACTC   -122
                                                                            ------
                !-111      !-95
-121 TCTGCTAGCCATCCATTTGGAGAGCCATCCAGATAGTCTACAAGAAGCCGCTCTATGGCAATAGCAACATCATCAAGGACAAGCGTATAA   -32

 -31 AGACGAAGCCCGTCAAACTGGAAACCAGCACCATGAGCAGCACTGGTGTTGCAAGTAGCAGCACAACAGCCGAAGAGGATTGGCCCACGG   58
                                     MetSerSerThrGlyValAlaSerSerSerThrThrAlaGluGluAspTrpProThrA   (20)

  59 CCGTTGAGTTTGTGATTATGACAACGCCCGCAAGCGAATTGGAAGCCAGCACGGAAACCATTGGTAACAATGGCACCACCGAAACGACCG   148
     laValGluPheValIleMetThrThrProAlaSerGluLeuGluAlaSerThrGluThrIleGlyAsnAsnGlyThrThrGluThrThrV   (50)

 149 TTGGCGAGGCACCCATCATCGGATCGTCGGAAGGATCCACACGATCGATGGAGCCAACCACCGCGAGTCCGCTGATGAGCACAAACCCAT   238
     alGlyGluAlaProIleIleGlySerSerGluGlySerThrArgSerMetGluProThrThrAlaSerProLeuMetSerThrAsnProS   (80)

 239 CGAGCAGCAGCAGTCTGGTTAGCACCATTCCCTTGCCACCGACAGCGGGACTACATGCGCAGGATAATCAGCCAGTGCCGTGCACATGCG   328
     erSerSerSerSerLeuValSerThrIleProLeuProProThrAlaGlyLeuHisAlaGlnAspAsnGlnProValProCysThrCysG   (110)

 329 GCGTCTTCCTCTCCTCGCAAATCCCAAATGGCTTGCCGACAAAGCCACTTATCCACCAGGAATTGGATCATATGTTTCCCTGCAATGCCA   418
     lyValPheLeuSerSerGlnIleProAsnGlyLeuProThrLysProLeuIleHisGlnGluLeuAspHisMetPheProCysAsnAlaI   (140)

 419 TCGGTCGCAAGCAGTGTCAAACCAAATGCCTAGAGACGGTGAGTACTGGGGAAACGAGGAGGAAAACATCAGGAGAAGCGCTCTATAACT   508
     leGlyArgLysGlnCysGlnThrLysCysLeuGluThr                                                       (152)

 509 CACCAATTTCGTCCATTTTAGATCGTACAACATCTGCCGAATTCCGCAAATATAGTATGCTCCGCACTGGGTCACGATTGTCACAAGGAA   598
                         IleValGlnHisLeuProAsnSerAlaAsnIleValCysSerAlaLeuGlyHisAspCysHisLysGlu   (182)

 599 CGGGCCTATTTGTTCATCAAGAACTGTCACAATCAATGGGTTAATACCAATCTGCAGGCGGGCAGGGAGTACTGTTGTCGCCTCGGCTTC   688
     ArgAlaTyrLeuPheIleLysAsnCysHisAsnGlnTrpValAsnThrAsnLeuGlnAlaGlyArgGluTyrCysCysArgLeuGlyPhe   (212)

 689 CCTACCGTTGCCCATTGATGGTTAAGCACTGTGCAAATGAAATAAATATATGATATGAC  747
     ProThrValAlaHisEnd                ------           |(A)n                                      (217)
```

Fcp3C SEQUENCE. Accession, M18281 (DROVITB).

mapping and cDNA sequencing. The 3′ end was obtained from a cDNA sequence. There is one intron after the Thr-152 codon (*Fcp3C* Sequence) (Burke et al. 1987).

Developmental Pattern

Transcription occurs during vitellogenesis and is restricted to the follicle cells. RNA is first detectable in stage 9 egg chambers, it reaches a maximum during stages 10 and 11, and it is absent from stage 12 chambers (Burke et al. 1987).

References

Burke, T., Waring, G. L., Popodi, E. and Minoo, P. (1987). Characterization and sequence of follicle cell genes selectively expressed during vitelline membrane formation in *Drosophila Dev. Biol.* **124**:441–450.

Fargnoli, J. and Waring, G. L. (1982). Identification of vitelline membrane proteins in *Drosophila melanogaster. Dev. Biol.* **92**:306–314.

Gigliotti, S., Graziani, F., De Ponti, L., Rafti, F., Manzi, A., Lavorgna, G., Gargiulo, G. and Malva, C. (1989). Sex-, tissue-, and stage-specific expression of a vitelline membrane protein gene from region 32 of the second chromosome of *Drosophila melanogaster. Dev. Genet.* **10**:33–41.

Mindrinos, M. N., Scherer, L. J., Garcini, F. J., Kwan, H., Jacobs, K. A. and Petri, W. H. (1985). Isolation and chromosomal location of putative vitelline membrane genes in *Drosophila melanogaster. EMBO J.* **4**:147–153.

Petri, W. H., Wyman, A. R. and Kafatos, F. C. (1976). Specific protein synthesis in cellular differentiation. III. The eggshell proteins of *Drosophila melanogaster* and their program of synthesis. *Dev. Biol.* **49**:185–199.

Popodi, E., Minoo, P., Burke, T. and Waring, G. L. (1988). Organization and expression of a second chromosome follicle cell gene cluster in *Drosophila. Dev. Biol.* **127**:248–256.

Savant, S. S. and Waring, G. L. (1989). Molecular analysis and rescue of a vitelline membrane mutant in *Drosophila. Dev. Biol.* **135**:43–52.

Scherer, L. J., Harris, D. H. and Petri, W. H. (1988). *Drosophila* vitelline membrane genes contain a 114 base pair region of highly conserved coding sequence. *Dev. Biol.* **130**:786–788.

32

yellow: *y*

Chromosomal Location:
X, 1B1

Map Position:
1-0.0

Product

Unknown. It plays a role in the accumulation or deposition of melanins in larval and adult cuticles.

Structure

Several features of the *y* product are suggested by the predicted amino-acid sequence (*y* Sequence). A signal peptide-like segment is an indication that the protein is either secreted or incorporated into membrane. Two potential N-glycosylation sites (Asn-X-Thr/Ser) are present, occurring at Asn-144 and Asn-215. The widespread occurrence of Pro and Gly residues suggests that extensive regions of α-helix or β-pleated sheet do not occur (Geyer et al. 1986).

Mutant Phenotypes

Mutations are classified into two groups. Type 1 alleles are probably amorphs; they show a uniform absence of melanin (yellow color) in all structures. Type 2 alleles show the mutant phenotype in some structures (body cuticles, wing blades) and the wild-type appearance in others (denticles, bristles, sex combs).

Gene Organization and Expression

Open reading frame, 541 amino acids; expected mRNA length, 1,985 bases. Primer extension and the sequences of two cDNA clones were used to define the 5′ end. The 3′ end was obtained from a cDNA sequence. There is an intron in the Gly-80 codon (*y* Sequence) (Geyer et al. 1986).

y

```
-3042  GTCGACTATTAAATGATTATCGCCCGATTACCACATTGAGTGGTTTAAAATAGCCATAAAATATGCAACTGACGATGGCTTAAGATAAAT  -2953
-2952  ACGTCGCAGAGTCACTCATAAATTTCGAACGCAGCCCGCTGATTTACCTACCCCTCTAAACGATTCATAGTATATGTACGAGTATATCCA  -2863
-2862  CTAAGCTTTTTCGAGCACTGATTTTTTCGCTTGCACGAGACAAGTGCACCACCGCAATTGCAGGCAAATTATGTCTGAGGTAATGATTCC  -2773
-2772  GTTTCGTGCAAGATTACACAGAAATCAAATTACGACAACCTTTATTCAGTAAGCAAACAAAGCCTTTGTTGGCATCTAATTATTCCACTT  -2683
-2682  ATGGTTGCGATTTCGGGAGCTACAATCGGTTTTGGTTTAGTATATCTAGCGAGTTCCTTGGCGACATTTAAAATTTACAAATAAAGTTTC  -2593
-2592  TCTATTCAATCGGAACAGTGGAAATTGACTATTTTATTTATATTAATGAACTTATTTTTAATTTGGCTTAAGTTACTAAGGGGTACTAAT  -2503
-2502  AGTTTGAGCGCAGTGCATGTCATGGGGACATGTGCAATTGTGTGTAAACGGGAAGTGATCGCGGCCTTCCGAATTTGGCCATGCCAAATA  -2413
-2412  ATCCCAGCTCGAAAGGAGGGGACCCGGCGGTCAGGGCCATGGACATTGAACTTGAAAAAAAAAAAACACAAAAATATATAACACAAAACG  -2323
-2322  GAAAATGCTGTGTACCGCTTATGTTAGAGAAGTTGAGCAACGGGTTTTTCGTTTTGCAGTCACGATGGATTTCCAAATTAGTGTAGGAGG  -2233
-2232  GGGGAGGGGAGGGAGGGAGATAATGTCCAGGCTGCCATAAGTGGGGAATAAGGAAAATAAAACATGAAACACGGGTCGGGCAATGTCATG  -2143
-2142  CGGTATTCGGCTTTGCTTTCCGCCCAAGTTGAAGTGATCCTGTGTGTAAATAATGTCGAATGTTGCCGGTCGGTTGCATAAGCGTTGGTC  -2053
-2052  AATTATGGCCAAAGAGATCTGATTTGTGGAAGCTTTTTTTGACCACTTAGCGCGCTCCGCTGATGTTGTTTTGTTTTGTGCTGGGGCAGA  -1963
-1962  AAACTTGTTTCAATTATTGGGAAAAGTGCGTATAAATCATTGCCGCAAGCTCTGAAAAGCGAAAAAGAAAAACAGTAACCAAACAGACAA  -1873
-1872  ACGCAGCATTCCCCCACACAATTAAGCAAAAACTTGAAACAAGTCAATTCGAAAAAAATTATAGGTTCAACGGCTGCAGCGATCGCATCA  -1783
-1782  TTAGTTGCGTTTTTAGTAAATACACCATTTCATTACACAACACACACAATTAATTAATAAAACTGTACTGGTTATTTCAAGTGTGTCTTT  -1693
-1692  TAATAAGCCTGCCGATCGCAATAAATTCGAGCAGCATTGCCGGTAATTTTGTGCAACATATTTTTCGATTGCCACACCGTGTTTGTTTAT  -1603
-1602  TTTTTCTGTGGGTGCAATGATTTAGAATGCGGGCAAGGGATCAAGTTGAACCACTTCTAAGAAAAAATAGACATTGCATAAATGATATAG  -1513
-1512  AGTCCAAAAACTACACCAATTCAATAGCAGTAATGGTTACATTAGCTTTGAAATTGTTTTTAGACATCCGAAGAAATAAGATTAAATTTA  -1423
-1422  AACGGCATTCTTTAATTTGTATTTTAATATTTTGAGAGGTTTTCCTTATTTAAAGTGTAGATTATTGAGGATTAATGCAATACCACTTTA  -1333
-1332  CCTGCGGAGGTCGTAAAACGTATTTTTACCCATTTGCATGTTTATTATGCGTGTCGCTGGTTGTTTACTTTACTTAAGTTTTGCAATTTT  -1243
-1242  TTCTTTAGCAAGCAGGTGCATTTGGGCCAAGAGATATATGCGATCGCTTTCGGTTCGAATTTTTAACATTTACTTGCGGCGATGGTCATT  -1153
-1152  AGAGCATTACCCACTTAGGGCACCCCCAACATCCAGTTGATTTTCAGGGACCACAATATTTTAAATAACAGCTAGTGGAATTACCTAAAA  -1063
-1062  GCGCTTTCGTTCCTTTTTGAAATTTTATGTAACACTCAATTATATTTATGTATATGTATGCTCAAAATCACCTGCCAATAACTAGCGGAA  -973
 -972  ACCAAATATTTGACCCTCAGTGAATTGTGAATCATCGGTGACGCCCAATCGAAATCCAATCCTAAGCAATTGAAACGAGCACGAGTTCCA  -883
                 ||gypsy=2
 -882  ATTTAATAGTATACAAGGAAACACCTGCTTTAAATACTCTACATAGTACACGTTATAATAACGATTTATTTGATATTTCTGGATTTTTGT  -793
 -792  CTGCATGTATTTCATATAATATTGATTTGATTTTTTTAATGAATTGAACTAAAAAATCATATTAGAACATTTTTTGCAGTCGCCGATAAA  -703
```

(continued)

```
-702  GATGAACACAGTTCTCAGAACACAACTGTCATGTATTAAGCTTTCAGATTTTCAGAAATTTGGAGAGCAATGCATTCTATGCACGAGCCT  -613

-612  CCTGGCCTTACAATTTACTTGTTTGAAATTAGATCGTCAAATAAAGTCCCTAAAATTAAATAAATAGTAGTCACAACTTTAAAATAGGTC  -523

-522  TTAATCTTTTAGGGTACCGAAAGGTATTTCGGCACAAATCAGCGCAGTTTTAAATGTCGATGAAGGCCAAAAATCATACCAAACCCAGCG  -433

-432  AAAGGTGATGTCTGACTCATTAAATTGGGGGATTCGAGTGTATTTATTAAACATGCGTGAAAATCAATCATGGAAGACAAAACGCAAAGT  -343

-342  TGGCCGATCTATGGGAACAGCATAAGCCACCTGATTACCCGAACACTGAACCACCCGAATCACTAAAACCACCGAAGTTGGCGCGCGCCT  -253

                                                                                      -->-170
-252  TCGTTTTCATTTTCATTGGCCTGTCTTCGTCTTCGGAGAAAAAAACCTTCATATAAAACGCGGCCGACATATTATGGCCACCAGTCGTTA  -163
                 ----                                   -----

                                                                           ||P=76d28
-162  CCGCGCCACGGTCCACAGAAGAGGATTAAAAAAATATCACACAGCCGAAGGCTAGAGAAGAACCCCCTATAGCTGAACATATATAAACAA  -73

 -72  ATATATTTTTTTTTATTGCCAACACACTTTGGCTTAAGTGTTAAGAGTGATTGTCAGCTTAGAGCTAAGTGCAATGTTCCAGGACAAAGG  17
                                                                                MetPheGlnAspLysGl  (6)

  18  GTGGATCCTTGTGACCCTGATCACCTTGGTGACGCCGTCTTGGGCTGCTTACAAACTTCAGGAGCGATATAGTTGGAGCCAGCTGGACTT  107
      yTrpIleLeuValThrLeuIleThrLeuValThrProSerTrpAlaAlaTyrLysLeuGlnGluArgTyrSerTrpSerGlnLeuAspPh  (36)

 108  TGCTTTCCCGAATACCCGACTAAAGGACCAAGCTCTGGCTAGTGGAGATTATATTCCGCAAAATGCTCTACCTGTTGGAGTCGAACACTT  197
      eAlaPheProAsnThrArgLeuLysAspGlnAlaLeuAlaSerGlyAspTyrIleProGlnAsnAlaLeuProValGlyValGluHisPh  (66)

 198  TGGCAATCGGTTATTCGTCACTGTTCCCCGCTGGCGTGATGGTAAGTGGAAGTTAAATATGAAGCCCTTGGGGAGATCGTAAATGGGACA  287
      eGlyAsnArgLeuPheValThrValProArgTrpArgAspG                                                  (80)

 288  TTCTTACTTAGGGCATCAGAGATATCTGATTGAGTGGTTGACAGTTTTATATGGCTTGTTTGACATGATGTAAAAACACAAAATTCATTT  377

 378  AGTTTAGGTATTCGAAATAAGAGCTTGTTATTTATTTTAGAATTTGGAGAACATTTTTTTGTCTTTCTACCCTTCTTAGAAAATAATATT  467

 468  GTTTTGTACAATTTAATTTTAAACTAGTACAGACGAAAAATGTATTTTTTTATTTGTATGCCTTTTCACCATTTTGGCAGAGGAATAAAT  557

 558  ATGACAATATATTTTGAGAGCACCCTCATGTAAAGGTTTTAGCGTGGCGACCTCTCATAAATCCGGTTGGTACCTGCGCGTTATTTTAAC  647

 648  ATTTTAAACAATTAACCGTTGTAAAATCGAAGCCAATAGCATGGCATTGGCTTTATACTGTATTAAATTGTATTATATTACCATCCGAAT  737

 738  TGTAAAGACTTCTTCAGGGCCGCCACATAGAAATGGAAATCCAATCACAAACAATAACTTATGGCATTAGCTATTAACCGACGATTAGCT  827

 828  GTCAGTTCAACAAATGTAAGAGTGGCGAAATGTTTAAATGCGAAGGCATTGTTCTGTGACTCACGTTTTATTATTAATCACACAAATGAT  917

 918  TTTGCTTCAAAATTATTTGGCTTACACAATAATCAAAATTTTTATGAAATAGTTGAAACACAAAACTAGGAAATTTTAAAAAGCAATGAA  1007

1008  ACTAAAAAACCCAATTGTTAAGATTATATGATGCGCATACAAATACTTCAGTACGTCTAGGAATGCTTTCGATGATTGATTAGTTTTTAT  1097

1098  GCATGGCTTACAATTGGTATTTACACAGAAAAACACGGCTGTATCGATTCAAAATGCGATGTTAATAAATTTTGTACATATGTTCTTAAG  1187

1188  CAGTCCGAAACACCCAAACTTCTGACTAAACTTAAAAAAACACGCTCTTCAAGGATATCTTAATGTCACTTATAATGTAATGGTATTGTA  1277

1278  TTATTAAGTATATCAAAATATTCTGGCCCAGCCTTGAGGTCTCTTTTTAAAAAGATATCGACTGACTACCTCCAGTCAATGAAATAATAG  1367

1368  CCCAGAAGGCCGAATCGGCAAAAAATAAACCCCAAGTTACGGCAAACAAAACATAGTGAAAGTTGTGGCAAAGTGGAACATTTAAAGGCA  1457
```

1458 TGCTTCAATGGCCATCGAAGCAAATCAATTAGTCAAAGCAAATCGGTAGTGGCAACAACAGGCTACAGAATACCTATAAGTGACAGTTAT 1547

1548 GGGGTATGATTAATTATAAATATTATCATTGACCACCAATGCTGGGCTCAATTGGAAAAACTATTCTATGAAGATTTGAGTAAATAAATT 1637

1638 TTGATTTAAAAAAGCCCATGGTTATCGCGACAACTAGCTACGGGACAAGATTACTGTTTAAAATCAAGTGTGAAATATCAAAATCAAAAT 1727

1728 CGGATTCCGATCGGGAAGTTGTATCCGATTCTGAAACTAAAACACAGAATTGCCAACATTTTCCGATATCGACTCAGCTCACGTATTTCA 1817

1818 TACAGATTCATTAGGCCACCAGCCATTGAATAATATACCCCAGTCAATTGAGCTACTCGATAGTTGATCAACTTAGCTTTTGTCAACGAG 1907

1908 TGAACGCATAAACTACTACATCAACGATATTTGCGGCCCATTCCAAGCTAAAAGTTCATCTTAATTACAAATAAGATTAGAAAAAATATC 1997

1998 TGAATGAAAAAAATGTTGAGACATATTTCTTTGGAAAAGGAGAACCTCAAGACAGTCGAAAAAATTGTTTACAATGAAAATGTTGAAAAT 2087

2088 CATGAAGCAGATAAATCTGTCAGTTGCGAGGTTTTAGGACTGAAAGAGCACATGTCAAAATATAAATTTGTTCAAATACTTTATATTTGA 2177

2178 CTGAATTAGATTGTTATTTTAAAAGTTATGAATTAAATAAAGATTGAAAGGTGCATTATGCTCAAATGTATATTTATCGCAACCCCCGGT 2267

2268 TACTTTGTAAAGCAAAAACGCCTGGTTTGATTTTTAAGAAGATGGGTCGGTAAATCGATAAAAGCTATATTTTCTGGTCGTTGCAGTCTC 2357

2358 ACTCGCCTGCTATAAAAACATTAAAAGTTCCCAGAAACAATAAATGTCTTTAAATTCAATTAACGAAGAAATAAAGAAGGAAAAGAACTG 2447

2448 GAGCGGAAATCGGTCGAAATACTGCCAATGGCCACATATACATTTAACAGCGATATATGGTATACATATTGATAATGATGTCAGACGCAA 2537

2538 TTGCTTCAGACGGCTAATGACATCGCAAATTGCACGCAACTTGCAATAGTGCCAATTATGACTGAAGTACATATAGCCGGGGATCTTTTA 2627

2628 ACAATAAACTTCCAGTAGATGTACAAGCAGAAAAAAGAGCCATTAGCACGGCAGTTACCATTGCTTATGATTCCTTGTGTCCAAAATAAT 2717

2718 GACAAATAGGTATATAAATAATTAAATGCCAAACATAAGCGATTCTAATTTACCTTTACATCTGTATGCATTTACATATTATCCAGAAAA 2807

2808 CAGACAGCGATAACTTGCAACATTGCTTAGTATAATAATCCAAAGAAGGAATTTAGGCAGAAATTCCAGTTAATTAAATATTCAAAACAA 2897

2898 ACTTTATTTAGTGCCTCAATAATAGTTTGGCCCTGCTAATTCTCCTATTTTATTTTTTAGGGATTCCGGCCACTCTGACCTATATAAACA 2987
lyIleProAlaThrLeuThrTyrIleAsnM (90)

2988 TGGACCGCAGTTTGACGGGTTCACCGGAGCTAATTCCGTATCCAGATTGGCGCTCAAATACAGCTGGAGATTGCGCCAACAGTATTACCA 3077
etAspArgSerLeuThrGlySerProGluLeuIleProTyrProAspTrpArgSerAsnThrAlaGlyAspCysAlaAsnSerIleThrT (120)

3078 CTGCCTACCGCATTAAAGTGGATGAGTGTGGTCGGCTGTGGGTTTTGGACACTGGAACCGTGGGCATCGGCAATACCACCACTAATCCGT 3167
hrAlaTyrArgIleLysValAspGluCysGlyArgLeuTrpValLeuAspThrGlyThrValGlyIleGlyAsnThrThrThrAsnProC (150)

3168 GCCCCTATGCGGTAAATGTCTTTGACTTGACCACGGATACGCGAATTCGGAGATACGAGCTACCTGGCGTGGACACAAATCCAAATACTT 3257
ysProTyrAlaValAsnValPheAspLeuThrThrAspThrArgIleArgArgTyrGluLeuProGlyValAspThrAsnProAsnThrP (180)

3258 TCATAGCTAACATTGCCGTGGATATAGGCAAAAATTGCGATGATGCATATGCCTATTTTGCCGATGAATTGGGATACGGCTTGATTGCTT 3347
heIleAlaAsnIleAlaValAspIleGlyLysAsnCysAspAspAlaTyrAlaTyrPheAlaAspGluLeuGlyTyrGlyLeuIleAlaT (210)

3348 ACTCCTGGGAACTGAACAAGTCCTGGAGATTCTCGGCACATTCGTATTTTTTCCCCGATCCATTGAGGGGCGATTTCAATGTCGCTGGTA 3437
yrSerTrpGluLeuAsnLysSerTrpArgPheSerAlaHisSerTyrPhePheProAspProLeuArgGlyAspPheAsnValAlaGlyI (240)

3438 TTAACTTCCAATGGGGCGAGGAGGGTATATTTGGTATGTCCCTTTCGCCCATTCGATCGGATGGTTATCGTACCCTGTACTTTAGTCCGT 3527
leAsnPheGlnTrpGlyGluGluGlyIlePheGlyMetSerLeuSerProIleArgSerAspGlyTyrArgThrLeuTyrPheSerProL (270)

(continued)

```
3528 TAGCAAGTCATCGACAATTTGCCGTATCCACGAGGATTTTGAGGGATGAAACCAGGACGGAAGATAGCTATCATGACTTTGTTGCCTTAG 3617
     euAlaSerHisArgGlnPheAlaValSerThrArgIleLeuArgAspGluThrArgThrGluAspSerTyrHisAspPheValAlaLeuA (300)

3618 ATGAACGGGGTCCAAACTCCCATACCACTTCACGTGTGATGAGCGATGATGGAATTGAGCTGTTCAATTTAATAGATCAAAATGCAGTGG 3707
     spGluArgGlyProAsnSerHisThrThrSerArgValMetSerAspAspGlyIleGluLeuPheAsnLeuIleAspGlnAsnAlaValG (330)

3708 GTTGCTGGCACTCATCAATGCCGTACTCACCGCAATTTCATGGCATTGTGGATCGCGATGACGTTGGCTTAGTTTTTCCGGCCGATGTGA 3797
     lyCysTrpHisSerSerMetProTyrSerProGlnPheHisGlyIleValAspArgAspAspValGlyLeuValPheProAlaAspValL (360)

3798 AAATTGATGAGAACAAAAACGTTTGGGTTCTATCCGATAGGATGCCCGTTTTCTTGCTGTCTGACTTGGATTATTCAGATACTAATTTCC 3887
     ysIleAspGluAsnLysAsnValTrpValLeuSerAspArgMetProValPheLeuLeuSerAspLeuAspTyrSerAspThrAsnPheA (390)

3888 GAATTTACACGGCTCCCTTGGCCACTTTAATTGAGAATACTGTGTGTGATTTGAGGAATAACGCCTATGGGCCGCCAAATACCGTTTCAA 3977
     rgIleTyrThrAlaProLeuAlaThrLeuIleGluAsnThrValCysAspLeuArgAsnAsnAlaTyrGlyProProAsnThrValSerI (420)

3978 TACCAAAACAAGCCGTTTTGCCAATGGGTCCACCGTTATATACGAAACAATATCGTCCTGTCTTGCCACAGAAACCTCAGACCAGCTGGG 4067
     leProLysGlnAlaValLeuProMetGlyProProLeuTyrThrLysGlnTyrArgProValLeuProGlnLysProGlnThrSerTrpA (450)

4068 CTTCCTCGCCGCCTCCTCCAAGTCGCACTTATTTGCCCGCCAATTCAGGCAATGTAGTCTCCAGTATTAGTGTCTCTACAAATTCTGTGG 4157
     laSerSerProProProProSerArgThrTyrLeuProAlaAsnSerGlyAsnValValSerSerIleSerValSerThrAsnSerValG (480)

4158 GTCCTGCAGGAGTGGAGGTGCCAAAGGCCTATATTTTCAACCAGCACAACGGCATAAATTACGAGACAAGTGGTCCCCATCTATTTCCCA 4247
     lyProAlaGlyValGluValProLysAlaTyrIlePheAsnGlnHisAsnGlyIleAsnTyrGluThrSerGlyProHisLeuPheProT (510)

4248 CCCATCAACCCGCCCAACCGGGTGGCCAGGATGGTGGGTTAAAAACTTATGTGAATGCCCGCCAATCTGGGTGGTGGCATCATCAGCATC 4337
     hrHisGlnProAlaGlnProGlyGlyGlnAspGlyGlyLeuLysThrTyrValAsnAlaArgGlnSerGlyTrpTrpHisHisGlnHisG (540)

4338 AAGGTTAACATAATCCTACACACGGTACTTGGGTATATTCTCACACACTCGATTGATGTAAAGAATATTTAAAGACAACAACATAGGGCA 4427
     lnGlyEnd                                                                                   (541)

4428 ACAGCGGTTAAAAAAACCACATGACGTATGAGCAAGTGGCAAATCAATACTTTATCTAGTTATGTTAAGCAAAAAATAACAATAAATCAA 4517
                                                                                    ------
4518 CTTTTTTTTGAAGGTTAAGAGTTTACGCAATTTTCTTGAGCGGAAAAAGCGGAAAAAAATGTAAGTATGC 4586
                   |(A)n
```

y SEQUENCE. Strain, Canton S. Accession, X04427 (DROYELLOW) and X06481 (DROYELL5). An insertion of the transposable element *gypsy* following the A at −870 causes the mutation y^2. Mutations y^{76d28} and $y^{1\#7}$ are both caused by insertion of a *P* element at the same site in the leader, but the insertions are in opposite orientations.

Most type 1 mutations occur in the transcribed region of the gene and likely result in non-functional *y* product (Chia et al. 1986; Geyer et al. 1986). Mutation y^{76d28} is the result of a *P* element insertion in the leader (*y* Sequence). In this insertion, *P* is transcribed in the opposite orientation from *y* and the RNA produced is derived from the *y* promoter. Some of that RNA includes both *y* and *P* sequences and is not functional. In a small fraction of the RNA, however, splicing of most of the *P* sequences takes place through the use of cryptic splice signals in the *y* leader and the *P* element. This processed RNA codes for a small amount of normal *y* product that is responsible for a hypomorphic phenotype.

Mutations in *suppressor of sable* (*su*(*s*)) leads to increased accumulation of processed RNA and a more complete restoration of the normal phenotype (Geyer et al. 1991). This mechanism of suppression is similar to that observed in some *v* mutations.

y is less than 1 kb from *achaete*, centromere distal and transcribed toward the centromere (Fig. 1.1).

Developmental Pattern

There are two broad peaks of expression, one beginning late in embryonic development (16–20 h) and lasting until the second larval instar, the other during the middle pupal stages, about 48 h after pupariation. Gene expression is detectable in epidermal structures in which pigmentation will develop (Parkhurst and Corces 1986; Martin et al. 1989).

Promoter

Analysis of 5′ deletions by germ line transformation identified 2,873 bp upstream of the transcription initiation site (up to −3,042) that are sufficient for full expression of *y*. The region between −3,042 and −2,038 controls expression in the wing blade and the adult abdominal cuticle. The region between −2,038 and −665 contains a *cis*-acting regulatory signal that also contributes to expression in the adult abdominal cuticle. Deletions that leave only 495 bp of the promoter region cause yellow body and wing blades but pigmented larval mouth parts and denticle belts and adult bristles and sex combs. The segment between −665 and 166 upstream of the transcription initiation site seems to control expression in larval mouth parts and denticle belts, and the segment between 166 and 95 appears to include elements that contribute to *y* expression in larval structures as well as elements that determine expression in the adult tarsal claws and sex combs. With 95 bp of the 5′ region left, only bristles are pigmented normally (Geyer and Corces 1987; Martin et al. 1989).

The long intron contains enhancer-like sequences that seem to be responsible for increased transcript levels; they act in a position-independent manner (Geyer and Corces 1987; Martin et al. 1989).

Most type 2 mutations, including y^2, are associated with rearrangements in the 5′ region of the gene; these seem likely to affect the regulation of *y* transcription (Chia et al. 1986; Geyer et al. 1986).

References

Chia, W., Howes, G., Martin, M., Meng, Y. B., Moses, K. and Tsubota, S. (1986). Molecular analysis of the *yellow* locus of Drosophila. *EMBO J.* **5**:3597–3606.

Geyer, P. K. and Corces, V. G. (1987). Separate regulatory elements are responsible for the complex pattern of tissue-specific and developmental transcription of the *yellow* locus in *Drosophila melanogaster*. *Genes Dev.* **1**:996–1004.

Geyer, P. K., Spana, C. and Corces, V. G. (1986). On the molecular mechanism of gypsy-induced mutations at the *yellow* locus of *Drosophila melanogaster. EMBO J.* **5**:2657–2662.

Geyer, P. K., Chien, A. J., Corces, V. G. and Green, M. M. (1991). Mutations in the *su*(*s*) gene affect RNA processing in *Drosophila melanogaster. Proc. Natl Acad. Sci.* (*USA*) **88**:7116–7120.

Martin, M., Meng, Y. B. and Chia, W. (1989). Regulatory elements involved in the tissue-specific expression of the *yellow* gene of Drosophila. *Mol. Gen. Genet.* **218**:118–126.

Parkhurst, S. M. and Corces, V. G. (1986). Interactions among the gypsy transposable element and the *yellow* and *suppressor of Hairy-wing* loci in *Drosophila melanogaster. Mol. Cell. Biol.* **6**:47–53.

33

The Yolk Protein Gene Family: *Yp1, Yp2, Yp3*

Chromosomal Location:			**Map Position:**
Yp1	X,	8F-9B	1-30
Yp2	X,	8F-9B	1-29.5
Yp3	X,	12B-C	1-44

Products

Yolk proteins 1, 2 and 3 (YP1, YP2, YP3) of 46, 45 and 44 kD, respectively; also known as vitellogenins, when circulating in the hemolymph, and vitellins, when deposited in the oocyte.

Structure

YP precursors contain signal peptides that are cleaved before secretion (Warren et al. 1979). Other post-translational modifications include the sulfation of Tyr residues (Tyr-172 in YP2) (Baeuerle and Huttner 1985; Baeuerle et al. 1988), glycosylation and phosphorylation (Minoo and Postlethwait 1985; Brennan and Mahowald 1982).

Judging from the predicted amino-acid sequence, the three yolk proteins have only moderate similarity (Fig. 33.1); sequence identities are 48–53% in pairwise comparisons over the whole lengths of the proteins and 73% if the comparisons are restricted to the C-terminal one-third (Hung and Wensink 1983; Garabedian et al. 1987; Yan et al. 1987).

The yolk proteins of higher dipterans seem to be related to the triacylglycerol lipase family of proteins rather than to the vitellogenins of vertebrates, nematodes and lower insects, which have a different common evolutionary origin (Terpstra and Geert 1988). Comparison to the yolk proteins of the Mediterranean fruit fly, *Ceratitis capitata*, shows that the most conserved region extends from residue 202 to 427 of YP1; in this segment there is 40% identity between the two species and 40% of the substitutions are conservative. In terms

```
    1                                                  50                                                  100
Yp1 MNPMRVLSLL ACLA.VAALA KP....NGRM DNSVNQALKP SQWLSGSQLE AIPALDDFTI ERLENMNLER GAELLQQVYH LSQIHHNVEP NY..VPSGIQ
Yp2 MNPLRTLCVM ACLLAVAMGN PQSGNRSGRR SNSLDNVEQP SNWVNPREVE ELPNLKEVTL KKLQEMSMEE GATLLDKLYH LSQFNHVFKP DYTPEPSQIR
Yp3 MMSLRICLLA TCLL.VAAHA SK........ .DASNDRLKP TKWLTATELE NVPSLNDITW ERLENQPLEQ GAKVIEKIYH VGQIKHDLTP SFVPSPSNVP
CON M---R----- -CL--VA--- ---------- ---------P --W------E --P-L---T- --L-----E- GA------YH --Q--H---P -----PS---

    101                                                150                                                 200
Yp1 VYVPKPNGDK TVAPLNEMIQ RLKQKQNFGE DEVTIIVTGL PQTSETVKKA TRKLVQAYMQ RYNLQQQRQH GKNGNQDYQD QSNEQRKNQR TSSEEDY...
Yp2 GYIVGERGQK IEFNLNTLVE KVKRQQKFGD DEVTIFIQGL PETNTQVQKA TRKLVQAYQQ RYNLQP.... ......YETT DYSNEEQSQR SSSEEQQTQR
Yp3 VWIIKSNGQK VECKLNNYVE TAKAQPGFGE DEVTIVLTGL PKTSPAQQKA MRRLIQAYVQ KYNLQQLQ.. ...KNAQEQQ QQLKSSDYDY TSSEEAADQ.
CON -------G-K ----LN---- --K----FG- DEVTI---GL P-T-----KA -R-L-QAY-Q -YNLQ----- ---------- ---------- -SSEE-----

    201                                                250                                                 300
Yp1 .SEEVKNAKT QSGDIIVIDL GSKLNTYERY AMLDIEKTGA KIGKWIVQMV NELDMPFDTI HLIGQNVGAH VAGAAAQEFT RLTGHKLRRV TGLDPSKIVA
Yp2 RKQNGEQDDT KTGDLIVIQL GNAIEDFEQY ATLNIERLGE IIGNRLVELT NTVNVPQEII HLIGSGPAAH VAGVAGRQFT RQTGHKLRRI TALDPTKIYG
Yp3 ....WKSAKA ASGDLIIIDL GSTLTNFKRY AMLDVLNTGA MIGQTLIDLT N.KGVPQEII HLIGQGISAH VAGAAGNKYT AQTGHKLRRI TGLDPAKVLS
CON ---------- --GD-I-I-L G--------Y A-L-----G- -IG------- N----P---I HLIG----AH VAG-A----T --TGHKLRR- T-LDP-K---

    301                                                350                                                 400
Yp1 KSKNTLTGLA RGDAEFVDAI HTSVYGMGTP IRSGDVDFYP NGPAAGVPGA SNVVEAAMRA TRYFAESVRP GNERSFPAVP ANSLQQYKQN DGFGKRAYMG
Yp2 KPEERLTGLA RGDADFVDAI HTSAYGMGTS QRLANVDFFP NGPSTGVPGA DNVVEATMRA TRYFAESVRP GNERNFPSVA ASSYQEYKQN KGYGKRGYMG
Yp3 KRPQILGGLS RGDADFVDAI HTSTFAMGTP IRCGDVDFYP NGPSTGVPGS ENVIEAVARA TRYFAESVRP GSERNFPAVP ANSLKQYKEQ DGFGKRAYMG
CON K----L-GL- RGDA-FVDAI HTS---MGT- -R---VDF-P NGP--GVPG- -NV-EA--RA TRYFAESVRP G-ER-FP-V- A-S---YK-- -G-GKR-YMG

    401                                                450
Yp1 IDTAHDLEGD YILQVNPKSP FGRNAPAQKQ SSYHGVHQAW NTNQDSKDYQ *..
Yp2 IATDFDLQGD YILQVNSKSP FGRSTPAQKQ TGYHQVHQPW RQSSSNQGSR RQ*
Yp3 LQIDYDLRGD YILEVNAKSP FGQRSPAHKQ AAYHGMHHAQ N*........ ...
CON -----DL-GD YIL-VN-KSP FG---PA-KQ --YH--H--- ---------- ---
```

FIG. 33.1. Amino-acid sequence comparison of the three yolk proteins. The CON(sensus) sequence indicates positions in which there is identity in all three sequences.

of gene organization, the Mediterranean fruit fly has two types of yolk protein genes, one type with only one intron, as in *Drosophila Yp1* and *Yp2*, and the other with two introns at the same positions as in *Drosophila Yp3* (Rina and Savakis 1991).

Function

Yolk proteins are the main protein component of the yolk platelets stored in mature oocytes.

Mutant Phenotype

Mutation *Yp3*^{S1} occurs in the signal peptide (*Yp3* Sequence) and blocks normal processing and secretion; as a consequence, YP3 fails to accumulate in oocytes (see below). Viability and fertility are normal, suggesting that *Yp3* has a redundant function (Liddell and Bownes 1991).

Tissue Distribution

YP synthesis occurs only in adult females. The proteins are barely detectable in newly eclosed females, but the rate of synthesis increases steadily during the first 24 h after eclosion. The main sites of synthesis are the fat body and the follicle cells. In female fat bodies, YP can reach 20–30% of newly made proteins, and all three YPs are produced in comparable amounts. YPs are secreted into the hemolymph and then pinocytosed by the maturing oocytes. Follicle cells of stages 9 and 10 egg chambers also actively synthesize YPs; these are transferred to the oocyte through the intercellular matrix, without entering the hemolymph. Follicle cells contribute a significant proportion of YP1 and YP2, but YP3 synthesis is under-represented by four-fold in these cells (Brennan et al. 1982; Bownes 1986 and references therein). Synthesis of YPs is under hormonal control: 20-hydroxyecdysone stimulates fat bodies to synthesize all three YPs; juvenile hormone stimulates synthesis in fat bodies and ovaries, but the effect is more pronounced on YP1 and YP2 than on YP3 (Jowett and Postlethwait 1980).

Organization and Expression of the Cluster

Yp1 and *Yp2* are separated by 1,228 bp and transcribed divergently; *Yp3* is several hundred kb closer to the centromere (Fig. 33.2).

Developmental Pattern

Transcription is limited to ovaries and fat bodies of adult females (Garabedian et al. 1985). Expression of *Yp1* and *Yp2* occurs, in general, in follicle cells lining

Yp3

```
-800  TTAATCTTTTTGGTGATGTTGCCTATGTTTTGATTGAGCTCATCATTTTAGCAGTTGCTATGCTTTTGCATATATAAATATAATGCATTC   -711

-710  ACCTGGCGGCTGGTCATTGATTCCAATTTGGCCGGCTTCCAATCGCTGGAGGTCAATGCCGGGTCACACCAGTTTCTCACTTGACGCAGG   -621

-620  TGTTGCAAGTTTGTTGCCAGTTCAATTCTAATCAAGGGATCTGCACAAGTTGTTTCAATCAATCCGTACTAGAATACATTTTAAGTGCAG   -531

-530  AGAACAAAAATTTGCATTACTTTGGGAATTATATGCATAAATCTGTAAGTGTCGTTAAAACCAAATGATAGTGATGATACAAATATATCA   -441

-440  CGATGCAATACTACTAGTGGTCAACGATTTTCCAATAATCTAAATCTTAACATTTTATGAATGGATTTTTTTTTGCACATTTTTTGCCAA   -351

-350  GTGTGAAGAGGTTCAAAAACCTTAGTGCGATAAGAGAACTAAATGGTTGGCAAACACACACACATGTGAAATAAATCCGGCTATTTGCAA   -261

-260  TCAATTTTCCCTTGACTTGCACTTTATACACCGGCGACAGATCAGCAGAACGAAAGGGGTGGGGAAAAAACTGGAAGCCTAGACAGCCGA   -171

-170  CAACGACGACAACGACGACGACGACGACGACTTCCTGTGGTCAGCAGAAAATCGCTGGCAGTGCGCTATCGGGAATCGGAGCTATATAAG   -81
                                                                                       -------

                          --> -56
-80   CCAGAGATGGGGCTGAAGGAAGCCATCAAACGTCGTTTAGCGTTTGGCCCTGATCTGATTCAATTCCGGATTTGCACCAAAATGATGAGT   9
                                                                                     MetMetSer   (3)

                A=S1
10    CTAAGGATTTGCCTGCTGGCCACCTGCCTCCTGGTGGCGGCCCATGCCTCCAAGGATGCCTCCAATGACCGACTGAAGCCGACCAAGTGG   99
      LeuArgIleCysLeuLeuAlaThrCysLeuLeuValAlaAlaHisAlaSerLysAspAlaSerAsnAspArgLeuLysProThrLysTrp   (33)
                Asp                                     |

100   CTGACCGCCACCGAGCTGGAGAACGTGCCCTCCCTCAACGACATCACCTGGGAGCGTTTGGAGAATCAGCCGCTGGAGCAGGGCGCCAAG   189
      LeuThrAlaThrGluLeuGluAsnValProSerLeuAsnAspIleThrTrpGluArgLeuGluAsnGlnProLeuGluGlnGlyAlaLys   (63)

190   GTGATCGAGAAGATCTGTGAGTAGAAACCGATGTTGCTGGAAATCTCCAGAGATAACCTCCTTGTGAATCACACCTAGACCACGTTGGCC   279
      ValIleGluLysIleT                                                                yrHisValGlyG   (73)

280   AAATCAAGCACGATCTGACCCCCAGCTTTGTGCCCAGCCCGAGCAATGTGCCCGTCTGGATTATCAAGTCCAATGGACAGAAGGTTGAGT   369
      lnIleLysHisAspLeuThrProSerPheValProSerProSerAsnValProValTrpIleIleLysSerAsnGlyGlnLysValGluC   (103)

370   GCAAGTTGAACAACTATGTGGAGACGGCCAAGGCACAGCCCGGATTCGGCGAGGATGAGGTCACCATTGTCCTGACTGGTCTGCCCAAGA   459
      ysLysLeuAsnAsnTyrValGluThrAlaLysAlaGlnProGlyPheGlyGluAspGluValThrIleValLeuThrGlyLeuProLysT   (133)

460   CCAGCCCCGCTCAGCAGAAGGCCATGCGCAGGTTGATCCAGGCCTACGTCCAGAAGTACAACCTCCAGCAGCTGCAGAAGAACGCCCAGG   549
      hrSerProAlaGlnGlnLysAlaMetArgArgLeuIleGlnAlaTyrValGlnLysTyrAsnLeuGlnGlnLeuGlnLysAsnAlaGlnG   (163)

550   AGCAGCAGCAGCAGCTCAAGAGCAGCGACTACGACTACACCAGCAGCGAGGAGGCCGCTGACCAATGGAAATCCGCCAAGGCTGCCAGCG   639
      luGlnGlnGlnGlnLeuLysSerSerAspTyrAspTyrThrSerSerGluGluAlaAlaAspGlnTrpLysSerAlaLysAlaAlaSerG   (193)

640   GCGATTTGATCGTAAGTTGGTCGCATTCCTATATTTCATAATTAAACGTGTACATATGGATATTTATGAAATTCAAATTGCAGATCATTG   729
      lyAspLeuIle
                                                                                      IleIleA   (199)

730   ACCTCGGCTCCACCCTGACCAACTTCAAACGCTACGCGATGCTGGATGTTCTGAACACCGGCGCCATGATCGGCCAGACCCTGATCGATC   819
      spLeuGlySerThrLeuThrAsnPheLysArgTyrAlaMetLeuAspValLeuAsnThrGlyAlaMetIleGlyGlnThrLeuIleAspL   (229)

820   TGACCAACAAGGGTGTGCCCCAGGAGATCATCCATCTGATCGGCCAGGGAATCAGCGCCCATGTGGCCGGAGCTGCTGGCAACAAGTACA   909
      euThrAsnLysGlyValProGlnGluIleIleHisLeuIleGlyGlnGlyIleSerAlaHisValAlaGlyAlaAlaGlyAsnLysTyrT   (259)

910   CCGCCCAAACCGGACACAAGCTGCGCCGCATCACCGGTCTGGATCCCGCCAAGGTGCTGTCCAAGCGTCCCCAGATCCTGGGTGGTCTGT   999
      hrAlaGlnThrGlyHisLysLeuArgArgIleThrGlyLeuAspProAlaLysValLeuSerLysArgProGlnIleLeuGlyGlyLeuS   (289)
```

```
1000  CCCGCGGCGATGCTGACTTCGTTGATGCCATTCACACATCGACCTTCGCCATGGGCACGCCCATCCGTTGCGGCGATGTTGACTTCTACC  1089
      erArgGlyAspAlaAspPheValAspAlaIleHisThrSerThrPheAlaMetGlyThrProIleArgCysGlyAspValAspPheTyrP  (319)

1090  CCAACGGACCGTCCACCGGTGTTCCCGGCTCCGAGAATGTGATCGAGGCTGTGGCCCGTGCCACCCGTTACTTTGCCGAGTCTGTGCGTC  1179
      roAsnGlyProSerThrGlyValProGlySerGluAsnValIleGluAlaValAlaArgAlaThrArgTyrPheAlaGluSerValArgP  (349)

1180  CCGGTAGCGAGCGCAATTTCCCCGCCGTTCCGGCCAACTCGCTGAAGCAGTACAAGGAGCAGGATGGCTTTGGCAAGCGCGCCTACATGG  1269
      roGlySerGluArgAsnPheProAlaValProAlaAsnSerLeuLysGlnTyrLysGluGlnAspGlyPheGlyLysArgAlaTyrMetG  (379)

1270  GTCTCCAGATCGACTACGATCTGCGCGGTGACTACATCTTGGAGGTCAACGCCAAGAGCCCCTTCGGTCAGCGCAGCCCTGCCCACAAGC  1359
      lyLeuGlnIleAspTyrAspLeuArgGlyAspTyrIleLeuGluValAsnAlaLysSerProPheGlyGlnArgSerProAlaHisLysG  (409)

1360  AGGCCGCCTACCATGGCATGCACCACGCCCAGAACTAGAGCGCCCATGGCCACGCCCCCTGGTTACCAGGGACGTTCGATCGTCACGCAC  1449
      lnAlaAlaTyrHisGlyMetHisHisAlaGlnAsnEnd                                                     (420)

1450  TTTCTGATAATCAGAAAATAAAAACCCGGAATGCGTAGTTTAGCTTAGAAGTTTCATCAAACAATCAAAAAAGAAAAATCTATAAAATCC  1539

1540  CATAAAAATAAAAGCTGCAAATTTTCGAAAAGTCAAGTTTTTAATAGCAATAGCAATGGTTATTCTGGATTGGATTCTAACTTTTATGG  1629
         -----                    |(A)n

1630  TATTAAAAAACACACACAAGAATTTGCTGGGCACATTTTTAGGCACCCCTTCTGAAGTAAATAGAAAAATTTCCGAAAATATACATATTT  1719

1720  AACATAGTAAATCGGCCAAACAACTTAAATGAGCTAATAATAAAAAGATAAATGCATATATCACAGGTGATCTTAAGCAGATGCTTAACC  1809

1810  AAAAAACAACACGATAAATAAAGCAAACAAAAAGTGCCTAAAATACAATTATGACACCTAATGAAAGGTACACGAAAGAAAATGTAGATA  1899

1900  ATAAATAAACTGAAAAGAAATTAGGAATAACTCATAAAAATCAAAATTTAGAAAACTGTGCAGCTTGGTATTTACTAGCACCCTAGATGC  1989

1990  TTAACAGGATTGCGAAGTTGGGATGGAAATACGCACAACGAGATGGATGCATTGAGTGGGCGGAAGTGAGAGTGAGGCAACTAGTGTCCG  2079

2080  TTGCCACTTGATGTGCACTCAATTAAAACTTGCATTCGGTTTATCGTTAGTGACTACTCGTTCAAAAATCACTGGGCAACCTGTGTAAAC  2169

2170  TCAATTGTTCCTTACAGTTTTGGGACATGCGCGGTGTAAATGTCAAAGTTGAACTTTATCAAATGCAATAGACAAACTAGAAAGGGCAGC  2259

2260  GAAAACAGCAGAGTCGAAAATAGAGCGAGATAGGGAGCTGGAGTGACAGGAGCGGAATGACAACAGTTGGCGTCTTTTGTTTGTGCATGT  2349

2350  CGTGACATGTTTGCTTTGACTCTGACCGAACGGAATGCGCCGTTAAGCTT  2399
```

Yp3 SEQUENCE. Strain, Canton S. Accession, M15898 (DROYP3) and X04754 (DROYP3G) as corrected near the transcription initiation site by Liddell and Bownes (1991). The vertical line at Ala-19 marks the putative cleavage site of the signal peptide.

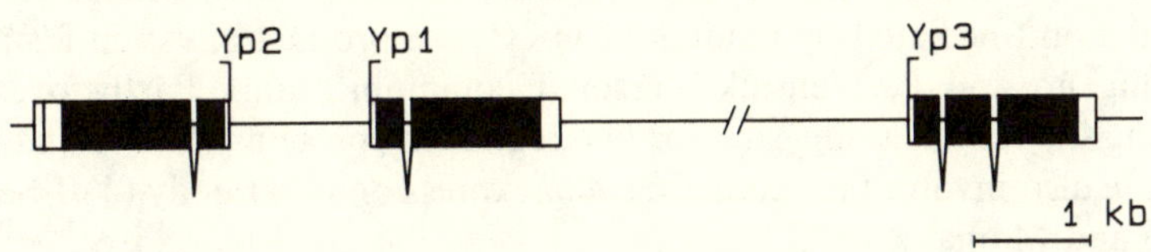

FIG. 33.2. *Yp* cluster, centromere to the right. Note that *Yp3* is many kbs from *Yp1*/*Yp2* and that the direction of transcription of the three genes relative to the centromere is not known

the maturing oocyte (stages 8–10) but not in the nurse cells (Logan and Wensink 1990) (see *Yp1 Promoter*).

Yp1

Gene Organization and Expression

Open reading frame, 439 amino acids; expected mRNA length, 1,559 bases, in agreement with northern analysis. S1 mapping, primer extension and sequence features were used to define the 5′ end. The 3′ end was obtained from S1 mapping. There is one intron in the Tyr-74 codon (*Yp1* Sequence) (Hung and Wensink 1981; Hovemann and Galler 1982).

Promoter

There is evidence that the 1,228-bp segment separating *Yp1* and *Yp2* includes two *cis*-acting regulatory elements, one for ovarian and the other for fat body expression; these two elements control both *Yp1* and *Yp2*. The two genes were cloned separately into *P* elements; this split the 1,228-bp segment leaving 886 associated with *Yp1* and the remaining 342 with *Yp2*. In germline transformants, the fragment with *Yp1* was expressed only in fat bodies and the one with *Yp2* only in ovaries (Garabedian et al. 1985).

Fat Body Enhancers Deletion mapping and ligation of fragments to a heterologous promoter (*Hsp70*) and a reporter gene (*lacZ*) showed further that 125 bp of the 886-bp segment (from −378 to −253 in the *Yp1* Sequence) was sufficient for stage-, sex- and tissue-specific expression in adult female fat bodies. This regulatory segment of DNA acts relatively independently of its orientation and distance from the genes, and it acts on both *Yp1* and *Yp2* (Garabedian et al. 1986; K. Coschigano and P. Wensink, personal communication). The rest of that segment, from −942 to −378 contains a weaker fat body enhancer (P. Wensink, personal communication).

Sex-specificity of expression seems to be controlled by the *doublesex* (*dsx*) gene products; these bind to three sites in the fat body enhancer, and all three binding sites contain sequences related to CTACAAAGT (Burtis et al. 1991). Binding sites A and B (between −378 and −253) direct male-specific repression (mediated by binding of DSX^{M}, the product of *dsx* in males) and female-specific stimulation (mediated by binding of DSX^{F}, the product of *dsx* in females) (K. Coschigano and P. Wensink, personal communication). Partly overlapping binding site A are binding sites for two regulatory proteins (AEF-1 and C/EBP) that are also involved in regulating *Adh* expression in fat body (*Yp1* Sequence) (Falb and Maniatis 1992).

Ovarian Enhancers Expression of *Yp1* and *Yp2* in ovarian follicle cells is controlled by an enhancer, oe1, located in the interval between −1,242 and

Yp1

```
-1453  GCTGCTCCACATTGTCCAGGGAGTTGGATCGGCGACCGGAACGGTTACCAGACTGGGGATTACCCATGGCGACCGCCAGAAGGCAGGCCA  -1364
                                                                                    |-
                                                                                          <--Yp2
-1363  TAACGCAAAGGGTGCGCAGAGGATTCATTGTGGCTTCCAAGTTCGACTTTTTCAGACACCGTACCAAATTGTACTGCATGCCCACTGCTG  -1274
                             Met(Yp2)                                                    -|oe2

-1273  CGACTCAATGCATTTTATACCCCTTGGAATCGGTAGTCTATACACACTATAATGCACGCGCCGGAAGCAATTGATTTCAGCAACCGATTT  -1184
                                     |-

-1183  CTGGATCAGCACAAATGCATTGATTCGCAGCGTCAGTGATTTTGCAACACTTCTGATGAGCTCTAAAATTTCGTTCCCCTTTTTTTTTTT  -1094

-1093  TTTTTTTTGGTTATTAAGTATCCATCGGGTAACAGGTAATGGGAAACTTCTTTAACCAGCACTTTCATAACATAAACAAAAGGTGGTCTG  -1004

-1003  GCCATTAAGGGGCTTGACAGTGGGGGCACGACTTGAACTCATGCACAGGTCAAGATAAAGCTTTTGTTTGAAAAAAATATTTGGCAATTT  -914
                                                                            -|oe1

 -913  TGTGAAATTTATGCAACTATTTAAGTGTTTGCCAAAAGAATTGTCTAAATTGTTCTATAAGCAGATAACACTTTCAGGGAAATGCAAAAT  -824

 -823  AAATATATTATAAATTATAATATTATAAATATAAATATTTACATCTATCGAAATATACATATATTTTTAATAAGTAGAATGAGTTACATG  -734

 -733  AAATAGCATCGATAAGATCATATATTATAAAACGAATCCCGGATATTAAAATAGAATCTCCTTGAAAAACGTTTCCCCTGAATCAATTCA  -644

 -643  TTTCTAAAGTCCAAAAACAAATATAATCTTACTATCTTGCCTTGGAAACTACAAACATTCCATACTTTTCGTATCAATGGCAAACATCTA  -554

 -553  GGAATCAATGAACTGTATCGGCCTTGAATTGAAAATGCAAAATTATGGACTTTTAATTAAGCAGAAGAAAAGTGCCAAATATAAATCTAC  -464

 -463  TTATAAACAAAAAAAATCAATAAAATGTTGTATATAATAACCAACTAATGCCCATGTTAGATCTATATTTTATGCATTTATTTGATCAAA  -374

 -373  TCCGGTGCACAACTACAATGTTGCAATCAGCGGAGCCTACAAAGTGATTACAAATTAAAATAATCAGGCGGCAGCAGGTGCTGCTAAGTC  -284
           -------------------A   ------------------------------B        --------------
         ------------------aef-1
                   ---------------------c/ebp

 -283  ATCAGTGGGGTCAGCTATAGGTAGGCCCCGTGTCTATTTTGTATGTATACAATTTATTCCGCTATCGATAGCATATACACTCGATCCGAT  -194
       ----C

 -193  TCCCAGGCACCCGAAAACCCTTACTCAGCACAAGTGACCGATTAAGGCCTGAGCCAGCGAAAAGCAAGTCGGAAAATGGGAAATCGCTCA  -104

                                                  -->-57
 -103  GCGTAAATTGTGGTATATAAACCACCATCGTTGGATTTGGAAGGCCAGTTCAACTCACTCAGTGTTGAAGTCGCATCCGCAGGACCAAAT  -14
                   ------

  -13  CCCAAATCCGAACCATGAACCCCATGAGAGTGCTGAGCCTTCTGGCTTGCTTGGCGGTCGCCGCCTTGGCCAAGCCCAATGGCCGTATGG  76
                      MetAsnProMetArgValLeuSerLeuLeuAlaCysLeuAlaValAlaAlaLeuAlaLysProAsnGlyArgMetA  (26)
                                                                                     |

   77  ACAACTCCGTCAACCAGGCATTGAAGCCGTCGCAGTGGCTCTCCGGATCCCAGCTGGAGGCCATTCCCGCCCTCGACGATTTCACCATTG  166
       spAsnSerValAsnGlnAlaLeuLysProSerGlnTrpLeuSerGlySerGlnLeuGluAlaIleProAlaLeuAspAspPheThrIleG  (56)

  167  AGCGTCTGGAGAACATGAACCTGGAGCGTGGCGCCGAGCTGCTGCAGCAAGTCTGTGAGTAATCCTAGATGCAGATAAAAAAAAAAAAAA  256
       luArgLeuGluAsnMetAsnLeuGluArgGlyAlaGluLeuLeuGlnGlnValT                                        (74)
```

(*continued*)

```
257  AAACATCGAATATTCTATGGAATATATATATCCTTTGTAGACCACCTGTCGCAGATCCACCACAACGTTGAGCCCAACTATGTGCCCAGC   346
                                                               yrHisLeuSerGlnIleHisHisAsnValGluProAsnTyrValProSer   (90)

347  GGCATCCAGGTCTATGTGCCCAAGCCCAATGGTGACAAGACCGTTGCTCCCCTGAACGAGATGATCCAGCGCCTGAAGCAGAAGCAGAAC   436
     GlyIleGlnValTyrValProLysProAsnGlyAspLysThrValAlaProLeuAsnGluMetIleGlnArgLeuLysGlnLysGlnAsn   (120)

437  TTTGGTGAGGATGAGGTGACCATCATTGTGACCGGACTGCCCCAGACCAGCGAGACCGTGAAGAAGGCGACCAGGAAGCTGGTTCAGGCT   526
     PheGlyGluAspGluValThrIleIleValThrGlyLeuProGlnThrSerGluThrValLysLysAlaThrArgLysLeuValGlnAla   (150)

527  TACATGCAGCGCTACAATCTGCAGCAGCAGCGCCAGCACGGCAAGAACGGCAACCAGGACTACCAGGATCAGAGCAACGAACAGAGGAAG   616
     TyrMetGlnArgTyrAsnLeuGlnGlnGlnArgGlnHisGlyLysAsnGlyAsnGlnAspTyrGlnAspGlnSerAsnGluGlnArgLys   (180)

617  AACCAGAGGACCAGCAGCGAGGAGGACTACAGCGAGGAGGTTAAGAACGCCAAGACCCAAAGCGGCGACATCATTGTGATCGATTTGGGC   706
     AsnGlnArgThrSerSerGluGluAspTyrSerGluGluValLysAsnAlaLysThrGlnSerGlyAspIleIleValIleAspLeuGly   (210)

707  TCCAAGCTGAACACCTATGAGCGTTATGCCATGCTCGACATTGAGAAGACCGGCGCCAAGATCGGCAAGTGGATCGTCCAGATGGTCAAC   796
     SerLysLeuAsnThrTyrGluArgTyrAlaMetLeuAspIleGluLysThrGlyAlaLysIleGlyLysTrpIleValGlnMetValAsn   (240)

797  GAGTTGGACATGCCCTTCGATACCATTCACCTGATTGGCCAGAATGTGGGTGCCCATGTTGCCGGTGCCGCTGCCCAGGAATTCACCCGT   886
     GluLeuAspMetProPheAspThrIleHisLeuIleGlyGlnAsnValGlyAlaHisValAlaGlyAlaAlaAlaGlnGluPheThrArg   (270)

887  CTCACCGGACACAAGCTGCGCCGTGTCACCGGTCTGGATCCCTCCAAGATCGTGGCCAAGAGCAAGAACACCCTGACCGGTCTGGCTCGC   976
     LeuThrGlyHisLysLeuArgArgValThrGlyLeuAspProSerLysIleValAlaLysSerLysAsnThrLeuThrGlyLeuAlaArg   (300)

977  GGTGATGCTGAATTCGTTGACGCCATCCACACCTCGGTCTACGGCATGGGCACCCCCATCCGCTCCGGTGATGTTGACTTCTATCCCAAT   1066
     GlyAspAlaGluPheValAspAlaIleHisThrSerValTyrGlyMetGlyThrProIleArgSerGlyAspValAspPheTyrProAsn   (330)

1067 GGACCTGCCGCCGGTGTTCCCGGAGCCAGCAACGTGGTGGAGGCCGCCATGCGTGCCACCCGCTACTTCGCCGAGTCCGTGCGTCCCGGA   1156
     GlyProAlaAlaGlyValProGlyAlaSerAsnValValGluAlaAlaMetArgAlaThrArgTyrPheAlaGluSerValArgProGly   (360)

1157 AACGAGAGGAGCTTCCCCGCCGTGCCAGCCAACTCCCTGCAGCAGTACAAGCAGAACGATGGATTCGGCAAGCGTGCCTACATGGGCATC   1246
     AsnGluArgSerPheProAlaValProAlaAsnSerLeuGlnGlnTyrLysGlnAsnAspGlyPheGlyLysArgAlaTyrMetGlyIle   (390)

1247 GATACCGCTCACGATCTCGAGGGTGACTACATTCTGCAGGTGAACCCCAAGTCTCCTTTCGGCCGCAACGCACCCGCCCAGAAGCAGAGC   1336
     AspThrAlaHisAspLeuGluGlyAspTyrIleLeuGlnValAsnProLysSerProPheGlyArgAsnAlaProAlaGlnLysGlnSer   (420)

1337 AGCTACCACGGTGTCCACCAGGCGTGGAACACCAACCAGGACAGCAAGGACTACCAGTAAGGATGAGTCTGCTTACTCTGGACACCTGGA   1426
     SerTyrHisGlyValHisGlnAlaTrpAsnThrAsnGlnAspSerLysAspTyrGlnEnd                                 (439)

1427 ATGGCAACTACCAAACAACCACCCAACCACACAAACACTGTAGTCCCTAAGTTGAACCCATATTGGCCCTTTTCTTGAGATTACCTAAAC   1516

1517 ATTTAACGAGCACATCGCGAAATTCAGCAAATAAACGCTCGATAAAGAGCTTAAAAATATCTATTTTGTTTATCTTAAATCATTTAGGAA   1606
                                     ------                                     |(A)n

1607 CTATAATAGTCTAATAGATCATCCCAAAAAAAAGGGAACAAAATCAAAAGTAAATATCGTAGTTTGGTTTTGTAAACTTAGATTTATTTT   1696

1697 ATTGTTGTCGGTGTTTTTGTGG  1718
```

Yp1 Sequence. Strain, Canton S. Accession, V00248 (DMYOLK), X01524, J01157 and M11170 (DROYP12). The segment between −1,453 and −1,282 corresponds to the reverse complement of *Yp2*: sites of transcription and translation initiation are indicated. The vertical line at Ala-19 marks the putative cleavage site of the signal peptide. A, B and C indicate the footprints produced by the *dsx* products in the main fat body enhancer; and aef-1 and c/ebp are footprints of fat-body specific proteins.

Yp2

-761 GAAAAGTATGGAATGTTTGTAGTTTCCAAGGCAAGATAGTAAGATTATATTTGTTTTTGGACTTTAGAAATGAATTGATTCAGGGGAAAC -672

-671 GTTTTTCAAGGAGATTCTATTTTAATATCCGGGATTCGTTTTATAATATATGATCTTATCGATGCTATTTCATGTAACTCATTCTACTTA -582

-581 TTAAAAATATATGTATATTTCGATAGATGTAAATATTTATATTTATAATATTATAATTTATAATATATTTATTTTGCATTTCCCTGAAAG -492

-491 TGTTATCTGCTTATAGAACAATTTAGACAATTCTTTTGGCAAACACTTAAATAGTTGCATAAATTTCACAAAATTGCCAAATATTTTTTT -402

-401 CAAACAAAAGCTTTATCTTGACCTGTGCATGAGTTCAAGTCGTGCCCCCACTGTCAAGCCCCTTAATGGCCAGACCACCTTTTGTTTATG -312

-311 TTATGAAAGTGCTGGTTAAAGAAGTTTCCCATTACCTGTTACCCGATGGATACTTAATAACCAAAAAAAAAAAAAAAAAAAAGGGGAACGA -222

-221 AATTTTAGAGCTCATCAGAAGTGTTGCAAAATCACTGACGCTGCGAATCAATGCATTTGTGCTGATCCAGAAATCGGTTGCTGAAATCAA -132

-->-53

-131 TTGCTTCCGGCGCGTGCATTATAGTGTGTATAGACTACCGATTCCAAGGGGTATAAAATGCATTGAGTCGCAGCAGTGGGCATGCAGTAC -42

-41 AATTTGGTACGGTGTCTGAAAAAGTCGAACTTGGAAGCCACAATGAATCCTCTGCGCACCCTTTGCGTTATGGCCTGCCTTCTGGCGGTC 48
MetAsnProLeuArgThrLeuCysValMetAlaCysLeuLeuAlaVal (16)

49 GCCATGGGTAATCCCCAGTCTGGTAACCGTTCCGGTCGCCGATCCAACTCCCTGGACAATGTGGAGCAGCCCAGCAACTGGGTCAACCCA 138
AlaMetGlyAsnProGlnSerGlyAsnArgSerGlyArgArgSerAsnSerLeuAspAsnValGluGlnProSerAsnTrpValAsnPro (46)

139 CGTGAAGTCGAGGAGCTGCCCAACCTGAAGGAGGTTACCCTTAAGAAGCTGCAGGAGATGAGCATGGAGGAGGGCGCTACGCTGTTGGAC 228
ArgGluValGluGluLeuProAsnLeuLysGluValThrLeuLysLysLeuGlnGluMetSerMetGluGluGlyAlaThrLeuLeuAsp (76)

229 AAGCTCTGTAAGTTCAAGGATCTCTAAAAGTTCTACCAATCATGTTATATTTACACGCACTATCCTATCCCGCAGACCATCTGTCCCAGT 318
LysLeuT yrHisLeuSerGlnP (84)

319 TCAACCATGTCTTCAAGCCCGATTACACCCCGGAACCCAGCCAGATCAGGGGCTACATTGTCGGCGAGCGCGGCCAGAAGATCGAGTTCA 408
heAsnHisValPheLysProAspTyrThrProGluProSerGlnIleArgGlyTyrIleValGlyGluArgGlyGlnLysIleGluPheA (114)

409 ACCTGAACACTTTGGTGGAGAAGGTTAAGCGCCAGCAGAAGTTCGGCGACGATGAGGTCACCATCTTCATCCAGGGCCTGCCCGAGACCA 498
snLeuAsnThrLeuValGluLysValLysArgGlnGlnLysPheGlyAspAspGluValThrIlePheIleGlnGlyLeuProGluThrA (144)

499 ACACCCAAGTGCAGAAGGCTACCAGGAAGCTGGTGCAGGCCTACCAGCAGCGTTACAACCTCCAGCCCTATGAGACCACCGACTACTCCA 588
snThrGlnValGlnLysAlaThrArgLysLeuValGlnAlaTyrGlnGlnArgTyrAsnLeuGlnProTyrGluThrThrAspTyrSerA (174)

589 ACGAGGAGCAGAGCCAGAGGAGTTCCAGCGAGGAGCAGCAAACGCAGCGCAGGAAGCAGAACGGTGAACAGGATGATACCAAGACCGGAG 678
snGluGluGlnSerGlnArgSerSerSerGluGluGlnGlnThrGlnArgArgLysGlnAsnGlyGluGlnAspAspThrLysThrGlyA (204)

679 ACCTGATTGTGATCCAGCTGGGCAATGCCATCGAGGACTTTGAGCAGTACGCCACCCTGAACATTGAGCGTCTGGGCGAGATCATTGGCA 768
spLeuIleValIleGlnLeuGlyAsnAlaIleGluAspPheGluGlnTyrAlaThrLeuAsnIleGluArgLeuGlyGluIleIleGlyA (234)

769 ACCGTCTGGTTGAGCTGACCAACACCGTGAACGTGCCCCAGGAGATCATCCATCTGATTGGCTCTGGACCCGCTGCCCACGTTGCCGGAG 858
snArgLeuValGluLeuThrAsnThrValAsnValProGlnGluIleIleHisLeuIleGlySerGlyProAlaAlaHisValAlaGlyV (264)

859 TGGCTGGACGCCAGTTCACCCGTCAGACCGGACACAAGTTGCGCCGCATCACCGCCCTGGACCCCACTAAGATCTACGGCAAGCCCGAGG 948
alAlaGlyArgGlnPheThrArgGlnThrGlyHisLysLeuArgArgIleThrAlaLeuAspProThrLysIleTyrGlyLysProGluG (294)

(*continued*)

```
949  AGAGGCTGACCGGGCTGGCCCGTGGTGATGCTGACTTCGTTGATGCCATCCACACCTCCGCCTACGGCATGGGTACCAGCCAGCGATTGG  1038
     luArgLeuThrGlyLeuAlaArgGlyAspAlaAspPheValAspAlaIleHisThrSerAlaTyrGlyMetGlyThrSerGlnArgLeuA  (324)

1039 CCAACGTGGACTTCTTCCCCAACGGACCCTCGACCGGAGTGCCCGGAGCCGATAATGTCGTTGAGGCCACCATGCGTGCCACCCGCTACT  1128
     laAsnValAspPhePheProAsnGlyProSerThrGlyValProGlyAlaAspAsnValValGluAlaThrMetArgAlaThrArgTyrP  (354)

1129 TCGCCGAGTCTGTGCGTCCTGGAAACGAGAGGAACTTCCCCTCCGTGGCCGCCAGCTCGTACCAGGAGTACAAGCAGAACAAGGGCTATG  1218
     heAlaGluSerValArgProGlyAsnGluArgAsnPheProSerValAlaAlaSerSerTyrGlnGluTyrLysGlnAsnLysGlyTyrG  (384)

1219 GCAAGCGCGGATACATGGGCATCGCCACCGATTTCGATCTGCAGGGCGATTACATTCTGCAGGTGAACTCCAAGAGCCCCTTCGGCAGGA  1308
     lyLysArgGlyTyrMetGlyIleAlaThrAspPheAspLeuGlnGlyAspTyrIleLeuGlnValAsnSerLysSerProPheGlyArgS  (414)

1309 GCACTCCCGCCCAGAAACAGACCGGCTACCACCAGGTCCACCAGCCCTGGCGCCAGTCCTCCTCCAACCAGGGTTCCCGCCGTCAGTAGA  1398
     erThrProAlaGlnLysGlnThrGlyTyrHisGlnValHisGlnProTrpArgGlnSerSerSerAsnGlnGlySerArgArgGlnEnd  (442)

1399 TCATCGCACAGTGATCCATCGATGACAACCAGATCGCACACCCCTCATGCGAGCGAACCACTCCAGCCCATCCTCATCCAGCAGAACCCT  1488

1489 CTGCCAGTTGCATCCACTACGATTAGTTAGCTTTGTTTTTTTAACTCACAATAAAAAACGTTTGCATTTTTAAACATTCTAAAGAGTTCA  1578
                                                ------                          |(A)n

1579 GTTCAATATCGGAAAAAAACCCCAGTTCAATTTACAATAAAAACAATTGCTTATGTCGAAATATTTGAGAGTTCCAAATGCTCCTTATAT  1668
                                  ------                                    |(A)n

1669 AAAAATATCCAAAACCAAATTATGCAATGCCACTGAGGCCATAAAAGAAGCACACAACAAACATTTGGGT  1738
```

Yp2 SEQUENCE. Strain, Canton S. Accession, X01524, J01157 and M11170 (DROYP12). The vertical line at Gly-19 marks the putative cleavage site of the signal peptide.

−942 in the *Yp1* Sequence, between 43 and 343 bp upstream of the transcription initiation site of *Yp2* (Logan et al. 1989). oe1 is composed of multiple parts, each controlling the expression in various subsets of follicle cells (Logan and Wensink 1990).

oe2, located between −1,389 and −1,284 (the first 105 bp of the first exon of *Yp2*), is also necessary for expression of *Yp1* in ovaries (Logan et al. 1989).

Other Regulatory Elements Another *cis*-acting regulatory region was identified by its ability to bind YPF1, a heterodimer with subunits of 85 and 69 kD, very specifically and very tightly ($K_D < 5 \times 10^{-16}$). This element occurs in the translated region of *Yp1* (between positions 82 and 126 in the *Yp1* Sequence) and is necessary for *Yp1* transcription (Mitsis and Wensink 1989a, 1989b).

Yp2

Gene Organization and Expression

Open reading frame, 442 amino acids; expected mRNA lengths, 1,546 or 1,630 bases depending on which of two polyadenylation sites is used. S1 mapping and primer extension were used to define the 5′ ends. The 3′ ends were obtained by S1 mapping. There is an intron in the Tyr-79 codon (*Yp2* Sequence) (Hovemann and Galler 1982; Hung and Wensink 1983).

Promoter

See discussion of the *Yp1* promoter, above.

Yp3

Gene Organization and Expression

Open reading frame, 420 amino acids; expected mRNA length, 1,488 bases. S1 mapping and cDNA sequencing were used to define the 5′ and 3′ ends. There are two introns, at Tyr-69 and after Ile-196 (*Yp3* Sequence) (Garabedian et al. 1987; Yan et al. 1987; Liddell and Bownes 1991).

References

Baeuerle, P. A. and Huttner, W. B. (1985). Tyrosine sulfation of yolk proteins 1, 2 and 3 in *Drosophila melanogaster*. *J. Biol. Chem.* **260**:6434–6439.

Baeuerle, P. A., Lottspeich, F. and Huttner, W. B. (1988). Purification of yolk protein 2 of *Drosophila melanogaster* and identification of its site of tyrosine sulfation. *J. Biol. Chem.* **263**:14925–14929.

Bownes, M. (1986). Expression of the genes coding for vitellogenin (yolk protein). *Ann. Rev. Entomol.* **31**:507–531.

Brennan, M. D. and Mahowald, A. P. (1982). Phosphorylation of the vitellogenin polypeptides of *Drosophila melanogaster*. *Insect Biochem.* **12**:669–673

Brennan, M. D., Weiner, A. J., Goralski, T. J. and Mahowald, A. P. (1982). The follicle cells are a major site of vitellogenin synthesis in *Drosophila melanogaster*. *Dev. Biol.* **89**:225–236.

Burtis, K. C., Coschigano, K. T., Baker, B. S. and Wensink, P. C. (1991). The doublesex proteins of *Drosophila melanogaster* bind directly to a sex-specific yolk protein gene enhancer. *EMBO J.* **10**:2577–2582.

Falb, D. and Maniatis, T. (1992). A conserved regulatory unit implicated in tissue-specific gene expression in *Drosophila* and man. *Genes Dev.* **6**:454–465.

Garabedian, M. J., Hung, M-C. and Wensink, P. C. (1985). Independent control elements that determine yolk protein gene expression in alternative *Drosophila* tissues. *Proc. Natl Acad. Sci.* (*USA*) **82**:1396–1400.

Garabedian, M. J., Shepherd, B. M. and Wensink, P. C. (1986). A tissue specific transcription enhancer from the *Drosophila* yolk protein 1 gene. *Cell* **45**:859–867.

Garabedian, M. J., Shirras, A. D., Bownes, M. and Wensink, P. C. (1987). The nucleotide sequence of the gene coding for *Drosophila melanogaster* yolk protein 3. *Gene* **55**:1–8.

Hovemann, B. and Galler, R. (1982). Vitellogenin in *Drosophila melanogaster*: a comparison of the *Yp1* and *Yp2* genes and their transcription products. *Nucl. Acids Res.* **10**:2261–2274.

Hung, M-C. and Wensink, P. C. (1981). The sequence of the *Drosophila melanogaster* gene for yolk protein 1. *Nucl. Acids Res.* **9**:6407–6419.

Hung, M-C. and Wensink, P. C. (1983). Sequence and structure conservation in yolk proteins and their genes. *J. Mol. Biol.* **164**:481–492.

Jowett, T. and Postlethwait, J. H. (1980). The regulation of yolk polypeptide synthesis in *Drosophila* ovaries and fat bodies by 20-hydroxyecdysone and a juvenile hormone analog. *Dev. Biol.* **80**:225–234.

Liddell, S. and Bownes, M. (1991). Characterization, molecular cloning and sequencing of Yp3-S1, a fertile yolk protein 3 mutant in *Drosophila. Mol. Gen. Genet.* **228**:81–82.

Logan, S. K., Garabedian, M. J. and Wensink, P. C. (1989). DNA regions that regulate the ovarian transcriptional specificity of *Drosophila* yolk protein genes. *Genes Dev.* **3**:1453–1461.

Logan, S. K. and Wensink, P. C. (1990). Ovarian follicle cell enhancers from the *Drosophila* yolk protein genes: different segments of one enhancer have different cell-type specificities that interact to give normal expression. *Genes Dev.* **4**:613–623.

Minoo, P. and Postlethwait, J. H. (1985). Processing and secretion of a mutant yolk polypeptide in *Drosophila. Biochem. Genet.* **23**:913–932.

Mitsis, P. G. and Wensink, P. C. (1989a). Identification of yolk protein factor 1. A sequence-specific DNA-binding protein from *Drosophila melanogaster. J. Biol. Chem.* **264**:5188–5194.

Mitsis, P. G. and Wensink, P. C. (1989b). Purification and properties of yolk protein factor 1. A sequence-specific DNA-binding protein from *Drosophila melanogaster. J. Biol. Chem.* **264**:5195–5202.

Rina, M. and Savakis, C. (1991). A cluster of vitellogenin genes in the Mediterranean fruit fly *Ceratitis capitata*: sequence and structural conservation in Dipteran yolk proteins and their genes. *Genetics* **127**:769–780.

Terpstra, P. and Geert, A. B. (1988). Homology of *Drosophila* yolk proteins and the triacylglycerol lipase family. *J. Mol. Biol.* **202**:663–666.

Warren, T. G., Brennan, M. D. and Mahowald, A. P. (1979). Two processing steps in the maturation of vitellogenin polypeptides in *Drosophila melanogaster. Proc. Natl Acad. Sci.* (*USA*) **76**:2848–2852.

Yan, Y. L., Kunert, C. J. and Postlethwait, J. H. (1987). Sequence homologies among the three yolk polypeptide (*Yp*) genes in *Drosophila melanogaster. Nucl. Acids Res.* **15**:67–85.

II

II

34

Size Variations Among the Elements that Constitute the Genes of *Drosophila* (Leader, coding region 3′ untranslated region, exons, introns)

The discussion in this chapter centers around two questions: (1) what are the size ranges of the various elements that constitute a functional gene? and (2) is there a correlation between the size of one element and the size of another.

The data analyzed in this chapter are derived from 73 of the genes presented in Part I. Because 12 of those 73 genes have multiple transcripts, they encompass a total of 87 transcripts. Two partly overlapping datasets can be examined: Dataset A includes all 87 transcripts, but elements shared by different transcripts of the same gene are considered only once. For example, if two transcripts of a gene differ only with respect to the poly(A) site, both 3′ untranslated regions (3′ UTR) are included in the analysis, but the leader is counted only once, since it is the same for both transcripts. Dataset B includes only one representative from each family of related genes or from the group of multiple transcripts of a given gene. In this case the sample is reduced to 40 "unrelated" transcripts. The size of a few elements were found to be outside the expected size range suggested by statistical analyses and so were excluded from the analysis. These elements are the 3′ UTR of *bsg25D* II and the leader, 3′ UTR and introns of *Ubx*.

Coding Regions and Untranslated Regions

The questions posed in the first paragraph are discussed as they apply to those parts of the gene that give rise to the mature mRNA: the leader, the coding region and the 3′ untranslated regions (the size of these regions in bp will be represented by the symbols *Leader*, *CR* and *3′UTR*, respectively, and *mRNA* will be used for *mRNA* size. These elements are often encoded by segments in more than one exon; however, because they are the constitutive parts of the mature message, they will be considered here as units. The size and position of exons and introns will be discussed in the next section.

Size Distribution

Table 34.1 covers Dataset A and lists 87 transcripts arranged in order of increasing *CR*, values are given for *Leader*, *CR*, *3′UTR*, *mRNA*, and the fraction

TABLE 34.1. Dataset A

	Gene	Leader	CR	3′UTR	mRNA	CR/mRNA
*	Mtn	124	123	140	387	0.32
	Mto	144	132	100	376	0.35
	CecA1	73	192	81	346	0.55
*	CecA2	81	192	81	354	0.54
	CecB	71	192			
	Sgs7	32	225	61	319	0.71
	Sgs8	32	228	92	353	0.65
	CytC2	44	318	311	673	0.47
*	CytC1	68	327	212	607	0.54
*	Hspg2	60	336	69	465	0.72
	Hspg2 d	182	336	103	622	0.54
	Lcp3	45	339			
	S15	45	348	126	519	0.67
	Vm32E	29	351	54	434	0.81
	Lcp2	42	381			
*	Lcp1	42	393			
*	Rp49	9	402			
	JanA	60	408	241	661	0.62
	Cp16	46	417	52	515	0.81
*	JanB	100	423	56	579	0.73
*	Vm26A1	81	426	122	629	0.68
*	Sgs5	33	492	129	653	0.75
	Vm26A2	62	507	56	625	0.81
*	Hspg3	168	510	301	979	0.52
	Cp18	44	519	86	649	0.80
*	Cp19	45	522	86	653	0.80
*	Hsp22	251	525	181	957	0.55
	Hsp23	112	561	201	874	0.64
	ASC-ac	63	606	243	912	0.66
*	Ddc-Cc	200	612	376	1,188	0.52
	Hsp26	184	627	138	949	0.66
	Hsp27	119	642			
*	Fcs3C	111	654	41	786	0.83
	Hspg1	93	717			
*	Ddc-Cs	353	738	605	1,696	0.44
	Adh d	123	771	173	1,067	0.72
*	Adh p	70	771	173	1,014	0.76
*	ASC-lsc	27	774	383	1,184	0.65
*	Cp36	31	861	112	1,004	0.86

TABLE 34.1. *Continued*

	Gene	Leader	CR	3′UTR	mRNA	CR/mRNA
	Cp38	77	921	293	1,290	0.71
	Sgs3	29	924	164	1,117	0.83
*	h alpha1	491	1,014	830	2,335	0.43
	h alpha2	295	1,014	830	2,139	0.47
*	ASC-sc	117	1,038	283	1,438	0.72
*	Ubx IVa	966	1,041	2,100	4,106	0.25
*	Sryb	144	1,056	99	1,299	0.81
*	Act5C I	155	1,131	184	1,560	0.73
	Act5C II	155	1,131	543	1,919	0.59
	Act5C III	119	1,131	184	1,524	0.74
	Act5C IV	119	1.131	543	1,883	0.60
	Act42A	102	1,131			
	Act79B	147	1,131			
	Act87E I	82	1,131	355	1,568	0.72
	Act87E II	82	1,131	367	1,580	0.72
	Act88F	95	1,131			
*	eve	94	1,131	191	1,416	0.80
	Ubx Ia	966	1,143	986	3,096	0.37
*	ftz	70	1,242			
*	Yp3	59	1,260	168	1,490	0.85
*	kni	271	1,290	507	2,068	0.62
	Sry d	67	1,293	104	1,464	0.88
	Yp1	61	1,320	181	1,562	0.85
	Yp2(I)	51	1,329	166	1,546	0.86
	Yp2(II)	51	1,329	250	1,630	0.82
*	EF-1AF2	138	1,389	1,030	2,558	0.54
	EF-1AF1	80	1,392	582	2,054	0.68
*	Kr(I)	185	1,401	265	1,851	0.76
	Kr(II)	185	1,401	633	2,219	0.63
*	Ddc I	197	1,428	298	1,923	0.74
*	Pgd	35	1,446	178	1,659	0.87
	ASC-ase	456	1,461	346	2,263	0.65
*	Amy	33	1,485	83	1,601	0.93
*	Ddc-DoxA	90	1,485	82	1,657	0.90
*	bcd	169	1,485	817	2,471	0.60
	Ddc II	232	1,533	298	2,064	0.74
	Ddc-amd	150	1,533	99	1,782	0.86
*	Sry a	43	1,593	226	1,862	0.86
*	y	171	1,626	188	1,985	0.82
*	prd	245	1,842	330	2,417	0.76
	Hsp70C1d	242	1,926			
*	Hsp70A7d	246	1,932	210	2,388	0.81
*	bsg25D I	296	2,226	198	2,720	0.82
	bsg25D II	296	2,226	2,227	4,749	0.47
	hb d	510	2,277	561	3,348	0.68
*	hb p	161	2,277	561	3,000	0.76
*	otu2	122	2,436	486	3,045	0.80
	otu1	171	2,562	486	3,220	0.80

Asterisks mark transcripts included in dataset B.

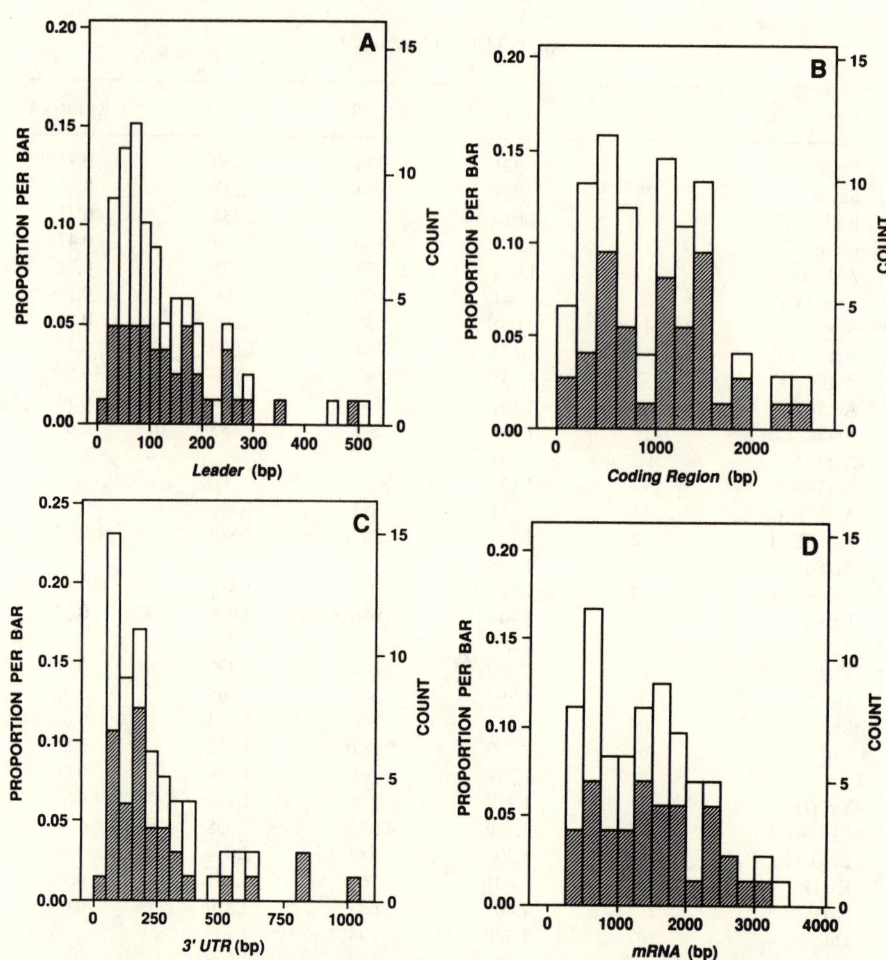

FIG. 34.1. Frequency distributions of size classes of: mRNA, coding regions, leaders and 3′ UTRs. Open bars represent Dataset A, shaded bars represent Dataset B. The "Count" scale measures the absolute number of cases in each class and it applies to both datasets. The "proportion per bar" scale measures the fraction of the total in each class and it applies only to dataset A. The transcripts for *Ubx* and *bsg25D* II were excluded.

	N	*Min.*	*Max.*	*Mean*	*St. Dev.*
(A) Leader:					
Dataset A	79	9	510	127	103
Dataset B	39	9	491	137	101
(B) Coding Region:					
Dataset A	76	123	2,562	959	589
Dataset B	39	123	2,436	1,036	575
(C) 3′ UTR:					
Dataset A	66	41	1,030	250	206
Dataset B	34	41	1,030	260	236
(D) mRNA:					
Dataset A	72	318	3,348	1,412	773
Dataset B	39	354	3,044	1,517	770

of the mature mRNA represented by *CR*. Fig. 34.1 shows frequency distributions for the size of these elements. For both datasets, *mRNA* and *CR* are broadly distributed; 90% of all *mRNA* values lie between 350 bp and 2,500 bp and 90% of all *CR* values are between 120 bp and 1,600 bp. For Dataset B, the *Leader* profile also forms a broad shoulder, but the *3′UTR* distribution is more skewed toward the smaller sizes. Both variables seem to have a threshold at the smaller end of the distribution, *Leader* at about 30 bp (with only 9 bp, *RP49* has the smallest leader) and *3′UTR* at about 50 bp (no *3′UTR* is smaller than 40 bp). Among the longer elements is found the leader of *Ubx* (966 bp) (excluded from the data in Fig. 34.1), which may contain a functional open reading frame (the leader associated with this secondary open reading frame is only 12 bp). The 3′ UTR of some *Ubx* and *bsg* transcripts are also outside the size normal range at approximately 2,100 and 2,200 bp, respectively.

Size Correlations

When regression analyses were applied to Dataset A, significant correlations were observed for several pairs of variables (*3′UTR vs CR*, *Leader vs 3′UTR*, etc.). However, many of these correlations were probably due to the inclusion of multiple members of the same family of transcripts. When the analysis was carried out using Dataset B, most of the correlations disappeared; the exceptions are as follows (Table 34.2):

1. There was a highly significant correlation ($p < 0.001$) between *Leader* and *3′UTR*. Even when a single representative from each family of transcripts was considered, 31% of the variability in *Leader* was associated with changes in *3′UTR* ($r^2 = 0.31$) (Fig. 34.2).

2. *Leader* ($r^2 = 0.23$), *CR* ($r^2 = 0.86$) and *3′UTR* ($r^2 = 0.42$) were correlated to *mRNA*. This is as would be expected since the last variable is the sum of the first three.

TABLE 34.2. Size correlations for various pairs of genetic elements from Dataset B

	Leader	*CR*	*3′UTR*	*Exon1*	*Exon2*	*LastExon*	*mRNA*	*Intron1*	*Intron2*
Leader		NS	***	NS	NS	**	***	*	NS
CR			NS	NS	NS	***	***	*	NS
3′UTR				NS	NS	***	***	NS	NS
Exon1					NS	NS	NS	NS	NS
Exon2						NS	NS	NS	NS
LastExon							***	*	NS
mRNA								*	NS
Intron1									NS
Intron2									

The significance of each correlation is indicated by asterisks: *, $p < 0.05$; **, $p < 0.01$; ***, $p < 0.001$; NS, indicates that the correlation is not significant. *Exon1*, *Exon2* and *Intron2* are not correlated with any of the variables. *Exon1* is not correlated with the number of exons either.

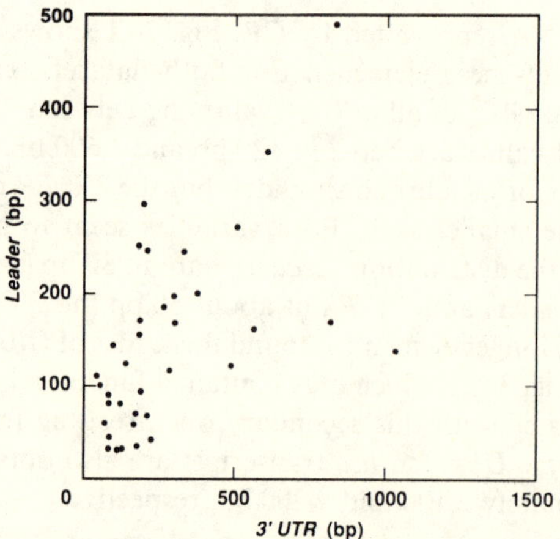

FIG. 34.2. Plot of leader size as a function of 3′ UTR size for Dataset B (*Ubx* was excluded). Regression analysis is actually not permissible on the raw data because there is lack of variance homogeneity. To obviate this problem a logarithmic transformation was applied and significant correlation was observed between the transformed variables.

Introns and Exons

The Number of Introns

Fig. 34.3 shows the frequency distribution of transcripts according to the number of introns. For genes with 0 to 3 introns, there was no statistically significant correlation between number of introns and *Leader*, *CR*, *3′UTR* or

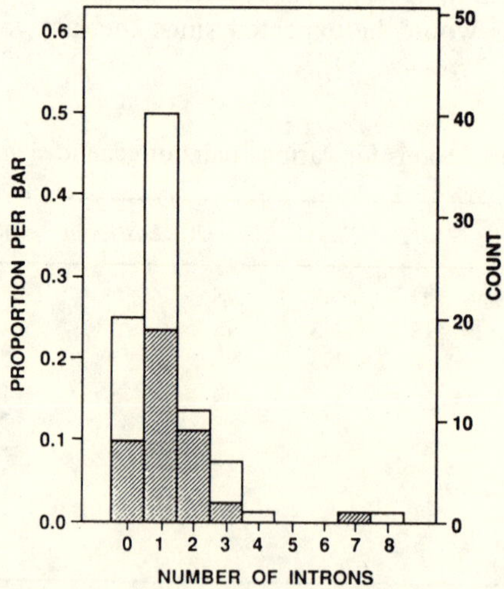

FIG. 34.3. Frequency distribution of transcripts classified according to the number of introns. See Fig. 34.1 legend.

TABLE 34.3. The size of the leader, the coding region, the 3′ UTR and the mRNA in genes with various number of introns

	Number of introns			
	0	*1*	*2*	*3*
Leader	125	108	173	144
	(20, 24)	(40, 15)	(11, 45)	(6, 25)
3′UTR	203	222	338	281
	(17, 24)	(31, 32)	(10, 90)	(6, 113)
CR	881	880	1,164	837
	(20, 129)	(40, 92)	(10, 166)	(6, 275)
mRNA	1,167	1,306	1,647	1,244
	(17, 142)	(31, 142)	(10, 234)	(6, 336)

Mean size in bp. Numbers in parentheses indicate the number of observations and the standard error of each mean.

mRNA (Table 34.3). There was no correlation either between number of introns and exon sizes.

The Size of Exons

In order to study the size distribution of exons, the last exons of all the genes were classified in a single category. The remaining exons were classified as exon 1, exon 2, exon 3, etc., starting at the 5′ end; the size of the corresponding exons are designated *LastExon*, *Exon1*, *Exon2*, and *Exon3*. No significant differences were found among *Exon1*, *Exon2* and *Exon3*; meaningful comparisons among higher numbered exons were not possible because they are so few in numbers. *Exon1*, *Exon2* and *Exon3* (*UpstreamExons*), however, are significantly smaller than *LastExon*. The frequency distribution of *UpstreamExons* is shown in Fig. 34.4; the most frequent size class, is between 50 and 150 bp. *LastExon* shows a much broader distribution (Fig. 34.5).

In order to evaluate the frequency with which leader introns occur, a plot was prepared of the frequency distribution of genes according to the position of the first intron. As Fig. 34.6 shows, the distribution is fairly uniform around the AUG codon; i.e., there is no obvious cluster of genes possessing a leader intron. It would appear that there is a preferred location for the first intron in the neighborhood of the AUG codon, and whether it occurs to its right or to its left is a question of chance. For Dataset A, the position of the first intron is centered around the origin of translation, with more than 50% of transcripts having the first intron within 50 bp on either side of the AUG. For dataset B, however, the peak is not quite so sharp, and it is centered 50 bp downstream of the AUG codon. This preference for the first intron to be near the translation initiation site may be a simple coincidence of the average sizes of leaders and first exons, or it may be determined by certain sequence characteristics of that region. That the first explanation is most likely the correct one is suggested by

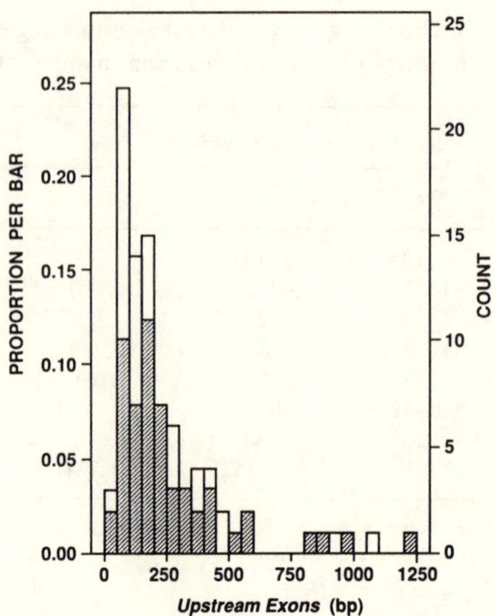

FIG. 34.4. Frequency distribution of upstream exons classified according to size. See Fig. 34.1 legend. *Ubx* was excluded.

	N	*Min.*	*Max.*	*Mean*	*St. Dev.*
Dataset A	89	22	1,245	245	240
Dataset B	55	22	1,245	264	242

the fact that leader introns seem to be more common among genes with longer leaders (Table 34.4).

The Size of Introns

The size distribution of introns appears to be uniform across the various classes of introns (intron 1, intron 2, etc.), and values were pooled for Fig. 34.7A and B: 47% of all introns fall in the size class 50–75 bp; and 24% are between 60 and 70 bp. However, introns that are many thousands of bp long also occur, as in the case of *Ubx*.

Size Correlations Dataset B was used to estimate correlations between various pairs of variables with the following results (Table 34.2):

1. *Exon1* and *Exon2* were independent of the size of the mature mRNA, but *LastExon* was highly correlated to *mRNA* ($r^2 = 0.64$).

2. As might have been expected from the *LastExon/mRNA* correlation, *CR* ($r^2 = 0.55$) and *3'UTR* ($r^2 = 0.22$) were correlated with *LastExon*. *Leader* was also correlated with *LastExon* ($r^2 = 0.32$). This might not have been expected except for the observation that *Leader* was correlated with *3'UTR*, as was mentioned in the previous section.

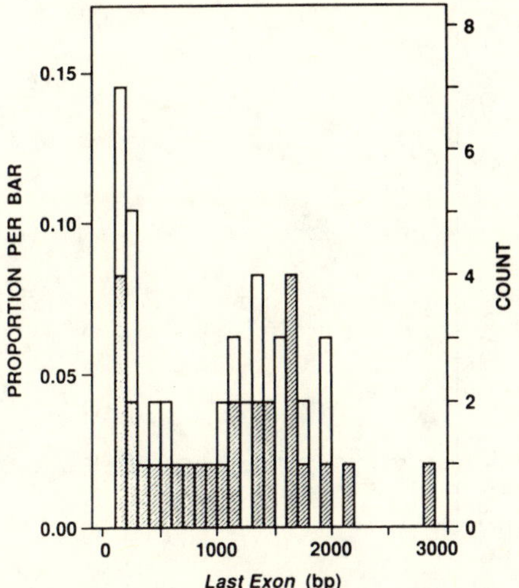

FIG. 34.5. Frequency distribution of the last exons classified according to size. See Fig. 34.1 legend. *Ubx* and *bsg* II were excluded.

	N	*Min.*	*Max.*	*Mean*	*St. Dev.*
Dataset A	50	124	2,856	1,046	713
Dataset B	39	124	2,856	1,109	698

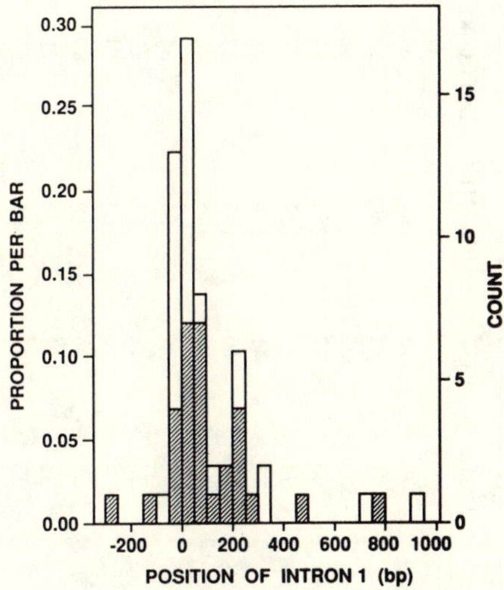

FIG. 34.6. Frequency distribution of transcripts classified acccording to the position of the first intron. Position 0 marks the AUG codon. See Fig. 34.1 legend. *Ubx* was excluded.

TABLE 34.4. Genes in dataset B classified by the size of intron 1

Gene	*Leader*	3′UTR	EXON1	INTRON1	LASTEX	mRNA	CR	*X(INT1)*	*EXON > 1*	*INTRON > 1*
Adh p	70	173	169	65	440	1,014	771	99	405	70
CecA2	81	81	180	58	174	354	192	99		
Ddc-Cs	353	605	53	62	1,643	1,696	738	−300		
Ddc-DoxA	90	82	254	61	1,403	1,657	1,485	164		
eve	94	191	233	71	1,183	1,416	1,131	139		
Fcs3C	111	41	568	73	218	786	654	457		
Hspg2	60	69	181	67	124	465	336	121	160	72
JanB	100	56	152	58	146	579	423	52	156, 125	57, 61
Lcp1	42		54	64			393	12		
Pgd	35	178	43	75	1,392	1,659	1,446	8	224	1,419
Rp49	9		102	59			402	93		
S19	45	86	60	89	593	653	522	15		
S36	31	112	79	91	925	1,004	861	48		
Sgs5	33	129	301	56	164	653	492	268	188	60
Sryb	144	99	217	68	1,082	1,299	1,056	73		
Yp3	59	168	264	62	843	1,490	1,260	205	383	72

Amy	33	83				1,601	1,485			
ASC-sc	117	283				1,438	1,038			
CytC1	68	212				607	327			
Hsp22	251	181				957	525			
Hsp70A7d	246	210				2,388	1,932			
Hspg3	168	301				979	510			
Sry a	43	226				1,862	1,593			
Vm26A1	81	122				629	426			
Act5C I	155	184	147	1,667	1,413	1,560	1,131	−8		
bcd	169	817	334	559	1,102	2,471	1,485	165	76, 959	40, 513
bsg25D I	296	198	528	776	1,947	2,720	2,226	232	245	1,168
Ddc I	197	298	191	869	1,646	1,923	1,428	−6	86	1,029
Ddc-Cc	200	376	401	360	787	1,188	612	201		
EF-1AF2	138	1,030	22	1,245	1,390	2,558	1,389	−116	87, 853, 206	450, 456, 78
ftz	70		827	150			1,242	757		
h alpha1	491	830	590	1,021	1,649	2,335	1,014	99	96	136
hb p	161	561	144	283	2,856	3,000	2,277	−17		
Kr(I)	185	265	222	372	1,629	1,851	1,401	37		
kni	271	507	271	733	1,719	2,068	1,290	0	78	214
Mtn	124	140	146	265	241	387	123	22		
otu2	122	486	118	537	663	3,045	2,436	−4	233, 98, 155, 396, 137, 1,245	62, 67, 57, 53, 583, 68
prd	245	330	312	356	2,105	2,417	1,842	67		
y	171	188	409	2,719	1,576	1,985	1,626	238		

X(INT1) indicates the position of the first intron relative to the translation initiation site. EXON > 1 and INTRON > 1 contain the values of all exons and introns between the first and last ones. The top panel includes Class I genes, the bottom panel, Class II and the middle panel, Class III.

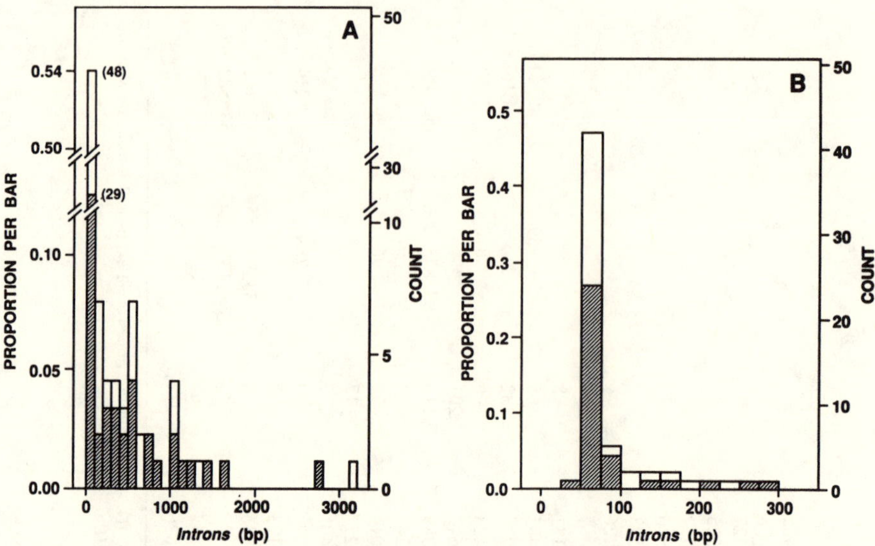

FIG. 34.7. Frequency distribution of introns classified according to size. See Fig. 34.1 legend. Panel B includes introns between 0 and 300 bp only. *Ubx* was not included.

3. Surprisingly, *Intron1* (the size of intron 1) had a significant, if not very strong, correlation with the size of several other elements. Naturally, some of these multiple associations may not have been independent of each other; i.e., *Intron1* might be correlated to *mRNA* because it was correlated to *LastExon*, which is in turn a determinant of *mRNA*.

Classes of Genes

The correlation between *Intron1* and *mRNA* ($r^2 = 0.19$) was not due to a smooth relationship between the two variables but rather to the fact that all introns 1 of small size were associated with mRNAs smaller than 1.7 kb while most larger introns were associated with mRNAs of more than 1.7 kb (Fig. 34.8). The same phenomenon explains the correlations between *Intron1* and the other variables and the correlation between *Leader* and *3'UTR*. In other words, all of the unexpected size correlations that were found are ascribable to the fact that genes with introns can be classified into two groups: class I (those having a first intron of less than 100 bp) and, class II (those having a first intron of more than 100 bp). As a group, class I genes have significantly smaller *Leader*, *3'UTR* and *CR* than class II genes. Within each class, none of the size correlations exist (Fig. 34.9, Table 34.4 and Table 34.5).

What is the biological or molecular significance of two such distinct classes of genes? One possibility is that some of the larger introns may contain segments important for the control of gene expression. Several instances of regulatory

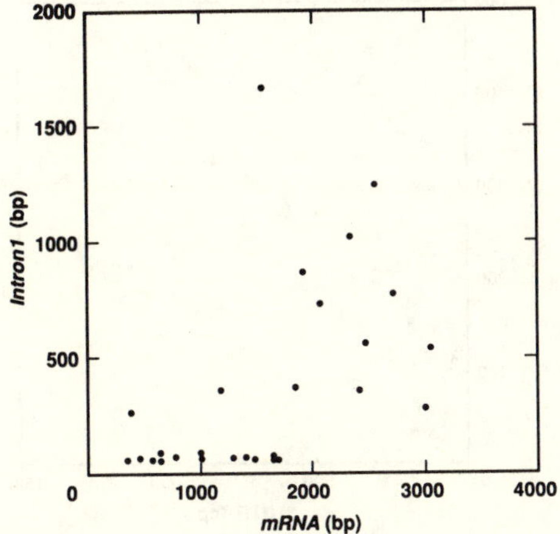

FIG. 34.8. Plot of intron 1 size as a function of mRNA size for Dataset B. Regression analysis is actually not permissible on the raw data because there is lack of variance homogeneity. To obviate this problem a logarithmic transformation was applied, and a significant correlation was observed between the transformed variables.

TABLE 34.5. The size of the leader, the coding region and the 3′ untranslated region in three classes of genes

	Class I	*Class II*	*Class III*
Leader	85	200	126
	(16, 80)	(15, 100)	(8, 87)
CR	760	1,435	980
	(16, 408)	(15, 614)	(8, 621)
3′UTR	148	444	202
	(14, 140)	(14, 278)	(8, 74)

Mean size in bp. Numbers in parentheses indicate the number of observations and the standard error of each mean. Analysis of variance indicates that in each case, Class II means are significantly different from Class I and Class III means ($p = 0.05$).

sequences in transcribed but non-coding regions of genes have been documented (see, for example, *bcd*, *ftz*, *Hsp70*, *Pgd*, *Ubx*). But why should the presence of such regulatory elements be associated exclusively with larger coding regions? Alternatively, the explanation may rest entirely with the mechanics of mRNA transcription and processing (see Chapter 35). Another possible explanation, albeit one that does not seem to be borne out by the data, is that genes within each class are more closely related to one another than to genes of the other class and that the correlations presented here are just a consequence of "family resemblance".

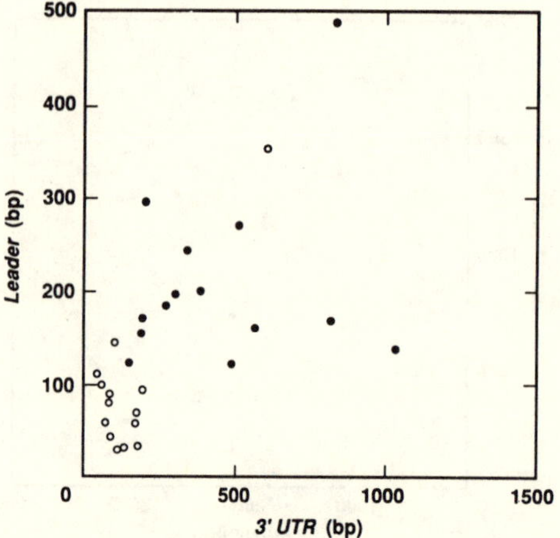

FIG. 34.9. Plot of leader size as a function of 3′ UTR size for Class I (○) and Class II (●) genes in Dataset B. Regression analysis within each class showed no significant correlation between the variables.

In addition to the two classes of genes treated heretofore, there is a third class, those without introns. The mean values for *Leader*, *CR* and *3′UTR* in intronless genes fall in between the values for class I and class II genes. Statistically, however, those values are significantly smaller than the values for class II genes, and not significantly different from the values for class I (Tables 34.4 and 34.5).

35

Messenger RNA splicing signals in *Drosophila* genes

Stephen M. Mount
Department of Biological Sciences
Columbia University
New York, NY 10027

This chapter provides a general description of introns in *Drosophila* genes, with emphasis on the genetic information responsible for the correct specification of boundaries between introns and exons. The problem of locating introns within unannotated DNA sequences is posed by any large genomic sequencing project, and provides a perspective for discussing the information that specifies their removal. I want to stress, however, that there may not be a single set of rules that can identify all introns in all tissues. Certainly, it has become clear that the rules for locating introns will differ between species, such as flies and humans, in different taxonomic classes. Here, I will attempt to describe in general terms both what is known about how introns are recognized by the splicing machinery, and how an investigator might go about identifying introns within the sequence of his favorite *Drosophila* gene. Ultimately, such searches will be carried out by computer. Most current software, however, is designed specifically or primarily for species other than *Drosophila* (one exception is the program GM (Fields and Soderlund 1990), which accepts organism-specific consensus matrices and codon asymmetry tables). I am currently developing computational applications of the ideas described here, and interested readers are encouraged to consult current releases of the electronic *Drosophila Information Newsletter.*

The Mechanism of Splicing

To understand how genetic information specifies the removal of introns, one must understand splicing at the level of biochemical mechanism. To date, the biochemistry of splicing has been studied in extracts from HeLa cells or yeast (reviewed by Smith et al. 1989; Green 1991; Guthrie 1991). However, *Drosophila* is becoming increasingly important to the study of messenger RNA splicing, primarily because of extremely promising genetic systems bearing on the regulation of alternative splicing (Laski et al. 1986; Boggs et al. 1987; Chou et al. 1987; Zachar et al. 1987; Bell et al. 1988; Nagoshi et al. 1988; Pongs et al. 1988; Schwartz et al. 1988; Siebel and Rio 1989; Collier et al. 1990; Geyer et al. 1991; Pret and Searles 1991; McAllister 1992; Steinhauser and Kalfayan 1992; Hazelrigg, unpublished results). Extracts from *Drosophila* cells or embryos that are capable of accurate and efficient removal of introns from RNA substrates have been described (Rio 1988; Hodges and Bernstein 1992; Guo et al. 1992), and are certain to be used increasingly. However, the HeLa *in vitro* system will just as certainly continue to provide the biochemical paradigm, and most of the information in this section pertains to results derived using extracts from HeLa cells.

The Chemistry of Splicing

The removal of introns from messenger RNA precursors occurs in a series of two cleavage–ligation reactions, each involving transesterification at a splice site phosphate (Fig. 35.1A). Thus, messenger RNA splicing resembles the

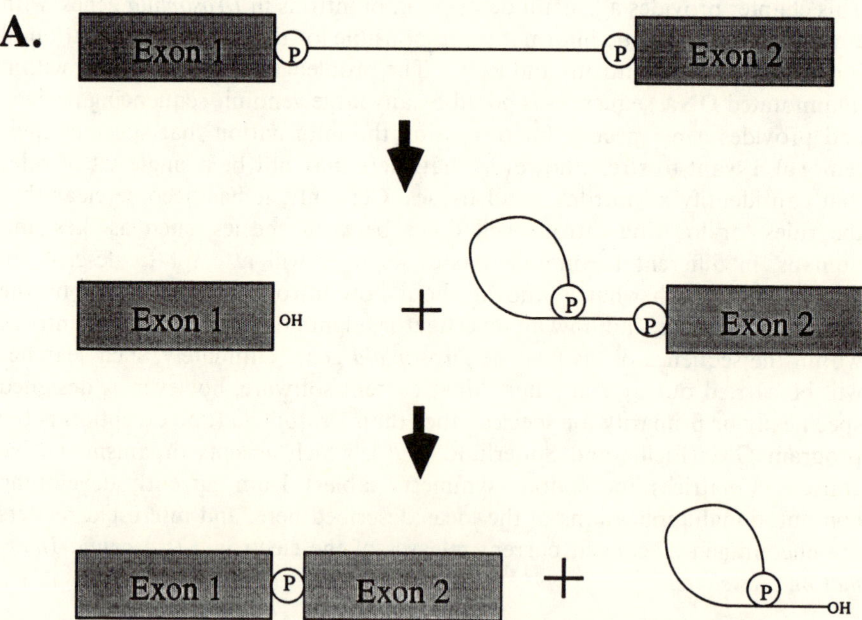

FIG. 35.1. Overview of the splicing mechanism. (a) Each of the chemically distinct steps in the splicing process is indicated. The first phosphotransfer reaction joins the 5′ phosphate of the intron to a 2′ hydroxyl group within the intron, resulting in a free upstream exon and a lariat intermediate. The second step of the splicing reaction joins the now free 3′ hydroxyl of the upstream exon to the phosphate at the 3′ splice site. (*continued*)

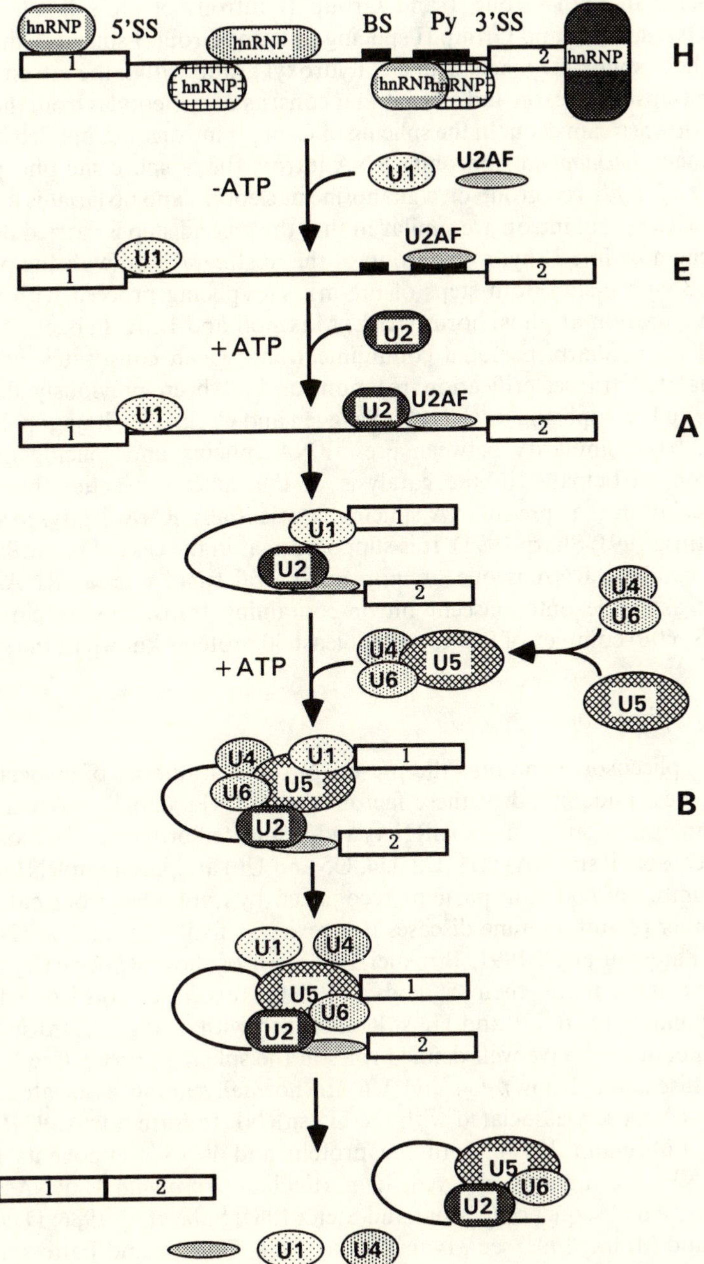

FIG. 35.1 (*continued*). Overview of the splicing mechanism. (B) Spliceosome assembly involves the ordered addition of snRNPs and protein factors. The generally recognized series of steps in HeLa nuclear extract spliceosome assembly are shown and the complexes named (see text).

splicing of both Group I and Group II introns of the self-splicing type. In mRNA splicing and Group II splicing (but not Group I splicing), the phosphate at the 5′ splice site reacts with a 2′ hydroxyl group within the intron, resulting in a free upstream exon and a lariat that consists of nucleotides from the intron and the downstream exon. In the splicing of Group I introns, exemplified by the *Tetrahymena thermophila* ribosomal RNA intron, the 5′ splice site phosphate reacts with a 3′ hydroxyl group on a guanosine nucleotide, and no lariat is formed. These three classes of intron are similar in that the second step is carried out by attack of the now free 3′ hydroxyl group of the upstream exon with the phosphate at the 3′ splice site. Both steps of pre-mRNA splicing proceed with inversion of configuration at phosphorus (K. L. Maschoff and R. A. Padgett, M. J. Moore and P. A. Sharp, personal communication), which constitutes evidence for a concerted transesterification reaction, as had been previously described for Group I self-splicing introns (McSwiggen and Cech 1989; Rajagopal et al. 1989). The basic similarity between pre-mRNA splicing and splicing in which the intron participates in the catalysis of the splicing reaction has led to the speculation that pre-mRNA splicing is essentially RNA-catalyzed (Cech 1986; Guthrie 1991; Sharp 1991). It is supposed that in the case of pre-mRNA splicing the catalytic RNA is one or more of several small nuclear RNAs (snRNAs) that assemble onto nascent intron-containing transcripts as part of a large (40S–60S) complex of RNAs with at least 30 proteins known as the spliceosome.

The Spliceosome

The spliceosome contains the pre-mRNA and a number of associated factors. The best understood of these factors are snRNPs (small ribonucleoproteins), complexes of one or more snRNAs and associated proteins. The most abundant spliceosomal snRNAs (U1, U2, U4, U5 and U6) are present in RNPs containing a number of common proteins recognized by antibodies from patients with a number of autoimmune diseases (for reviews of snRNPs and snRNP proteins, see Paterson et al. 1991; Birnstiel 1988). All of these RNAs carry a trimethyl guanosine cap at their 5′ ends, with the exception of U6, which has a monomethyl cap. U1 and U2 snRNPs, each with a single U snRNA, are most abundant, and have well-defined roles in the splicing process (see Fig. 35.1 and the discussion below). U4 and U6 are normally found associated in a single snRNP, loosely associated with the U5 snRNP to form a tri-snRNP (Beherens and Lührmann 1991). Both the protein and RNA components of these U snRNPs are highly conserved. In particular, *Drosophila* U RNAs are highly conserved in sequence (Mount and Steitz 1981; Saba et al. 1986; Das et al. 1987; Lo and Mount 1991; see Mylinski et al. 1984; Guthrie and Patterson 1988; and Reddy and Busch 1988; for overviews of snRNA conservation). Furthermore, it is generally possible to make a one-to-one correspondence between HeLa cell and *Drosophila* snRNP proteins on the basis of mobility and antigenicity (Paterson et al. 1991), and those proteins involved in splicing whose sequences have been determined in *Drosophila* as well as in vertebrates are also highly conserved (Mancebo et al. 1990; Harper et al. 1992; Zahler et al. 1992).

A considerable number of specific interactions among various components of the spliceosome and the splicing substrate occur prior to the first step of splicing. Green (1991) divides spliceosome assembly into four steps: the U1 snRNP-binding reaction, the U2 snRNP binding reaction, the entry of the U4/U5/U6 tri-snRNP and the loss of U4 snRNP from the spliceosome (Fig. 35.1). A number of intermediates in this process can be separated on non-denaturing gels (Konarska and Sharp 1987) or on sizing columns (Michaud and Reed 1991), and some of the intermediate complexes have been named (Fig. 35.1). Prior to its assembly with spliceosomal components, the pre-mRNA can be found associated with heterogeneous nuclear ribonucleoprotein (hnRNP) proteins both *in vivo* (Dreyfuss 1986) and *in vitro* (Bennett et al. 1992). This early complex, known as the H complex, contains different hnRNP proteins on different substrates. A second complex, known as the E complex, consists of stably bound U1 snRNP, and can assemble in the absence of ATP (Michaud and Reed 1991). Subsequent addition of the U2 snRNP (which associates with the branchpoint) requires ATP and results in the formation of the A complex. A pre-existing complex of U4, U5 and U6 is added to the A complex to form the B complex. Then, the U4 snRNP (without U6) is either lost from the spliceosome (Lamond et al. 1988; Yean and Lin 1991) or destabilized (Blencowe et al. 1989), and splicing follows. Splice site recognition by snRNPs has recently been reviewed by Steitz (1992).

Recognition of 5′ Splice Sites

A 5′ splice site that conforms to the consensus sequence MAG|GURAGU (M = A or C; R = A or G), within which the underlined GU dinucleotide is invariant, is generally required for splicing (Aebi et al. 1986; Green 1986; Smith et al. 1989). The 5′ splice site is recognized by the U1 snRNP (Mount et al. 1983; Black et al. 1985) via base-pairing with the 5′ end of U1 RNA (Zhuang and Weiner 1986; Séraphin et al. 1988; Siliciano and Guthrie 1988), as originally proposed by Lerner et al. (1980) and by Rogers and Wall (1980). The 5′ splice site is probably also recognized by additional factors (Siliciano and Guthrie 1988; Bruzik and Steitz 1990; Seraphin and Rosbash 1990; Stolow and Berget 1991), including the U5 snRNP (Newman and Norman 1991), which appears to recognize the exonic portions of both the 5′ and the 3′ splice sites (Newman and Norman 1992). The G at intron position 1 is required for the second step of splicing as well as for the first; mutations in this position can result in accumulation of lariat intermediates in both yeast (Newman et al. 1985; Vijayraghavan et al. 1986) and mammalian (Aebi et al. 1986) systems. Thus, it appears that nucleotides at the 5′ splice site are recognized multiple times in the course of a single splicing event, and this may help to explain the observation that consensus sequences for the 5′ splice site are highly conserved between species (Mount 1982; Shapiro and Senapathy 1987; Jacob and Gallinaro 1989; Fields 1990; Mount et al. 1992). It is of particular interest to this discussion that the *Drosophila* matrix is remarkably similar to those obtained from mammalian introns (Table 35.1).

TABLE 35.1. 5′ splice site sequences

Drosophila (frequencies, as percentages).

	−5	*−4*	*−3*	*−2*	*−1*	*1*	*2*	*3*	*4*	*5*	*6*	*7*	*8*
A	33	34	37	52	9	0	0	60	71	9	11	39	27
C	24	21	29	15	8	0	0	1	9	2	14	13	21
G	14	23	15	11	71	100	0	35	9	82	6	19	20
T	29	22	19	21	12	0	100	4	11	6	68	29	32
consensus:			M	A	G	G	T	R	A	G	T	W	

Total (all species, dominated by mammals).

	−3	*−2*	*−1*	*1*	*2*	*3*	*4*	*5*	*6*
A	32	60	9	0	0	59	71	7	16
C	37	13	5	0	0	3	9	6	16
G	18	12	79	100	0	35	11	82	18
T	13	15	7	0	100	3	9	6	50
consensus:	M	A	G	G	T	R	A	G	T

Drosophila 5′ splice site scoring table. Scores were calculated according to Hertz et al. (1990).

	−3	*−2*	*−1*	*1*	*2*	*3*	*4*	*5*	*6*	*7*
A	0.6	1.1	−1.4	−5.7	−5.7	1.3	1.5	−1.4	−1.1	0.6
C	0.2	−0.7	−1.6	−4.7	−4.7	−4.1	−1.4	−3.1	−0.8	−0.9
G	−0.7	−1.1	1.5	2.0	−5.7	0.5	−1.4	1.7	−1.9	−0.4
T	−0.4	−0.2	−1.0	−5.7	2.0	−2.5	−1.1	−2.0	1.5	0.2

Recognition of Branchpoints, Pyrimidine Tracts, and 3′ Splice Sites

3′ Splice sites conform to the consensus sequence YAG|G and are typically found at the site of the first AG dinucleotide downstream of the branchpoint. Mammalian branchpoints fit the consensus sequence UNCURAC (in which branch formation occurs at the underlined A) and usually reside between 18 and 38 nucleotides upstream of the 3′ splice site (Noble et al. 1988; Reed and Maniatis 1988; Nelson and Green 1989). Between the branchpoint and the 3′ splice site is a pyrimidine-rich region. The way in which sequences at the 5′ splice site, the branchpoint, the pyrimidine-rich stretch, and the 3′ splice site act together in mammalian splicing to specify intron boundaries has been investigated in detail and much is known of the factors that recognize these sites (Reed and Maniatis 1988; Smith et al. 1989; reviewed in Smith et al. 1989; Green 1991). The branchpoint is recognized by the U2 snRNP via base pairing (Parker and Patterson 1987; Nelson and Green 1989; Wu and Manley 1989; Zhuang et al. 1989; Zhuang and Weiner 1989). However, binding of the

U2 snRNP to the branchpoint requires a number of factors, including the U1 snRNP (Zillman et al. 1987; Ruby and Abelson 1988; Séraphin et al. 1988; Barabino et al. 1990) and U2AF, a factor that binds to the pyrimidine-rich stretch (Ruskin et al. 1988; Zamore and Green 1991).

There exists considerable evidence supporting the proposal that after a branchpoint has been selected (and possibly, but not necessarily, after the first step of splicing) a 3′ splice site is selected at the first AG dinucleotide downstream of the branch. This model is supported by the result, observed in both yeast (Rymond and Rosbash 1985) and HeLa cell extracts (Smith et al. 1989), that the first step of splicing can proceed without an AG dinucleotide if certain conditions are met (see below). In particular, Reed (1989) has divided introns into two categories based on the relative importance of the branchpoint and the pyrimidine tract, and finds that a tract of 14 pyrimidines is sufficient to confer AG-independent splicing. In any event, the lack of AG dinucleotides in the region between the branchpoint and the 3′ splice site (Mount 1982; Shapiro and Senapathy 1987; Gelfand 1989) is suggestive of some sort of microscanning model, as was noted very early (Mount 1982). Consistent with this, mutational analysis indeed indicates that the first AG downstream of such a branchpoint is used as the 3′ splice site (Langford and Gallwitz 1983; Smith et al. 1989). In the mammalian case (Smith et al. 1989), CAG, UAG or AAG, introduced between the branchpoint and the genuine 3′ splice site, were found to "capture" splicing, but GAG in the same position prevented splicing altogether, a result that is consistent with the lack of any recorded 3′ splice sites with the sequence GAG.

Recently, Reich et al. (1992) observed that compensatory changes in U1 RNA can suppress mutations in the AG at the 3′ splice site in *Schizosaccharomyces pombe*, indicating base pairing between U1 and the 3′ splice site prior to the first step of splicing. Thus, U1 RNA interacts with both splice sites prior to the first step of splicing, at least for some introns (possibly all those introns that require the 3′ splice site AG to complete the first step of splicing). This division of introns into categories based on AG-dependence can be extended to include a third category: those introns that do not require U1 at all (Bruzik and Steitz 1990). Thus, it is becoming apparent that the relative contributions of particular factors to intron recognition may vary among introns.

Species-specificity of Splicing Signals

Although it is now clear that mRNA splicing is carried out by a universally conserved fundamental mechanism, it does not follow that there is conservation of splicing signals. In fact, both *in vivo* and *in vitro* systems splice introns derived from other phyla either inaccurately or not at all, and those interested in the expression of genes in *Drosophila* must keep in mind that there is no counterpart in *Drosophila* to the wealth of information available about splicing signals in yeast and mammalian cells. However, judicious consideration of *Drosophila* intron sequences, the small but growing database of experimental results obtained in *Drosophila*, and selected results from other species, allows a good understanding of *Drosophila* splicing signals. In this section, I will review what is known about variation between species with respect to the nature and relative contribution of various splicing signals.

Exon Definition and Intron Retention

What happens when a splice site is defective? Naively, one would think that the splice site would be ignored, resulting in retention of the intron whose excision is dependent upon that splice site (intron inclusion—Fig. 35.2). Alternatively, if there is information elsewhere that indicates that a splice should take place within any given region, then another site may be used for the splice (cryptic sites, Fig. 35.2). This result can also be explained by competition between the two sites—either alone would be sufficient to compel a splice, but the stronger site is better at recruiting factors that result in a commitment to splicing. In fact, the result of many mutations in mammalian splice sites is skipping of an entire exon that includes the affected splice site (Mitchell et al. 1986; reviewed in Robberson et al. 1990; see exon skipping, Fig. 35.2). This implies that exons, rather than introns, are recognized as a unit. Such results from mutational analyses have been used, in combination with results from the study of complex assembly *in vitro* on model substrates, including an association of U1 snRNP with the 3′ half of introns (Zillman et al. 1987), to advance a theory of exon definition (Robberson et al. 1990). Exon definition implies that the productive assembly of spliceosomal components at splice sites is dependent upon the presence of functional sequences at both ends of each exon. This phenomenon has now been well documented experimentally (Talerico and Berget 1990; Grabowski et al. 1991). The strength of the 5′ splice site at the 3′ end of the internal exon has been shown to be critical for the efficiency of splicing in a manner that is independent of the strength of the upstream 5′ splice site (Grabowski et al. 1991), implying that two U1 snRNPs, interacting with two distinct binding sites (5′ splice site sequences) are critical for splicing.

Drosophila has relatively shorter introns (Hawkins 1988; Bingham et al. 1988; Mount et al. 1992), and relatively longer exons (Hawkins 1988; Maroni, this volume, Chapter 34), than do mammalian species, and exon definition may play a correspondingly smaller role in the determination of splicing patterns. Consistent with this, a two-intron adenovirus test substrate (an exon of 94 nucleotides flanked by an upstream intron of 120 nucleotides and a downstream intron of 89 nucleotides) reveals species-specific behavior when tested in splicing extracts from *Drosophila* and human cells. Mutation of the 5′ splice site results in exon skipping in splicing extracts from HeLa cells (Talerico and Berget 1990), but intron inclusion in extracts from *Drosophila* cells (M. Talerico and S. Berget, personal communication). Intron inclusion has also been observed in response to similar mutations in a two-intron *Drosophila* substrate from the *zeste* gene assayed in *Drosophila* extracts (M. Talerico and S. Berget, personal communication). Finally, there are hints that intron inclusion may be

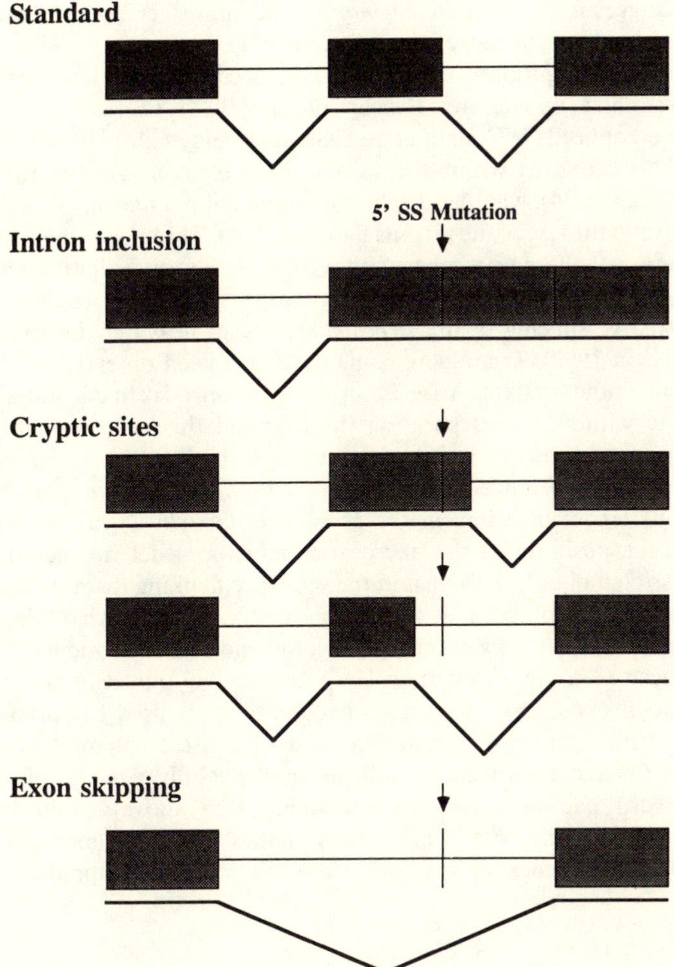

FIG. 35.2. Exon-skipping, intron inclusion, and the use of cryptic splice sites as responses to inactivation of a splice site. The splicing pattern of a typical gene segment including three exons and two introns is depicted in the top cartoon ("standard"), and altered patterns of splicing that can result from a mutation at one splice site (here the a mutation at the 5′ splice site) are shown below: activation of cryptic splice sites, intron inclusion, exon skipping. Most alternative splicing can be explained in terms of one of these three responses to an inactivated (or activated) site.

accompanied by a greater stability of intron-containing RNA *in vivo*. The classical *in vivo* result of 5′ splice site mutations in vertebrate systems is no RNA. In contrast, flies carrying 5′ splice site mutations have been observed to accumulate intron-containing RNA (S. Wasserman, personal communication; unpublished data from the author's laboratory). These results are all consistent with the suggestion that in *Drosophila* the intron, rather than the exon, is the unit of recognition during spliceosome assembly.

However, recognition of exons probably occurs as well. The term "microexon" was first applied by Beachy et al. (1985) to two, 51 nucleotide, alternatively spliced (O'Conner et al. 1988; Kornfeld et al. 1989), exons in *Ubx*. In fact, these exons are within the normal range of exon sizes (see Maroni, this volume, Chapter 34); what led to their being called microexons was their small size relative to the size of the introns flanking them. This makes them candidates for regulation of alternative splicing by regulation of exon definition/recognition, as is the case in the sex-specific autoregulation of *Sexlethal* (Bell et al. 1988), and alternative splicing of the *Drosophila* myosin heavy chain gene (Hodges and Bernstein 1992). True microexons have also been observed in *Drosophila* genes. One rather striking case is an exon of only six nucleotides that lies somewhere within 26 kb separating the first and third exons of the *invected* gene (Coleman et al. 1987). To my knowledge, the location of this microexon has never been ascertained. McAllister et al. (1992) describe two nine-nucleotide microexons whose inclusion in the *Drosophila fasciclin I* gene is variable. In this case, the positions of the microexons have been determined (they reside within a stretch of only 2.7 kb), and the sequence flanking them provides a clue as to how exons of such a small size might be recognized. Each of the microexons is preceded by a long stretch (160 and 120 nucleotides) of sequence with reduced G content (less than 10%) and no AG dinucleotides. Thus, it is possible to propose that these microexons are recognized by formation of a complex at the microexon 5′ splice site and a site greater than 100 nucleotides upstream. Once commitment to splicing (and possibly removal of the downstream exon) had occurred, microscanning (see above) could locate the appropriate 3′ splice site. This model makes the experimentally testable prediction that microexons will generally use remote branchpoints.

Introns with High A + T Content

Animal introns are not properly recognized in transfected plant cells (Weibauer et al. 1988). This is despite the observation that splice site consensus sequences are fairly similar between plants and animals (Brown 1986; Goodall and Filopowicz 1991; White et al. 1992). It appears that the relative A + T-richness of plant introns is critical to their proper recognition (Goodall and Filopowicz 1989), an effect that is more pronounced in introns from dicots than in introns from monocots. The upshot of considerable mutational analyses (assayed by transfection into tobacco (dicot) protoplasts) is that these cells will recognize as an intron almost any sequence that is extremely A + T-rich and is flanked by appropriate, short, consensus sequences (Goodall and Filopowicz 1991); branchpoint and pyrimidine tract sequences are not important to splicing. As would be predicted from the foregoing, deletion of intron sequences so as to move the boundary between A + T-rich and flanking sequences is sufficient to activate cryptic 3′ splice sites that lie in the vicinity of the new boundary (Lou et al. 1992). *Drosophila* also has introns that are significantly richer in A + T than are flanking exons (65% versus 48%; Mount et al. 1992), but the possible contribution of A + T content (or of critical subsequences composed of A and T) to intron recognition has not been demonstrated experimentally. In fact, a survey of base composition in introns versus exons (Csank et al. 1990) reveals that mammals are unique among species surveyed in their lack of a significant difference between introns and exons with respect to A + T content, and the yeast *Saccharomyces cerevisiae* is among species with the smallest difference in A + T content between introns and exons. Thus, a contribution of A + T content to the recognition of introns may be general, but poorly described because of the choice of experimental organisms by the pre-mRNA splicing community.

Variation in Intron Size

Hawkins (1988) and Bingham et al. (1988) were the first to note that there are considerable differences between *Drosophila* and other species (notably mammals) with respect to the size of introns. Specifically, approximately half of all sequenced *Drosophila* introns are less than 80 nucleotides, with a modal length between 60 and 65 nucleotides (Mount et al. 1992). Thus, the typical *Drosophila* intron is smaller than all but a few mammalian introns (Hawkins 1988; Ge et al. 1990), and shorter than the length of approximately 80 nucleotides generally required for efficient splicing in mammalian cells (Wieringa et al. 1984; Ruskin et al. 1985). This strongly suggests species specificity in the recognition of introns (as opposed to the idea that, although smaller than most mammalian introns, the many short *Drosophila* introns would nevertheless be recognized by a mammalian splicing system). An experimental demonstration of species specificity with respect to size requirements has been obtained recently by Guo et al. (1992), working with a short (74 nucleotide) *Drosophila* intron that was properly recognized in homologous (*Drosophila* Kc cell), but not heterologous (HeLa cell) nuclear extracts. An even more extreme situation exists in *C. elegans*, where intron lengths of less than 50 nucleotides are common (Blumenthal and Thomas 1988). Consistent with these observations, a *C. elegans* intron of 53 nucleotides was efficiently spliced in HeLa cell nuclear extracts only when expanded to 84 nucleotides (Ogg et al. 1990).

The distance between the 5′ splice site and the branchpoint and the distance between the branchpoint and the 3′ splice site are presumably subject to different constraints, so it is of interest to know in which portion of the intron species-specific length preferences reside. The distribution of sequences that resemble the branchpoint, for example CTAA, within small introns indicates that distances between the branchpoint and the 3′ splice site in *Drosophila* are very similar to those found in mammals, but 5′ splice site to branchpoint distances are often shorter (typically 38–43 nucleotides; Mount et al. 1992). In the case of the *white* second intron, the experimentally observed branchpoint used by Kc cell extracts is 42 nucleotides from the 5′ splice site (Guo et al. 1992). This 5′ splice site to branchpoint distance is considerably less than that in mammalian introns. For example, manipulation of this distance in the small-t intron (Fu et al. 1988) indicated that the wild-type distance in that case (48 nucleotides) is minimal; an intron with a distance of 46 nucleotides showed no splicing, while a distance of 53 nucleotides showed significantly increased splicing. In another study, Smith and Nadal-Ginard (1989) found 51 nucleotides too short, but 59 sufficient. In addition, a distance of 49 nucleotides between the 5′ splice site and branchpoint was found too short to allow U4,U5,U6 tri-snRNP binding to an adenovirus E1A pre-mRNA *in vitro* (Himmelspach et al. 1991).

Intriguingly, although the distance between branchpoint and 3′ splice site is conserved between mammals and fruit flies, it is not conserved in all species with small introns. For example, Prabhala et al. (1992) examined sequence data for introns from *Schizosaccharomyces pombe*, and found that over half of *S. pombe* introns appear to have less than 10 nucleotides between the branchpoint and the 3′ splice site. In the nematode *C. elegans*, there is a modal intron size of roughly 45 nucleotides (Blumenthal and Thomas 1988), smaller than all but the very smallest *Drosophila* introns. Because no *C. elegans* branchpoint consensus can be discerned, it is unclear which half of the intron has altered size constraints.

The smallest intron in the *Drosophila* data set examined by Mount et al. (1992) is 51 nucleotides. However, a 36 nucleotide intron is described in the *vasa* gene by Lasko and Ashburner (1988), and the set of genes in this atlas contains a 40 nucleotide intron in the *bicoid* gene. In the latter case, the unusually small size may be due to overlap between the branchpoint and 3′ splice site (CTTATCAG|A incorporates both the CTAAT branchpoint consensus and the YAG|G 3′ splice site consensus). The distance between the putative branchpoint and the 3′ splice site in this case would be only four nucleotides, which is an extremely short distance. However, there are a number of *Drosophila* introns with alternative 3′ splice sites bearing a similar relationship. One site is very close to a putative branchpoint and the other is about 15 nucleotides further on. In the *bcd* example, the downstream 3′ splice site corresponds to a short intron that is normal in every respect save one—the occurrence of an AG dinucleotide relatively close to the 3′ splice site, at position −15. Other examples of this arrangement are the two 3′ splice sites in the *Sxl* male-specific exon (Bell et al. 1988), where a branchpoint consensus is 10 nucleotides upstream of one 3′ splice site and 28 nucleotides upstream of another; and Hrb98DE, where the corresponding distances are 5 and 17 nucleotides (Haynes et al. 1990).

The short but relatively constant distance between the 5′ splice site and branchpoint of small *Drosophila* introns raises the possibility of direct contact between complexes at the 5′ splice site and at the branchpoint. A model that incorporates the information summarized here is described in Fig. 35.3. Fig. 35.3A depicts the mechanism of splicing of large introns in *Drosophila* and follows information from splicing in HeLa cells presented in greater detail in Fig. 35.1. Fig. 35.3B presents a model involving direct interaction between a complex at the 5′ splice site and a complex at the 3′ splice site, indicating how accommodation to a shorter distance between the 5′ splice site and branchpoint might lead to a novel, and possibly species-specific, interaction between complexes at the two sites.

Variation in Branchpoint Recognition

Most mammalian introns are not spliced in yeast cells (Beggs et al. 1980; Langford and Gallwitz 1983). This is due, at least in part, to the fact that yeast introns almost always use the precise sequence UACUAAC as a branchpoint, and this sequence is the primary determinant of yeast 3′ splice site selection (Jacquier et al. 1985; Newman et al. 1985; Parker and Guthrie 1985). In contrast, the branchpoint sequence of mammalian introns has greater flexibility (Keller and Noon 1984; Ruskin et al. 1984; Zeitlin and Efstratiatis 1984; Konarska et al. 1985; Reed and Maniatis 1988; Nelson and Green 1989) and the pyrimidine-rich stretch is relatively more important (Frendeway and Keller 1985; Reed 1989). This dichotomy is nicely illustrated by differences in the sequences required for the first step of the splicing reaction to take place in the absence of the second. In yeast, UACUAAC is sufficient (Rymond and Rosbash 1985), while the mammalian splicing machinery demands a significant stretch of pyrimidines for AG-independent splicing (Reed 1989; Smith et al. 1989).

Branchpoint recognition has not been carefully examined in *Drosophila.* In one case, *in vitro* splicing was observed to proceed, using a non-consensus branchpoint, when the wild-type site was mutated (Guo et al. 1992). In addition, the pyrimidine-rich stretch that is so prominent in the literature on mammalian intron splicing (cited above) is absent in a large fraction of *Drosophila* introns, implying that branchpoint recognition must occur in the absence of a significant pyrimidine tract. For example, 49% of short *Drosophila* introns lack even a single stretch of 12 nucleotides including 10 pyrimidines in the region between −50 and −3 relative to the 3′ splice site. In the −26 to −5 region, the average content of pyrimidines in a mammalian intron is 72%. In *Drosophila*, this number is 66%. Perhaps more striking is the observation that A residues are actually more common than C residues (25 versus 22%). This region is high in T (44%) and low in G (9%). In fact, 78% of all *Drosophila* introns in that data set have TTT somewhere in the region between −35 and −3. However, there are introns with few Ts in the region, but no introns that are G-rich in this region. Counting the number of G residues in a 25 nucleotide window adjacent to 3′ splice sites leads to rather striking results; only three out of 205 3′ splice sites have more than five Gs in this region, and the average intron has less than three (Fig. 35.4A). Of 205 3′ splice sites in the data set, only seven have more Gs in an adjacent exonic 25 nucleotide window than in this window. Thus, this region carries a lot of information that can be used to predict the location of a 3′ splice site.

How is this region recognized by the splicing machinery in *Drosophila*? U2AF recognizes the pyrimidine tract and promotes recognition of the branchpoint by U2 in mammalian splicing, and U2AF activity has been found in *Drosophila* extracts (Zamore and Green 1991). It is possible that pyrimidine-poor *Drosophila* introns are indeed recognized by U2AF; the *Drosophila*

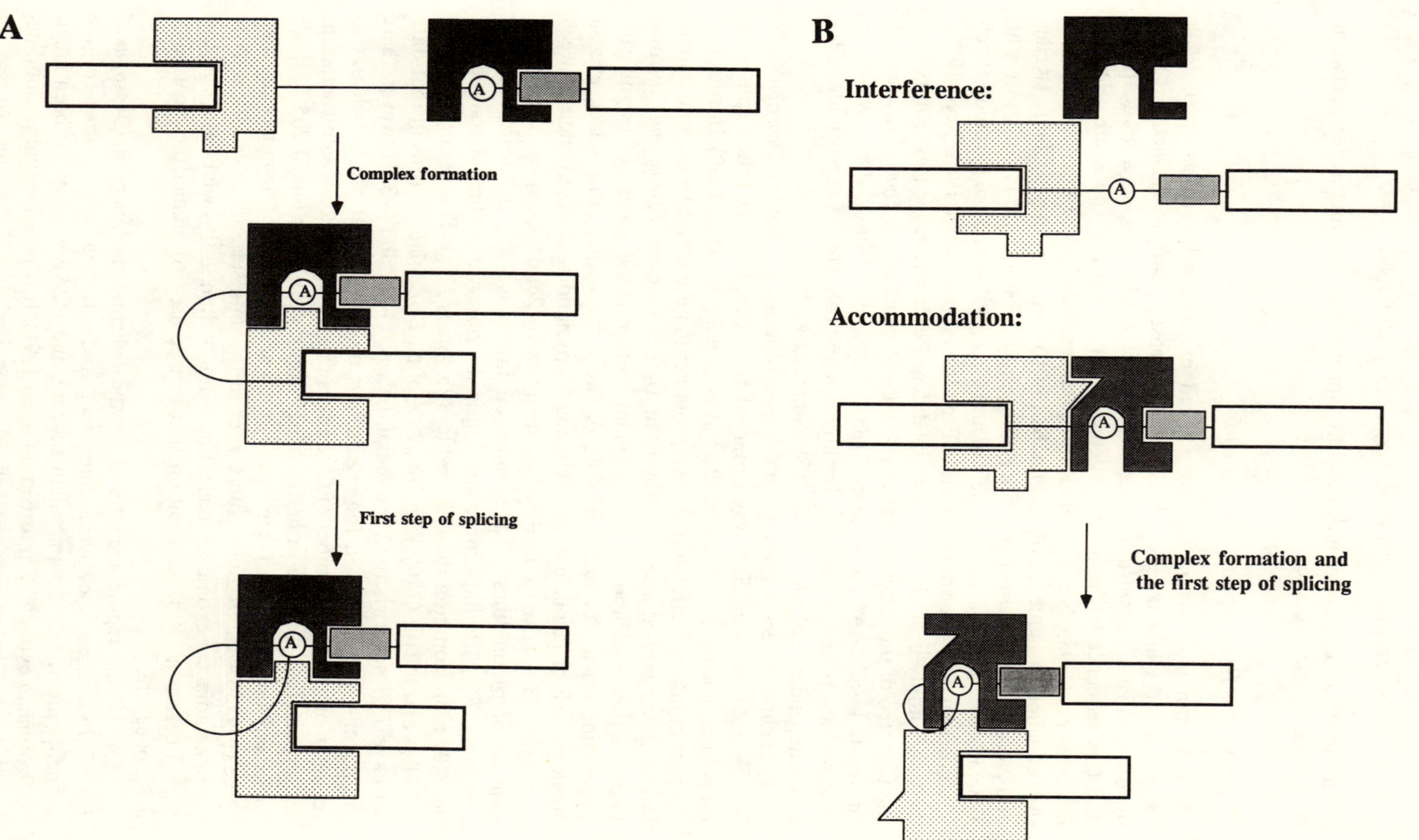
A
Complex formation
First step of splicing
B
Interference:
Accommodation:
Complex formation and
the first step of splicing

homologue of U2AF could simply have an altered sequence specificity, and bind to G-poor rather than pyrimidine-rich regions. Another possibility is that other factors are involved.

A Strategy for the Identification of *Drosophila* Introns

This section is written for the *Drosophila* geneticist or developmental biologist who has just sequenced his favorite gene, but has not yet isolated cDNAs, or would like to assess the likelihood that additional spliced RNAs exist, and if so, which splice sites are likely to be used. The ideas presented here are being developed into computer programs that can be used by those involved in any large scale *Drosophila* genomic sequencing project.

Splice sites conform well to a consensus, and the identification of potential splice sites on the basis of conformity to frequency matrices, such as those in Tables 35.1–35.3, generated by tabulating known splice sites, would appear to be a straightforward matter. However, there is no universally accepted method for weighting specific nucleotides within such matrices. Lear et al. (1990) experimentally determined the strength, relative to a reference site, of 37 actual 5′ splice sites within a common defined sequence context by HeLa cell transfection assays, and then compared those results to a compilation of primate data (Shapiro and Senapathy 1987) using a number of distinct scoring schemes. These techniques, including the increasingly applied "Senapathy score," calculated according to Shapiro and Senapathy (Shapiro and Senapathy 1987), performed comparably (giving coefficients of correlation between measured strength and score of between 0.68 and 0.76. However, it is entirely possible that scoring schemes not examined in that paper would yield better results. In particular, the log-likelihood scoring technique derived from information theory and embodied in the programs CONSENSUS and PATSER

FIG. 35.3. A model for the splicing of small introns in *Drosophila.* Many *Drosophila* introns are smaller than the minimum size recognized by the mammalian splicing apparatus (Bingham et al. 1988; Hawkins 1988; Mount et al. 1992), and there are indications of a distinct mechanism for the splicing of small introns in *Drosophila* (Guo et al. 1992). (A) depicts the mechanism of splicing of large introns in *Drosophila,* following Fig. 35.1. Boxes around the splice sites and branchpoint represent complexes formed from splicing factors. (B) presents a model of direct contact between complexes at the 5′ splice site and at the branchpoint that may explain those observations. "Interference" refers to the effect of too short a 5′ splice site to branchpoint distance on mammalian splicing, as described by Himmelspach et al. (1991), who observed a lack of complexes involving U4/U5/U6 or U2. "Accommodation" depicts the idea that *Drosophila* spliceosomal components may have evolved to be compatible with assembly on introns with shorter 5′ splice site to branchpoint distances, and indicates how that might have led to the observation of a preferred 5′ splice site to branchpoint distance of approximately 40 nucleotides in small *Drosophila* introns (Mount et al. 1992).

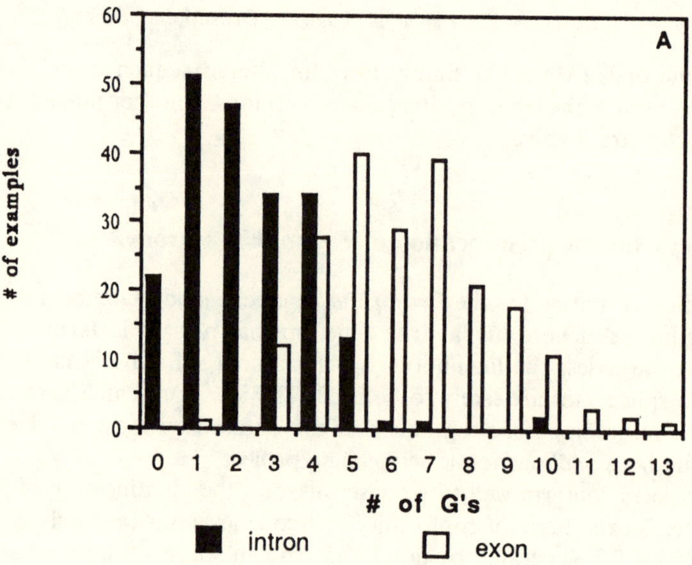

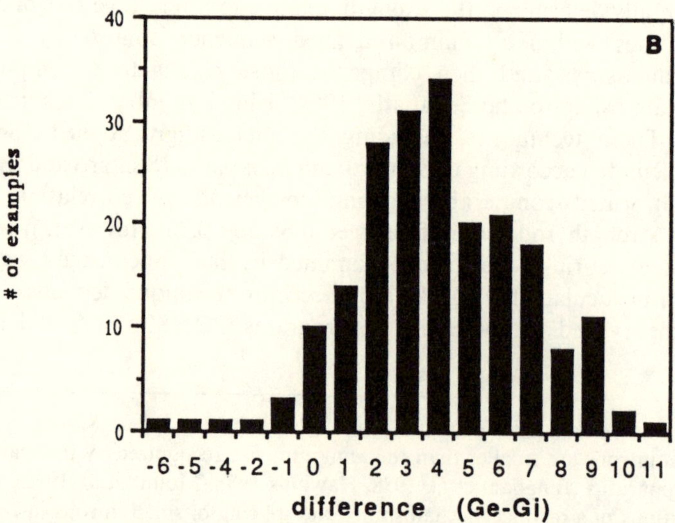

FIG. 35.4. Distribution of G content in 25 nucleotide windows flanking 3′ splice sites. (A) The number of Gs in a 25 nucleotide window within the intron and adjacent to the 3′ splice site (positions −29 to −5) was determined, and the number of examples in a dataset of 205 3′ splice sites with a given number of Gs is indicated by black bars. Note that there are relatively few cases of introns with more than 4 Gs in this window, and only three cases of introns with more than 5 Gs in this window. The number of examples with a given number of Gs in a 25 nucleotide window in the adjacent exon is indicated by white bars. (B) The difference between the number of Gs in a 25 nucleotide window within the intron is subtracted from the number of Gs in a 25 nucleotide window within the flanking exon for each of 205 *Drosophila* 3′ splice sites, and the distribution of results is plotted. Note that very few (seven out of 205) cases of 3′ splice sites have more Gs in the flanking exon than in the intron.

TABLE 35.2. 3′ splice site sequences

Drosophila.

	−14	−13	−12	−11	−10	−9	−8	−7	−6	−5	−4	−3	−2	−1	1	2	3
A	21	21	22	20	19	19	24	19	10	11	28	5	99	0	33	17	18
C	21	23	16	24	24	37	28	36	28	20	23	68	0	0	15	21	32
G	8	10	9	9	10	6	11	6	5	4	23	0	0	100	34	19	25
T	49	45	53	47	47	39	37	40	57	64	25	27	0	0	18	43	25
	T	T	T	T	T	Y	Y	Y	T	T		C	A	G	R	T	

Total.

	−14	−13	−12	−11	−10	−9	−8	−7	−6	−5	−4	−3	−2	−1	1
A	11	11	10	8	11	10	11	11	7	8	25	3	100	0	27
C	29	33	30	30	32	34	37	38	39	36	26	75	0	0	14
G	14	12	10	10	9	11	10	9	7	6	26	1	0	100	49
T	46	44	50	52	48	45	42	43	47	51	23	21	0	0	10
	T	Y	Y	Y	Y	Y	Y	Y	Y	Y		C	A	G	G

Drosophila 3′ splice site scoring table. Scores were calculated according to Hertz et al. (1990).

	−22	−21	−20	−19	−18	−17	−16	−15	−14	−13	−12	−11	−10
A	0.5	0.5	0.1	0.3	0.0	0.2	−0.2	0.1	−0.2	−0.2	−0.2	−0.3	−0.4
C	−0.6	−0.6	−0.3	−0.3	−0.2	−0.3	−0.3	−0.4	−0.2	−0.1	−0.6	0.0	0.0
G	−1.1	−1.3	−1.1	−1.2	−0.9	−1.5	−1.8	−0.8	−1.5	−1.3	−1.4	−1.4	−1.3
T	0.6	0.6	0.8	0.7	0.7	0.8	1.1	0.7	1.0	0.9	1.1	0.9	0.9

	−9	−8	−7	−6	−5	−4	−3	−2	−1	1	2	3
A	−0.4	0.0	−0.4	−1.3	−1.2	0.2	−2.2	2.0	−5.7	0.4	−0.5	−0.5
C	0.6	0.2	0.5	0.2	−0.3	−0.1	1.5	−4.7	−5.7	−0.7	−0.2	0.4
G	−2.0	−1.2	−2.0	−2.1	−2.4	−0.1	−5.7	−5.7	2.0	0.5	−0.4	0.0
T	0.6	0.6	0.7	1.2	1.4	0.1	0.1	−5.7	−5.7	−0.4	0.8	0.0

(Hertz et al. 1990) was not tested. Tables 35.1–35.3 include scoring matrices in addition to frequency matrices for splice sites and branchpoints. Scores were calculated from the data set used by Mount et al. (1992) using the formula $\log_2(4[N_b + 1]/N + 1)$, where N_b is the frequency of a particular base and N is the number of examples that contribute to the matrix (Hertz et al. 1990).

It should be kept in mind that one cannot know the splicing pattern of a gene without looking at the mRNA from that gene. This is primarily because there are exceptions to each of the various features common to most *Drosophila* introns or splice sites. However, there are multiple signals involved in the specification of most introns, and these signals are usually in agreement. Furthermore, biochemical information summarized in the preceding section may serve as guidelines.

TABLE 35.3. Branchpoint consensus

Mammalian examples. Actual numbers, from Nelson and Green (1989).

						BP		
A	3	10	0	8	10	29	1	2
C	8	9	20	6	4	1	15	11
G	5	7	3	0	13	0	7	2
T	15	5	8	17	4	1	8	16
Consensus:	T	N	C	T	R	A	C	Y
Yeast sequence:	T	A	C	T	A	A	C	T

Branchpoints determined for *Drosophila* introns in homologous extracts.

ftz:	A	G	C	T	A	A	C	C	Rio (1988)
white:	T	C	T	T	A	A	T	A	Guo et al. (1992)
Myosin HC exon 19:	T	T	T	T	A	A	T	C	Hodges and Bernstein (1992)
Myosin HC exon 19:	A	A	C	T	A	A	T	T	Hodges and Bernstein (1992)
Myosin HC exon 6:	T	C	C	T	A	A	T	G	Hodges and Bernstein (1992)

Drosophila branchpoint matrix* as determined by CONSENSUS (percentages).

A	36	41	7	20	92	86	1
C	1	10	42	10	2	4	10
G	8	3	2	9	5	1	1
T	55	45	48	60	0	9	88
Consensus:	W	W	Y	T	A	A	T

Drosophila branchpoint scoring table. Scores were calculated according to Hertz et al. (1990) using a weighted average of the matrix above and that in Mount et al. (1992).

A	0.6	0.7	−2.3	−0.2	1.7	1.8	−2.0
C	−1.8	−1.8	1.1	−0.7	−1.7	−2.2	−1.2
G	−1.6	−1.4	−1.2	−1.9	−1.9	−4.3	−4.7
T	0.9	0.8	0.4	1.2	−4.7	−1.7	1.7

Alternative *Drosophila* branchpoint scoring table. Scores were calculated according to Hertz et al. (1990) using the five experimentally determined branchpoints listed above.

A	1.0	0.4	−0.6	−0.6	2.0	2.0	−0.6
C	−0.6	1.0	1.4	−0.6	−0.6	−0.6	0.4
G	−0.6	0.4	−0.6	−0.6	−0.6	−0.6	−0.6
T	1.4	0.4	1.0	2.0	−0.6	−0.6	1.7

* Different in detail from that reported in Mount et al. because a different (uniform) *a priori* base composition was assumed.

Splice sites will also generally occur at boundaries between DNA with roughly 50% A + T content (exons) and DNA with higher A + T content (introns). Introns in non-coding regions may be an exception to this rule. However, G content in the vicinity of 3′ splice sites is an extremely reliable predictor (Fig. 35.4; see below). Splice sites will generally conform to the matrices given in Tables 35.1 and 35.2. If the exons are coding, an open reading frame will generally be continued across the splice sites. Size is also an important clue—over half of the *Drosophila* introns that occur in GenBank are between 50 and 80 nucleotides in length, with the majority of those being between 60 and 66.

Identification of 5′ Splice Sites

5′ splice sites are best identified by the invariant GT and a consensus matrix. The *Drosophila* 5′ splice site matrix determined by Mount et al. (1992) is presented in Table 35.1. This matrix or one like it can be considered universal, reports of minor differences between species (Jacob and Gallinaro 1989) or between introns of different sizes (Fields 1990) notwithstanding. One example of such a difference is the greater frequency of T at position 6 in *Drosophila* as opposed to mammalian introns (68% versus roughly 50%). When the matrix shown in Table 35.1 is used to calculate scores for each of the 205 5′ splice sites in the dataset used, the average score obtained is 6.8. Eighty per cent of actual sites score over 5.0, 95% score over 3.0 and only three sites have negative scores.

The Branchpoint, G-poor Region, and 3′ Splice Site

Two scoring matrices for the *Drosophila* branchpoint are given in Table 35.3. One matrix is based on the five branchpoints that have been experimentally determined in a homologous extract, and one is based on a weighted average of matrices derived using the program CONSENSUS (Hertz et al. 1990; Mount et al. 1992). Given the low number of experimentally determined branchpoints, it is unclear which scoring matrix is preferable, or, indeed, how much importance should be attached to finding a match to the branchpoint matrix.

A G-poor region should be found between the branchpoint and the 3′ splice site, and is an extremely useful tool for locating 3′ splice sites within unannotated sequence. Note that a large portion of this G-poor region is incorporated in the 3′ splice site scoring matrix given in Table 35.2.

In summary, the strategy I propose is a simple one. First, determine open reading frames and plot A + T content and G content across the gene. Then look for splice sites and branchpoints using that information and the scoring matrices given in Tables 35.1–35.3, bearing in mind the size constraints and preferences described above. I am currently developing computational applications of the ideas described here, and interested readers are encouraged to consult current releases of the electronic *Drosophila Information Newsletter* for information about the availability of software. To add your name to the Newsletter distribution list, send e-mail to LISTSERV@IUBVM.UCS.INDIANA.EDU with the message "SUB DIS-L *Your-real-name*." Statistics cited in this chapter were derived by the author and Lonny Sorkin using the data set described in Mount et al. (1992).

Acknowledgements

I thank Lonny Sorkin, a Columbia University Undergraduate Research Fellow, for help with numerous computational aspects of this work, particularly Fig. 35.4 and calculation of the scoring chart in Tables 35.1–35.3. I also thank Owen White for his interest, Susan Berget and Bruce Baker for comments on a draft of this review, and Jian Wu for Fig. 35.1B. This chapter is informed by a series of "Recognizing Genes" workshops at the Aspen Center for Physics. The author's research has been supported by NIH grant 37991, a NSF Presidential Young Investigator Award, and by Basic Research Grant 0347 from the March of Dimes.

References

Aebi, M., Hornig, H., Padgett, R. A., Reiser, J. and Weissman, C. (1986). Sequence requirements for splicing of higher eukaryotic pre-mRNA. *Cell* **47**:555–565.

Barabino, S. L., Blencowe, B. J., Ryder, U., Sproat, B. S. and Lamond, A. I. (1990). Targeted snRNP depletion reveals an additional role for mammalian U1 snRNP in spliceosome assembly. *Cell* **63**:293–302.

Beachy, P. A., Krasnow, M. A., Gavis, E. R. and Hogness, D. S. (1985). Segmental distribution of bithorax complex proteins during *Drosophila* development. *Nature* **313**:545–551.

Beggs, J. D., Berg, J. v. d., Ooyen, A. v. and Weissman, C. (1980). Abnormal expression of a chromosomal rabbit β-globin gene in *Saccharomyces cerevisiae. Nature* **283**:835–840.

Beherens, S-E. and Lührmann, R. (1991). Immunoaffinity purification of a [U4/U6.U5] tri-SnRNP from human cells. *Genes Dev.* **5**:1439–1452.

Bell, L. R., Maine, E. M., Schedl, P. and Cline, T. W. (1988). *Sex-lethal*, a *Drosophila* sex determination switch gene, exhibits sex-specific RNA splicing and sequence similarity to RNA binding proteins. *Cell* **55**:1037–1046.

Bennett, M., Pinol-Roma, S., Staknis, D., Dreyfuss, G. and Reed, R. (1992). Differential binding of heterogeneous nuclear ribonucleoproteins to mRNA precursor prior to spliceosome assembly in vitro. *Mol. Cell. Biol.* **12**:3165–3175.

Bingham, P. M., Chou, T-B., Mims, I. and Zachar, Z. (1988). On–off regulation of gene expression at the level of splicing. *Trends Genet.* **4**:134–138.

Birnstiel, M. L. (ed.) (1988). *Structure and Function of Major and Minor Small Nuclear Ribonucleoprotein Particles* (Berlin: Springer-Verlag).

Black, D. L., Chabot, B. and Steitz, J. A. (1985). U2 as well as U1 small nuclear ribonucleoproteins are involved in pre-messenger RNA splicing. *Cell* **42**:737–750.

Blencowe, B. J., Sproat, B. S., Tyder, U., Barabino, S. and Lamond, A. I. (1989). Antisense probing of the human U4/U6 snRNP with biotinylated 2′O-Me RNA oligonucleotides. *Cell* **59**:531–539.

Blumenthal, T. and Thomas, J. (1988). Cis and trans mRNA splicing in *C. elegans. Trends Genet.* **4**:305–308.

Boggs, R. T., Gregor, P., Idriss, S., Belote, J. and McKeown, M. (1987). Regulation of sexual differentiation in *D. melanogaster* via alternative splicing of RNA from the *transformer* gene. *Cell* **50**:739–747.

Brown, J. W. S. (1986). A catalogue of splice junction and putative branch point sequences from plant introns. *Nucl. Acids Res.* **14**:9549–9559.

Bruzik, J. P. and Steitz, J. A. (1990). Spliced leader RNA sequences can substitute for the essential 5′ end of U1 RNA during splicing in a mammalian in vitro system. *Cell* **62**:889–899.

Cech, T. R. (1986). The generality of self-splicing RNA: relationship to nuclear mRNA splicing. *Cell* **44**:207–210.

Chou, T., Zachar, Z. and Bingham, P. M. (1987). Developmental expression of a regulatory gene is programmed at the level of splicing. *EMBO J.* **7**:4095–4104.

Coleman, K. G., Poole, S. J., Weir, M. P., Soeller, W. C. and Kornberg, T. (1987). The *invected* gene of *Drosophila*: sequence analysis and expression studies reveal a close kinship to the engrailed gene. *Genes Dev.* **1**:19–28.

Collier, V. L., Kronert, W. A., O'Donnell, P. T., Edwards, K. A. and Bernstein, S. I. (1990). Alternative myosin hinge regions are utilized in a tissue-specific fashion that correlates with muscle contraction speed. *Genes Dev.* **4**:885–895.

Csank, C., Taylor, F. M. and Martindale, D. W. (1990). Nuclear pre-mRNA introns: analysis and comparison of intron sequences from Tetrahymena thermophila and other eukaryotes. *Nucl. Acids Res.* **18**:5133–5141.

Das, G., Henning, D. and Reddy, R. (1987). Structure, organization, and transcription of *Drosophila* U6 small nuclear RNA genes. *J. Biol. Chem.* **262**:1187–1193.

Dreyfuss, G. (1986). Structure and function of nuclear and cytoplasmic ribonucleoprotein particles. *Ann. Rev. Cell Biol.* **2**:459–498.

Fields, C. (1990). Information content of *Caenorhabditis elegans* splice site sequences varies with intron length. *Nucl. Acids Res.* **18**:1509–1512.

Fields, C. A. and Soderlund, C. A. (1990). gm: a practical tool for automating DNA sequence analysis. *Comput. Appl. Biosci.* **6**:263–270.

Frendeway, D. and Keller, W. (1985). The stepwise assembly of a pre-mRNA splicing complex requires U-snRNPs and specific intron sequences. *Cell* **42**:355–367.

Fu, X-Y., Colgan, J. and Manley, J. L. (1988). Multiple cis-acting sequence elements are required for efficient splicing of simian virus 40 small-t antigen pre-mRNA. *Mol. Cell. Biol.* **8**:3582–3590.

Ge, H., Noble, J., Colgan, J. and Manley, J. L. (1990). Polyoma virus small tumor antigen pre-mRNA splicing requires cooperation between two 3′ splice sites. *Proc. Natl Acad Sci.* (*USA*) **87**:3338–3342.

Gelfand, M. S. (1989). Statistical analysis of mammalian pre-mRNA splicing sites. *Nucl. Acids Res.* **17**:6369–6382.

Geyer, P., Chien, A. J., Corces, V. G. and Green, M. M. (1991). Mutations in the *su*(*s*) gene affect RNA processing in *Drosophila melanogaster*. *Proc. Natl Acad. Sci.* (*USA*) **88**:7116–7120.

Goodall, G. J. and Filopowicz, W. (1989). The AU-rich sequences present in the introns of plant nuclear pre-mRNAs are required for splicing. *Cell* **58**:473–483.

Goodall, G. J. and Filopowicz, W. (1991). Different effects of intron nucleotide composition and secondary structure on pre-mRNA splicing in monocot and dicot plants. *EMBO J.* **10**:2635–2644.

Grabowski, P. J., Nasim, F-U. H., Kuo, H-C. and Burch, R. (1991). Combinatorial splicing of exon pairs by two-site binding of U1 small nuclear ribonucleoprotein particle. *Mol. Cell. Biol.* **11**:5919–5928.

Green, M. R. (1986). Pre-mRNA splicing. *Ann. Rev. Genet.* **20**:671–708.

Green, M. R. (1991). Biochemical mechanisms of constitutive and regulated pre-mRNA splicing. *Annu. Rev. Cell Biol.* **7**:559–600.

Guo, M., Lo, P. and Mount, S. (1993). Species-specific signals for the splicing of a short *Drosophila* intron *in vitro*. *Mol. Cell. Biol.* **13**:1104–1118.

Guthrie, C. (1991). Messenger RNA splicing in yeast: clues to why the spliceosome is a ribonucleoprotein. *Science* **253**:157–163.

Guthrie, C. and Patterson, B. (1988). Spliceosomal snRNAs. *Ann. Rev. Genetics* **22**:387–419.

Harper, D. S., Fresco, L. D. and Keene, J. D. (1992). RNA binding specificity of a *Drosophila* snRNP protein shares sequence homology with mammalian U1-A and U2-B″ proteins. *Nucl. Acids Res.* **20**:3645–3650.

Hawkins, J. D. (1988). A survey on intron and exon lengths. *Nucl. Acids Res.* **16**:9893–9905.

Haynes, S. R., Raychaudhuri, G. and Beyer, A. L. (1990). The *Drosophila* Hrb98DE locus encodes four protein isoforms homologous to the A1 protein of mammalian heterogeneous nuclear ribonucleoprotein complexes. *Mol. Cell. Biol.* **10**:316–323.

Hertz, G. Z., Hartzell, G. W. H., III and Stormo, G. D. (1990). Identification of consensus patterns in unaligned DNA sequences known to be functionally related. *Comput. Appl. Biosci.* **6**:81–92.

Himmelspach, M., Gattoni, R., Gerst, C., Chebli, K. and Stevenin, J. (1991). Differential block of U small ribonucleoprotein particle interactions during in vitro splicing of adenovirus E1A transcripts containing abnormally short introns. *Mol. Cell. Biol.* **11**:1258–1269.

Hodges, D. and Bernstein, S. I. (1992). Suboptimal 5′ and 3′ splice sites regulate alternative splicing of *Drosophila melanogaster* myosin heavy chain transcripts in vitro. *Mech. Dev.* **37**:127–140.

Jacob, M. and Gallinaro, H. (1989). The 5′ splice site: phylogenetic evolution and variable geometry of association with U1 RNA. *Nucl. Acids Res.* **17**:2159–2180.

Jacquier, A., Rodriguez, J. R. and Rosbash, M. (1985). A quantitative analysis of the effects of 5′ junction and TACTAAC box mutants and mutant combinations on yeast mRNA splicing. *Cell* **43**:423–430.

Keller, E. B. and Noon, W. A. (1984). Intron splicing: a conserved internal signal in introns of animal pre-mRNAs. *Proc. Natl Acad. Sci.* (*USA*) **81**:7417–7420.

Konarska, M. M., Grabowski, P. J., Padgett, R. A. and Sharp, P. A. (1985). Characterization of the branch site in lariat RNAs produced by splicing of mRNA precursors. *Nature* **313**:552–557.

Konarska, M. M. and Sharp, P. A. (1987). Interactions between small nuclear ribonucleoprotein particles in formation of spliceosomes. *Cell* **49**:763–774.

Kronfeld, K., Saint, R. B., Beachy, P. A., Harte, P. J. and Peattie, D. A. (1989). Structure and expression of a family of *Ultrabithorax* mRNAs generated by alternative splicing and polyadenylation in *Drosophila*. *Genes Dev.* **3**:243–258.

Lamond, A. I., Konarska, M. M., Grabowski, P. J. and Sharp, P. A. (1988). Spliceosome assembly involves binding and release of U4 small nuclear ribonucleoprotein. *Proc. Natl. Acad. Sci.* (*USA*) **85**:411–415.

Langford, C. J. and Gallwitz, D. (1983). Evidence for an intron-contained sequence required for the splicing of yeast RNA polymerase II transcripts. *Cell* **33**:7–19.

Laski, F., Rio, D. and Rubin, G. (1986). Tissue specificity of *Drosophila* P element transposition is regulated at the level of mRNA splicing. *Cell* **44**:7–19.

Lasko, P. F. and Ashburner, M. (1988). The product of the *Drosophila* gene *vasa* is very similar to eukaryotic initiation factor-4A. *Nature* **335**:611–617.

Lear, A. L., Eperon, L. P., Wheatley, I. M. and Eperon, I. C. (1990). Hierarchy for 5′ splice sites preference determined in vivo. *J. Mol. Biol.* **211**:103–115.

Lerner, M. R., Boyle, J. A., Mount, S. M., Wolin, S. L. and Steitz, J. A. (1980). Are snRNPs involved in splicing? *Nature* **283**:220–224.

Lo, P. C. H. and Mount, S. M. (1990). *Drosophila melanogaster* genes for U1 snRNA variants and their expression during development. *Nucl. Acids Res.* **18**:6971–6979.

Lou, H., McCullough, A. J. and Schuler, M. A. (1993). Expression of Maize Adh 1 intron mutants in tobacco nuclei. *The Plant J.* **3**.

Mancebo, R., Lo, P. C. H. and Mount, S. M. (1990). Structure and expression of the *Drosophila melanogaster* gene for the U1 small nuclear ribonucleoprotein particle 70K protein. *Mol. Cell. Biol.* **10**:2492–2502.

McAllister, L., Rehm, E. J., Goodman, G. S. and Zinn, K. (1992). Alternative splicing of microexons creates multiple forms of the insect cell adhesion molecule fasciclin I. *J. Neurosci.* **12**:895–905.

McSwiggen, J. A. and Cech, T. R. (1989). Stereochemistry of RNA cleavage by the *Tetrahymena* ribozyme and evidence that the chemical step is not rate-limiting. *Science* **244**:679–683.

Michaud, S. and Reed, R. (1991). An ATP-independent complex commits pre-mRNA to the mammalian spliceosome assembly pathway. *Genes Dev.* **5**:2534–2546.

Mitchell, P. A., Urlaub, G. and Chasin, L. (1986). Spontaneous splicing mutations at the dihydrofolate reductase locus in Chinese hamster ovary cells. *Mol. Cell. Biol.* **6**:1926–1935.

Mount, S. M. (1982). A catalogue of splice junction sequences. *Nucl. Acids Res.* **10**:459–472.

Mount, S. M., Burks, C., Hertz, G., Stormo, G. D., White, O. and Fields, C. (1992). Splicing signals in *Drosophila*: intron size, information content, and consensus sequences. *Nucl. Acids Res.* **20**:4255–4262.

Mount, S. M., Petterson, I., Hinterberger, M., Karmas, A. and Steitz, J. A. (1983). The U1 small nuclear RNA-protein complex selectively binds a 5′ splice site in vitro. *Cell* **33**:509–518.

Mount, S. M. and Steitz, J. A. (1981). Sequence of U1 RNA from Drosophila melanogaster: implications for U1 secondary structure and possible involvement in splicing. *Nucl. Acids Res.* **9**:6351–6368.

Myslinski, E., Branlant, C., Weiben, E. D. and Pederson, T. (1984). The small nuclear RNAs of *Drosophila. J. Mol. Biol.* **180**:927–945.

Nagoshi, R. N., McKeown, M., Burtis, K. C., Belote, J. M. and Baker, B. S. (1988). The control of alternative splicing at genes regulating sexual differentiation in *D. melanogaster*. *Cell* **53**:229–236.

Nelson, K. K. and Green, M. R. (1989). Mammalian U2 snRNP has a sequence-specific RNA-binding activity. *Genes Dev.* **3**:1562–1571.

Newman, A. J., Lin, R-J., Cheng, S-C. and Abelson, J. (1985). Molecular consequences of specific intron mutations on yeast mRNA splicing in vivo and in vitro. *Cell* **42**:335–344.

Newman, A. J. and Norman, C. (1991). Mutations in yeast U5 snRNA alter the specificity of 5′ splice site cleavage. *Cell* **65**:115–123.

Newman, A. J. and Norman, C. (1992). U5 snRNA interacts with exon sequences at the 5′ and 3′ splice sites. *Cell* **68**:743–754.

Noble, J. C. S., Prives, C. and Manley, J. L. (1988). Alternative splicing of SV40 early pre-mRNA is determined by branch site selection. *Genes Dev.* **2**:1460–1475.

O'Conner, M. B., Binari, R., Perkins, L. A. and Bender, W. (1988). Alternative RNA products from the Ultrabithorax domain of the bithorax complex. *EMBO J.* **7**:435–445.

Ogg, S. C., Anderson, P. and Wickens, M. P. (1990). Splicing of a *C. elegans* myosin pre-mRNA in a human nuclear extract. *Nucl. Acids Res.* **18**:143–149.

Parker, R. and Guthrie, C. (1985). A point mutation in the coserved hexanucleotide at a yeast 5′ splice junction uncouples recognition, cleavage and ligation. *Cell* **41**:107–118.

Parker, R. and Patterson, B. (1987). Architecture of fungal introns: implications for spliceosome assembly. In *Molecular Biology of RNA: New Perspectives*, eds M. Inouye and B. S. Dudock (New York, Academic Press), pp. 133–149.

Paterson, T., Beggs, J., Finnegan, D. and Luhrmann, R. (1991). Polypeptide components of *Drosophila* small nuclear ribonucleoprotein particles. *Nucl. Acids Res.* **19**: 5877–5882.

Pongs, O., Kecskemethy, N., Muller, R., Krah-Jentgens, I., Baumann, A., Kiltz, H. H., Canal, I., Llamazares, S. and Ferrus, A. (1988). Shaker encodes a family of putative potassium channel proteins in the nervous system of *Drosophila. EMBO J.* **7**:1087–1096.

Prabhala, G., Rosenberg, G. H. and Käufer, N. F. (1992). Architectural features of pre-mRNA introns in the fission yeast *Schizosaccharomyces pombe. Yeast* **8**:171–182.

Pret, A. M. and Searles, L. (1991). Splicing of retrotransposon insertions from transcripts of the *Drosophila* melanogaster vermilion gene in a revertant. *Genetics* **129**:1137–1145.

Rajagopal, J., Doudna, J. A. and Szostak, J. W. (1989). Stereochemical course of catalysis by the *Tetrahymena* ribozyme. *Science* **244**:692–694.

Reddy, R. and Busch, H. (1988). *Structure and Function of Major and Minor Small Nuclear Ribonucleoprotein Particles* (Berlin: Springer-Verlag), pp. 71–79.

Reed, R. (1989). The organization of 3′ splice-site sequences in mammalian introns. *Genes Dev.* **3**:2113–2123.

Reed, R. and Maniatis, T. (1988). The role of mammalian branchpoint sequences in pre-mRNA splicing. *Genes Dev.* **2**:1268–1276.

Reich, C. I., VanHoy, R. W., Porter, G. L. and Wise, J. A. (1992). Mutations at the 3′ splice site can be suppressed by compensatory base changes in U1 snRNA in fission yeast. *Cell* **69**:1159–1169.

Rio, D. C. (1988). Accurate and efficient pre-mRNA splicing in *Drosophila* cell-free extracts. *Proc. Natl Acad. Sci.* (*USA*) **85**:2904–2909.

Robberson, B. L., Cote, G. L. and Berget, S. M. (1990). Exon definition may facilitate splice site selection in RNAs with multiple exons. *Mol. Cell. Biol.* **10**:84–94.

Rogers, J. and Wall, R. (1980). A mechanism for RNA splicing. *Proc. Natl Acad. Sci.* (*USA*) **77**:1877–1879.

Ruby, S. W. and Abelson, J. (1988). An early hierarchic role of U1 small nuclear ribonucleoprotein in spliceosome assembly. *Science* **242**:79–85.

Ruskin, B., Greene, J. M. and Green, M. R. (1985). Cryptic branch point activation allows accurate in vitro splicing of human β-globin intron mutants. *Cell* **52**:207–219.

Ruskin, B., Krainer, A. R., Maniatis, T. and Green, M. R. (1984). Excision of an intact intron as a noval lariat structure during pre-mRNA splicing in vitro. *Cell* **38**:317–331.

Ruskin, B., Zamore, P. D. and Green, M. R. (1988). A factor, U2AF, is required for U2 snRNP binding and splicing complex assembly. *Cell* **52**:207–219.

Rymond, B. C. and Rosbash, M. (1985). Cleavage of 5′ splice site and lariat formation are independent of 3′ splice site in yeast mRNA splicing. *Nature* **317**:735–737.

Saba, J. A., Busch, H., Wright, D. and Reddy, R. (1986). Isolation and characterization of two full-length *Drosophila* U4 small nuclear RNA genes. *J. Biol. Chem.* **261**:539–542.

Schwartz, T. L., Temple, B. L., Papazian, D. M., Jan, Y. N. and Jan, L. Y. (1988). Multiple potassium-channel components are produced by alternative splicing at the *Shaker* locus in *Drosophila. Nature* **332**:740.

Séraphin, B., Kretzner, L. and Rosbash, M. (1988). A U1 snRNA:pre-mRNA base pairing interaction is required early in yeast spliceosome assembly but does not uniquely define the 5′ splice site. *EMBO J.* **7**:2533–2538.

Séraphin, B. and Rosbash, M. (1990). Exon mutations uncouple 5′ splice site selection from U1 snRNA pairing. *Cell* **63**:619–629.

Shapiro, M. B. and Senapathy, P. (1987). RNA splice junctions of different classes of eukaryotes: sequence statistics and functional implications in gene expression. *Nucl. Acids Res.* **15**:7155–7174.

Sharp, P. A. (1991). Five easy pieces. *Science* **254**:663.

Siebel, C. W. and Rio, D. C. (1989). Regulated splicing of the *Drosophila* P transposable element third intron in vitro: somatic repression. *Science* **248**:1200–1208.

Siliciano, P. G. and Guthrie, C. (1988). 5′ splice site selection in yeast: genetic alterations in base-pairing with U1 reveal additional requirements. *Genes Dev.* **2**:1258–1267.

Smith, C. W. J. and Nadal-Ginard, B. (1989). Mutually exclusive splicing of a-tropomyosin exons enforced by an unusual lariat branch point location: implications for constitutive splicing. *Cell* **56**:749–758.

Smith, C. W. J., Patton, J. G. and Nadal-Ginard, B. (1989). Alternative splicing in the control of gene expression. *Ann. Rev. Genet.* **23**:527–577.

Smith, C. W. J., Porro, E. B., Patton, J. G. and Nadal-Ginard, B. (1989). Scanning from an independently specified branch point defines the 3′ splice site of mammalian introns. *Nature* **342**:243–247.

Steinhauer, W. R. and Kalfayan, L. J. (1992). A specific *ovarian tumor* protein isoform is required for efficient differentiation of germ cells in *Drosophila* oogenesis. *Genes Dev.* **6**:233–243.

Steitz, J. A. (1992). Splicing takes a holiday. *Science* **257**:888–889.

Stolow, D. T. and Berget, S. M. (1991). Identification of nuclear proteins that specifically bind to RNAs containing 5′ splice sites. *Proc. Natl Acad. Sci.* (*USA*) **88**:320–324.

Talerico, M. and Berget, S. M. (1990). Effect of 5′ splice site mutations on splicing of the preceding intron. *Mol. Cell. Biol.* **10**:6299–6305.

Vijayraghavan, U., Parker, R., Tamm, J., Limura, Y., Rossi, J., Abelson, J. and Guthrie, C. (1986). Mutations in conserved intron sequences affect multiple steps in the yeast splicing pathway, particularly assembly of the spliceosome. *EMBO J.* **5**:1683–1695.

Weibauer, K., Herrero, J-J. and Filopowicz, W. (1988). Nuclear pre-mRNA processing in plants: distinct modes of 3′ splice site selection in plants and animals. *Mol. Cell. Biol.* **8**:2042–2051.

White, O., Soderland, C., Shanmugam, P. and Fields, C. (1992). Information contents and dinucleotide compositions of plant intron sequences vary with evolutionary origin. *Plant Molec. Biol.* **19**:1057–1064.

Wieringa, B., Hofer, E. and Weissmann, C. (1984). A minimal intron length but no specific internal sequence is required for splicing the large rabbit β-globin intron. *Cell* **37**:915–925.

Wu, J. and Manley, J. L. (1989). Mammalian pre-mRNA branch site selection by U2 snRNP involves base pairing. *Genes Dev.* **3**:1553–1561.

Yean, S-L. and Lin, R-Y. (1991). U4 small nuclear RNA dissociates from a yeast spliceosome and does not participate in the subsequent splicing reaction. *Mol. Cell. Biol.* **11**:5571–5577.

Zachar, Z., Chou, T. B. and Bingham, P. M. (1987). Evidence that a regulatory gene autoregulates splicing of its transcript. *EMBO J.* **6**:4105–4111.

Zahler, A. M., Lane, W. S., Stolk, J. A. and Roth, M. B. (1992). SR proteins: a conserved family of pre-mRNA splicing factors. *Genes Dev.* **6**:837–847.

Zamore, P. D. and Green, M. R. (1991). Biochemical characterization of U2 snRNP auxiliary factor: an essential pre-mRNA splicing factor with a novel intramolecular distribution. *EMBO J.* **10**:207–214.

Zeitlin, S. and Efstratiatis, A. (1984). In vivo splicing products of the rabbit β-globin gene. *Cell* **39**:589–602.

Zhuang, Y., Goldstein, A. M. and Weiner, A. M. (1989). UACUAAC is the preferred branch site for mammalian mRNA splicing. *Proc. Natl Acad. Sci.* (*USA*) **86**:2752–2756.

Zhuang, Y. and Weiner, A. M. (1986). A compensatory base change in U1 snRNA suppresses a 5′ splice site mutation. *Cell* **46**:827–835.

Zhuang, Y. and Weiner, A. M. (1989). A compensatory base change in human U2 snRNA can suppress a branch site mutation. *Genes Dev.* **3**:1545–1552.

Zillman, M., Rose, S. D. and Berget, S. M. (1987). U1 small nuclear ribonucleoproteins are required early during spliceosome assembly. *Mol. Cell. Biol.* **7**:2877–2883.

36

Translation Start Sites and mRNA Leaders

Douglas R. Cavener and Beth A. Cavener

Vanderbilt University
Department of Molecular Biology
Nashville, TN 37235

Introduction

A prototypical eukaryotic mRNA is often described as having a short (less than 100 nt) 5′ untranslated leader sequence upstream of start codon containing a good consensus sequence (Lewin 1990). Translation initiation from such a mRNA follows the scanning model whereby: (1) a complex of proteins including the cap-binding protein (eIF-4E) associates with the 5′ cap of the mRNA; (2) this complex in turn facilitates binding of the preinitiation complex (40S ribosomal subunits + eIF-2-GTP-tRNA^{met}); (3) the preinitiation complex scans the mRNA searching for the start codon (the first AUG encountered in the prototypical mRNA); (4) the large ribosomal subunit (60S) joins the 40S subunit beginning translation (Kozak 1989). The first *Drosophila* mRNAs characterized (e.g., *Adh*, larval cuticle proteins, and the glue proteins) fit the eukaryotic prototype. However, in more recent years an increasing number of eukaryotic mRNAs have been discovered that contain unusual features. First, approximately 9% of the characterized vertebrate mRNAs contain long leader sequences with upstream open reading frames (Kozak 1987). The presence of upstream open reading frames present a dilemma for the scanning model. If the ribosome engages translation of an upstream open reading frame, terminates, and then dissociates from the mRNA, how is translation of the major coding region achieved? Kozak demonstrated two possible solutions. The scanning preinitiation complex can ignore an AUG codon in the leader if it is in a poor context for initiation or the ribosomes can engage translation of the URF, terminate, resume scanning (presumably in the form of the small ribosomal subunit), reload initiation factors, and reinitiate translation downstream at the start codon for the major coding region (Kozak 1989). Recently, Macejak and Sarnow (1991) have demonstrated a more radical solution: cap-independent, internal binding of the ribosome. Under the "internal initiation" model, the ribosome can bind downstream of any offending URFs and then traverse the remaining leader sequence to the major start codon.

A reanalysis of the translation start site consensus sequence has also altered our view of the prototypical mRNA. Kozak (1984) initially argued that the sequence CCACCAUGG was the eukaryotic consensus sequence for translation initiation and showed the sequences that departed markedly from this consensus reduced translation initiation of the rat preproinsulin mRNA (Kozak 1986). However, similar experiments in yeast failed to show significant reduction in translation initiation from start codons with a "poor context" (Baim and Sherman 1988). Studies on the start codon context of the *Drosophila Adh* gene showed a significant effect of context intermediate to that observed in the rat and yeast studies (Feng et al. 1991). A further complication of the start codon context came from the finding that Kozak's consensus sequence was not based upon explicit quantitative criteria and did not represent a true consensus sequence for any major eukaryotic group (Cavener 1987; Cavener and Ray 1991). Moreover, various eukaryotic groups exhibit somewhat different consensus sequence for the translation initiation site. For example, yeast mRNAs exhibit relatively high frequencies of U at -2 and -1; Kozak had shown that Us at these positions were rare in vertebrates mRNA and detrimental to translation of the rat preproinsulin mRNA. Only the presence of A or G at the -3 position is a consensus throughout eukaryotes (Cavener and Ray 1991).

Data Acquisition and Analysis

We compiled the following data for *Drosophila* mRNAs: (1) length of the leader sequence; (2) method of determining the extent of the leader sequence; (3) the number of upstream start codons (uAUG); and (4) the start codon context from positions -6 to $+4$ for the major translation start sites and for a random sample of the uAUGs. Initially, most of the mRNA sequences were identified in GenBank Release 69 using the INTERBAS computer program (Cavener and Ray 1991). Sequences reported recently in several journals were added to this list. In the vast majority of cases the start codons are readily discernible from the GenBank records. However, information regarding the leader sequence is almost always inaccurate and/or incomplete in GenBank. In many cases the extent of the leader sequence has not been determined empirically. Consequently we examined the primary literature reporting each of the 403 mRNAs listed in Table 36.1 in order to ascertain the method for mapping the leader sequence and to verify the map features of the GenBank records. Since the 5′ end of the leader sequence is defined by the presumptive start site of transcription, the precise limits of the leader sequence are only known in cases where extensive transcript mapping experiments have been conducted. Ideally, this involves a combination of comparing cDNA sequences with genomic DNA sequences, primer extension and nuclease protection experiments. For the majority of *Drosophila* mRNAs these data are incomplete. Typically, the extents of mRNA sequences are inferred only from the analysis of the longest cDNA

TABLE 36.1. Leader lengths (nt), number of upstream AUGs, and translation start site sequences from -6 to $+4$

File	*Encoded protein*	*Method*	*Leader lengths*	*Number of uAUGs*	*Start site*
5HTR	5HT serotonin receptor	b	894	15	CUGCUG**AUG**G
6DHR	RAD6 homolog	c	88	0	UGAAAA**AUG**U
ABDA	abd-A, abdominal-A, homeotic	b	668	0	AGCAAG**AUG**U
ABDBP3	abd-b, abdominal-B, P3	a	?	?	CCCGUC**AUG**C
ACHE	ace, acetylcholinesterase	b	993	6	UAUUCA**AUG**C
ACHRR	acetylcholinesterase receptor	b	254	6	AAAAUC**AUG**G
ACHRX	muscarinic acetylcholine receptor	b	32	0	CCGGCG**AUG**A
ASC1	ac, achaete	e	63	0	CUUAAA**AUG**G
ACS2	sc, scute	e	117	0	GUGUUA**AUG**A
ACT42A	actin 42A	e	102	0	UACAAA**AUG**U
ACT5CX	actin 5C	e	156	0	UACAAA**AUG**U
ACT79B	actin 79B	c	149	0	CCAAAC**AUG**U
ACT87EA	actin 87E	e	82	0	GCCAAG**AUG**U
ACT88F	actin 88F	c	187	4	GCCAAG**AUG**U
ADF1A	adf-1 transcription factor	b	312	0	AUUGAG**AUG**G
ADHa	Adh, alcohol dehydrogenase distal protein	e	123	0	GUCACC**AUG**U
ADHb	Adh, alcohol dehydrogenase proximal protein	e	70	0	GUCACC**AUG**U
AFLL	arf-like, GTP binding protein	b	118	0	GUCAUC**AUG**G
ALSR	acetylcholine receptor alpha	b	1,282	7	CCUAAG**AUG**G
AMA	ama, amalgam	e	235	0	CCAGAC**AUG**G
AMYAG1	amy, amylase	c	35	0	AUCAUC**AUG**U
ANNX	annexin	b	90	0	UGCAUA**AUG**G
ANP*	andropin	c	37	0	CUAGUU**AUG**A
ANTCA	Dfd, deformed homeotic	b	490	4	UCCGUC**AUG**A
ANTCF	ftz, fushi tarazu	c	120	1	UCCGAU**AUG**G
ANTPa	Antp, antennapedia P1 mRNA	e	1,527	8	GCCACG**AUG**A
AnTPb	Antp, antennapedia P2 mRNA	e	1,729	15	GCCACG**AUG**A
ANTPS2	position-specific antigen 2	b	258	5	GACAAA**AUG**A
APRT	adenine phosphoribosyltransferase	b	89	0	AGAAAA**AUG**A
ARMa	armadillo E16	e	135	1	ACCAAG**AUG**A
ARMb	armadillo E9	e	170	0	ACCAAG**AUG**A
ARR	arrestin-1	c	120	0	UCCAAA**AUG**G
ARRA	arrestin-2	c	116	0	UCCAAA**AUG**G
ASCA	T3 of achaete-scute	c	27	0	AUUACC**AUG**A
ASCB	T8 of achaete-scute	a	?	?	UUUGGC**AUG**C
ASE	ase, asense	c	456	8	UUAAUU**AUG**G
ATPA	Da-47, Na^+/K^+ ATPase alpha subunit	b	12	0	AAUAAC**AUG**G
AWDR	awd, abnormal wing disc	e	25	0	GCGACA**AUG**G
B52*	B52 protein, NHCP	b	55	0	GUUAUC**AUG**G
BAM	bam, bag-of-marbles	c	186	1	AGAAUA**AUG**C
BCD16	bic, bicoid	b	169	2	GGGAAA**AUG**G
BICD	bic^D bicaudal-D	b	131	0	AUCAUC**AUG**U

(continued)

TABLE 36.1. *Continued*

File	*Encoded protein*	*Method*	*Leader lengths*	*Number of uAUGs*	*Start site*
BJ1G	BJ1, chromatin-binding protein	b	210	0	GCUAAA**AUG**C
BJ6	no-on transient A, Bj6	b	76	0	UAAAAA**AUG**G
BR*	br, broad	b	386	0	AUCGAG**AUG**G
BROWN	bw, brown	b	268	1	CUCGAA**AUG**C
BSG25D	bsg25D, blastoderm	e	296	1	CGGAUA**AUG**G
BX189A	pH189A ORF, BX-C	a	?	?	UCCUAA**AUG**U
BX189B	ph189B ORF, BX-C	c	1,019	5	UACCCG**AUG**G
BX200	pH200 gene, BX-C	c	494	1	UACAGA**AUG**G
C1A9	NHC, non-histone chromosomal protein	b	349	4	AACAAA**AUG**G
CACTTR	choline actyltransferase	c	406	0	GCGAAC**GUG**G
CADA1a	cad, caudal zygotic	e	460	4	CCAGCC**AUG**G
CADA1b	cad, caudal maternal	e	301	3	CCAGCC**AUG**G
CAIM1	calmodulin	b	85	0	ACAAAA**AUG**G
CAPKCA	cAMP-dep protein, kinase catalytic	a	?	?	UCCAAG**AUG**G
CATHPO	catalase	b	87	1	AGCAAA**AUG**G
CCG	Cc gene, Ddc region	a	?	?	AGGAUA**AUG**G
CDC2P24	cdc2 homolog	b	55	0	UAAAUU**AUG**G
CHAB	potassium channel protein	b	406	5	GGUGGC**AUG**G
CHORS16	chorion, s16	?	46	0	AAAAAA**AUG**U
CHORS3	chorion, s36	c	31	0	GGCAAC**AUG**C
CHORS3	chorion, s38	e	77	0	GACAAG**AUG**A
CHORSGa	chorion, S18-1	c	44	0	CUCAGA**AUG**A
CHORSGb	chorion, S15-1	c	45	0	CUCACC**AUG**A
CHORSGc	chorion, S19-1	c	45	0	AUAGCC**AUG**A
CID	ciD, cubitus interruptus dominant	b	415	6	AAUGAA**AUG**G
CLARET	claret non-disjunctional^{+}	a	?	?	UUGGCG**AUG**G
CNC	cnc, segmentation protein	b	94	0	UGUCGC**AUG**G
COPO1	chaoptin	e	255	0	AGCAAA**AUG**G
CRN*	crn, crooked neck, cell cycle	b	80	0	CACAGC**AUG**G
CRPA	crumbs protein	b	213	4	GCGAUC**AUG**G
CSG	Cs, Ddc region	a	?	?	GAUUCG**AUG**U
CSKA	casein kinase II alpha	b	258	0	AGAAAA**AUG**A
CSKB	casein kinase II beta	b	22	0	AUCAAA**AUG**A
CSPAA	cysteine-string protein 29	b	150	0	AUCAGG**AUG**A
CSTAA	ctr, concertina	b	133	1	CCAGCG**AUG**U
CTCL1	cuticle protein I	c	42	0	GCGAAU**AUG**U
CTCL2a	cuticle protein II	f	42	0	GCCAAC**AUG**U
CTCL2b	cuticle protein III	c	45	0	AUCAAA**AUG**U
CTCL2c	cutical protein IV	f	45	0	GUCAAA**AUG**U
CUT	cut	b	268	4	CCACGA**AUG**C
CYCA	cyclin A	b	296	5	CGCACC**AUG**G
CYCC*	cyclin	b	93	0	UACGAA**AUG**G
CYCDC3	cytochrome c, DC3	a	?	?	UCCAAG**AUG**G
CYCDC4	cytochrome c, DC4	a	?	?	UCCAUA**AUG**G
CYCLB	cyclin B	b	123	0	AUCAAA**AUG**G
CYP1	cyp-1 protein, cyclophilin	a	?	?	UCAAAG**AUG**A

TABLE 36.1. *Continued*

File	Encoded protein	Method	Leader lengths	Number of uAUGs	Start site
D1DE	insulin-degrading enzyme	b	297	1	CCCAAGAUGA
D1P	chromosomal protein D1	b	227	0	AGAGAAAUGG
DA2	D alpha-2 protein, D'2	b	492	1	GUCACCAUGG
DC1AB	DC1, putative protein kinase	b	92	3	GCUGUUAUGA
DC2	DC2, putative protein kinase	b	894	1	ACAGCGAUGU
DCKA	calmodulin-dependent protein kinase	b	250	0	AUCGCGAUGG
DCO	cAMP-protein kinase catalytic subunit	c	828	2	UCCAAGAUGG
DDC a	Ddc, dopa decarboxylase CNS	c	233	0	UCUGAAAUGA
DDC b	Ddc, hypoderm form	c	197	0	AUCGACAUGG
DDY3	Ddyn3, dynamin shibire locus	b	394	2	GCCGCAAUGG
DDYN4	Ddyn4, dynamin shibire locus	b	51	0	GCCGCAAUGG
DEC1A	dec-1 chorion-1 fc125	e	75	0	UACAGGAUGA
DELTA	D1, delta, neurogenic (DLG)	b	141	0	AUAAACAUGC
DFUR1	dfur1, furin-type protein	b	104	0	CCCACAAUGA
DG1A1	cGMP-dependent protein kinase	c	108	1	GGCAGAAUGG
DG2T1A3	cGMP-dependent protein kinase	e	97	0	GCCUGGAUGC
DG2T2A	cGMP-dependent protein kinase	e	776	9	UUCGUAAUGA
DG2T2B	cGMP-dependent protein kinase	e	338	1	UUCGUAAUGA
DGHTRL	da, daughterless	b	212	2	GCUGAAAUGG
DIPT	diptericin	b	24	0	ACUGAGAUGC
DLGA	discs-large tumor suppressor	b	380	3	UGCGAUAUGA
DMYD	Dmyd, myogenic	b	262	2	UGAAAAAUGA
DNC	dnc, dunce	b	363	4	AGUCUUAUGA
DORSAL	dl, dorsal	b	274	2	CACAUAAUGU
DOXA2	A2 comp. of diphenol oxidase	c	90	0	UACAAAAUGA
DPPC	dpp, decapentaplegic	b	1,187	6	GCGACCAUGC
DRCII1	II-cAMP-dependent protein kinase regulatory subunit	e	402	1	AGCGAAAUGG
DRCIV1	IV-cAMP-dependent protein kinase regulatory subunit	c	182	1	AGCCCGAUGC
DRICI1	I-cAMP-dependent protein kinase regulatory subunit	e	565	3	UACCACAUGU
DSK	sulfated tyrosine kinin	a	?	?	CUGUUUAUGC
DSX*	doublesex, male and female	e	1,020	9	GGAAUCAUGG
E74A	E74A, ecdysone inducible	e	1,891	17	UCAGCGAUGC
E74B	E74B, ecdysone inducible	e	793	6	UGCAAAAUGA
E75A	E75A, ecdysone inducible	e	380	3	AGCAAAAUGU
E75B	E75B, ecdysone inducible	e	284	3	UCAAAUAUGG
EAG	putative potassium channel protein	b	463	2	GGCAAAAUGC

(*continued*)

TABLE 36.1. *Continued*

File	Encoded protein	Method	Leader lengths	Number of uAUGs	Start site
EAST	easter, putative serine protease	b	203	0	ACGAAA**AUG**C
ECR*	EcR, ecdysone receptor	b	1,068	11	CAGAGG**AUG**A
EDG78A	EDG-78 cuticle protein	c	76	0	AUCAUC**AUG**U
EDG84A	EDG-84 cuticle protein	c	61	0	AUCAGC**AUG**U
EDG91B	EDG-1, cuticle protein	c	34	0	AUCGCA**AUG**G
EF1AF1	elongation factor, F1	c	80	0	UCCAAC**AUG**G
EF1AF2	elongation factor, F2	c	139	0	GCAAGG**AUG**G
EF2A	translation elongation factor 2	b	72	0	UCCAAA**AUG**G
EFSII	RNA pol II elongation factor	b	236	0	GCCAAA**AUG**A
EGFRA	epidermal growth factor receptor homolog	b	84	0	GAUAUC**AUG**A
EGFRB	epidermal growth factor receptor homolog	b	22	0	GCAACA**AUG**C
EIF2AL*	eIF-2 alpha subunit	a	?	?	UUUAAC**AUG**G
EIF2BE*	eIF-2 beta subunit	b	>99	1	GACACA**AUG**G
EIP28G	ecdysone inducible protein	e	65	1	GAAAUC**AUG**U
ELAVK	elav protein	b	491	1	AAAACA**AUG**G
ELF1	Elf1, DNA binding protein	b	920	7	CGUAUA**AUG**U
EMC	emc, extramacrochaetae	c	258	0	UCCAGA**AUG**A
ENGM	en, engrailed	b	168	0	AAACCA**AUG**G
ENHSPA	E(spl), enhancer of split	b	222	0	AACAAC**AUG**U
ESPLM4	E(spl), m4 transcription unit	f	79	0	AUCAUC**AUG**U
ESPLM5	E(spl), m5 transcription unit	c	84	0	UACAAA**AUG**G
ESPLM7a	E(spl), m7 transcription unit	f	128	0	CACACA**AUG**G
ESPLM7b	E(spl), m8 transcription unit	f	96	0	ACAAAA**AUG**G
EST6	Est-6, esterase-6	b	24	0	AGCAAC**AUG**A
EVE	eve, even skipped	c	94	0	CCAAAC**AUG**C
F1GA	F1 50kd protein	b	200	2	UCCAAC**AUG**G
FCN	fasciclin I	b	174	0	GCUAAA**AUG**C
FCNIII	fasciclin III	b	582	2	AAAAUC**AUG**U
FKH	fork head	b	707	3	GACAUC**AUG**C
FMRF	FMRFamide	b	18	1	GCCUUG**AUG**U
FOS*	fos homolog	b	772	5	GCAACA**AUG**A
FPS85D	dfps 85D	b	243	2	AGCAUC**AUG**G
FRZAC2	frizzled, AC2	b	709	8	UCCAAA**AUG**U
FS1YA	fs(1)Ya, nuclear env.	b	23	0	AGGUGU**AUG**U
FSHA	fsh membrane protein A	c	662	3	ACCACC**AUG**U
GADPH1	GAPDH-1	e	62	0	UCAGCC**AUG**U
GADPH2	GAPDH, glyceraldehyde-3-phosphate dehydrogenase	c	49	0	UUAACC**AUG**U
GART	Gart	c	160	0	GGAAUU**AUG**U
GART p	pcp, pupal cuticle gene Gart	b	33	0	GACACC**AUG**U
GIAA	guanine nucleotide binding, regulatory subunit	b	441	3	CACAAG**AUG**A
GLDGMC	Gld, glucose dehydrogenase	e	344	0	AUCAAC**AUG**U
GLUEDA	Glued	b	360	6	UCCUCC**AUG**A
GNBPSA1	guanine nucleotide binding protein alpha	b	486	3	GCUGCG**AUG**G
GOALB*	G-o-alpha-like protein	b	519	2	CGCACC**AUG**G

TABLE 36.1. *Continued*

File	Encoded protein	Method	Leader lengths	Number of uAUGs	Start site
GPAMA*	G protein alpha mRNA type a	b	189	0	ACCACA**AUG**G
GPDHA	Gpdh, glycerol-3-phosphate dehydrogenase	e	136	0	CAAAAU**AUG**G
GTUB	gamma-tubulin	b	196	1	ACCACA**AUG**C
HAIRR	h, hairy	c	492	0	ACCGAA**AUG**G
HBGa	hunchback, maternal mRNA	c	511	1	GCCAAG**AUG**C
HBGb	hunchback, zygotic mRNA	c	165	0	GCCAAG**AUG**C
HELI	RNA helicase	b	33	0	UGAAUA**AUG**A
HGSG2	heat-shock 2, male specific	e	60	0	ACUACA**AUG**G
HISH1	histone, H1	c	36	0	AAAAAG**AUG**U
HLI*	HL, putative troponin I	b	134	0	CUCAAA**AUG**G
HMGCO	HMG CoA reductase	b	572	2	GCAGCC**AUG**A
HOXH20	H2.0 homeobox	b	205	0	CGGACA**AUG**U
HP1	Hp-1	c	169	0	ACAAAA**AUG**G
HRB87F*	Hrb87F, A/B hnRNP protein	c	132	0	GAGAGA**AUG**G
HREC2C	putative steroid hormone receptor	b	198	1	CCCAGG**AUG**G
HSC7A1	cognate of hsp70	a	?	?	GCCGAC**AUG**C
HSP1	heat-shock protein 1	c	94	0	GUGAAA**AUG**U
HSP22G	hsp22, heat-shock protein	e	253	0	ACUACA**AUG**C
HSP27G	hsp27, heat-shock protein	e	121	0	UCAAAA**AUG**U
HSP4	hsp23	e	111	0	ACAAAA**AUG**
HSP7A2	hsp70	e	244	0	CACACA**AUG**C
HSP83A	hsp83	c	148	0	UUGCAG**AUG**C
HSPG3	heat-shock gene 3 from 67B	e	168	0	AGUAAA**AUG**C
HSPHEX	heat-shock transcription factor	b	228	0	CACUUU**AUG**U
IMP	IMP-E2, ecdysone inducible	b	75	0	GCGAUA**AUG**A
INT1HO	Dint-1	b	417	7	GCAAUA**AUG**G
INVR	invected	e	294	3	AAACUG**AUG**U
JUN	dJRA/Djun, jun homolog	b	207	0	GCAAAC**AUG**A
K10G	K10 putative DNA-binding protein	e	191	0	CCUGCA**AUG**G
KINHCA	kinesin heavy chain	b	320	1	UAAGCA**AUG**U
KINLA	nod, kinesin-like protein	b	71	1	AUCUGC**AUG**G
KNIRPS	knirps	b	270	0	UUCCAG**AUG**A
KNR1	knirps-related protein	b	516	4	ACCAUA**AUG**A
KR	krueppel	d	185	1	UUGUUG**AUG**U
L2AMD	alpha-methyldopa hypersen.	b	150	0	AGCGGU**AUG**G
LA9	LAP, DNA-binding protein	b	435	8	GUCAAA**AUG**G
LABG1	labial F24	d	239	0	GACAAU**AUG**A
LAMB1	laminin B1	b	423	5	AUCGAG**AUG**U
LAMB2	laminin B2	b	227	2	CCCACC**AUG**A
LAMDMO	laminin, nuclear	b	130	0	GUGAAC**AUG**U
LAMIN	lamin	c	148	1	GUGAAC**AUG**U
LARM	DLAR, protein tyrosine phosphatase	b	117	0	GAAAUA**AUG**G
LETHAL	lethal(1)2cb sarcoplasmic actinin	b	66	0	CACAAG**AUG**A

(*continued*)

TABLE 36.1. *Continued*

File	Encoded protein	Method	Leader lengths	Number of uAUGs	Start site
LGL2	lethal(2) giant (L2GLR)	b	474	6	CCAAUU**AUG**U
LOD*	lodestar, nucleotide triphosphate binding	b	84	1	CUAAAA**AUG**U
LSP1A5	Lsp larval serum protein, alpha	c	88	0	UCCAGG**AUG**A
LSP1B	Lsp larval serum protein, beta	c	85	0	GUCAAC**AUG**A
LSP1C	Lsp larval serum protein, gamma	c	82	0	CCAAGG**AUG**A
MACE	muscarinic acetylcholine receptor	b	293	3	UCCGUC**AUG**G
MAP205	205kd microtubule-associated protein	e	420	0	UAAAGG**AUG**G
MASTER	mastermind	c	753	8	GCAUUU**AUG**G
MET	Met, metallothionein	b	123	0	AUCAAG**AUG**C
METO	Met, metallothionein	b	69	0	UACAAG**AUG**G
MEX1A	mex1	c	76	0	AUCACC**AUG**U
MLE*	mle, maleless	b	79	0	CUAAGA**AUG**G
MOV34	Mov34 protein	b	111	0	ACAAAC**AUG**C
MP20	mp20, muscle-specific protein	c	70	0	UCAAAC**AUG**U
MPP1	patched (PTCR)	b	772	7	ACCAUA**AUG**G
MSP316	msP316 male-specific protein	c	34	0	AUCAAC**AUG**G
MST355a	msP355a male-specific protein	c	22	0	CUCGAA**AUG**A
MST355b	msP355b male-specific protein	c	25	0	UCCACA**AUG**A
MYBDR	D-myb oncogene homolog	b	605	7	CUUAAG**AUG**G
MYHB	myosin heavy chain	e	113	0	AGCAAG**AUG**C
MYL	myosin light chain	b	43	0	GACAAA**AUG**G
MYLA	myosin light chain 2	c	66	0	AGCACC**AUG**G
MYONMAa	non-muscle myosin heavy chain	c	93	0	AAACAA**AUG**A
MYONMAb	non-muscle 2nd start codon	c	228	1	GCCAAA**AUG**U
MYSP	myospheroid	b	93	0	AAAGCC**AUG**A
NCDA	ncd, non-claret disjunctional	b	65	0	UUGGCG**AUG**G
NEU*	neu, neuralized	b	273	2	ACUACC**AUG**G
NEUROT	neurotactin	b	508	1	GACAAU**AUG**G
NINAA	ninaA	a	?	?	AAAAUC**AUG**A
NINAC	ninaC	c	146	1	UAAGUC**AUG**A
NORPA	norpa, phospholipase C	b	652	5	GCAAUA**AUG**A
NOS*	nanos	c	261	1	UUCGCC**AUG**U
NOTCH1	Notch, ectodermal determinant	c	865	8	AACAAA**AUG**C
NRGAA	neuroglian	b	27	0	ACCAAA**AUG**U
NUMB	numb	b	791	5	ACAGGC**AUG**G
OPSA	ninaE, opsin	c	170	2	AACACA**AUG**G
OPSAA	Rh2, opsin	e	37	0	CUGAGC**AUG**G
OSKAR	oskar	c	15	0	CAAGCG**AUG**G
OTEDA	otefin	b	75	1	GCCAAA**AUG**C
OTUA	ovarian tumor (OTU)	c	154	1	GUCGCC**AUG**G
PABP	poly(A)-binding protein PABP	b	132	0	CCAAAU**AUG**G
PAH	pah, phenylalanine hydroxylase	b	84	0	GUGAAA**AUG**U
PCGENE	Pc, polycomb	b	109	0	UUAAAA**AUG**A
PCNA	proliferation cell nuclear antigen	d	89	0	UUCAAC**AUG**U

TABLE 36.1. *Continued*

File	Encoded protein	Method	Leader lengths	Number of uAUGs	Start site
PEP*	pep, protein on ecdysone puffs	b	217	0	AAAAUUAUGG
PEPCK	PEPCK, phosphoenolpyruvate carboxylase	v	29	0	AACAAAAUGC
PERA	per, period	c	368	1	AGCACCAUGG
PKC53E	protein kinase C 53E	b	62	0	CUUUUAAUGG
PKC98F	protein kinase C 98F	b	398	7	GACCUCAUGC
PKCR	protein kinase C	b	886	17	GCAACAAUGU
PLC21A	plc-21, phospholipase c	b	824	11	GUGAGGAUGA
PMSH2	msh-2	b	289	0	GCGAGGAUGU
PN*	pn, prune	b	70	0	CUGGUAAUGG
POLO*	polo, putative protein kinase	b	219	1	AGCAAGAUGG
PP1A	phosphatase 1 alpha	b	129	0	GCAAACAUGG
PRD	paired	b	245	1	GAAACUAUGA
PROS*	prospero, axonal growth regulation	b	301	0	GGCUUCAUGA
PROS281	proteasome subunit	b	60	0	AACAAGAUGU
PROS29	proteasome subunit	b	77	1	UUAGCAAUGG
PROS35	proteasome 35kd	b	70	0	AAAGUCAUGU
PTPM	tyrosine phosphatase DPTP	b	54	1	CAAGCCAUGG
R118C	intronic R1 gene 18C	b	117	0	UGCAAAAUGA
RAB3	rab3, neuronal GTP-binding protein	b	586	3	GAUAAAAUGG
RAFPO	raf, proto-oncogene	b	84	0	GAACUAAUGG
RAS1	Dras1, proto-oncogene	b	167	0	AGCCAAAUGA
RAS21	Dras2, proto-oncogene	b	184	3	CUUAUAAUGU
RAS3	Dras3, proto-oncogene	b	57	0	GCCAGCAUGC
RCC1*	RRC1, regulator chromatin condensation	b	211	0	GCUAAAAUGC
RDG*	rdgB, retinal degeneration	b	180	2	GUCAACAUGC
REF2P	ref(2)p, sigma rhabdovirus multiplication	e	371	0	GCGAAAAUGC
RGPS14a	rp14, ribosomal protein S14 A	c	29	0	CCCAGAAUGG
RGPS14b	rp14, ribosomal protein S14 B	c	34	0	UGCAGAAUGG
RH3A	Rh3, opsin	e	22	0	CGGAGCAUGG
RH4A1	Rh4, opsin	c	87	0	ACCGAUAUGG
RM62RH	rm62, RNA helicase	b	482	2	GGAGUAAUGG
RNP70K	U1 70K snRNP	c	208	1	CACAAAAUGA
RNPOL2	RP140, RNA polymerase II 140 kilodalton subunit	c	168	4	AUUCAGAUGU
RP128	RNA polymerase III 128 kilodalton subunit	a	?	?	AACGAAAUGG
RP135	RNA polymerase III 135 kilodalton subunit	c	98	0	UACAACAUGC
RP21C	rp21C, A-type ribosomal protein	b	48	0	UUCGACAUGU
RP49	rp49, ribosomal protein 49	d	9	0	UUCAAGAUGA
RPA1R	rpA1, ribosomal protein	e	89	0	UUAAACAUGC

(*continued*)

TABLE 36.1. *Continued*

File	Encoded protein	Method	Leader lengths	Number of uAUGs	Start site
RPII	RNA polymerase II, 215 kilodalton subunit	d	435	3	ACCAGG**AUG**A
RPL1R	ribosomal protein L1	b	69	0	ACGAAA**AUG**A
RPS17	rp17, ribosomal protein S17	c	56	0	AACAUA**AUG**G
RRP1	Rrp1, strand transferase	b	132	0	UCCAUA**AUG**C
RUD1	rudimentary	e	11	0	UCCAAU**AUG**G
RUNTR	runt, segmentation gene	e	252	0	UACGAG**AUG**C
S12*	1(3)S12	a	?	?	UGCAGC**AUG**G
S1C4	beta-amyloid-like	b	152	0	CGAACA**AUG**U
S2ZSTM	suppressor-2 of zeste	b	149	1	AGAAAG**AUG**C
S59	S59 homeo box	b	67	0	CCAAAA**AUG**G
SAD	sad, nicotinic acetylcholine receptor	b	343	1	GUCACC**AUG**G
SAL	spalt	b	50	0	GCCACG**AUG**A
SAS	sas, putative cell adhesion receptor	b	44	0	ACCAAA**AUG**C
SCAa	sca, scabrous, 1st putative start	c	321	1	GUGUGA**AUG**A
SCAb	sca, scabrous, 2nd, in-frame start	c	396	2	GCAACA**AUG**G
SD*	Sd, segregation distortion	b	121	2	CGAGGC**AUG**U
SER2a	serine protease SER1	c	24	0	AACAAG**AUG**A
SER2b	serine protease SER2	c	24	0	ACCAAG**AUG**A
SERCA	sarcoplasmic/endoplasmic reticulum Ca^{2+}-ATPase	b	32	0	AUCAAG**AUG**G
SEV	sevenless	c	229	2	GCCUCG**AUG**A
SGG	shaggy	b	280	0	GUUACG**AUG**A
SGS378a	Sgs-3, salivary gland protein	c	29	0	AAAAAC**AUG**A
SGS378b	Sgs-7, salivary gland protein	c	33	0	AGAACC**AUG**A
SGS378c	Sgs-8, salivary gland protein	c	33	0	ACAACC**AUG**A
SGS4C1	Sgs-4, salivary gland protein	e	13	0	GUCAAG**AUG**C
SGS5	Sgs-5, salivary gland protein	d	33	1	UACGAC**AUG**U
SHAKE2	shaker	b	269	1	GCCAAG**AUG**A
SHAKE3	shaker, larval	b	72	2	GCCUGU**AUG**G
SINA	seven in absentia	e	903	11	CUUCCA**AUG**U
SING2	sn, singed	b	739	1	AGCACC**AUG**A
SLIT	slit	f	314	1	GCCACA**AUG**G
SNAIL	snail	b	163	0	UCAAAA**AUG**G
SNAKE	snake	b	78	2	AAUAGA**AUG**A
SOD	Sod, superoxide dismutase	b	68	0	UUCGAA**AUG**G
SODCHA	para locus, sodium channel alpha	b	>271	4	UAGACA**AUG**A
SOL	sol, small optic lobes gene	b	263	0	CGCGCA**AUG**G
SPCA	alpha-spectrin	b	270	1	AGCGAA**AUG**G
SPERM	mst(3)gl-9, spermatogenesis	b	97	0	UUAAUC**AUG**U
SQH*	sqh, regulatory non-muscle myosin	c	221	0	GCAACC**AUG**U
SRC28C	Dsrc proto-oncogene	b	133	1	GGCAAC**AUG**A
SRCC	Dsrc proto-oncogene	a	?	?	UAAGCC**AUG**G
SRYG1a	serendipity, beta	e	145	0	GACUAG**AUG**A

TABLE 36.1. *Continued*

File	Encoded protein	Method	Leader lengths	Number of uAUGs	Start site
SRYG1b	serendipity, alpha	e	43	0	AACAGCA**UG**G
SRYG1c	serendipity, gamma	e	67	0	GGCGCAA**UG**G
STAUFEN	staufen	b	274	3	AAGAAAA**UG**C
STELL	stellate	e	30	0	GGCAACA**UG**U
STGA	string, cdc25	b	391	3	AACAAAA**UG**C
STIMG	stimulatory G protein	b	299	2	GCUGCGA**UG**G
SUHW	suppressor of hairy wing	b	59	0	ACCAACA**UG**A
SUSG	suppressor of sable	e	507	4	UCGAUAA**UG**U
SVP1	seven-up protein, svp type 1	b	450	3	GGCGUCA**UG**U
SWA*	swallow	c	39	0	AAAGCGA**UG**A
SX1PS11	sex-lethal	e	425	1	CAGGAUA**UG**U
SYT	synaptotagmin	b	359	0	AACAAAA**UG**C
TAC*	tachykinin-like receptor	b	258	1	GCAGCCA**UG**G
TCP1	T complex protein Tcp-1	b	42	0	AGGAAAA**UG**U
TER	terminus protein	c	154	0	UCAAUCA**UG**U
TFIID	TATA-box binding protein TFIID	c	173	1	UGUAAGA**UG**G
TGA	transformer, sex determination	c	70	0	UUUCCGA**UG**A
TKABL1	abl, tyrosine kinase abelson homolog	c	96	0	UGGCAAA**UG**G
TKO	tko, technical knock-out	b	171	0	GAGAGCA**UG**A
TLD*	tolloid, dorsal/ventral pattern	b	72	0	CACGCAA**UG**A
TLL	tailless	b	177	0	AUCGGUA**UG**C
TMLPA	serrate (SER)	b	433	3	CCCAGAA**UG**U
TOLL	toll	b	574	4	GACAACA**UG**A
TORSO	torso, tyrosine kinase	f	195	0	AGGAAAA**UG**C
TRA2Aa	tra-2, transformer "A" non-sex determination	e	186	1	AGCCAGA**UG**G
TRA2Ab	tra-2, transformer "B" non-sex determination	e	488	2	AUCACUA**UG**U
TRA2Ac	tra-2, transformer "C" male germline	e	503	1	GAACGAA**UG**C
TROIIN	tropomyosin II, non-muscle	b	435	0	ACAAAAA**UG**A
TROPI2	tropomyosin I	c	103	0	AACACCA**UG**G
TROPT	wupA, troponin-T	a	?	?	GUAGCCA**UG**U
TRP	trp protein	c	191	3	GCAGAUA**UG**G
TRPB	transient receptor pot	b	484	2	CGGAAGA**UG**G
TRYA	trypsin like, alpha	a	?	?	CCCAUCA**UG**U
TSH*	teashirt, ventral trunk development	b	1,008	9	UUAAAAA**UG**U
TTKFTZ	tramtrack (FTZF2)	b	251	3	CUCCCAA**UG**A
TU4A	TU-4 vitelline membrane	c	62	0	UCCGCAA**UG**G
TUBA1	alpha-tubulin-1	e	141	0	CUCAAUA**UG**G
TUBA2	alpha-tubulin-2	e	96	0	AUCAUCA**UG**G
TUBA3	alpha-tubulin-3	e	504	0	AUCAAUA**UG**C
TUBA4	alpha-tubulin-4	e	149	0	AAUAAAA**UG**G
TUBB2A	beta-tubulin-2	c	175	0	AUCAAAA**UG**C
TUBE	tube	b	193	2	AACACCA**UG**G

(*continued*)

TABLE 36.1. *Continued*

File	*Encoded protein*	*Method*	*Leader lengths*	*Number of uAUGs*	*Start site*
TWISTG	twist	e	159	0	CACCAA**AUG**A
TYRDROG	tyramine receptor (OCR)	b	312	4	GGAAAG**AUG**C
UB52AA	ubiquitin 52-AA extension protein	b	34	0	CGCAUU**AUG**C
UBIA	ubiquitin	e	139	6	UCCAAA**AUG**C
UBXG5	Ubx, ultrabithorax	e	697	2	CGUUCG**AUG**G
UROX	urate oxidase	e	33	0	GUCACA**AUG**U
VASA	vasa	b	131	1	AUCAAU**AUG**U
VERM	vermilion	e	57	0	UGCACC**AUG**A
VHATP	vacuolar H^+ ATPase	b	116	0	AGCAAA**AUG**U
VITA	vitelline membrane protein, 26A-1	c	81	0	ACCAAG**AUG**A
VITB	vitelline membrane protein, 3C-1	c	96	1	AGCACC**AUG**A
VMP	vitelline membrane protein	b	29	0	UUCAUC**AUG**C
WL	w, white	a	?	?	CCGGCA**AUG**G
XDH	ry, xanthine dehydrogenase	b	180	1	UUCACG**AUG**U
XR2C	xr2c, ultraspiracle	b	162	0	CCCAGG**AUG**G
YELLOW	y, yellow	c	171	0	AGTGCA**AUG**U
YOLK	yolk protein I	c	61	0	CGAACC**AUG**A
YP3	Yp3, yolk protein-3	c	59	0	ACCAAA**AUG**A
Z60MEX1a	z600	e	63	0	GUUAUU**AUG**U
Z60MEX1b	gld-F female specific	e	142	0	GUUAAG**AUG**G
Z60MEX1c	gld-M male specific	e	307	1	GUUAAG**AUG**G
ZPBA	trithorax	b	841	11	ACUAUU**AUG**G
ZESTE	zeste	d	964	4	ACUCAA**AUG**U
ZFH1	zinc finger homeobox protein 1	b	358	2	UUCCAA**AUG**U
ZFH2	zinc finger homeobox protein 2	b	369	4	UCUCCA**AUG**U
ZIPR	zipper	b	261	3	AGCACG**AUG**A

Methods used to map the leader sequence: a = none or ambiguous data; b = 5′ UTR of the longest cDNA; c = 5′ UTR of longest cDNA and primer extension data or nuclease protection; d = primer extension or nuclease protection (genomic sequence only); e = primer extension and nuclease protection along with cDNA and/or genomic sequence data; f = presence of consensus TATA sequence and *Drosophila* consensus transcription start site plus partial cDNA sequence of leader.

* An asterisk at the end of the file name indicates that this sequence was not included in GenBank Release 69. A temporary file name was assigned to such sequences by us.

Lower case letters in file names were used to uniquely identify multiple mRNAs present in a single GenBank file.

clone obtained from an exhaustive screening of a cDNA library. In many, if not most of these cases, the 5′ end of the longest cDNA is likely to correspond to the transcription start site. A terminal G residue is often found at the 5′ end of the cDNA that is not found at the corresponding position in the genomic sequence; it is thought that this G is copied from the 5′ methyl G cap that is added post-transcriptional to the 5′ end of eukaryotic mRNAs. Sequences from *Drosophila* species other than *D. melanogaster* were not included because their orthologous counterparts in *D. melanogaster* are almost always represented in the database. Many *Drosophila* genes contain multiple mRNAs arising from alternative promoters and/or RNA processing. In cases where alternative leader sequences have been clearly documented, more than one leader sequence is listed for a particular gene. If such alternative leaders share the same translation start site, the start context (−6 to +4) is listed for just one of the leader sequences. Several genes have been characterized by more than one research group and reported to GenBank. We have arbitrarily used the information and GenBank file name given for one of the duplicate entries if the data are equivalent. Where data are not equivalent for the same gene we have chosen the data which are most strongly supported by experimental evidence.

Leader Length and Upstream AUGs

Inspection of Table 36.1 reveals numerous mRNAs with leader sequences exceeding 100 nt and containing multiple upstream start codons (uAUGs). Indeed the average *Drosophila* leader sequence is 248 nt and the median is 156. The distribution of leader lengths is shown in Fig. 36.1. Forty-six of the leader sequences exceed 500 nt. The smallest size class (0–100 nt) is the largest containing 140 mRNA sequences. Many of the reported leader sequences are based upon the analysis of the longest cDNA obtained but may nonetheless

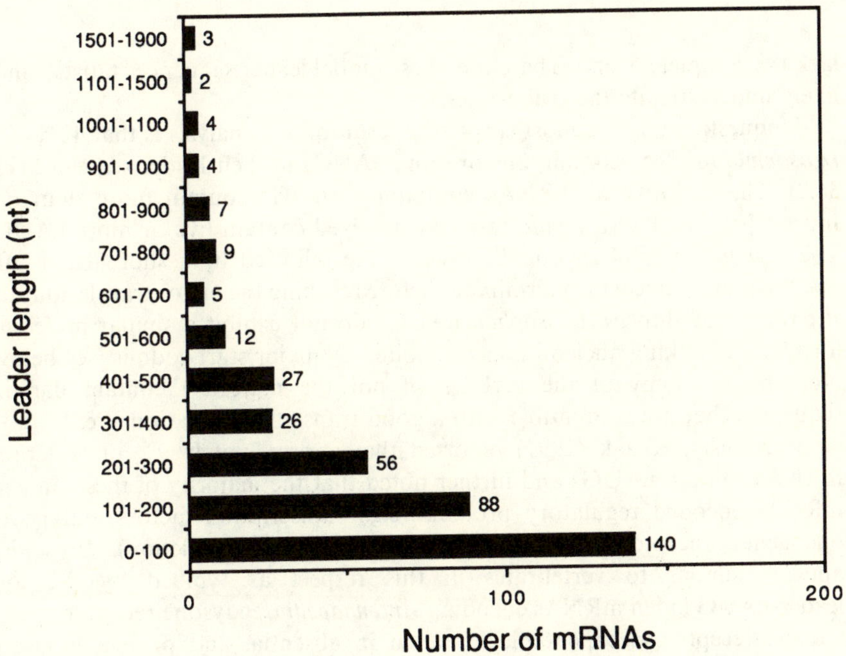

FIG. 36.1. Distribution of the number of nucleotides (nt) in the 5′ untranslated leader sequences of *Drosophila* mRNAs.

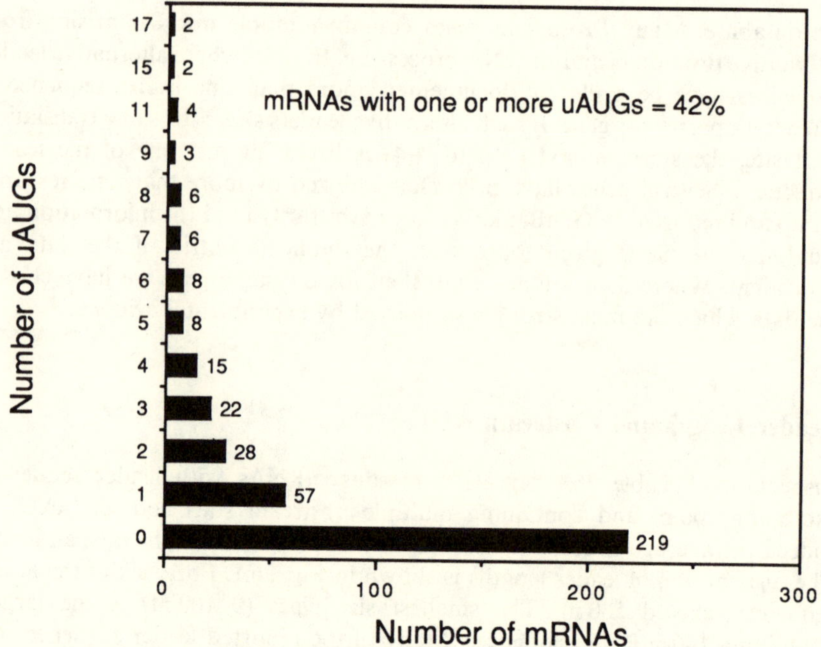

FIG. 36.2. Distribution of upstream AUGs in the 5′ untranslated leader sequences of *Drosophila* mRNAs.

lack the complete 5′ end. Therefore, these global leader sequence statistics most likely underestimate the true values.

Unquestionably the most surprising result of our analysis is that 42% of all *Drosophila* mRNAs contain one or more uAUGs in their leader sequence (Fig. 36.2). The majority of mRNAs containing uAUGs contain more than one. Indeed 10% of all *Drosophila* mRNAs surveyed contain five or more uAUGs. The vast majority of *Drosophila* uAUGs are followed by a short (ca. 1–100) open reading frame which terminates before reaching the major translation start site (data not shown). *Drosophila* uAUGs do not exhibit a similar preference for specific flanking nucleotides as exhibited by major start codons (see below). Nonetheless, many of the uAUGs (if not the majority) contain flanking sequences that are compatible with a good translation initiation site.

Previously, Kozak (1991) reported that approximately 9% of vertebrate mRNAs contain uAUGs and further noted that the majority of these unusual mRNAs encoded regulatory proteins (e.g., transcription factors and proto-oncogenes, receptors, and components of signal transduction). *Drosophila* appears similar to vertebrates in this respect as typified by the long leader-uAUG laden mRNAs encoding *Antennapedia*, ecdysone receptor, acetyl-choline receptor, decapentaplegic, seven in absentia, and protein kinase C (Table 36.1). In general long leader-uAUG laden mRNAs encode low abundant proteins, particularly as compared to very short-leader mRNAs encoding such proteins as the yolk, cuticle, and larval serum proteins. This dichotomy is consistent with the general finding that removal of long leader sequences typically increases translation initiation rates (e.g. Chinkers et al. 1989; Muller and Witte 1989). Long leader mRNAs may be tolerated as a consequence of the absence of natural selection to increase translation rates of proteins that are not needed in abundance. Alternatively, the presence of a long leader with multiple uAUGs may afford devices to regulate translation initiation. The paradigm par excellence in this regard is the yeast GCN4 gene. GCN4 mRNA is constitutively produced but the translation of GCN4 protein is highly regulated through the interaction of four upstream open reading frames, the scanning preinitiation complex, and some of the translation initiation factors that undergo changes in activity as a consequence of amino deprivation (Miller and Hinnebusch 1989; Ramirez et al. 1991; Dever et al. 1992). Whether other eukaryotic mRNAs that contain long leader and uAUGs are under similar control is unknown. For many of the long-leader *Drosophila* genes, mRNA and protein expression are temporally and spatially correlated suggesting the lack of translational regulation. However, it should be noted that translation initiation rates are almost never determined empirically. Consequently, the relative translation rates among different mRNAs cannot be compared at this time.

OH and coworkers have recently reported (OH et al. 1992; and personal communication) that the long-leader sequences of the *Antennapedia* and *Ultrabithorax* can promote internal ribosome binding in *Drosophila* cell culture. Since internal binding circumvents the requirement for the cap-binding protein (eIF-4E), it is likely that internal initiation also circumvents global translation control as mediated by altering the level and activity of eIF-4E. It will be interesting to see if *Antp* and *Ubx* use an internal mode of initiation in flies and whether internal initiation is used by other mRNAs with exceptionally long leader sequences.

Translation Start Sites

Table 36.2 presents an update of the translation start sites from positions −6 to +4 relative to the start codon. The 50/75 consensus rule (Cavener and Ray 1991) was used to assign consensus nucleotides. The derived consensus sequence, C/A A N N AUG has not appreciably changed with the doubling of the database from that reported by Cavener and Ray (1991). Since long leader mRNAs are thought to be poorer substrates for translation initiation, such mRNAs may on average contain a poorer fit to the consensus sequence. To examine this question the mRNA sequences listed in Table 36.1 were divided into two groups: long leaders, exceeding the median leader length and short leaders, less than the median leader length (Table 36.2). The short leader mRNAs do exhibit a significantly stronger preference for A or G at the critical −3 position as compared to the long leader mRNAs as might be expected. In addition, differences are observed between the short and long leader classes at

TABLE 36.2. Nucleotide frequencies flanking the start codons for the major protein coding regions and the start site consensus sequences

Total mRNA dataset.

	−6	−5	−4	−3	−2	−1	+1	+2	+3	+4
A	33	23	26	70	47	41	100	0	0	23
G	28	18	11	20	11	20	0	0	100	39
C	17	34	51	6	24	30	0	0	0	15
U	22	25	12	5	18	9	0	100	0	23
	a	c	C/A	A	a	a	A	U	G	g

Short leader mRNA dataset.

	−6	−5	−4	−3	−2	−1	+1	+2	+3	+4
A	34	22	29	77	51	39	100	0	0	26
G	24	16	9	15	14	17	0	0	100	35
C	18	30	54	3	18	35	0	0	0	14
U	23	32	8	4	16	9	0	100	0	25
	a	u	C/A	A	A	A/C	A	U	G	g

Long leader mRNA dataset.

	−6	−5	−4	−3	−2	−1	+1	+2	+3	+4
A	34	24	23	66	43	45	99	0	0	22
G	32	22	13	20	10	23	1	0	100	42
C	16	38	48	10	29	23	0	0	0	16
U	19	16	16	4	19	9	0	100	0	20
	a	c	c	A	a	a	A	U	G	g

Upstream AUGs (uAUGs).

	−6	−5	−4	−3	−2	−1	+1	+2	+3	+4
A	32	28	27	29	45	40	100	0	0	30
G	23	18	35	15	17	16	0	0	100	9
C	20	21	16	21	25	14	0	0	0	26
U	25	34	22	35	13	30	0	100	0	35
	a	u	g	u	a	a	A	U	G	u

The 50/75 Consensus Rule was applied: if the frequency of one nucleotide is greater than 50% and is greater than twice the frequency of the next highest nucleotide, it is assigned as the consensus and denoted as such with a capital letter (e.g., A). If the sum of the frequency of the two most frequent nucleotides is greater than 75% but neither meet the requirement for singular consensus, the two nucleotides are assigned as co-consensus nucleotides and denoted with capital letters (e.g., C/A). Lower case letters indicate the most frequent nucleotide at a particular position when no nucleotides meet the consensus criteria.

positions −4, −2 and −1. However, these are largely differences in the relative distribution of A and C, both favored at these positions in most eukaryotic groups (Cavener and Ray 1991). Overall, the differences between long and short leader mRNAs translation start sites are significant but minor. Among the long leader mRNAs occurs an exceptional GUG start codon for choline acetyltransferase. The data supporting the use of GUG as a start codon in this case are strong (Sugihara et al. 1990). Preliminary evidence indicates that the E74A gene uses CUG as a major alternative start codon (L. Boyd and C. Thummel, personal communication). Non-AUG start codons are likely to be more prevalent than currently recognized because most AUG translation start sites have not been empirically confirmed.

The sequence context flanking 151 upstream AUGs was examined in order to see if uAUGs lie in a poor context. It might be expected that uAUGs would exhibit a strong anti-consensus sequence to discourage translation initiation at these sites. However, the summary of these data in Table 36.1 indicates that uAUGs collectively neither show a good or poor fit to consensus relative to major translation start sites. At the critical −3 position A or G is found in 44% of the cases. The frequency of A at −2 and −1 is relatively high similar to major translation initiation sites. Some unique biases are observed including a relatively high frequency of G at −4 and a relatively low frequency of G at +4; just the opposite biases are seen for major translation initiation sites. One possible explanation for the lack of consensus either opposite or similar to major translation initiation sites is that uAUG context data may contain a mixture of uAUG which are either selected for or against as initiation sites depending upon the mRNA. This assumes that some uAUGs may be involved in translational regulation but that others are not. Overall, a large fraction of uAUGs would appear to be in a reasonably good context. How the scanning preinitiation complex traverses a leader sequence burdened with uAUGs is an interesting mechanistic and regulatory question.

A Caveat to Using the Translation Start Site Consensus

Comparing putative translation start sites of newly sequenced genes with the start site consensus sequence is a common practice. In some cases investigators have favored a downstream start codon over an in-frame upstream start codon based upon a better fit of the former to the consensus sequence. However, mutational analysis of the translation start site of *Adh* (Feng et al. 1991) and inspection of the diversity of start contexts in Table 36.1 demonstrates that start codons that exhibit a poor fit to the consensus can nonetheless serve as the major site of translation initiation. A good example of this is provided by the translation start site for *hsf* encoding the *Drosophila* heat-shock transcription factor (Table 36.1). The start codon context for *hsf* is UUUAUGU (Clos et al. 1990). Based upon mutational analysis and the consensus sequence, a UUUAUGU context is exceptionally poor. Changing the start codon context for *Adh* to this same sequence resulted in a 6–12-fold reduction in translation depending upon the developmental stage. However, an appreciable level of ADH protein was still observed in this mutant. Thus a "poor" context may reduce but not necessarily eliminate translation initiation. Kozak's studies on the rat preproinsulin mRNA have clearly indicated that a poor context reduces the probability that the ribosome will initiate at that particular site. If initiation does not occur at a particular start codon, the preinitiation complex resumes scanning, a process called leaky scanning. The overall effect of leaky scanning may be the use of multiple start codons, particularly when two start codons are in-frame and within close proximity. An important perspective to bear in mind when analyzing translation start sites is that the sequence context may be adapted to down-regulate the rate of translation initiation. Moreover, sequence context effects are likely to be developmentally dependent (Feng et al. 1991) as a function of changes in the concentration and activities of the translation initiation factors (particularly eIF-2). These considerations are also relevant to the presence of upstream start codons in the leader sequence which are either out of frame with the major coding region or followed by an in-frame termination codon.

Summary

Drosophila genes exhibit a diverse array of untranslated leader sequences and translation start sites. The presence of a long leader or multiple uAUGS or a poor sequence context surrounding the start codon should no longer be perceived as abnormal or unusual given the large fraction of *Drosophila* mRNAs which contain such features. In many cases these features will affect translation initiation rates. How they affect translation and what the physiological rationale is for these effects remain to be elucidated. Although it would appear that *Drosophila* mRNAs are more prone to long leaders and uAUGs than vertebrate genes, only a small fraction of all mRNAs have been characterized for either group. The current *Drosophila* and vertebrate databases are biased somewhat differently as a consequence of the types of genes and questions being analyzed using different systems. In particular the *Drosophila* database contains a larger fraction of genes encoding proteins that regulate development. Whether the current *Drosophila* database is a more representative sample than the vertebrate database is unknown. Fortunately our obsession for cloning and sequencing will eventually answer this question.

References

Baim, S. B. and Sherman, F. (1988). mRNA structures influencing translation in the yeast *Saccharomyces cerevisiae*. *Mol. Cell Biol.* **8**:1591–1601.

Cavener, D. R. (1987). Comparison of the consensus sequence flanking translational start sites in *Drosophila* and vertebrates. *Nucl. Acids Res.* **15**:1353–1361.

Cavener, D. R. and Ray, S. C. (1991). Eukaryotic start and stop translation sites. *Nucl. Acids Res.* **19**:3185–3192.

Chinkers, M., Garbers, D. L., Chang, M. S., Lowe, D. G., Chin, H. M., Goeddel, D. V. and Schulz, S. (1989). A membrane form of guanylate cyclase is an atrial natriuretic peptide receptor. *Nature* **338**:78–83.

Clos, J., Westwood, J. T., Becker, P. B., Wilson, S., Lambert, K. and Wu, C. (1990). Molecular cloning and expression of a hexameric *Drosophila* heat shock factor subject to negative regulation. *Cell* **63**:1085–1097.

Dever, T. E., Feng, L., Wek, R. C., Cigan, A. M., Donahue, T. F. and Hinnesbusch, A. G. (1992). Phosphorylation of initiation factor 2α by protein kinase GCN2 mediates gene-specific translational control oc *GCN4* in yeast. *Cell* **68**:1–20.

Feng, Y., Gunter, L. E., Organ, E. L. and Cavener, D. R. (1991). Translation initiation in *Drosophila* is reduced by mutations upstream of the AUG initiator codon. *Mol. Cell. Biol.* **11**:2149–2153.

Kozak, M. (1984). Compilation analysis of sequences upstream from the translational start site in eukaryotic mRNAs. *Nucl. Acids Res.* **12**:857.

Kozak, M. (1986). Point mutations define a sequence flanking the AUG initiator codon that modulates translation by eukaryotic ribosomes. *Cell* **44**:283–292.

Kozak, M. (1987). An analysis of 5′-noncoding sequences from 699 vertebrate messenger RNAs. *Nucl. Acids Res.* **15**:8125–8148.

Kozak, M. (1989). The scanning model for translation: an update. *J. Cell Biol.* **108**:229–241.

Kozak, M. (1991). An analysis of vertebrate mRNA sequences: Intimations of translational control. *J. Cell. Biol.* **115**:887–903.

Lewin, B. (1990). *Genes IV* (Oxford: Oxford University Press).

Macejak, D. G. and Sarnow, P. (1991). Internal initiation of translation mediated by the 5′ leader of a cellular mRNA. *Nature* **353**:90–94.

Miller, P. F. and Hinnebusch, A. G. (1989). Sequences that surround the stop codons of upstream open reading frames in GCN4 mRNA determine their distinct functions in translational control. *Genes Dev.* **3**:1217–1225.

Muller, A. J. and Witte, O. N. (1989). The 5′ noncoding region of the human leukemia-associated oncogene BCR/ABL is a potent inhibitor of in vitro translation. *Mol. Cell. Biol.* **9**:5234–5238.

OH, S. K., Scott, M. P. and Sarnow, P. (1992). Homeotic gene Antennapedia mRNA contains 5′-noncoding sequences that confer translational initiation by internal ribosome binding. *Genes Dev.* **6**:1643–1653.

Ramirez, M., Wek, R. C. and Hinnebusch, A. G. (1991). Ribosome association of GCN2 protein kinase, a translational activator of the GCN4 gene of *Saccharomyces cerevisiae. Mol. Cell. Biol.* **11**:3027–3036.

Sugihara, H., Andrisani, V. and Salvaterra, P. M. (1990). Drosophila choline acetyltransferase uses a non-AUG initiation codon and full length RNA is ineffectively translated. *J. Biol. Chem.* **265**:21714–21719.

37

Codon Usage

Paul M. Sharp and Andrew T. Lloyd

Department of Genetics
Trinity College
Dublin 2
Ireland

Introduction

In most genes in most species, alternative synonymous codons are not used in equal frequencies (Aota et al. 1988)—*Drosophila* is no exception (Ashburner et al. 1984; O'Connell and Rosbash 1984; Shields et al. 1988). In Table 37.1 we present the total codon usage for 438 *D. melanogaster* genes. This dataset was extracted from the GenBank/EMBL/DDBJ DNA sequence data library (GenBank release 71) using the ACNUC sequence retrieval software (Gouy et al. 1985), and screened to remove duplicates and/or multiple alleles (making particular use of FlyBase; Ashburner 1992); the genes are listed in Appendix A. As we discuss below, there is considerable heterogeneity of codon usage patterns among genes, and so the values in Table 37.1 must be taken only as an overall, or average, guide to *D. melanogaster* codon usage. This may be useful in the design of oligonucleotide probes, or in the assessment of whether a novel open reading frame is actually a coding sequence: many genes approximate to this pattern, but particular genes may differ quite markedly. Note also that genes from transposable elements have rather different patterns of codon usage from "chromosomal" genes (Shields and Sharp 1989), and so they are presented separately in Table 37.1 (to be discussed in more detail below). (Codon usage in the *Drosophila* mitochondrial genome is completely different from nuclear genes (Clary and Wolstenholme 1985), and will not be discussed further.)

Why is codon usage in *Drosophila* biased, and why does it vary among genes? Clearly, the pattern of synonymous codon usage in a gene must reflect the net result of past evolutionary pressures: the two main influences are natural selection (some codons may be translated more accurately and/or efficiently than synonyms encoding the same amino acid), and mutational biases (which may give rise to strongly biased codon usage even in the absence of any selection). Thus, in general terms, codon usage can be considered as the result

Table 37.1. Codon usage in *Drosophila melanogaster*

	Total		T.E.			Total		T.E.			Total		T.E.			Total		T.E.	
	N	*RSCU*	N	*RSCU*		N	*RSCU*	N	*RSCU*		N	*RSCU*	N	*RSCU*		N	*RSCU*	N	*RSCU*
Phe UUU	2,896	0.66	478	1.17	Ser UCU	1,745	0.49	226	0.95	Tyr UAU	2,610	0.69	341	0.96	Cys UGU	1,396	0.55	167	0.82
UUC	5,822	1.34	336	0.83	UCC	5,377	1.50	216	0.91	UAC	4,957	1.31	368	1.04	UGC	3,712	1.45	238	1.18
Leu UUA	922	0.26	430	1.30	UCA	1,857	0.52	297	1.25	ter UAA	220	1.60	2	0.60	ter UGA	75	0.55	3	0.90
UUG	3,787	1.05	319	0.96	UCG	4,705	1.31	153	0.64	ter UAG	117	0.85	5	1.50	Trp UGG	2,380	1.00	223	1.00
Leu CUU	1,959	0.54	364	1.10	Pro CCU	1,683	0.46	195	0.78	His CAU	2,814	0.78	295	1.01	Arg CGU	2,480	1.08	128	0.66
CUC	3,410	0.94	256	0.77	CCC	5,087	1.39	204	0.82	CAC	4,363	1.22	292	0.99	CGC	4,866	2.13	134	0.69
CUA	1,758	0.49	337	1.02	CCA	3,484	0.95	437	1.76	Gln CAA	3,984	0.55	580	1.28	CGA	1,964	0.86	197	1.02
CUG	9,835	2.72	283	0.85	CCG	4,436	1.21	160	0.64	CAG	10,537	1.45	328	0.72	CGG	1,941	0.85	103	0.53
Ile AUU	4,065	0.97	583	1.16	Thr ACU	2,232	0.60	344	1.04	Asn AAU	5,565	0.84	747	1.10	Ser AGU	2,704	0.75	255	1.07
AUC	6,523	1.56	349	0.69	ACC	6,152	1.67	307	0.93	AAC	7,692	1.16	614	0.90	AGC	5,174	1.44	283	1.19
AUA	1,947	0.47	581	1.15	ACA	2,773	0.75	513	1.55	Lys AAA	3,759	0.51	1,115	1.36	Arg AGA	1,041	0.45	405	2.09
Met AUG	6,366	1.00	335	1.00	ACG	3,611	0.98	163	0.49	AAG	11,038	1.49	519	0.64	AGG	1,440	0.63	195	1.01
Val GUU	2,757	0.72	290	1.07	Ala GCU	4,022	0.78	343	1.11	Asp GAU	7,274	1.05	497	0.97	Gly GGU	4,344	0.92	222	1.05
GUC	3,839	1.00	223	0.82	GCC	9,783	1.89	319	1.03	GAC	6,631	0.95	528	1.03	GGC	8,179	1.73	218	1.03
GUA	1,450	0.38	300	1.11	GCA	3,265	0.63	409	1.33	Glu GAA	4,778	0.58	808	1.30	GGA	5,208	1.10	294	1.39
GUG	7,234	1.89	271	1.00	GCG	3,601	0.70	163	0.53	GAG	11,655	1.42	435	0.70	GGG	1,140	0.24	113	0.53

"Total" indicates summed codon usage for 438 nuclear chromosomal genes (i.e., excluding transposable elements), a total of 264,421 codons. "T.E." indicates summed codon usage for 30 genes from 16 transposable elements (listed in Appendix 37.B), a total of 20,836 codons. Codon usage is presented as *N* (the observed number of occurrences) and RSCU (the relative synonymous codon usage, obtained by dividing *N* by the average value for the amino acid); the RSCU value is useful for comparing the level of bias among different amino acids, or among data sets of different sizes.

of a selection-mutation balance (Sharp and Li 1986; Bulmer 1991). However, while it is clear that selection among synonyms shapes codon usage in certain prokaryotes and unicellular eukaryotes (reviewed in Ikemura 1985; Andersson and Kurland 1990), it is not obvious how widespread selective codon usage may be in the genomes of multicellular organisms. In particular, it is not clear whether the long-term evolutionary effective population sizes of most multicellular species are large enough for the selective differences between alternative synonymous codons (which are expected to be very small) to overcome random genetic drift (Sharp 1989). In an earlier study (Shields et al. 1988), we concluded (somewhat to our surprise!) that the evidence suggests that codon usage in many *D. melanogaster* genes *is* influenced by natural selection. Here we briefly review that evidence, utilizing the much larger *D. melanogaster* gene sequence data set now available.

Codon usage variation among genes

Codon usage patterns vary considerably among *D. melanogaster* genes (Shields et al. 1988). To take an extreme example, 92% (33/36) of the Leu residues in the enolase phosphoglycerate hydrolase gene (*Eno*) are encoded by CUG; in contrast, the cubitus interruptus Zn finger gene (*ci*) uses this codon in only seven of 91 cases (8%); differences are also seen for all other 17 amino acids where there is a choice of codons. Under the selection–mutation balance model, two possible reasons for this variation stand out. If selection among synonymous codons for translational efficiency occurs in *D. melanogaster*, then the strength of selection is likely to vary among genes, depending on their level (and perhaps also tissue and developmental stage) of expression. For example, in *Escherichia coli* (Gouy and Gautier 1982) and *Saccharomyces cerevisiae* (Sharp et al. 1986) the strength of codon usage bias in a gene is very highly correlated with the level of gene expression. Alternatively, or perhaps additionally, genes may be affected by different mutational biases. For example, mammalian genes vary greatly in base composition (G + C content) at silent sites (and thus in codon usage) depending on the local base composition of the chromosome (Bernardi et al. 1985; Ikemura 1985); this variation can be most simply explained as variation in the mutation pattern around the genome (Filipski 1988; Sueoka 1988; Wolfe et al. 1989).

To elucidate the situation in *Drosophila*, the first step is to characterize the nature of the codon usage variation among genes. Since the codon usage pattern of each gene is a composite of 59 values (one for each codon, less Met, Trp and stop codons), it is necessary to use multivariate statistical analyses. In codon usage studies the most commonly used method is correspondence analysis (pioneered by Grantham et al. 1981). It is not appropriate to go into any details of the method here, except to say that it allows definition of the major trends among genes—see Grantham et al. (1981) or Shields et al. (1988) for more discussion of this method. We applied this approach to 84 genes (Shields et al. 1988) and have also used it on a data set of 438 genes here. In each case, the major variation in synonymous codon usage among genes is found to be strongly associated with G + C content at silent sites (GC_S): genes at one end of the trend have relatively unbiased codon usage, while genes at the other end of the trend have very highly biased codon usage, and high values of GC_S.

This seems very like the situation found with mammalian (e.g., human) genes, but in fact there are several important differences (Shields et al. 1988). Some of these become apparent from a comparison of GC_S, the G + C content at silent third positions of codons (i.e., excluding Met and Trp) in a gene, and GC_I, the G + C content in the introns of a gene. First, in *D. melanogaster* (unlike humans) GC_S is not strongly correlated with GC_I. Second, GC_S values are generally much higher than GC_I values, particularly in genes with very biased codon usage. It is also noticeable that GC_S becomes reduced in pseudogenes (Shields et al. 1988; Moriyama and Gojobori 1992). Most of the *D. melanogaster* genes studied have been mapped, and there is no obvious relationship between GC_S and map position, although local variations in base composition on the scale that they are thought to occur in the human genome would be difficult to detect at this level. However, it is clear that neighboring genes can have quite different GC_S values. For example, in the highly biased alcohol dehydrogenase gene GC_S is 0.77, but in the relatively unbiased Adh-related gene (less than 300 bp away; Kreitman and Hudson 1991) GC_S is only 0.53. It is also noticeable that in human genes the trend in G + C content is due to similar changes in the frequency of both C and G, but in *D. melanogaster* the major trend is more specifically (though not exclusively) due to a change in the frequency of C. Thus, the major variation in GC_S in *Drosophila* does not appear to be due to regional chromosomal base composition differences.

On the other hand, the main trend in codon usage differences among genes may be correlated with expression level. We might expect that the highly biased genes at the G + C-rich extreme would be those under the most selection pressure, particularly as their GC_S values are the most different from noncoding DNA (i.e., introns). Of course, in a multicellular organism with a complex series of developmental stages, it is rather more difficult to quantify "expression level" than it would be in *E. coli* or yeast. Nevertheless, the G + C-rich genes do seem to include many genes that can be identified as highly expressed. For example, one is alcohol dehydrogenase: *Adh* mRNA "accounts for about 1–2% of the translational activity of mRNA from adult flies" (Benyajati et al. 1980), and must be considered a highly expressed gene. Others at this extreme include *Yp1* and *Yp2* encoding yolk proteins, the nine ribosomal protein genes in the data set, and genes encoding actins and cuticle proteins; all such genes were considered by O'Connell and Rosbash (1984) to be "abundantly expressed".

"Optimal" Codons in *Drosophila melanogaster*

If it is true that the major trend among genes in codon usage is associated with expression-level-mediated selection on codon usage, then contrasting the codon usage patterns for the genes at either end of this trend should reveal which particular codons for each amino acid are favoured. Codon usage in 10% of genes at each extreme of this trend (as identified by multivariate statistical analysis) is presented in Table 37.2. There are 23 codons used with (significantly) higher frequency in the highly biased group of genes: 22 of these are here defined as "optimal" codons, the exception being GGU (for Gly), where the difference in RSCU values is small (though significant at the 5% level). These optimal codons are G + C-rich: of the 22, 15 end in C and six end in G—only one ends in U (CGU) and none end in A. Interestingly, CGU appears to be an optimal Arg codon in many other species (Sharp et al. 1992).

A simple measure of the strength of species-specific codon usage bias is given by the frequency of optimal codons (F_{op}) in a gene (Ikemura 1985). We define a F_{op} for *D. melanogaster* as the number of occurrences of these 22 optimal codons (Table 37.2), divided by the total number of occurrences of codons for these 18 amino acids (i.e., excluding Met and Trp codons). (Calculation of F_{op} values is an option in the FORTRAN program CODONS (Lloyd and Sharp 1992), which is available from the authors on request.) F_{op} values for these 438 genes are given in Appendix A, and they range from 0.22 (*Scr* encoding the sex combs reduced homeobox protein) to 0.88 (*Lsp1-b* encoding β-larval serum protein).

While we have already alluded to the difficulties in discussing absolute expression levels, it is nevertheless possible to compare F_{op} values among genes whose relative expression levels have been described. There are two cytochrome c genes, and it is known that *Cyt-c2* ($F_{op} = 0.77$) "is expressed at much higher levels than" *Cyt-c1* ($F_{op} = 0.57$) (see Limbach and Wu 1985); among four α-tubulin genes, the transcript of *Tuba84D* (α-1) ($F_{op} = 0.79$) is "much more abundant" than that of *Tuba84E* (α-2) ($F_{op} = 0.69$) (see Kalfayan and Wensink 1982); of two elongation factor 1α genes, expression of *Ef1a100E* ($F_{op} = 0.76$) is "generally markedly stronger" than that of *Ef1a48D* ($F_{op} = 0.71$) (see Hovemann et al. 1988); and there are two lysozyme genes, *LysP* ($F_{op} = 0.63$) whose expression was only detected in adults, and whose expression in the adult "was low compared to that of *LysD*" ($F_{op} = 0.70$) (see Kylsten et al. 1992). In some cases these differences in F_{op} values are quite small—the genes' similarity in sequence may reflect quite recent gene duplication events; nevertheless, the differences are all in the direction predicted.

In Table 37.2, it is interesting to note that the highly biased (and highly expressed?) genes favour the most A + T-rich stop codon (UAA), even though the rest of their codons are generally G + C-rich. The highly biased genes also appear to avoid UGA, which is more common in the low bias genes. This is reminiscent of the pattern of stop codon usage in genes of high and low expression in *E. coli*, *Bacillus subtilis*, and yeast (Sharp et al. 1992), and lends further credence to the idea that codon usage in *D. melanogaster* is influenced by natural selection.

Transposable Element Genes

Codon usage in the open reading frames (ORFs) of the various transposable elements (TEs) found in the *D. melanogaster* genome (see Appendix B) is

TABLE 37.2. Codon usage in high and low bias genes in *D. melanogaster*

	High		Low			High		Low			High		Low			High		Low	
	N	*RSCU*	N	*RSCU*		N	*RSCU*	N	*RSCU*		N	*RSCU*	N	*RSCU*		N	*RSCU*	N	*RSCU*
Phe UUU	37	0.16	541	1.14	Ser UCU	73	0.51	407	0.99	Tyr UAU	59	0.25	365	1.03	Cys UGU	18	0.19	206	0.81
UUC*	421	1.84	406	0.86	UCC*	412	2.86	476	1.16	UAC*	420	1.75	345	0.97	UGC*	169	1.81	305	1.19
Leu UUA	6	0.04	334	0.93	UCA	13	0.09	423	1.03	ter UAA	32	2.29	24	1.76	ter UGA	0	0.00	10	0.73
UUG	101	0.60	503	1.40	UCG*	176	1.22	323	0.79	ter UAG	10	0.71	7	0.51	Trp UGG	147	1.00	196	1.00
Leu CUU	39	0.23	370	1.03	Pro CCU	44	0.30	298	0.87	His CAU	46	0.35	368	1.06	Arg CGU*	227	2.20	204	0.99
CUC*	149	0.88	190	0.53	CCC*	385	2.61	320	0.94	CAC*	215	1.65	324	0.94	CGC*	327	3.17	225	1.10
CUA	17	0.10	282	0.79	CCA	85	0.58	474	1.39	Gln CAA	44	0.16	699	0.97	CGA	16	0.16	264	1.29
CUG*	703	4.16	475	1.32	CCG	75	0.51	276	0.81	CAG*	501	1.84	743	1.03	CGG	13	0.13	146	0.71
Ile AUU	147	0.56	576	1.31	Thr ACU	76	0.38	390	0.96	Asn AAU	73	0.25	793	1.10	Ser AGU	17	0.12	426	1.04
AUC*	633	2.43	340	0.77	ACC*	637	3.21	425	1.05	AAC*	513	1.75	645	0.90	AGC	174	1.21	404	0.99
AUA	2	0.01	401	0.91	ACA	25	0.13	507	1.25	Lys AAA	38	0.09	775	0.99	Arg AGA	4	0.04	220	1.07
Met AUG	301	1.00	549	1.00	ACG	55	0.28	301	0.74	AAG*	813	1.91	798	1.01	AGG	32	0.31	172	0.84
Val GUU	129	0.54	446	1.18	Ala GCU	267	0.87	558	1.20	Asp GAU	284	0.77	903	1.29	Gly GGU	316	1.19	426	1.02
GUC*	346	1.46	300	0.79	GCC*	854	2.79	539	1.15	GAC*	457	1.23	493	0.71	GGC*	505	1.91	450	1.08
GUA	19	0.08	278	0.73	GCA	41	0.13	491	1.05	Glu GAA	75	0.17	885	1.10	GGA	230	0.87	619	1.49
GUG*	453	1.91	492	1.30	GCG	62	0.20	279	0.60	GAG*	809	1.83	728	0.90	GGG	7	0.03	169	0.41

"High" and "Low" denote groups of genes with high and low codon usage bias; they are the 10% of genes at each extreme of the major codon usage trend among genes (identified by multivariate statistical analysis). Twenty-two codons defined as "optimal" (see text) are indicated by *. The High and Low groups each comprise 44 genes, and total 13,374 and 26,307 codons, respectively. *N* and RSCU are explained in Table 37.1.

different, overall, from that of "chromosomal" genes (Table 37.1). The TE ORFs are more similar to the low bias genes than the high bias genes (Table 37.2), and exhibit very little evidence of selected codon usage.

However, as with chromosomal genes, codon usage varies greatly among TE ORFs: in general, ORFs from the same TE have rather similar codon usage patterns, but ORFs from different TEs have different codon usage patterns (this is apparent, to some extent, in the GC_S values in Appendix B, but see Shields and Sharp (1989) for more details). This observation is most simply explained if the TEs have been subject to different mutational biases, and we consider two possible scenarios. Since TEs appear to have been subject to occasional horizontal transfer among species, their base composition could reflect different mutation biases in different previous host genomes. However, it seems rather more likely that the differences reflect current/ongoing differences in mutation pattern. For many TEs, movement around the genome involves an RNA intermediate which is then subject to a (quite highly error prone) reverse transcription process. The different TEs have reverse transcriptases which differ considerably in their primary amino-acid sequences (Xiong and Eickbush 1990), and it is quite likely that each reverse transcriptase has a slightly different error propensity which leads to different mutational spectra, and ultimately to different base composition and codon usage (Shields and Sharp 1989).

Conclusions

We have concluded above that *Drosophila melanogaster* genes are subject to different levels of codon selection. This seems to be corroborated by the observation that silent sites in genes with high codon usage bias have diverged to a lesser extent between *D. melanogaster* and other related species (e.g., *D. simulans* and *D. pseudoobscura*), suggesting that there is more constraint on codon usage in the highly biased genes (Sharp and Li 1989). In a recent examination of silent site base composition and substitution rates, Moriyama and Gojobori (1992) suggested that the variation in each can be explained by mutational biases, in a manner consistent with the situation in mammalian genes (Wolfe et al. 1989). However, we have outlined many discrepancies between the observations relating to *Drosophila* and mammals which make a similar explanation unlikely. We have detailed some cases where it seems that the strength of codon usage bias can be correlated with the level of gene expression—it will be of particular interest to investigate whether any of the heterogeneity in codon usage among genes can be related to the genes' tissue or time of expression. Certainly, while we have discussed the major pattern of codon usage variation among genes, we do not exclude the possibility that there are other (as yet undefined) trends which explain some further part of the heterogeneity in these data.

Another question of interest concerns the extent to which a similar pattern is found in other species of *Drosophila*. Codon usage differs among *Adh* genes derived from various *Drosophila* species (Starmer and Sullivan 1989).

Interestingly, Moriyama and Gojobori (1992) reported that in the *Adh* gene of Hawaiian *Drosophila*, GC_S is low and the silent substitution rate is high (see also Thomas and Hunt 1991); these two observations can be consistently explained if codon selection has been relaxed in that lineage, due possibly to a small effective population size caused by several bottleneck events. It will be interesting to examine to what extent (and to ask why) codon usage patterns generally vary among *Drosophila* species.

Acknowledgements

This is a publication from the Irish National Centre for Bioinformatics. We are particularly grateful to Michael Ashburner for supplying FlyBase, and to Ken Wolfe for his comments. This study was supported by EOLAS grant SC/91/603.

References

Andersson, S. G. E. and Kurland, C. G. (1990). Codon preferences in free-living micro-organisms. *Microbiol. Rev.* **54**:198–210.

Aota, S-I., Gojobori, T., Ishibashi, F., Maruyama, T. and Ikemura, T. (1988). Codon usage tabulated from the GenBank genetic sequence data. *Nucl. Acids Res.* **16**:r315–r402.

Ashburner, M. (1992). FlyBase—a *Drosophila* genetic database (version 9206). Available electronically from the FTP.BIO.INDIANA.EDU and EMBL-HEIDELBERG.DE fileservers.

Ashburner, M., Bodmer, M. and Lemeunier, F. (1984). On the evolutionary relationships of *Drosophila melanogaster. Dev. Genet.* **4**:295–312.

Benyajati, C., Wang, N., Reddy, A., Weinberg, E. and Sofer, W. (1980). Alcohol dehydrogenase in Drosophila: Isolation and characterization of messenger RNA and cDNA clone. *Nucl. Acids Res.* **8**:5649–5656.

Bernardi, G., Olofsson, B., Filipski, J., Zerial, M., Salinas, J., Cuny, G., Meunier-Rotival, M. and Rodier, F. (1985). The mosaic genome of warm-blooded vertebrates. *Science* **228**:953–958.

Bulmer, M. (1991). The selection-mutation-drift theory of synonymous codon usage. *Genetics* **129**:897–907.

Clary, D. O. and Wolstenholme, D. R. (1985). The mitochondrial DNA molecule of *Drosophila yakuba*: nucleotide sequence, gene organization, and genetic code. *J. Mol. Evol.* **22**:252–271.

Filipski, J. (1988). Why the rate of silent codon substitutions is variable within a vertebrate's genome. *J. Theoret. Biol.* **134**:159–164.

Gouy, M. and Gautier, C. (1982). Codon usage in bacteria: correlation with gene expressivity. *Nucl. Acids Res.* **10**:7055–7074.

Gouy, M., Gautier, C., Attimonelli, M., Lanave, C. and di Paola, G. (1985). ACNUC—a portable retrieval system for nucleic acid sequence databases: logical and physical designs and usage. *Comp. Appl. Biosci.* **1**:167–172.

Grantham, R., Gautier, C., Gouy, M., Jacobzone, M. and Mercier, R. (1981). Codon catalog usage is a genome strategy modulated for gene expressivity. *Nucl. Acids Res.* **9**:r43–r74.

Hovemann, B., Richter, S., Walldorf, U. and Czielpluch, C. (1988). Two genes encode related cytoplasmic elongation factors 1-α (EF-1) in *Drosophila melanogaster* with continuous and stage specific expression. *Nucl. Acids Res.* **16**:3175–3194.

Ikemura, T. (1985). Codon usage and tRNA content in unicellular and multicellular organisms. *Mol. Biol. Evol.* **2**:13–34.

Kalfayan, L. and Wensink, P. C. (1982). Developmental regulation of *Drosophila* α-tubulin genes. *Cell* **29**:91–98.

Kreitman, M. and Hudson, R. R. (1991). Inferring the evolutionary histories of the *Adh* and *Adh-dup* loci in *Drosophila melanogaster* from patterns of polymorphism and divergence. *Genetics* **127**:565–582.

Kylsten, P., Kimbrell, D. A., Daffre, S., Samakovlis, C. and Hultmark, D. (1992). The lysozyme locus in *Drosophila melanogaster*: different genes are expressed in midgut and salivary glands. *Mol. Gen. Genet.* **232**:335–343.

Limbach, K. J. and Wu, R. (1985). Characterization of two *Drosophila melanogaster* cytochrome c genes and their transcripts. *Nucl. Acids Res.* **13**:631–644.

Lloyd, A. T. and Sharp, P. M. (1992). CODONS: A microcomputer program for codon usage analysis. *J. Heredity* **83**:239–240.

Moriyama, E. N. and Gojobori, T. (1992). Rates of synonymous substitution and base composition of nuclear genes in Drosophila. *Genetics* **130**:855–864.

O'Connell, P. and Rosbash, M. (1984). Sequence, structure, and codon preference of the *Drosophila* ribosomal protein 49 gene. *Nucl. Acids Res.* **12**:5495–5513.

Sharp, P. M. (1989). Evolution at "silent" sites in DNA. In *Evolution and Animal Breeding*, eds. W. G. Hill and T. F. C. Mackay (Wallingford: C.A.B. International), pp. 23–32.

Sharp, P. M., Burgess, C. J., Cowe, E., Lloyd, A. T. and Mitchell, K. J. (1992). Selective use of termination codons and variations in codon choice. In *Transfer RNA in Protein Synthesis*, eds. D. L. Hatfield, B. J. Lee and R. M. Pirtle (Boca Raton, FL: CRC Press), pp. 395–420.

Sharp, P. M. and Li, W-H. (1986). An evolutionary perspective on synonymous codon usage in unicellular organisms. *J. Mol. Evol.* **24**:28–38.

Sharp, P. M. and Li, W-H. (1989). On the rateof DNA sequence evolution in *Drosophila*. *J. Mol. Evol.* **28**:398–402.

Sharp, P. M., Tuohy, T. M. F. and Mosurski, K. R. (1986). Codon usage in yeast: cluster analysis clearly differentiates highly and lowly expressed genes. *Nucl. Acids Res.* **14**:5125–5143.

Shields, D. C. and Sharp, P. M. (1989). Evidence that mutation patterns vary among *Drosophila* transposable elements. *J. Mol. Biol.* **207**:843–846.

Shields, D. C., Sharp, P. M., Higgins, D. G. and Wright, F. (1988). "Silent" sites in *Drosophila* genes are not neutral: evidence of selection among synonymous codons. *Mol. Biol. Evol.* **5**:704–716.

Starmer, W. T. and Sullivan, D. T. (1989). A shift in the third-codon-position nucleotide frequency in alcohol dehydrogenase genes in the genus *Drosophila*. *Mol. Biol. Evol.* **6**:546–552.

Sueoka, N. (1988). Directional mutation pressure and neutral evolution. *Proc. Natl Acad. Sci.* (*USA*) **85**:2653–2657.

Thomas, R. H. and Hunt, J. A. (1991). The molecular evolution of the alcohol dehydrogenase locus and the phylogeny of Hawaiian *Drosophila*. *Mol. Biol. Evol.* **8**:687–702.

Xiong, Y. and Eickbush, T. H. (1990). Origin and evolution of retroelements based upon their reverse transcriptase sequences. *EMBO J.* **9**:3353–3362.

Wolfe, K. H., Sharp, P. M. and Li, W-H. (1989). Mutation rates differ among regions of the mammalian genome. *Nature* **337**:283–285.

Appendix 37.A: Codon Usage Bias in *D. melanogastar* Genes

Gene	*Function/Product*	*Map*	*AA*	GC_S	GC_I	F_{op}	*Acc.#*
Abd-A	abdominal-A: homeodomain TF	3-58.8	330	0.72		0.62	X54453
Abd-B	abdominal-B: homeodomain TF	3-58.8	491	0.73		0.62	X16134
Abl	abl-oncogene analog: Tyr kinase	3-[44]	1,520	0.69	0.40	0.57	M19692
ac	achaete: T5 AHLH protein	1-0.0	201	0.52		0.41	M17120
Ace	acetylcholinesterase	3-52.5	649	0.74		0.64	X05893
Acp70A	male accessory gland protein	3-[40]	55	0.47		0.37	M21201
Acr60C	muscarinic acetylcholine receptor C	2-[107]	788	0.79		0.62	M23412
Acr64B	nicotinic acetylcholine receptor D	3-[8]	521	0.72		0.62	X04016
Acr96Aa	nicotinic acetylcholine receptor B	3-[83]	567	0.75		0.66	X07194
Acr96Ab	nicotinic acetylcholine receptor E	3-[83]	535*	0.70		0.61	X52274
Act5C	actin	1-[14]	376	0.78		0.78	K00667
Act42A	actin	2-[55.2]	376	0.68		0.64	K00670
Act57A	actin	2-[92]	376	0.76		0.79	K00673
Act79B	actin	3-[47]	376	0.82	0.33	0.80	M18829
Act87E	actin	3-[53]	376	0.76		0.77	K00674
Act88F	actin	3-57.1	376	0.80	0.48	0.79	M18830
Actn	sarcomeric α actinin	1-[0.5]	895	0.81		0.76	X51753
ade3	glycinamide ligase	2-[22]	434	0.65	0.42	0.55	J02527
Adfl	Adh distal factor 1: AHLH protein	2-[56]	253	0.74		0.67	M37787
Adh	alcohol dehydrogenase	2-50.1	256	0.81	0.39	0.77	J01066
Adhr	alcohol dehydrogenase related	2-50.1	272	0.53		0.43	
Ald	fructose-1,6-biphosphate aldolase	3-91.5	363	0.82	0.40	0.82	M76409
ama	amalgam protein	3-[47.5]	333	0.77	0.28	0.66	M23561
amd	α-methyl-dopa hypersensitivity	2-53.9	510	0.67	0.38	0.58	X04695
Amy-d	α-amylase 1	2-77.7	494	0.88		0.82	X04569
AnnIX	annexin IX	3-[70]	296*	0.87		0.81	M34068
AnnX	annexin X	1-[64]	321	0.87		0.78	M34069
annon-77F	histone-like protein	3-[46]	215	0.56	0.42	0.48	X16962
Anr	andropin: male-specific protein	3-[101]	57	0.46	0.26	0.39	X16972
Antp	antennapedia: homeodomain TF	3-47.5	378	0.75		0.63	X03791
Appl	β-amyloid-like gene	1-0.0	886	0.71		0.63	J04516
Aprt	adenine phosphoribosyltransferase	3-1.5	183	0.64		0.57	M18432
arl	arf-like: GTP-binding protein	3-[43]	180	0.81	0.40	0.74	M61127
arm	armadillo	1-[0.4]	843	0.64	0.47	0.57	X54468
Arr1	arrestin A/phosphorestin II	2-[53]	364	0.76	0.32	0.71	M30177
Arr2	arrestin B/phosphorestin I	3-[26]	401	0.76	0.34	0.70	M32141
ase	asense: T8 AHLH protein	1-0.0	396	0.52		0.41	X12550
Atpa	Na/K-ATPase α subunit	3-[70]	1,038	0.70		0.66	X14476

(*continued*)

APPENDIX 37.A. *Continued.*

Gene	*Function/Product*	*Map*	*AA*	GC_S	GC_I	F_{op}	*Acc.#*
awd	abnormal wing discs	3-[105]	153	0.89		0.86	X13107
B	Bar: homeodomain protein	1-57.0	543	0.70		0.56	M73079
bam	bag-of-marbles	3-[85]	442	0.64	0.33	0.54	X56202
bcd	bicoid: homeodomain TF	3-[47.5]	489	0.66		0.54	X14458
Bd	Beaded: EGF-like transmembrane P	3-92.5	1,408	0.66		0.55	X56811
BicD	bicaudal D α-helical coiled coil	2-52.9	782	0.67		0.59	M31684
Bj1	chromatin-binding protein	3-[20]	547	0.68	0.40	0.58	X58530
boss	bride of sevenless: transmembrane P	3-[89]	896*	0.61	0.35	0.53	X55887
br	broad: Zn finger protein	1-[0.4]	704	0.80		0.67	X54664
brm	brahma: homeotic regulator	3-43.0	1,638	0.62		0.52	M85049
Bsg25D	blastoderm-specific transcript	2-[16]	741	0.68	0.43	0.57	X04896
bw	brown	2-104.5	675	0.81		0.68	M20630
cad	caudal: homeodomain TF	2-[54]	472	0.74	0.35	0.58	M21070
Cal	calmodulin	2-[64]	152	0.67	0.38	0.63	X05951
Cam	CAM-kinase type II α	4-[3]	490	0.32		0.28	M74583
Cat	catalase	3-[45]	506	0.76		0.68	X52286
cdc2	protein kinase	2-[40]	297	0.54		0.45	X57485
cdc2c	cdc2c protein kinase	3-[68]	314	0.60		0.53	X57486
CecA1	cecropin A1	3-[101]	63	0.56	0.34	0.52	X16972
CecA2	ceceopin A2	3-[101]	63	0.51	0.31	0.52	X16972
CecB	cecropin B	3-[101]	63	0.61	0.26	0.56	X16972
CecC	cecropin C	3-[101]	63	0.64	0.26	0.59	Z11167
Cf1a	chorion transcription factor 1α	3-[22]	549	0.65		0.54	X58435
Cf2	chorion transcription factor 2	2-[15]	235*	0.72		0.62	X53380
Cg25C	collagen α-1 type IV	2-[15]	1,775	0.52		0.47	M23704
Cha	choline acetyltransferase	3-64.6	728*	0.66		0.55	M13219
chi	chickadee: profilin	2-[18]	126	0.74		0.66	M84528
chp	chaoptin: cell surface glycoprotein	3-[102]	1,134	0.72		0.64	M19017
ci	cubitis interruptus: Zn finger P	4-0.0	1,377	0.30		0.23	X54360
CkIIa	casein kinase II α subunit	3-[47]	336	0.37		0.34	M16534
CkIIb	casein kinase II β subunit	1-[36]	215	0.55		0.51	M16535
cnc	cap-n-collar: AHLH protein	3-81.2	533	0.64		0.52	M37495
Cp15	chorion protein S15	3-[26]	115	0.65	0.54	0.64	X02497
Cp16	chorion protein S16	3-[26]	138	0.69	0.40	0.66	X16715
Cp18	chorion protein S18	3-[26]	172	0.70	0.29	0.67	X02497
Cp19	chorion protein S19	3-[26]	173	0.74	0.45	0.71	X02497
Cp36	chorion protein S36	1-[23]	286	0.73	0.47	0.67	X05245
Cp38	chorion protein S38	1-[23]	306	0.64	0.36	0.60	X05245
crb	crumbs: transmembrane protein	3-82	2,139	0.63		0.55	M33753
Csp	cysteine-string protein 29	3-[47]	223	0.65		0.54	M63008
ct	cut: homeodomain protein TF	1-20.0	2,175	0.61		0.49	X07985
cta	concertina: G-protein-α1-like	2-[54.8]	457	0.36		0.31	M63651
CycA	cyclin A	3-[36]	491	0.66		0.56	M24841
CycB	cyclin B	2-[101]	530	0.71		0.61	M33192
CycC	cyclin C	3-[55]	267	0.73		0.64	X62948
Cyp1	cyclophilin-1	—	165	0.82		0.77	M62398
Cyt-c1	cytochrome c DC3	2-[52]	105	0.65		0.57	X01761
Cyt-c2	cytochrome c DC4	2-[52]	108	0.81		0.77	X01760

APPENDIX 37.A. *Continued.*

Gene	*Function/Product*	*Map*	*AA*	GC_S	GC_I	F_{op}	*Acc.#*
D1	chromosomal protein D1	3-[49]	355	0.61		0.54	J04725
da	daughterless: AHLH protein	2-41.3	710	0.70		0.57	J03148
Dbp73D	D-E-A-D box protein 73D	3-[44]	572	0.51	0.31	0.44	M74824
Ddc	dopa decarboxylase	2-53.9	508	0.71	0.37	0.62	X04661
dec1	defective chorion-1	1-20.8	1,123	0.54		0.45	M35887
Dfd	deformed: homeodomain TF	3-47.5	590	0.64		0.50	X05136
Dhod	dihydroorotate dehydrogenase	3-48.0	51	0.50		0.46	X17297
dim	didymous: homeodomain protein	2-[46]	475	0.70		0.56	M65016
disco	disconnected: Zn finger protein	1-53.1	568	0.68		0.56	X56232
Dl	delta: EGF-transmembrane	3-66.2	833	0.67		0.57	Y00222
dl	dorsal: embryonic polarity	2-52.9	678	0.66		0.56	M23702
Dlar	protein Tyr phosphatase	2-[52]	2,029	0.65		0.56	M27700
dlg1	discs-large: guanate-cyclase-like	1-34.8	960	0.64		0.53	M73529
DmsII	RNA pol II elongation factor	—	313	0.77		0.70	X53670
dnc	dunce: cAMP phosphodiesterase	1-3.9	584	0.58		0.48	X55167
Dox-A2	diphenol oxidase A2	2-53.9	494	0.70	0.41	0.62	M63010
dpp	decapentaplegic	2-4.0	588	0.74		0.60	M30116
Dpt	diptericin	2-[87]	106	0.57		0.47	M55432
Dromsopa	CAX (opa) repeat	3-[47]	69	0.78		0.75	X56491
Dsk	sulfated tyrosine-kinin	3-[47.1]	128	0.46		0.36	J03957
ea	easter: serine protease	3-57	392	0.76		0.64	J03154
eag	ether-a-gogo: K^+ channel protein	1-50.0	1,174	0.65		0.53	M61157
EcR	ecdysone receptor	2-[55.2]	878	0.68		0.56	M74078
Edg78E	pupal cuticle protein	3-[47]	122	0.77	0.46	0.73	M71247
Edg84A	pupal cuticle protein	3-[47.5]	188	0.55	0.38	0.49	M71249
Edg91	pupal cuticle protein	3-[62]	159	0.48	0.51	0.42	M71250
Ef1a100E	elongation factor 1-α F1	3-[102]	463	0.76	0.44	0.76	X06869
Ef1a48D	elongation factor 1-α F2	2-[64]	462	0.79	0.44	0.71	X06870
Ef2b	elongation factor 2	2-[54.6]	844	0.66		0.64	X15805
Egon	embryonic-gonad: Zn finger protein	3-[47]	373	0.68		0.54	X16631
Eip71CD	ecdysone-induced protein	3-[42]	255	0.64	0.36	0.55	X04024
Eip74EF	ecdysone-induced protein	3-[45]	883	0.72		0.56	X15087
Eip75B	ecdysone-induced protein	3-[45]	1,443	0.73		0.59	X15586
elav	embryonic lethal, abnormal vision	1-[0.0]	483	0.68		0.56	M21152
emc	extramacrochaetae protein	3-0.0	199	0.75		0.63	M31902
en	engrailed: homeodomain EF	2-62.0	60*	0.88	0.43	0.71	X01765
Eno	enolase phosphoglycerate hydrolase	2-[3]	433	0.87		0.86	X17034
esg	escargot: Zn finger protein	2-[51]	470	0.66		0.56	M83207
E(spl)	enhancer of split	3-89.1	186	0.70		0.66	X16553
Est6	esterase 6	3-35.9	544	0.50	0.25	0.42	J04167
EstP	esterase P	3-35.9	544	0.42	0.32	0.35	M33780
Ets2	ets-oncogene analog	2-[100]	159*	0.60	0.22	0.51	M20408
eve	even-skipped: homeodomain TF	2-[59]	376	0.78	0.41	0.68	M14767
Fas1	fasciclin I	3-[59]	652	0.69	0.41	0.61	M32311
Fas2	fasciclin II	1-[6]	811	0.62		0.55	M77165
Fas3	fasciclin III	2-[53]	508	0.70		0.61	M27813
Fcp3C	vitelline membrane protein 3C-1	1-[1.5]	210	0.58	0.45	0.42	M18281

(continued)

APPENDIX 37.A. *Continued.*

Gene	Function/Product	Map	AA	GC_S	GC_I	F_{op}	Acc.#
fkh	fork head: DNA-binding protein	3-95	510	0.78		0.57	J03177
Fmrf	FMRFamide polyprotein	2-[59]	342	0.74		0.62	J03232
Fps85D	fps-oncogene analog: P Tyr kinase	3-[49]	803	0.69		0.61	X52844
fs(1)h	FS: bromodomain membrane protein	1-21	2,038	0.64		0.52	M23221
fs(1)K10	FS: DNA-binding protein	1-0.5	463	0.64	0.40	0.53	X12836
fs(1)Ya	FS: nuclear envelope protein	1-[1.5]	708	0.75		0.62	M38442
ft	fat: cadherin-like protein	2-12.0	5,147	0.58		0.49	M80537
ftz	fushi tarazu: homeodomain TF	3-47.5	413	0.77	0.29	0.67	X00854
ftz-f1	ftz transcription factor 1	3-[45]	1,043	0.64		0.52	M63711
Fur1	furin-1: serine protease	—	899	0.66		0.54	X59384
fz	frizzled: transmembrane protein	3-41.7	581	0.66		0.53	X54646
Gapdh1	glyceraldehyde-3-phosphate DH 1	2-[57]	332	0.83		0.80	M11254
Gapdh2	glyceraldehyde-3-phosphate DH 2	1-[51]	332	0.75		0.72	M11255
Gb13F	G protein b subunit	1-[51]	340	0.66		0.58	M22567
Gld	glucose dehydrogenase	3-48	612	0.67	0.40	0.58	M29298
Glu-RII	glutamate receptor II	2-[17]	906	0.68		0.59	M73271
G-oa47A	G-protein 0α subunit	2-[60]	354	0.64		0.57	M86660
Gpdh	glycerol-3-phosphate dehydrogenase	2-17.8	362*	0.75	0.39	0.69	X61224
Gprk1	G-protein coupled receptor kinase 1	2-[55.1]	700	0.31		0.23	M80493
Gprk2	G-protein coupled receptor kinase 2	3-[102]	427	0.74		0.65	M80494
grh	grainy head: AHLH TF	2-86	1,063	0.69		0.57	X15657
gro	groucho: G-protein b-subunit-like	3-89.1	719	0.67		0.57	M20571
G-sa60A	G-protein Sa-60A	2-[106]	385	0.45	0.42	0.37	M33998
Gst	glutathione S-transferase 1-1	3-[51]	209	0.88		0.83	X14233
gt	giant: AHLH (Leu zipper)	1-1.0	448	0.70	0.49	0.59	X61148
h	hairy: AHLH	3-26.5	337	0.76		0.67	X15905
H2.0	homeodomain P 2.0 TF	2-[20]	410	0.68		0.57	Y00843
hb	hunchback: Zn finger protein	3-48.3	758	0.71	0.40	0.60	Y00274
His1	histone H1	2-[54.6]	255	0.48		0.41	X14215
His2A	histone H2A	2-[54.6]	124	0.54		0.54	X14215
His2AvD	histone H2A variant	3-[91]	134*	0.44		0.34	X07485
His2B	histone H2B	2-[54.6]	123	0.63		0.53	X14215
His3	histone H3	2-[54.6]	136	0.57		0.55	X14215
His4	histone H4	2-[54.6]	103	0.55		0.54	X14215
HmG-CoAR	3-OH-3-Methylglutaryl CoA reductase	3-[81]	916	0.62		0.55	M21329
HmgD	high mobility group protein D	2-[99]	112	0.74		0.65	M77023
Hrb87Fa	RNA-binding protein	3-[54]	386	0.64	0.33	0.60	X59691
Hsc70-1	heat-shock protein cognate 1	3-[41]	68*	0.67	0.42	0.58	J01085
Hsc70-2	heat shock protein cognate 2	3-[52]	68*	0.78	0.29	0.66	K01297
Hsc70-4	heat shock protein cognate 4	3-[57]	651	0.79		0.78	M36114
Hsp22	heat shock protein 22 kD	3-[28]	174	0.77		0.68	J01098
Hsp23	heat shock protein 23 kD	3-[28]	186	0.75		0.69	J01100
Hsp26	heat shock protein 26 kD	3-[28]	208*	0.75		0.69	J01099
Hsp27	heat shock protein 27 kD	3-[28]	213	0.72		0.64	J01101
Hsp67Ba	heat shock protein	3-[28]	238	0.71		0.59	M26267

APPENDIX 37.A. *Continued.*

Gene	*Function/Product*	*Map*	*AA*	GC_S	GC_I	F_{op}	*Acc.#*
Hsp67Bb	heat shock protein	3-[28]	111	0.55	0.30	0.45	X07311
Hsp67Bc	heat shock protein	3-[28]	169	0.53		0.44	X06542
Hsp70A	heat shock protein 70 kD	3-[51]	643	0.75		0.68	J01103
Hsp70B	heat shock protein 70 kD	3-[51]	641	0.73		0.66	J01104
Hsp83	heat shock protein 83 kD	3-[5]	375*	0.77	0.35	0.76	K01685
Ide	insulin-degrading enzyme	—	990	0.63		0.54	M58465
ImpE2	ecdysone inducible gene E2	3-[6]	466	0.57		0.52	M55099
inaC	protein kinase C	2-82	700	0.53		0.48	J04845
Inr	insulin-like receptor b subunit	3-[70]	300*	0.56		0.46	M13568
Jra	jun-related AHLH (Leu zipper)	2-[59]	289	0.72		0.64	M36181
Kin	kinesin heavy chain	2-[76]	975	0.74		0.66	M24441
Klp54D	kinesin-like protein (KLP1)	2-[80]	133*	0.65		0.50	M74427
Klp61F	kinesin-like protein (KLP2)	3-[0]	130*	0.55		0.49	M74428
Klp64D	kinesin-like protein (KLP4)	3-[19]	129*	0.55		0.48	M74430
Klp67A	kinesin-like protein (KLP3)	3-[27]	118*	0.69		0.56	M74429
Klp68D	kinesin-like protein (KLP5)	3-[36]	123*	0.63		0.50	M74431
Klp98A	kinesin-like protein (KLP6)	3-[98]	95*	0.66		0.53	M74432
kni	knirps: steroid receptor P family	3-[46]	429	0.75	0.42	0.62	X13331
knrl	knirps-related protein	3-[46]	647	0.58		0.49	X14153
Kr	Krüppel: Zn finger protein	2-107.6	467	0.54	0.33	0.44	X03414
Kr-h	Kr homolog: Zn finger protein	2-[20]	79*	0.83		0.74	M14940
l(1)sc	lethal at scute: T3 AHLH protein	1-0.0	257	0.59		0.51	X12549
l(2)37Cc	mitochondrial protein	2-53.9	203	0.73	0.42	0.61	X04227
l(2)gl	lethal giant larvae: transmembrane P	2-0.0	1,160	0.36		0.28	X05426
lab	labial: homeodomain TF	3-[47.5]	495*	0.71		0.57	X13103
Lam	nuclear lamin	2-[17]	621	0.79		0.73	X07278
LanA	laminin A chain	3-[21]	1,951*	0.60		0.52	M75882
LanB1	laminin B1 chain	2-[24]	1,787	0.61		0.53	M19525
LanB2	laminin B2 chain	3-[28]	1,639	0.68		0.61	M25063
Lcp1	cuticle protein I	2-[58]	130	0.70	0.47	0.69	J01080
Lcp2	cuticle protein II	2-[58]	126	0.66	0.43	0.65	J01081
Lcp3	cuticle protein III	2-[58]	112	0.77	0.45	0.73	J01081
Lcp4	cuticle protein IV	2-[58]	111	0.74	0.40	0.72	J01081
lds	lodestar: DEAH-family NTP-binding	3-47.8	974	0.59		0.49	X62629
Lsp1-a	α larval serum protein	1-39.5	70*	0.83	0.48	0.76	X03872
Lsp1-b	β larval serum protein	2-1.9	100*	0.91	0.47	0.88	X03873
Lsp1-g	gamma larval serum protein	3-[0]	105*	0.69	0.42	0.64	X03874
LvpD	larval visceral protein	2-[58]	508	0.62		0.55	V00204
LvpH	larval visceral protein	2-[58]	521	0.70	0.31	0.65	V00204
LvpL	larval visceral protein	2-[58]	505	0.71	0.45	0.65	V00204
LysD	lysozyme	3-[0]	140	0.76		0.72	X58382
LysP	lysozyme	3-[0]	141	0.73		0.65	X58382
M(2)21C	ribosomal protein 21C	2-0.0	112	0.82		0.81	Y00504
M(3)67C	ribosomal protein S17	3-28.9	131	0.84	0.37	0.84	M22142
M(3)99D	ribosomal protein rp49	3-[101]	133	0.78	0.47	0.72	X00848
mam	mastermind: neurogenic protein	2-70.3	1,596	0.67		0.58	X54251
Map205	microtubule-associated 205 kD	3-[105]	1,163	0.45		0.37	X54061

(continued)

APPENDIX 37.A. *Continued.*

Gene	*Function/Product*	*Map*	*AA*	GC_S	GC_I	F_{op}	*Acc.#*
Mdr49	P-glycoprotein (drug resistance)	2-[67]	1,302	0.67		0.60	M59076
Mdr65	P-glycoprotein (drug resistance)	3-[21]	1,302	0.51		0.43	M59077
me31B	maternal expression: DEAD-helicase	2-[37]	459	0.46		0.41	M59926
mex1	midgut expression 1	3-[42]	83	0.79	0.33	0.70	M63626
Mhc	myosin heavy chain	2-52.2	1,962	0.77	0.40	0.76	M61229
Mlc1	myosin light chain 1	3-[98]	155	0.75		0.70	K01567
Mlc2	myosin light chain 2	3-[101]	222	0.73	0.45	0.70	M11947
mle	male-less: DEAH-family helicase	2-55.2	1,293	0.53		0.46	M74121
mod	modulo: DNA-binding protein	3-[102]	544	0.49		0.40	X15702
Mov34	Mov34	2-[106]	338	0.78		0.70	M64643
Mp20	muscle-specific protein 20	2-[68]	184	0.82	0.34	0.83	Y00795
msh1	muscle homeodomain 1	3-[100]	61*	0.43		0.31	M38582
Mst26Aa	male accessory gland	2-[20]	264	0.41	0.32	0.33	Y00219
Mst26Ab	male accessory gland	2-[20]	90	0.51	0.36	0.43	Y00219
Mst87F	sperm protein	3-[45]	56	0.47	0.29	0.45	Y00831
Mst95E	male-specific protein msp316	3-[81]	52	0.39	0.31	0.29	M32022
mys	myospheroid: integrin b-subunit	1-[21]	846	0.70		0.61	J03251
N	notch: transmembrane protein	1-3.0	2,703	0.63	0.42	0.52	M16152
nau	nautilus: AHLH protein	3-[81]	332	0.62		0.50	X56161
ncd	non-claret disjunctional	3-100.7	700	0.72	0.39	0.63	X52814
ninaA	ninaA: transmembrane protein	2-1.4	237	0.80	0.39	0.68	M22851
ninaC	ninaC: protein kinase	2-[22]	1,501	0.62	0.32	0.54	J03131
ninaE	opsin-R1/R6	3-66.4	373	0.79	0.33	0.71	K02315
NK1	NK-1 homeodomain TF	3-[72]	659	0.70		0.54	X55393
NK2	NK-2 homeodomain TF	1-[0.0]	158*	0.66		0.55	M27290
NK3	NK-3 homeodomain TF	3-[72]	194*	0.77	0.38	0.64	M27291
nod	kinesin-like protein	1-36	666	0.64		0.52	M36195
nonA	RNA-binding protein	1-52.3	700	0.55		0.48	X55902
norpA	phospholipase C-b-type	1-6.5	1,095	0.67		0.59	J03138
nos	nanos: posterior determinant	3-66.2	401	0.68	0.37	0.55	M72421
Nrq	neuroglian Ig-like	1-23.6	1,239	0.65		0.57	M28231
Nrt	neurotactin: Ser protease-like TMP	3-[44]	846	0.64		0.55	X53837
numb	numb	2-[35]	556	0.66		0.55	M27815
oc	ocelliless: homeodomain TF	1-23.1	671*	0.65		0.47	X58983
Ocr	octopamine receptor	3-[100]	601	0.80		0.69	M60789
ogre	optic ganglion reduced	1-18.8	362	0.79		0.71	X61180
omb	optomotor-blind	1-7.5	974	0.63		0.48	M81796
osk	oskar: maternal effect	3-48.4	606	0.63	0.27	0.51	M65178
Ote	otefin: nuclear envelope protein	2-[86]	406	0.62		0.51	X17495
otu	ovarian tumors	1-22.7	811	0.55		0.46	X13693
pAbp	poly(A)-binding protein	2-[80]	574	0.70		0.65	M38019
Pah	phenylalanine-4-hydroxylase	—	453	0.61		0.52	M32802
para	paralytic: Na-channel α subunit	1-52.1	1,820*	0.56		0.49	M32078
Pcna	proliferating cell nuclear antigen	2-[88]	260	0.79	0.30	0.72	M33950
Pcp	pupal cuticle protein	2-[22]	184	0.70	0.51	0.60	J02527
pcx	pecanex transmembrane protein	1-0.9	2,483	0.56	0.50	0.45	M74329
Pep	protein on ecdysone puffs: Zn finger	3-[45]	716	0.67		0.64	X56689

APPENDIX 37.A. *Continued.*

Gene	Function/Product	Map	AA	GC_S	GC_I	F_{op}	Acc.#
Pepck	phosphoenolpyruvate carboxykinase	—	647	0.74		0.69	Y00402
per	period: biological clock protein	1-1.2	1,218	0.79	0.48	0.62	M30114
Pgd	6-phosphogluconate dehydrogenase	1-0.5	481	0.80	0.47	0.73	M80598
phl	pole-hole: raf-oncogene analog	1-[1]	666	0.64	0.31	0.51	X07181
Pig1	pre-intermoult gene 1	1-[3]	187	0.47		0.38	X15760
Pka-C1	cAMP-dependent protein kinase A	2-[34]	353	0.83		0.74	M18655
Pka-C2	cAMP-dependent protein kinase-B	3-[102]	354	0.72	0.46	0.65	X16960
Pka-C3	cAMP-dependent protein kinase-related	3-[43]	502	0.61		0.49	X16961
Pkc53E	protein kinase C 53E	2-[78]	639	0.61	0.26	0.55	X05283
Pkc98E	protein kinase C 98E	3-[99]	634	0.78		0.69	J04848
Pkg24A	cGMP-dependent protein kinase 24A	2-[9]	894	0.69		0.59	M30147
Plc21C	phospholipase C	2-[0.1]	1,312	0.63		0.52	M60453
polo	protein Ser/Thr kinase	3-46	576	0.74		0.65	X63361
Pp1-87B	protein-Ser/Thr phosphatase 1 α	3-[51]	302	0.77		0.71	X15583
PpY-55A	protein Ser/Thr phosphatase Y	2-[83]	314	0.53		0.46	Y07510
prd	paired: homeodomain TF	2-45	613	0.66	0.47	0.56	M14548
Prm	paramyosin	3-[26]	477	0.88		0.84	X62591
pros	prospero: homeodomain	3-[51]	1,407	0.71		0.60	M81389
Pros28	proteasome 28 kD subunit	—	249	0.72		0.70	M57712
Pros35	proteasome 35 kD subunit	3-[59]	279	0.63		0.57	X15497
Psc	posterior sex combs: Zn finger	2-67	1,603	0.56		0.45	X59275
Ptp	protein Tyr phosphatase	—	1,462	0.49		0.42	M27699
Ptp10D	protein Tyr phosphatase 10D	1-[36]	1,558	0.68		0.57	M80538
Ptp99A	protein Tyr phosphatase 99A	3-[100]	1,301	0.68		0.59	M81795
pum	pumilio	3-48.5	1,533	0.64		0.53	X62589
R	roughened: ras analog	3-1.4	184	0.84		0.75	M80535
r	rudimentary: dihydroorotase	1-54.5	2,236	0.64		0.54	X04813
Rab3	ras-related GTP-binding protein	2-[60]	220	0.74		0.64	M64621
Ras64B	GTPase ras-analog 2	3-[15]	187	0.79		0.71	K01962
Ras85D	GTPase ras-analog 1	3-[49]	189	0.72		0.66	K01960
Rdl	GABA-A receptor	3-[27]	606	0.54		0.46	M69057
ref(2)P	male fertility (Zn finger)	2-54.0	599	0.58	0.34	0.50	X16993
Rh2	rhodopsin-2	3-[65]	381	0.65	0.33	0.54	M12896
Rh3	rhodopsin-3	3-[67]	383	0.72		0.62	M17718
Rh4	rhodopsin-4	3-[44]	378	0.74		0.61	M17730
Rm62	DEAD-family helicase	3-[47.4]	575	0.72		0.68	X52846
RpA1	ribosomal protein A1	2-[78]	113	0.80		0.78	X05016
RpI135	RNA polymerase I 135 kD subunit	2-[0.1]	1,129	0.53	0.32	0.44	X17298
RpII140	RNA polymerase II 140 kD subunit	3-54	1,123	0.58	0.26	0.55	X05709
RpII215	RNA polymerase II 215 kD subunit	1-35.7	1,896	0.66	0.34	0.58	M27431
RpIII128	RNA polymerase III 128 kD subunit	2-	1,135	0.64		0.56	X58826

(*continued*)

APPENDIX 37.A. *Continued.*

Gene	Function/Product	Map	AA	GC_S	GC_I	F_{op}	Acc.#
RpL1	ribosomal protein L1	3-[98]	407	0.79		0.79	X13382
RpS14A	ribosomal protein S14 A	1-[21]	151	0.70	0.42	0.72	M21045
RpS14B	ribosomal protein S14 B	1-[21]	151	0.69	0.39	0.71	M21045
Rrp1	recombination repair protein	2-[6]	679	0.56		0.48	M62472
run	runt: ATP-binding protein	1-65	509	0.80		0.65	X56432
rut	rutabaga: adenylyl cyclase	1-46	2,248	0.70		0.59	M81887
ry	rosy: xanthine dehydrogenase	3-[52]	1,335	0.64	0.37	0.55	Y00308
sala	spalt accessory	2-44	142	0.28	0.25	0.26	X57474
sas	stranded at second	3-[47.5]	1,348	0.64		0.52	M68866
sc	scute: AHLH protein	1-0.0	345	0.51		0.43	M17119
sca	scabrous: fibrinogen-like	2-66.7	774	0.73		0.61	M60065
Scr	sex combs reduced: homeodomain TF	3-47.5	73*	0.36		0.22	X05228
sd	scalloped: DNA-binding protein	1-51.5	440	0.56		0.45	M83787
Sd	segregation distorter: Leu zipper	2-54	363	0.59		0.51	X60218
Ser99Da	serine protease 1	3-[101]	265	0.82		0.78	M24379
Ser99Db	serine protease 2	3-[101]	265	0.82		0.77	M24379
Ser99Dc	serine protease 3	3-[101]	61	0.54		0.51	M24380
sev	sevenless: protein Tyr kinase	1-33.4	2,554	0.66	0.37	0.54	J03158
sgg	shaggy: Ser/Thr kinase	1-1.3	514	0.57		0.49	X53332
Sgs4	salivary gland secretion	1-[3]	182*	0.47		0.37	X06565
Sgs5	salivary gland secretion	3-[60]	163	0.54	0.25	0.46	X04269
Sh	shaker K^+-channel	1-57.6	643	0.48		0.38	X07132
Shab	shaker cognate b	3-[3]	924	0.63		0.54	M32659
Shal	shaker cognate l	3-[46]	490	0.79		0.66	M32660
Shaw	shaker cognate w	2-[10]	498	0.71		0.62	M32661
shi	shibire: dynamin	1-51.5	836	0.56		0.50	X59448
sim	single-minded: AHLH protein	3-52.2	655*	0.73		0.61	M19020
sina	seven in absentia: nuclear protein	3-[44]	314	0.80		0.69	M38384
sli	slit: transmembrane protein	2-77	1,480	0.73		0.64	X53959
slo	slowpoke: Ca-activated K^+-channel	3-86	1,184*	0.59		0.50	M69053
sn	singed	1-21.0	512	0.75	0.37	0.65	X17549
sna	snail: Zn finger protein	2-51	390	0.73		0.64	Y00288
snk	snake: serine protease	3-52.1	435	0.67		0.59	X04513
snRNP27D	sn-ribonucleoprotein 70 kD	2-[21]	448	0.76	0.47	0.65	M31162
Sod	Cu-Zn superoxide dismutase	3-[34]	153	0.74		0.67	Y00367
sol	small optic lobes: Zn finger	1-[65]	1,597	0.73		0.60	M64084
Sos	son of sevenless: G-exchange	2-[48]	1,595	0.68		0.56	M83931
Spec-a	α-spectrin	3-[1.5]	2,415	0.76		0.72	M26400
SR55	Ser-Arg RNA-binding protein	3-[53]	350	0.66		0.65	X58720
Src29A	src-oncogene analog	2-[24]	590	0.63		0.55	M16599
Src64B	src-oncogene analog	3-[15]	552	0.74		0.65	M11917
Sry-a	serendipity α	3-[101]	530	0.70		0.60	X03121
Sry-b	serendipity β: Zn finger protein	3-[101]	351	0.88		0.78	X03121
Sry-d	serendipity δ: Zn finger protein	3-[101]	430	0.88	0.53	0.80	X03121
stau	staufen	2-83.5	1,026	0.61		0.47	M69111
Ste	stellate: casein kinase II-b-like	1-45.7	172	0.74	0.35	0.61	X15899
stg	string: Tyr phosphatase	3-[100]	479	0.78		0.70	M24909
su(f)	suppressor of forked	1-65.9	733	0.59	0.34	0.52	X62679

APPENDIX 37.A. *Continued.*

Gene	Function/Product	Map	AA	GC_S	GC_I	F_{op}	Acc.#
Su(H)	suppressor of Hairless: DNA-binding	2-50.5	550	0.65		0.56	X58393
su(Hw)	suppressor of Hairy wing	3-54.8	944	0.60	0.33	0.54	Y00228
su(s)	suppressor of sable: RNA-binding	1-0.0	1,334	0.59	0.38	0.49	M57889
Su(var)20	suppressor of variegation: DNA-binding	2-31.1	206	0.64	0.37	0.56	M57574
Su(z)2	suppressor of zeste-2: Zn finger	2-[67]	1,364	0.58	0.39	0.44	X56798
svp	seven-up: steroid receptor	3-[51]	543	0.75		0.65	M28863
Sxl	Sex-lethal: RNA-binding protein	1-19.2	366	0.58		0.52	M59448
Syt	synaptotagmin-p65	2-[7]	474	0.72		0.64	M55048
Takr86C	tachykinin-like receptor	3-[50]	504	0.71		0.60	M77168
Takr99D	tachykinin-like receptor	3-[101]	519	0.80		0.64	X62711
T-cp1	T complex protein 1 analog	3-[76]	557	0.72		0.69	M21159
term	terminus: Zn finger protein	3-[45]	428	0.82		0.73	M19140
TfIID	transcription factor IID	2-[99]	353	0.67		0.56	M38082
Tgfb-60A	TGF-b-like	2-[106]	455	0.86		0.75	M84795
tin	tinman: homeodomain TF	3-[72]	150*	0.74	0.29	0.65	M27292
tko	technical knockout: mt RP S12	1-1.0	140	0.79		0.68	M19494
Tl	Toll: transmembrane protein	3-91	1,097	0.66		0.53	M19969
tld	tolloid: bone morphogenetic P-1-like	3-85	1,057	0.61		0.53	M76976
tll	tailless: steroid receptor	3-102	452	0.73		0.64	M34639
Tm1	tropomyosin I	3-[55]	284	0.84	0.44	0.85	K02623
Tm2	tropomyosin II/troponin H	3-[55]	285	0.71		0.72	M15466
top	torpedo: protein Tyr kinase	2-[97]	174*	0.70		0.61	K03417
Top2	type II DNA topoisomerase	2-[54]	1,447	0.63	0.37	0.57	X61209
tor	torso: receptor Tyr kinase	2-57	923	0.60	0.33	0.51	X15150
tra2	transformer-2: RNA-binding protein	2-[71]	264	0.54		0.49	M23633
trp	serine protease	3-[100]	1,275	0.68	0.38	0.59	M34394
trx	trithorax: Zn finger protein	3-54.2	3,759	0.50		0.41	M31617
Try	trypsin-like Ser protease	2-[60]	256	0.73		0.67	X02989
tsh	teashirt: Zn finger protein	2-[54.8]	993	0.49		0.39	M57496
ttk	tramtrack: Zn finger protein	3-[102]	641	0.69		0.55	X17121
tub	tube: (dorso-ventral polarity)	3-[47.1]	462	0.59		0.44	M59501
Tuba67C	α-4 tubulin	3-[28]	462	0.72	0.30	0.64	M14646
Tuba84B	α-1 tubulin	3-[47.5]	450	0.79	0.39	0.79	M14643
Tuba84D	α-3 tubulin	3-[48]	450	0.79		0.79	M14645
Tuba85E	α-2 tubulin	3-[49]	449	0.74	0.32	0.69	M14644
Tubb60D	β-3 tubulin	2-[107]	454	0.88	0.42	0.80	M22335
Tubb85D	β-2 tubulin	3-48.5	446	0.72		0.66	M20420
Tubb97EF	β-1 tubulin	3-[92]	447	0.81		0.79	M20419
Tubg	gamma-tubulin	2-[6]	475	0.66		0.58	M61765
tud	tudor protein	2-97	2,515	0.54		0.47	X62420
tuf	tufted: transmembrane protein	2-59	1,286	0.78	0.35	0.65	M28999
twi	twist: AHLH protein	2-[102]	490	0.81	0.25	0.71	X12506
twn	twain: homeodomain protein	2-[46]	601	0.69		0.59	M65015
UbcD6	ubiquitin conjugating enzyme	3-[47.1]	151	0.43		0.38	M63792
Ubi-f	ubiquitin-RP hybrid	1-[17]	128	0.80		0.80	X53059

(*continued*)

Appendix 37.A. *Continued.*

Gene	*Function/Product*	*Map*	*AA*	GC_S	GC_I	F_{op}	*Acc.#*
Ubi-m	ubiquitin-RP S27A hybrid	—	156	0.71		0.70	M22536
Ubi-p	poly-ubiquitin protein	3-[6]	231	0.68		0.69	M22428
Ubx	Ultrabithorax: homeodomain TF	3-58.8	246	0.70	0.54	0.59	M24608
up	upheld: troponin-T	1-41.0	396	0.76		0.74	X54504
Uro	urate oxidase	2-[24]	352	0.70	0.22	0.62	X51940
usp	ultraspiracle: chorion 1 TF	1-[0.5]	508	0.77		0.65	X53417
uzip	unzipped	2-107.6	500	0.53		0.44	X07450
v	vermilion: tryptophan oxidase	1-33.0	379	0.68	0.39	0.61	M34147
vas	vasa: DEAD-family helicase	2-51	661	0.47	0.32	0.41	X12946
Vha	vacuolar H^+-ATPase 16 kD subunit	—	159	0.63		0.55	X55979
Vm26Aa	vitelline membrane protein 26Aa	2-[20]	168	0.74		0.72	M20936
Vm26Ab	vitelline membrane protein 26Ab	2-[20]	141	0.82		0.79	M18280
Vm32Ec	vitelline membrane protein 32Ec	2-[44]	116	0.59		0.49	M27647
Vm34Ca	vitelline membrane protein 34Ca	2-[47]	96*	0.72		0.65	X01802
w	white eye	1-1.5	687	0.71		0.58	X51749
wg	wingless: int1-oncogene analog	2-[22]	468	0.74		0.63	M17230
y	yellow body	1-0.0	541	0.46	0.33	0.35	X04427
yema	nuclein a DNA-binding protein	3-[99]	1,022	0.68	0.35	0.55	X63503
Yp1	yolk protein 1	1-30	442	0.80	0.38	0.76	X01524
Yp2	yolk protein 2	1-30	439	0.80	0.25	0.75	X01524
Yp3	yolk protein 3	1-44	420	0.80	0.37	0.75	M15898
z	zeste	1-1.0	575	0.68	0.38	0.58	Y00049
Z600	histone-like protein	3-[42]	90	0.68		0.63	X58286
zfh1	Zn-finger homeodomain protein 1	3-[102]	1,060	0.76		0.65	M63449
zfh2	Zn-finger homeodomain protein 2	4-[1]	3,005	0.43		0.35	M63450
zip	zipper: myosin heavy chain	2-[108]	1,972	0.63		0.56	M35012
—	65 kD protein phosphatase	—	591	0.67		0.62	M86442
—	retinal specific G-α protein	—	353	0.52	0.37	0.46	M58016
—	fushi tarazu repressor	—	641	0.69		0.56	M62856
—	Glu-tRNA aminoacyl synthetase	—	1,475	0.57		0.51	M74104
—	DNA polymerase	—	1,505	0.55	0.29	0.47	D90310
—	laminin receptor	—	253	0.88		0.83	M77133

Genes are presented in alphabetical order; gene names follow FlyBase (Ashburner 1992). Map is the genetically defined map location. AA is the length of the gene in codons, * indicates a partial gene sequence. GC_S is the G + C content at silent third positions of codons (i.e., excluding Trp, Met and stop codons); GC_I is the G + C content in introns. F_{op} is the frequency of optimal codons (see text for definition). Acc.# indicates the accession number allowing retrieval of the sequence from the GenBank/EMBL/DDBJ DNA sequence data library. Abbreviations: AHLH = amphipathic helix-loop-helix; DH = dehydrogenase; FS = female sterile; G = guanine; mt = mitochondrial; P = protein; RP = ribosomal protein; TF = transcription factor; TGF = transforming growth factor; TMP = transmembrane protein.

Appendix 37.B. Codon Usage Bias in *D. melanogaster* Transposable Elements

Family	*Element*	*Gene*	*AA*	GC_S	F_{op}	*Acc.#*
LINE-like:	F	NA binding	122	0.50	0.43	M17214
		RT	858	0.45	0.38	
	I	NA binding	429	0.38	0.36	M14954
		RT	1,086	0.41	0.37	
	Jockey	NA binding	583	0.38	0.32	M22874
		RT	916	0.46	0.37	
	DOC	NA binding	565	0.42	0.37	X17551
		RT	888	0.41	0.35	
	R1Dm	orf1	471	0.63	0.50	X51968
		RT	1,021	0.56	0.45	
	R2Dm	RT	1,057	0.46	0.36	X51967
Ty-like	Copia		1,409	0.28	0.23	X02599
	1,731	gag	273	0.49	0.35	X07656
		pol	982	0.50	0.39	
Retrovirus-like:	17.6	gag	445	0.30	0.27	X01472
		pol	1,058	0.33	0.28	
		env	472	0.28	0.26	
	297	gag	424	0.31	0.26	X03431
		pol	1,059	0.26	0.23	
		env	471	0.29	0.26	
	Gypsy	gag	451	0.52	0.43	M12927
		pol	1,035	0.54	0.45	
		env	509	0.51	0.44	
	412	gag	444	0.36	0.32	X04132
		pol	1,219	0.27	0.22	
Foldback:	FB4	orf	148	0.35	0.30	J01084
	FBw^c	orf1	633	0.39	0.29	X15469
		orf2	403	0.37	0.29	
P-like:	P element		751	0.38	0.31	V01520
	HOBO		644	0.32	0.26	M69216

Abbreviations: NA = nucleic acid; RT = reverse transcriptase. See also the footnote to Appendix 37.A.

APPENDIX

Early Stages of Embryonic Development

Many of the genes treated in Part I are expressed in early embryos. Four figures that summarize different aspects of the processes involved are presented in this appendix.

FIG. A.1. "Schematic drawing of the embryonic stages leading up to gastrulation in *D. melanogaster*" from Foe and Alberts (1983). "This figure is modified from Zalokar & Erk (1976) to show the correct times of appearance of pole and somatic buds and to indicate the cessation of division of the yolk nuclei. The number beside each embryo, which denotes its developmental stage, corresponds to the total number of nuclear division cycles undergone by the almost synchronously dividing embryonic nuclei. A stage begins with the start of interphase and ends with the conclusion of mitosis. Stage 1 is the fertilized zygote during its first interphase and mitosis. The subsequent stages, each of which corresponds to one complete nuclear division cycle (interphase plus mitosis), are numbered consecutively. Embryos are shown in longitudinal section and with their anterior ends at the top. They are depicted without vitelline membranes to emphasize the changes in surface morphology of the plasma membrane that surrounds the syncytial embryo. Solid black circles represent nuclei, stippled regions denote yolk, and non-textured regions denote the yolk-free regions of cytoplasm. As shown, when development begins there is a thin layer of yolk-free cytoplasm at the egg periphery (the 'periplasm'), and a yolk-free region of cytoplasm surrounding each nucleus (the 'protoplasmic islands'). For stages 1–5 all nuclei are indicated, even though they would not all normally be in the same plane. For stages 6–14, only a fraction of the embryonic nuclei is shown."

"Stages 1–7: The nuclei multiply exponentially in the central region of the egg."

"Stage 8: The majority of the still dividing nuclei, with their enveloping protoplasmic islands, have started their migration outwards, leaving the future yolk nuclei behind. These yolk nuclei will divide in approximate synchrony with the remaining nuclei in cycles 8–10, and thereafter cease dividing and become polyploid."

"Stage 9: Early in their 9th interphase, a few migrating nuclei appear in the posterior periplasm, creating there the posterior cytoplasmic protuberances called pole buds. At the end of this stage, these nuclei (like all others in the syncytium) enter into mitosis, thus doubling the number of pole buds." (*continued*)

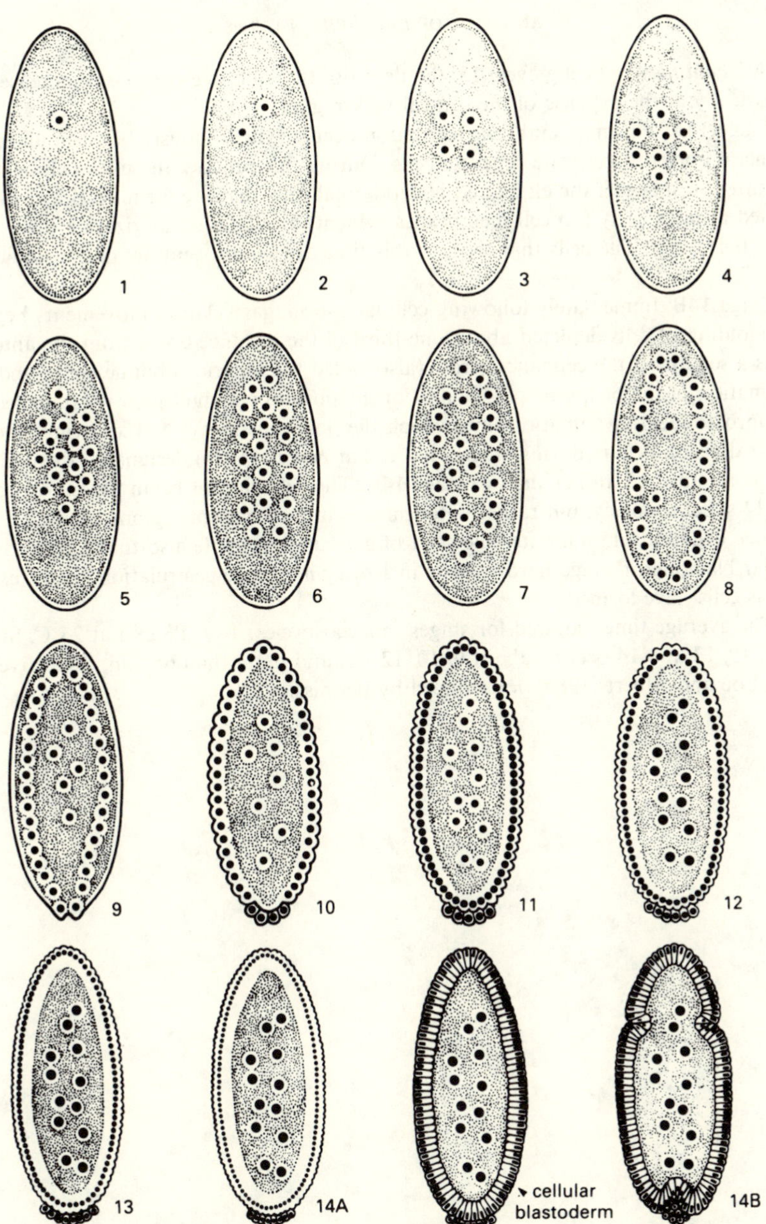

FIG. A.1 *continued.* "Stage 10: The remainder of the migrating nuclei appear in the periplasm at the beginning of their 10th interphase, organizing somatic buds over the entire embryonic surface. During mitosis of this cycle, the pole buds divide again and, nearly simultaneously, are pinched off from the syncytial embryonic mass to produce the pole cells; after this stage these cells, which are the potential germ cell progenitors, will continue to divide, but they lose mitotic synchrony with the embryonic syncytium."

"Stages 10–13: The syncytial nuclei in their somatic buds at the embryonic periphery divide with near synchrony. (*continued*)

FIG. A.1 *continued.* During cycle 13, the depth of the yolk-free periplasm increases dramatically at the expense of the central yolk region."

"Stage 14A: Plasma membrane formation occurs synchronously between all of the peripheral nuclei to generate separate cells. During this process, the nuclei elongate, matching the shape of the elongated blastodermal cells that are forming. Stage 14A is depicted at both early (no cell membranes evident) and late (cellularization just completed) times. The cells that form at this time are the progenitors of the somatic tissues."

"Stage 14B: Immediately following cellularization, gastrulation movements begin. The infolding of cells depicted about one-third of the distance down from the anterior pole is a section of the cephalic furrow (also called the anterior oblique cleft), and the invagination of the posterior pole is part of the posterior midgut furrow (all called the amnioproctodaeal invagination) into which the pole cells move. Not knowing when nuclear division occurs during stage 14, Zalokar & Erk (1976) designated the early gastrula as stage 15, rather than as stage 14B. The cells do not begin the mitosis of cycle 14 synchronously, but rather enter mitosis in a consistent region-specific sequence beginning 15 min after the start of gastrulation. Note also that a true 'cellular blastoderm' stage hardly exists in *Drosophila*, since gastrulation begins as soon as cells have formed."

"The average time required for stages (nuclear cycles) 1–9 is 8 min at 25°C. Stages 10, 11, 12, 13 and 14 occupy about 9, 10, 12, 21, and more than 65 min, respectively." From Foe and Albert (1983); reproduced by permission.

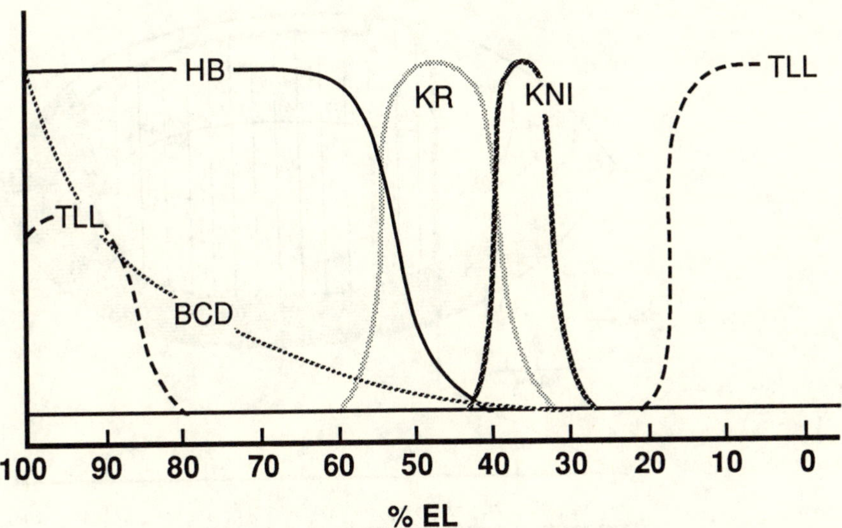

FIG. A.2. Main zones of expression of several gap genes along the antero–posterior axis of the egg (%EL) during blastoderm stage (modified from Hülskamp et al. 1990). 0% egg length = posterior pole to the right; the scale on the vertical axis is arbitrary and cannot be used to compare levels of gene products to each other. The horizontal line near the bottom of the graph represents the threshold of detection; it is meant to indicate that the presence, and effect, of some of these products may extend beyond the region of the embryo where they are detected. Localization of these products occurs at the mRNA level and it is due, at least in part, to the following interactions.

Maternal *bcd* RNA is anchored at the anterior pole by cytoskeletal elements. BCD stimulates transcription of *hb* thus limiting this RNA to the anterior half of the embryo. Low concentrations of BCD and HB stimulate transcription of *Kr* in the middle section of the embryo, while high concentrations repress it (thus defining the anterior border of the KR band). KR in turn represses *hb* thus defining this gene posterior border of expression, and it activates *kni* in a band immediately posterior to its own. Low to moderate concentrations of HB and TLL repress *kni* thus defining the anterior and posterior border of the KNI band.

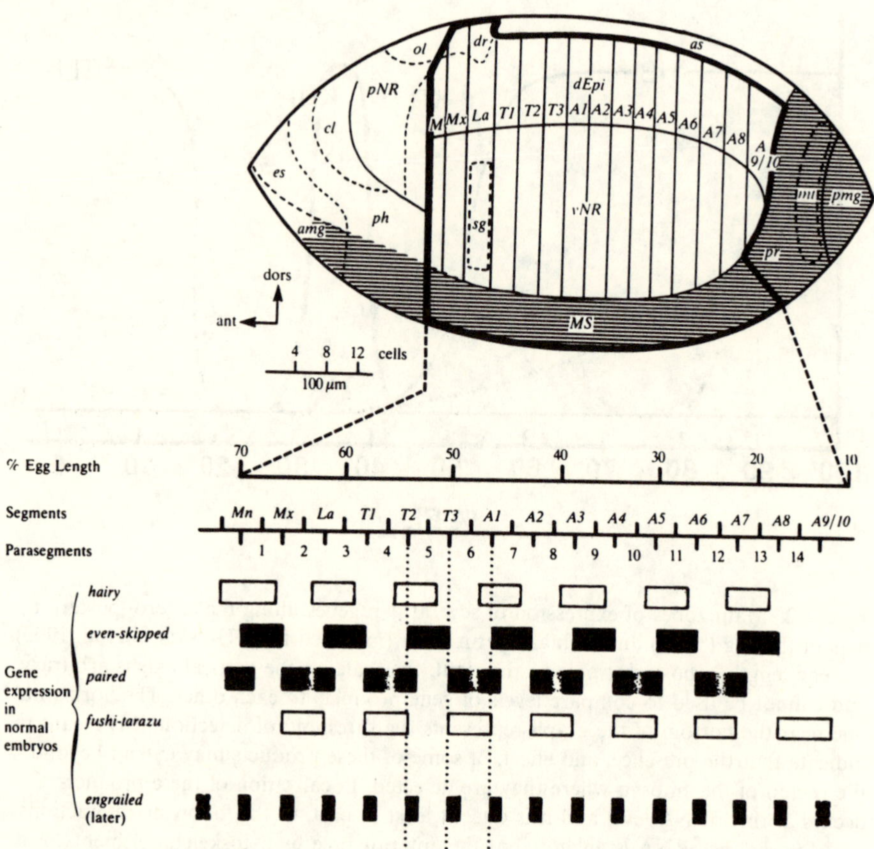

FIG. A.3. Top. "A fate map of the *Drosophila* blastoderm (from Campos-Ortega and Hartenstein, 1985)" as modified by Akam (1987). "The shape is a planimetric reconstruction of the blastoderm surface. All parts of the egg surface contribute to the embryo proper, except the narrow dorsal primordium for the amnioserosa (*as*). Hatched areas will invaginate at gastrulation. Cells that will generate metameric structures are enclosed by a thick line. Abbreviations: *amg*, anterior midgut; ant, anterior; *as*, amnioserosa; *cl*, clypeolabrum; *dEpi*, dorsal epidermis; dors, dorsal; *dr*, dorsal ridge; *es*, oesophagus; *mt*, Malpighian tubules; *MS*, mesoderm; *ol*, optic lobe; *ph*, pharynx; *pmg*, posterior midgut; *pNR*, procephalic neurogenic region; *pr*, proctodeum; *sg*, salivary gland; *vNR*, ventral neurogenic region; *M"* or *Mn*, "mandibular segment; *Mx*, maxillary segment; *La*, labial segment; *T1–T3*, thoracic segments; *A1–A10* abdominal segments."

Bottom. "Expression of segmentation genes in the *Drosophila* blastoderm: approximate registration of pair-rule stripes, *engrailed* expression and metameric units." Each segment is divided, by the parasegment line, into an anterior (A) and a posterior (P) compartment. "The patterns of expression are shown for four of the pair-rule genes at about cleavage stage 14A/B. The later patterns of *engrailed* expression have been projected onto the same diagram, even though at this mid-blastoderm stage hybridization reveals only a single well-defined *engrailed* stripe (stripe 2)."

"The bands of *engrailed* expression define P compartments and so lie at the anterior margin of each parasegment.

(*continued*)

FIGURE A.3 *continued.* Stripes of *even-skipped* and *fushi-tarazu* expression are each approximately four cells wide at mid-blastoderm, and appear to lie out of phase with each other. Double-labelling experiments in later embryos suggest that the anterior margins of both *ftz* and *eve* stripes coincide precisely with the *engrailed* stripes, and hence define parasegment boundaries (Lawrence et al. 1987). *hairy* stripes are about the same width, but are displaced slightly with respect to parasegments and overlap those of *ftz. paired* stripes are broader than a single metameric repeat, but the seven stripes split into fourteen before gastrulation." (This figure combines elements from Figs 1 and 4B from Akam (1987); reproduced by permission.) For a review and discussion see also Carroll (1990).

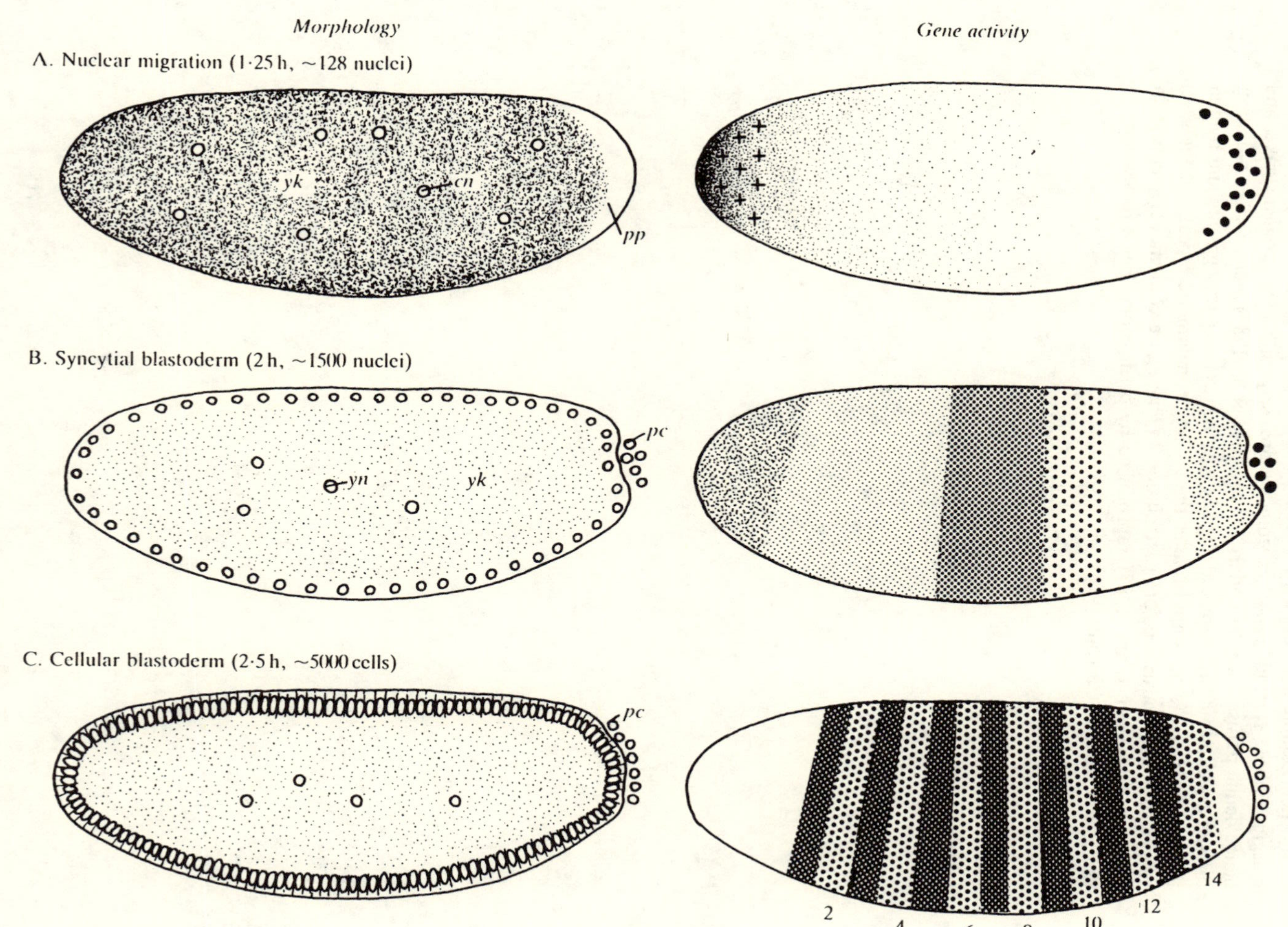
Morphology
Gene activity
A. Nuclear migration (1·25 h, ~128 nuclei)
yk
cn
pp
B. Syncytial blastoderm (2 h, ~1500 nuclei)
pc
yn
yk
C. Cellular blastoderm (2·5 h, ~5000 cells)
pc
2
4
6
8
10
12
14

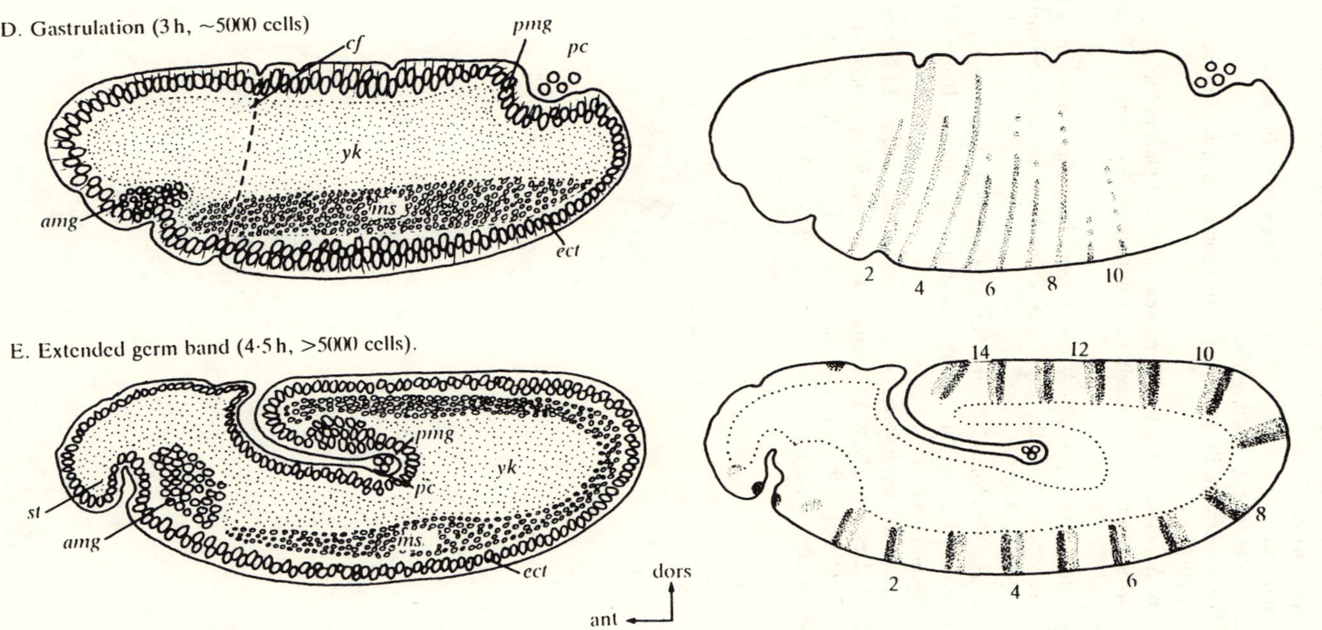

Fig. A.4. "Patterns of gene activity during early *Drosophila* development . . . Diagrams on the left show the morphology of stages during early embryogenesis. Corresponding panels on the right show patterns of gene activity established at the corresponding stages: A. Localized maternal determinants: *bicoid* RNA (crosses); polar granules (dots). The bicoid protein gradient is shown by shading. B. Gap gene expression: *hunchback*, *Krüppel*, *knirps* (shading, zones from anterior to posterior); *tailless* (stipple at both ends). C. Pair rule stripes: *even-skipped* (dark) and *fushi-tarazu* (light). D, E. Evolving pattern of segment polarity gene expression: *wingless* (dark) and *engrailed* (light)." By M. Akam, from *The Encyclopaedia of Molecular Biology* (Oxford: Blackwell Scientific Publications), reproduced by permission.

References

Akam, M. E. (1987). The molecular basis for metameric development in the *Drosophila* embryo. *Development* **101**:1–22.

Campos-Ortega, J. A. and Hartenstein, V. (1985). *The Embryonic Development of* Drosophila melanogaster. Berlin: Springer-Verlag.

Carroll, S. B. (1990). Zebra patterns in fly embryos: Activation of stripes or repression of interstripes? *Cell* **60**:9–16.

Foe, V. E. and Alberts, B. M. (1983). Studies of nuclear and cytoplasmic behaviour during the five mitotic cycles that precede gastrulation in *Drosophila* embryogenesis. *J. Cell Sci.* **61**:31–70.

Hülskamp, M., Pfeifle, C. and Tautz, D. (1990). A morphogenetic gradient of *hunchback* protein organizes the expression of the gap genes *Krüppel* and *knirps* in the early *Drosophila* embryo. *Nature* **346**:577–580.

Lawrence, P. A., Johnston, P., Macdonald, P. and Struhl, G. (1987). The *fushi tarazu* and *even-skipped* genes delimit the borders of parasegments in *Drosophila* embryos. *Nature* **328**:440–442.

Sonnenblick, B. P. (1950). The early embryology of *Drosophila melanogaster*. In *Biology of Drosophila*, ed. M. Demerec (New York: John Wiley and Sons), pp. 62–167.

Zalokar, M. and Erk, I. (1976). Division and migration of nuclei during early embryogenesis of *Drosophila melanogaster*. *J. Microbiol. Cell* **25**:97–106.

INDEX